Recent Advances in Poultry Management

Recent Advances in Poultry Management

Editors

George K. Symeon
Vassilios Dotas

MDPI Basel • Beijing • Wuhan • Barcelona • Belgrade • Novi Sad • Cluj • Manchester

Editors

George K. Symeon
Research Institute of Animal Science, ELGO-DEMETER
Giannitsa
Greece

Vassilios Dotas
Aristotle University of Thessaloniki
Thessaloniki
Greece

Editorial Office
MDPI
St. Alban-Anlage 66
4052 Basel, Switzerland

This is a reprint of articles from the Special Issue published online in the open access journal *Sustainability* (ISSN 2071-1050) (available at: https://www.mdpi.com/journal/sustainability/special_issues/poultry_management).

For citation purposes, cite each article independently as indicated on the article page online and as indicated below:

Lastname, A.A.; Lastname, B.B. Article Title. *Journal Name* **Year**, *Volume Number*, Page Range.

ISBN 978-3-0365-9606-8 (Hbk)
ISBN 978-3-0365-9607-5 (PDF)
doi.org/10.3390/books978-3-0365-9607-5

Contents

About the Editors

George K. Symeon

Dr. G. K. Symeon serves as an associate researcher at the Research Institute of Animal Science of the Hellenic Agricultural Organization DEMETER (ELGO-DIMITRA), in the field of animal production. He is a graduate of the Department of Animal Production of the Agricultural University of Athens, from where he received his Ph.D. in 2009. He has published more than 30 papers in peer-reviewed scientific journals and has participated in many national and international conferences. He specializes in poultry production and nutrition and has been a member of the research team of several national and international research programs. His scientific interests include poultry management, novel additives in poultry nutrition, developmental programming, thermal stress, and strategies to cope with climate change in animal production.

Vassilios Dotas

Dr. Dotas is actively involved in teaching, instructing eight undergraduate courses at the Aristotle University of Thessaloniki and one postgraduate course as an external partner at the International Hellenic University, all related to animal nutrition. He is a graduate of the Department of Animal Production of the Aristotle University of Thessaloniki, from where he received his Ph.D. in 2010. Dr. Dotas has received scholarships and awards of distinction from the Greek Ministry of Education Scholarship and the Research Committee of the Aristotle University of Thessaloniki. He has published more than 30 papers in peer-reviewed scientific journals and has participated in many national and international conferences. He specializes in animal nutrition and has been a member of the research team of several national research programs. Dr. Dotas is actively involved in teaching, instructing eight undergraduate courses and one postgraduate course at the aforementioned university, all related to animal nutrition. Additionally, he is the co-author of four books and serves as an active reviewer of several scientific journals. His scientific interests include dietary feed evaluation, the evaluation of the energy value and nutrient availability of various feeds, aromatic plants, and their essential oils in farm animals' nutrition, the production of functional foods through interventions in livestock nutrition, and the evaluation of the nutritive value of pasture, grazing capacity, and the intensity of pastures.

Preface

It is common knowledge among poultry experts that if the broiler did not exist, it should have been invented. Indeed, it is the perfect scientific tool, as it can be reared in large numbers and has a very short production cycle, allowing researchers to easily achieve significant power of study and conduct repeated experimental trials at minimal cost. Consequently, poultry research is thriving, which can be evidenced by the large annual number of remarkable scientific papers on poultry health, management, and nutrition.

On the other hand, commercial poultry production has proven to be perhaps the most successful sector of animal production. It has achieved this success not only by marketing a wide variety of products but also by demonstrating admirable adaptability to the concerns and attitudes of the modern consumer, along with high rates of incorporating research results into modern production. Nevertheless, there is still work to be done in terms of research and development, especially in addressing the modern challenges of climate change, welfare issues, and the overall sustainability of production.

The subject of this Special Issue is "Recent Advances in Poultry Management", and its aim is to contribute to research on the key aspects of poultry production, including (a) the structural characterization and clustering of production systems, (b) the application of innovative IT methods to housing and management practices, and (c) the novel trends in poultry nutrition and their effects on the quality of poultry meat and eggs. The significant number of manuscripts that have been submitted demonstrates that these goals remain relevant, and researchers worldwide are dedicated to providing advanced knowledge and applicable scientific results for everyday practice.

The Guest Editors would like to wholeheartedly thank all of the contributors to this Special Issue for their hard work and excellent collaboration throughout the manuscript review process. They would also like to thank the Assistant Editor, Ms. Sybil Han, for her tireless efforts and support. Editing this Special Issue has been a rewarding journey that has produced excellent papers and fruitful collaborations. Hopefully, the scientific audience will gain new insights from this Special Issue, contributing to the common goal of improving poultry production toward a viable and sustainable future.

George K. Symeon and Vassilios Dotas
Editors

sustainability

MDPI

Article

Typology, Structural Characterization and Sustainability of Integrated Broiler Farming System in Epirus, Greece

Vassilios Dotas [1,*], Dimitrios Gourdouvelis [1], Lampros Hatzizisis [2], Ioannis Kaimakamis [3], Ioannis Mitsopoulos [4] and George Symeon [5]

1 Department of Animal Production, Faculty of Agriculture, Aristotle University of Thessaloniki, 54124 Thessaloniki, Greece; dgourdou@agro.auth.gr
2 Faculty of Agriculture, University of Ioannina, 47100 Arta, Greece; lamprosxatz@yahoo.gr
3 Department of Animal Science, School of Agricultural Sciences, University of Thessaly, 41334 Larissa, Greece; ikaimaka@auth.gr
4 Department of Agriculture, School of Geosciences, International Hellenic University, 57400 Sindos, Greece; gmitsop@ihu.gr
5 Research Institute of Animal Science, Hellenic Agricultural Organization Demeter, 58100 Giannitsa, Greece; gsymewn@yahoo.gr
* Correspondence: vdotas@agro.auth.gr; Tel.: +30-2310991735

Citation: Dotas, V.; Gourdouvelis, D.; Hatzizisis, L.; Kaimakamis, I.; Mitsopoulos, I.; Symeon, G. Typology, Structural Characterization and Sustainability of Integrated Broiler Farming System in Epirus, Greece. *Sustainability* **2021**, *13*, 13084. https://doi.org/10.3390/su132313084

Academic Editor: Giuseppe Todde

Received: 8 November 2021
Accepted: 23 November 2021
Published: 26 November 2021

Abstract: The aim of this study was the detailed characterization of the existing zootechnical and financial management applied in broiler poultry farms in the Region of Epirus, Greece. The current situation was captured through the formation of a typology on the structural characterization of broiler farming system. The variables were recorded based on data from a stratified random sample according to Neyman's methodology of 110 poultry farms. In the typology, hierarchical cluster analysis was applied to identify differences between farms and to support most of this differentiation. Chebyshev distance was used to maximize the effect of the cluster elements distance, as well as Ward's clustering method, which aims to achieve greater homogeneity within the clusters. Bonferroni multiple comparison tests were used to evaluate the differences. Four clusters of different farm types were identified from the hierarchical cluster analysis. In conclusion, the production system of broiler farms in Epirus is intensive, especially in large farms that have made significant investments in fixed capital and implement successful management. However, the poultry sector in Epirus has further margin for improvement in both its productivity and profitability.

Keywords: broiler farms; farming system; hierarchical cluster analysis; sustainable development

1. Introduction

Over the past decades, the broiler industry in Greece has made tremendous progress to meet the growing demand for inexpensive and safe supply of high-quality meat. This development was combined with structural changes in the sector, characterized by the modernization of farming facilities, the intensification of the farming system and the concentration of poultry farms.

One such area with a high concentration of poultry activity and mainly broiler farming in Greece is the Region of Epirus (45% of the total national broiler production). The Region of Epirus is one of the 13 Regions of Greece. It occupies an area of 9203 km^2, and its population amounts to 336,856 inhabitants. The total generated GDP amounts to EUR 4499 million, which is 2.2% of the total GDP of Greece. Agriculture and livestock have a turnover of EUR 235.4 million and the food industry EUR 353.9 million [1]. The developed farming system of broiler production in the Region of Epirus is considered intensive, industrial, and large-scale according to the classification scale of Camargo-Barros et al. [2]. The Greek poultry industry is structured around a few large vertically integrated companies (i.e., "Pindos Poultry Co-operative", "Arta Poultry Co-operative" and "Nitsiakos S.A."

in the Region of Epirus) operating throughout the whole supply chain, from feed production and preparation to breeding, raising, and delivering the end products to retailers [3].

Large initial and operating costs, as well as economies of scale and more efficient capital investments, lead broiler farmers to establish larger scale commercial farms (i.e., flock sizes of more than 10,000 and up to 100,000 birds) and to participate into large enterprises [4,5]. The facilities are well equipped and relatively mechanized, including both semi-automatic and automatic equipment. The available automated systems include supply systems, water supply with nipples and central control for temperature/humidity, ventilation and lighting. Some systems have more extensive automation, including remote digital monitoring and computer controls [6].

Regarding Greece, the self-sufficiency rate in poultry meat reaches approximately 75%, and the corresponding imports in the year 2018 amounted to 82,000 tons [7]. During the same period, the corresponding exports of poultry meat to EU countries amounted to 22,000 tons of carcass weight. Covering national self-sufficiency in poultry meat makes it imperative to intensify the production system and increase productivity, reduce production costs, improve specific production indicators, and use innovative and sustainable practices with European legislation [8].

An effective tool that highlights how the particular structural characteristics of farms affect the efficient use of their own resources (land, labor and capital), but also the choice of actions and investments that can be made for their utilization, is the typology of farms [9]. The main objective of research focusing on the construction of a farm typology is to identify the farming system in the case study region [10]. A farm typology allows an overall assessment of farming activities not only in terms of farm structural characteristics (i.e., farm size, managerial ownership and farm intensity), but also in terms of farm economics [11]. Determining the types of poultry farms can be achieved with cluster analysis (C.A.) which is one of the most basic data analysis tasks. The aim of C.A. is the division of a total sample into clusters formed based on the similarity of their members [12]. Cluster analysis has been used in many cases to study the typology and productivity in broiler farming worldwide [13–18]. Corresponding studies, i.e., application of farm typology, regarding broiler farms in the direction of structural characterization, economic viability and future sustainability have not been conducted in Greece.

One of the most important concerns of the European Union is the achievement of sustainable development of human activities, such as the intensive broiler production [19]. Moreover, farmers evaluate their production practices mainly on the basis of economic efficiency, which generally requires large amounts of inputs, paying little attention to environmental pollution and future sustainability [20]. According to Vaarst et al. [21] the concept of 'sustainability' of a single agricultural sector is multi-dimensional, encompassing economic (production of valuable and cheap food), environmental (pollution and use of antibiotics), social (work conditions and animal welfare) and institutional (food chain governance) aspects. The economic impact of EU legislation on environmental protection, food safety and animal welfare and its implementation in the poultry sector has led to an increase in production costs estimated at 0.05 euros per kilogram (kg) of chicken live weight on average [22,23].

Poultry production has been found to be relatively environmentally friendly compared to the production of other terrestrial animals [24] because it has the best conversion rate of feed into human food and the smallest environmental footprint in terms of energy and water use per kilogram of meat produced [25]. As the production and transport of feeds contribute about 70% to the global warming potential of poultry systems, while manure management contributes to their eutrophication and acidification potential (nutrient waste) by about 40 and 60%, respectively [24–26], an important measure to reduce the impact of poultry activities is to improve feed efficiency by implementing better management practices or by using alternative feed additives (e.g., enzymes) and selected genetic material [19,21,26–28].

All of the above indicate that when using the C.A. to build a typology for such an intensified sector as broiler farming, all those economic and managerial indicators that contribute to the sustainable development and future viability of this sector should be taken into account. One of the most crucial factors affecting the profitability and economic viability of broiler farms is the feed conversion ratio (FCR) [29,30]. This is because the cost of feeds in broiler farming constitutes the largest percentage of variable costs, about 70% [31–33], and has an impact on the productivity and profitability of each farm, especially when economies of scale increase. It follows that the size of a farm and its financial viability are strongly related [34]. In terms of improving environmental sustainability, FCR is again the key factor both in reducing the environmental footprint of feedstuffs and in manure management due to reduced nutrient losses [24,27,28,35]. In addition, the technical and economic efficiency of broiler farms, according to Marcu et al. [36], is greatly influenced by the FCR and other productivity indexes included in the European Production Efficiency Factor (EPEF), which has been used by many researchers [37–41] to compare the results of broilers from different flocks and different regions. In the current study, this index (EPEF) will be used to confirm the results of the cluster analysis, by comparing the technical efficiency of the created clusters.

In summary, the aim of this study is the detailed characterization of the existing zootechnical, and financial management applied in broiler poultry farms in the Region of Epirus. The current situation was reflected in the formation of a typology on the structural characterization of the production of broiler farms. The analysis and interpretation of the common characteristics of the created clusters of farms will reflect the existing liabilities of the farming system but will also identify the key points that will be the lever of support for achieving social and economic development and sustainability of the sector.

2. Materials and Methods

2.1. Study Area

In the present study, the administrative Region of Epirus has been defined as the research area. This region includes a total of 4 Regional Units (R.U.); the R.U. of Ioannina and Arta, which have a large number of broiler farms, the R.U. of Preveza that participated in the sampling with a small number of respective farms and the R.U. of Thesprotia, where there were no industrial-sized broiler farms.

The capacity of the farms ranges from 5001 broilers to over 75,000 chickens. Therefore, the minimum size per production cycle carried out by each farm participating in the sample is 5001 chickens. The main reason that farms with less than 5001 chickens were not included in the original sample was that such small-scale farms do not participate in the existing vertical and integrated business or cooperative schemes operating in the Region.

The farms that formed the original sample cooperate either with private companies or are members of cooperatives that have systems for processing, slicing and marketing of chickens and their products, as well as modern slaughterhouses of high productivity with freezing, slicing, cooking and packing infrastructure. They also have broiler breeders' farms that supply eggs to privately owned hatcheries for the production of day-one chicks, as well as privately owned feedstuff mills.

The sampling framework was designed based on the statistical data of EL.STAT. [1]. According to the data obtained, the distribution of broiler farms in the Region of Epirus is presented in Table 1.

2.2. Data Collection and Sample Determination

The structured questionnaire method was used to collect the survey data, in a selected representative and proportional sample of statistically sufficient size. The survey, conducted in the year 2019, required the use of a questionnaire with a detailed record of the technical and economic data describing the structure of each farm. These are technical and economic characteristics related to both inputs (factors of production used) and outputs (final products) related to the broilers' farming.

Table 1. Distribution of broiler farms in the Region of Epirus.

Regional Unit	Number of Farms	Total Production (Number of Chickens/Year)
Ioannina	767	26,475,000
Arta	214	15,000,000
Preveza	23	700,000
Thesprotia	-	-
Total	1004	42,175,000

The initial number of 1004 farms was divided into 4 size categories according to the number of broilers reared per year, regardless of the annual placements. According to Unay-Gailhard et al. [10] and FAO [42], farm size (physical and economic) is one of the most important categorization variables in typologies because it determines significant differences between farms. The first category, M1, included farm sizes from 5001–25,000 chickens per production cycle, the second, M2, from 25,001–50,000, the third, M3, from 50,001–75,000 and the last, M4, more than 75,001 chickens.

The sampling method followed was Neyman's stratified sampling method [43,44]. The distribution of the sample into stratums is given by the following equation [45]:

$$n_h = n \cdot \frac{N_h S_h}{\sum\limits_{h=1}^{H} N_h S_h}$$

where:

n = sample size
n_h = sample size of stratum h
N = size of total (initial) sample
N_h = size of total sample concerning stratum h
S_h = the standard deviation of the variable in each stratum

$$S_h = \sqrt{\frac{1}{N_h - 1}\sum_{i=1}^{N_h}(y_i - \overline{y}_i)^2}$$

y_i = the stratification-related variable.

The final sample included 110 farms and constitutes 9.13% of the total broiler farms in the study area. The number of broilers per year in the final sample amounted to 3,241,480 chickens, which is 7.7% of the total population of the sampling frame. There is a sufficient sample, and therefore the estimates are representative and can be generalized. According to the above stratification methodology, the sample collected from all 3 Regional Units and the 4 size categories is presented in Table 2.

Table 2. Distribution of broiler farms in the 3 R.U. of Epirus, per size categories.

Regional Unit	Size Categories				Total
	M1	M2	M3	M4	
Arta	12	10	2	1	25
Ioannina	49	26	4	2	81
Preveza	2	1	1	-	4
Total	63	37	7	3	110

The random selection of farms in the final sample was done according to the range of each selection, using an initially random start number of each selection [46]. The selection range from which the successively selected farms were obtained through the 2-way stratification was calculated as the quotient in each *Nh* stratum to the number of sample members in each n_h stratum [47]. The calculation was performed using the OpenEpi Open-Source Calculator, Version 3.

2.3. Data Processing

The primary data were collected through questionnaires recording the main characteristics of each farm that concerned (a) the applied management practices and (b) the financial data. The data were encoded and integrated into tables, coded sheets of MS-Excel, in order to create a matrix-database for further statistical analysis with SPSS version 25 and R software by adding (adds-on) the Benchmarking Package in the MacOS High Sierra operating system.

Multivariate statistical analysis was used to process and interpret the data by creating separate groups of farms that apply a separate rearing model in order to identify differences between farms and to support most characteristics of this differentiation [48].

2.4. Hierarchical Cluster Analysis

The multivariate statistical technique used in the present study to determine the relationship of various quantitative parameter, was the hierarchical cluster analysis. The cluster analysis aims to create homogeneous (internally) groups of production systems, in terms of their technical and economic characteristics (variables), which differ significantly from each other. Groups are called classes or clusters. The analysis also tries to determine the number and composition of the groups. Cluster analysis is an exploratory method, and the variables used can be both quantitative and qualitative, only the latter require the use of specialized techniques for their integration and separation of clusters [49].

The steps followed for the cluster analysis of the sample of 110 broiler farms are described in detail by Hair et al. [50] and Everitt et al. [51]. Initially, the items to be grouped were selected (110 farms). The following variables were then selected to fulfill the objectives of the research:

- Number of production cycles per year;
- Feed conversion ratio (kg of total feed/kg of final body weight);
- Average days of a production cycle–length;
- Total feed consumption (kg/year);
- Mortality (%): the index was obtained by subtracting from the number of day-old chicks placed for fattening the final output related to the total number of chickens taken to the slaughterhouse;
- Gross farm profit (euros-€).

To address the missing values for the study, the missing data were replaced by applying the mean-based method. That is, the mean value of each group was calculated, and each missing variable received the mean value of the group to which it belongs.

One way to determine the degree or extent of similarity of two observations in the data is their distance. The Chebychev distance used in the study to classify farms maximizes the effect of the data distance and is expressed by the equation:

$$d_{ij} = \max_{i} \left| X_{ij} - X_{ik} \right|$$

The clustering of the farms was performed using the Ward's linkage method. The analysis was based on the estimation of the variance of the observations with the ultimate goal of estimating the distances between the clusters. Essentially, with the application of the method, the minimization of the variability between the two examined clusters that are formed in each successive stage of the hierarchical analysis was achieved.

A diagrammatic representation of a hierarchical classification method is the dendrogram on the basis of which the number of groups was determined [49]. In order to evaluate the a posteriori differences, multiple comparison tests were used through the Bonferroni test which follows the criterion of the least significant difference.

Summary cluster analysis is the organization of a collection of data or variables into classes based on a measure of similarity, in which special attention must be paid to the selection of variables that enter the analysis process [50,51].

2.5. European Production Efficiency Factor

To interpret the final result, the values of the variables in each group must be studied to assess whether the clusters actually exist or are simply the result of an algorithm. A useful evaluation tool that is directly related to the selected cluster classification variables is the European Production Efficiency Factor (EPEF) [36,39] given by the equation:

$$\frac{(100 - \text{ mortality})\ \% \ \times \ \text{final body weight (kg)}}{\text{FCR} \times \ \text{slaughter age of chicks (days)}} \times 100 = \text{EPEF}$$

3. Results

3.1. Hierarchical Cluster Analysis

The result of classification by applying the cluster analysis in the final sample of 110 broiler farms in the Region Epirus was the creation of the dendrogram (Figure 1) from which four (4) clusters of farms emerged. The distribution of farms by cluster and Regional Unit is given in Table 3.

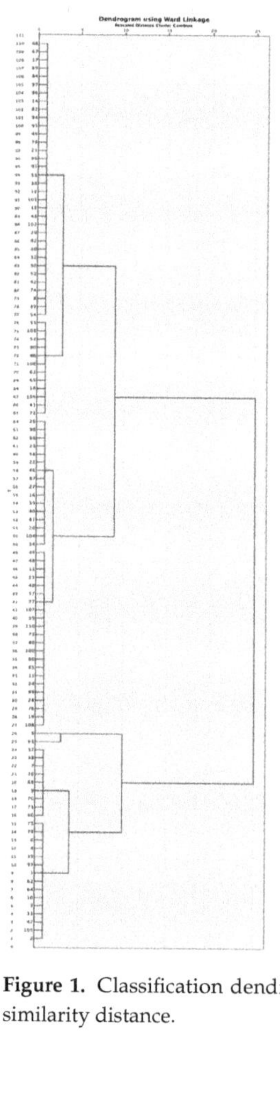

Figure 1. Classification dendrogram of 110 farms with Ward linkage technique and Chebychev similarity distance.

Table 3. Farm types per Regional Unit.

Cluster	Regional Unit			Total	%
	Arta	Ioannina	Preveza		
C1	11	35	0	46	41.8
C2	8	30	0	38	34.6
C3	1	1	0	2	1.8
C4	5	15	4	24	21.8
Total	25	81	4	110	100

The above table shows that the percentage of farms included in the 1st cluster amounts to 41.8% while the lowest frequency is shown by the 3rd cluster which includes only 2 farms. The 4th cluster includes farms with activity in all three Regional Units, with the Regional Unit of Ioannina dominating in number of farms and the other two being numerically equivalent.

3.2. Characteristics and Interpretation of Clusters—Structural Characterization

The formation of farm types was based on the aforementioned variables. Table 4 contains the averages by type of farm of the variables included in the analysis.

Table 4. Variable analysis of generated clusters.

Variable	Mean Value	C1	C2	C3	C4	p	SD
Number of farms (n)	110	46	38	2	24		
Production cycles/year/farm	5.07	5.22 ^{b}	4.58 ^{a}	6.00 ^{c}	5.50 ^{b}	<0.001	0.93
Feed conversion ratio	1.81	1.77 ^{a}	1.86 ^{a}	2.24 ^{b}	1.78 ^{a}	0.004	0.21
Annual rest period (days)	81.25	78.24	85.18	79.00	80.96	0.618	23.69
Consumed feed (tn)	673.42	565.42 ^{b}	271.25 ^{a}	4156.25 ^{d}	1226.97 ^{c}	<0.001	107.75
Mortality (%)	2.91	3.05 ^{c}	2.98 ^{bc}	1.56 ^{a}	2.65 ^{b}	0.046	0.35
Gross profit (€)	27,979	27,546 ^{c}	11,798 ^{b}	−202,419 ^{a}	73,628 ^{d}	<0.001	56,942
Farm size (broilers/production cycle)	29,468	26,847 ^{b}	14,118 ^{a}	127,000 ^{d}	50,666 ^{c}	<0.001	20,288
Total number of broilers/year	152,049	134,736 ^{b}	62,329 ^{a}	752,000 ^{d}	277,923 ^{c}	<0.001	120,700
Total live weight of broilers/year (tn/year)	372.11	323.47 ^{b}	147.76 ^{a}	1846.95 ^{d}	697.68 ^{c}	<0.001	304.16
Average days of a production cycle-length	45.26	45.17	44.94	42.50	46.20	0.349	3.50
Average slaughter weight (kg)	2.49	2.48	2.45	2.48	2.58	0.107	0.20
Average selling price/ kg of live weight (€)	1.11	1.11	1.09	1.12	1.12	0.137	0.02
Total production days/year	229.87	235.97 ^{b}	205.79 ^{a}	255.00 ^{c}	254.25 ^{c}	<0.001	45.73
Land (ha)	1.09	1.01 ^{b}	0.73 ^{a}	3.00 ^{d}	1.64 ^{c}	<0.001	0.92
Total Fixed Capital (€)	37,187	32,289 ^{b}	13,981 ^{a}	170,268 ^{d}	72,226 ^{c}	<0.001	30,500
Average Gross Revenue (€)	413,953	357,999 ^{b}	161,779 ^{a}	2,056,270 ^{d}	783,612 ^{c}	<0.001	340,681
Average Net Profit (€)	−16,678	−12,516 ^{b}	−9743 ^{c}	−366,669 ^{a}	−6469 ^{d}	<0.001	6411
Average Farm Income (€)	20,093	19,928 ^{c}	7357 ^{b}	−217,853 ^{a}	60,406 ^{d}	<0.001	55,583
Average Return on Capital (%)	18.90	−4.02 ^{b}	48.7 ^{d}	−20.5 ^{a}	12.4 ^{c}	0.018	10.6
Average production cost (€/kg of live weight)	0.82	0.82 ^{a}	0.83 ^{a}	0.94 ^{b}	0.81 ^{a}	0.014	0.12

^{a,b,c,d} Values with no common superscript in the same row are statistically different, at a significance level of $p \leq 0.05$, according to the results of the Bonferroni test.

The profile of each cluster was further investigated based on a broader framework of variables derived from field research based on the prevailing typology (4 clusters). In order to achieve optimal comparability and consequently an increased degree of interpretation of the results, the following ratio was used as a reference base:

4 LU (Livestock Units) = 1000 broilers,

Therefore, the quantities of the variables listed in Table 5 are expressed per 1000 reared chickens.

Table 5. Basic financial results per 4 LU (1000 broilers).

Variable	Mean Value	C1	C2	C3	C4
Number of farms (*n*)	110	46	38	2	24
Farm size (number of broilers)	29.47	26.85	14.12	127.00	50.67
Total number of broilers/year	152.05	134.74	62.33	752.00	277.92
Total Fixed Capital (€)	244.65	239.65	224.32	226.42	260.00
Average Gross Revenue (€)	2722.48	2656.97	2595.54	2734.40	2819.56
Average Net Profit (€)	−109.69	−92.90	−156.32	−487.59	−23.28
Average Farm Income (€)	132.15	147.90	118.04	−289.70	217.35

Cluster 1 (C1). Medium-size farms, with very good feed conversion rate ($n = 46$).

This cluster includes the largest percentage of farms in the sample. As noted, the capacity of C1 cluster farms amounts, on average, to 26,847 broilers (107.39 LU). In farms of this cluster, 5.22 production cycles are carried out per year. The characteristic feature of these farms is the achievement of the best FCR (1.77). However, they have the highest mortality rate (3.05%) compared to the other clusters.

The above type of farms achieves almost zero return in terms of invested capital, with the average gross profit amounting to EUR 27,546.46 per farm.

Cluster 2 (C2). Small farms, with mediocre feed conversion rate ($n = 38$).

The second cluster includes 38 farms characterized by low capacity per production cycle. Also, compared to the other clusters, the average number of production cycles per year (4.58) is significantly lower. In addition, they have a relatively high mortality rate (about 3%). Most of them are located within the geographical boundaries of the Regional Unit of Ioannina.

Farms of this type achieve the highest return on invested capital, which amounts approximately to 50%.

Cluster 3 (C3). Very large farms, with very bad feed conversion rate ($n = 2$).

This cluster includes 2 farms, in the Regional Units of Arta & Ioannina. The farms are characterized as very large in size as shown by the average production capacity. Based on the worst FCR among clusters (2.24), C3 shows a liability that can be interpreted in many ways, the main one focusing on the irrational use of the supplied feed or its possible loss during feeding. This is also reflected in high production cost per kg of live weight.

The above farming type shows the greatest economic losses as well as the lowest average return on invested capital.

Cluster 4 (C4). Large farms, with very good feed conversion rate ($n = 24$).

This cluster, which represents 21.82% of the farms in the sample, is characterized by large-scale farms with a capacity per production cycle of approximately 50,667 broilers. It also achieves good utilization of the supplied feed while producing broilers for slaughter with the highest average live weight (2.58 kg).

Farms of this type show the highest return on invested capital. The investment in fixed capital is much smaller compared to the other clusters (26.00 euros/1000 broilers). In addition, this specific farming type gives the highest gross income per 4 LU (2819.56 euros). According to the above results, the farms of this type are characterized as more economically efficient compared to other clusters.

The financial performance of each cluster for the key economic indexes is presented in Figure 2.

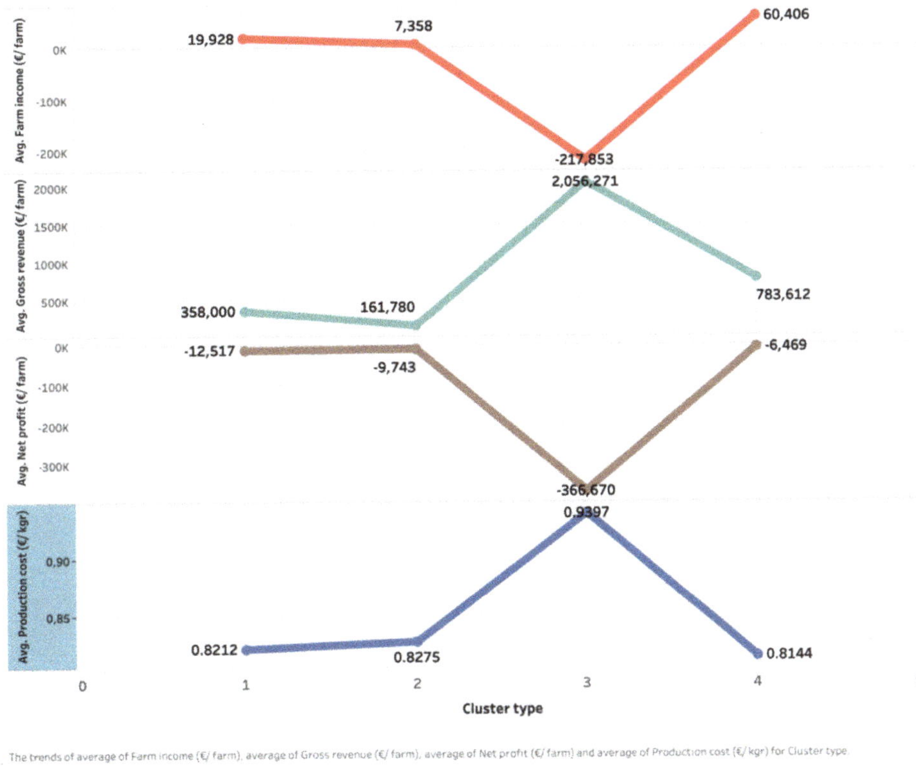

Figure 2. Financial performance of each cluster for the key economic indexes.

4. Discussion

The created clusters were characterized based on the average size and the average FCR after examining a wide range of variables that influenced the classification of the farms into one of the four clusters. From the results of the hierarchical cluster analysis it is clear that large farms of the C4 cluster, with their very good productive performance and the lowest average production costs, determine the path that will be followed by the broiler poultry industry in the Region of Epirus. Khan and Afzal [52] reported similar findings, namely that the highest net profit benefits were observed in large-scale broiler farms followed by medium and small farms. In this study, the results concerning the profitability of the farms revealed that they are negatively affected by the mortality of the chicks, something that is also evident from the classification of the farms in our study.

Emaikwu et al. [53] reports that the increase in farm size has positively affected the expansion of poultry activity, leading poultry farmers to focus on improving production management practices that allow for the efficient use of available resources and increased productivity. Based on the above, it is expected that the 2 very large farms of the C3 category will have remarkable financial results, as they manage to have 6 annual production cycles, an extremely low mortality rate (1.56%) and the shortest production period length (about 2 days earlier than the general average). The problem arises from the high feed consumption, which negatively affects both the production results (FCR = 2.24) and the economic results (production costs). Given that these are modern farms on the basis of their high fixed capital, in order to manage the existing problem, poultry farmers should focus on better management either by avoiding potentially extensive feed waste (e.g., poor feeder regulation) or by stricter control of inputs (quantities of feed).

The efficient, or inefficient, resource utilization of small-scale broiler farms, studied by Baba et al. [54], led them to conclude that underutilized production parameters should be increased, while the quantities of overused inputs should be more rationally managed in order to optimize their efficiency. Also, Souza et al. [13], using hierarchical cluster analysis to classify broiler farms by facility level and by checking the yield of the productivity index (PI), found that the larger farms with the largest poultry houses achieve higher performance.

Table 6 shows the average values recorded by the farms of the 4 clusters regarding the European Production Efficiency Factor (EPEF), reflecting their technical efficiency.

Table 6. Technical efficiency based on EPEF.

	Average	**C1 (*n* = 46)**	**C2 (*n* = 38)**	**C3 (*n* = 2)**	**C4 (*n* = 24)**
EPEF	295.10	300.73	284.36	256.44	305.29

From the data in the above table with the European Production Efficiency Factor (EPEF), it can be concluded that primarily the large farms, which constitute the C4 cluster, and secondarily the medium-sized farms of the most populous cluster, C1, seem to present an excellent evaluation of their technical parameters, which are consistent with their financial results.

The explanation for this situation is based on two components: (a) the large farms of C4 have significantly increased the economy of scale compared to C1 and even more importantly with the C2 farm cluster, which enters the EPEF calculation equation, and (b) the equation does not include the variable "yearly placements" (annual production cycles) that affects the total production volume, even more so as the size of the farms increases.

According to Wang et al. [55], the optimum return is achieved when the EPEF reaches the highest value, which in turn is achieved when the FCR and rearing duration reach the lowest value. However, the objective in broiler production is to increase the production volume per year [56], which is directly affected by the annual production cycles.

The effect of the number of yearly placements on key economic indexes is presented in Figure 3.

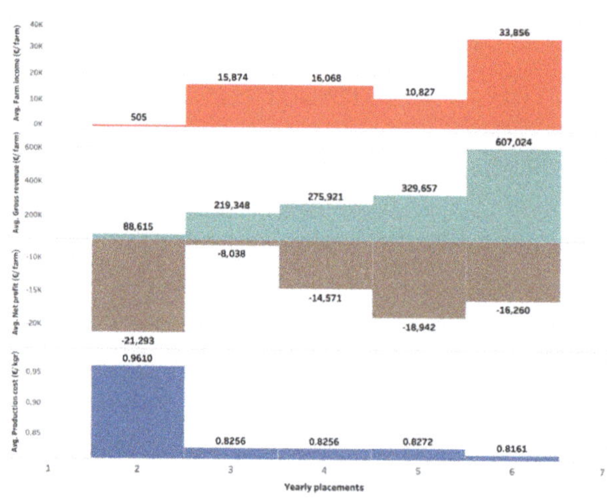

Figure 3. Effect of the number of yearly placements on key economic indexes.

In a similar study [38], the EPEF was found to be 300.6 when the fattening days are 46 and the annual production cycles are 6.07. According to these researchers, the EPEF increases with the reduction of fattening days as the FCR improves significantly, and there is potential for more yearly placements. Another parameter that significantly affects the efficiency of broiler farms and that enters the EPEF according to Goliomytis et al. [57] is the mortality rate, which for all clusters and especially for the C4 of large farms is extremely low (2.65%).

5. Conclusions

The size of the farm was a decisive parameter in the classification of farms by using hierarchical cluster analysis.

Although the C3 cluster of very large farms has been found to have good production characteristics, it cannot achieve profitability and economic sustainability if these farms do not improve their FCR. In other words, this is an example of inefficient management of modernized industrial-scale farms. Their inefficient way of operating does not allow their future financial viability and does not ensure the environmentally sustainable development of the sector.

The C2 cluster of small-scale farms is unable to follow the trend of our time to increase mass production, as these farmers do not seem willing to invest in fixed capital. Poultry farmers of this category (C2) do not have high fixed capital costs, as the facilities they use are old. However, they will have to adopt development strategies, based mainly on the modernization of current facilities, in order to improve the FCR and increase the annual number of production cycles. These farmers, who do not envision the sustainable development of their farms both from an economic and a socio-environmental point of view can be considered as conservative.

The C1 cluster of medium-sized farms seems to be on a path of modernization, but only in terms of improving their existing infrastructure, a remark confirmed by the relatively low fixed capital investment. Their size does not allow a significant increase in production volume even if their productivity parameters are intensified. In this case, too, the profile of poultry farmers needs to be investigated in order to interpret their reluctance/incapacity for further investments, so that this knowledge can be utilized by poultry companies and government agencies to create development funding tools appropriate to support the needs of the broiler poultry industry.

The already modernized large farms of the C4 cluster, in the current conditions, cannot achieve positive net profitability if their operating costs are not reduced, because the planning of production absorption by their organizations does not allow them to further increase production. This is also illustrated by two parameters in Table 4 related to the satisfactory, for the study area, status (FCR), namely the "average slaughter weight", which is 100 g higher than the overall average and the "average production cycle length" which is 1 day higher than the corresponding average. These observations suggest that in some production cycles it is difficult to absorb the production volume, thus affecting the efficiency (worse than expected FCR) of the C4 cluster. Finally, it seems that the C4 cluster is the model with the highest prospect of sustainable development of the industry.

The perspective of sustainable development of the sector in the Region of Epirus, an area remote from the urban centers, will contribute significantly to the social sustainability of the region, partially ensuring the employment of its population. In order to achieve sustainable development after the year 2011, a large co-operative in the area established a subsidiary whose main purpose is to utilize renewable resources for the energy requirements of their farmers and to implement an integrated waste management system that views poultry waste as an asset rather than a liability [3]. Similar actions should be adopted by the whole industry in the Region of Epirus, in order to develop into an even more environmentally friendly livestock activity, but also to reduce the production costs that will ensure the economic sustainability of broiler farming.

Author Contributions: Conceptualization, V.D.; methodology, L.H. and I.K.; software, D.G. and I.K.; validation, I.M. and G.S.; formal analysis, D.G., L.H. and I.K.; investigation, V.D. and L.H.; resources, V.D. and L.H.; data curation, L.H., I.K. and I.M.; writing—original draft preparation, D.G. and L.H.; writing—review and editing, V.D. and G.S.; visualization, D.G., I.M. and G.S.; supervision, V.D.; project administration, V.D.; funding acquisition, V.D. and L.H. All authors have read and agreed to the published version of the manuscript.

Funding: This research was funded by the Region of Epirus in the framework of the project with C.N. 80250 and title "Pilot study to investigate development strategies of the poultry industry in the Region of Epirus".

Data Availability Statement: All the relevant data are available in the paper.

Acknowledgments: We sincerely thank the poultry farmers who participated in the survey and answered the questions asked to them through personal interviews.

Conflicts of Interest: The authors declare no conflict of interest.

References

1. Elstat, Hellenic Statistical Authority. 2018. Available online: www.statistics.gr (accessed on 26 March 2019).
2. Camargo-Barros, G.S.; Zen, S.D.; Piedade-Bacchi, M.R.; Galvão de Miranda, S.H.; Narrod, C.; Tiongco, M. Policy, technical, and environmental determinants and implications of the scaling-up of swine, broiler, layer and milk production in Brazil. In *Annex V, Final Report of IFPRI-FAO Livestock Industrialization Project: Phase II*; International Food Policy Research Institute: Washington, DC, USA, 2003.
3. Iliopoulos, C.; Theodorakopoulou, I. Measuring and Communicating the True Value of Membership: The Case of the Pindos Poultry Co-operative. In *Sustainable Co-operative Enterprise: Case Studies of Organisational Resilience in the Co-operative Business Model*; Chapter 17; Mazzarol, T., Reboud, S., Limnios, E.M., Eds.; Edward Elgar Publishing: Cheltenham, UK, 2014; pp. 223–239.
4. Burgos, S.; Hong Hanh, P.T.; Roland-Holst, D.; Burgos, S.A. Characterization of Poultry Production Systems in Vietnam. *Int. J. Poult. Sci.* **2007**, *6*, 709–712. [CrossRef]
5. Keramidou, I.; Mimis, A.; Pappa, E. Performance evaluation of the poultry sector in Greece. *J. Food Agric. Environ.* **2011**, *9*, 431–437.
6. Koutoulis, K. Present and future of poultry industry in Greece. In Proceedings of the 12th Pan-Hellenic Veterinary Congress, Athens, Greece, 5–8 April 2012.
7. Eurostat Trade Statistics. 2018. Available online: https://ec.europa.eu/eurostat/web/main/data/database (accessed on 26 March 2019).
8. Eurostat. *Sustainable Development in the European Union*; Key Messages; Office for Official Publications of the European Communities: Luxembourg, 2015. Available online: https://ec.europa.eu/eurostat/documents/3217494/6975281/KS-GT-15-001-EN-N.pdf (accessed on 26 March 2019).
9. Dormady, N.; Roa-Henriquez, A.; Rose, A. The Resilience of the Firm. A Production Theory Approach. *Int. J. Prod. Econ.* **2017**, *208*, 446–460. [CrossRef]
10. Unay-Gailhard, I.; Balmann, A.; Appel, F. Dimensions of SURE-Farm Farm Typology for Farm Resilience Assessments. In *Working Paper (727520)*; Leibniz Institute of Agricultural Development in Transition Economies (IAMO): Madrid, Spain, 2018.
11. Andersen, E. Regional Typologies of Farming Systems Contexts, SEAMLESS Report No.53, SEAMLESS 2010 Integrated Project, EU 6th Framework Programme, Contract no. 010036-2. Available online: http://www.seamless-ip.org/Reports/Report_53_PD4.4.3.pdf (accessed on 26 March 2019).
12. Kruskal, J. The Relationship between Multidimensional Scaling and Clustering. In *Classification and Clustering*; Van Ryzin, J., Ed.; Academic Press Inc.: New York, NY, USA, 1977; pp. 17–45.
13. Souza, F.; Milani, A.; Furlan, R.; Nogueira, W.; Fernandez Alarcon, M.F.; Almeida, E. Cluster Analysis to Study the Typology and Productivity in Broiler Facilities. In Proceedings of the International Conference of Agricultural Engineering (CIGR-Ageng2012), Valencia, Spain, 6–12 July 2012.
14. Worapol, A.; Manakant, I.; Jaroon, W.; Thanapol, N.; Surangkanang, C.; Prayat, S.; Komvut, T.; Neti, J.; Khlahan, S.; Tawinwaang, T. Poultry Production Clusters (PPCs) after AI Outbreaks in Thailand: Past, Present and Future Direction. *Int. J. Poult. Sci.* **2012**, *11*, 541–550.
15. Thevenot, A.; Aubin, J.; Tillard, E.; Vayssieres, J. Accounting for farm diversity in Life Cycle Assessment studies—The case of poultry production in a tropical island. *J. Clean. Prod.* **2013**, *57*, 280–292. [CrossRef]
16. Rymuza, K.; Grużewska, A.; Biesieda-Drzazga, B. Multidimensional analyses in studies on broiler chicken production results. *Europ. Poult. Sci.* **2014**, *78*, 22.
17. Gocsik, E.; Brooshooft, S.D.; De Jong, I.C.; Saatkamp, H.W. Cost efficiency of animal welfare in broiler production systems: A pilot study using the Welfare Quality® assessment protocol. *Agric. Syst.* **2016**, *146*, 55–69. [CrossRef]
18. Roskam, J.L.; Oude Lansink, A.G.J.M.; Saatkamp, H.W. The relation between technical farm performance and antimicrobial use of broiler farms. *Poult. Sci.* **2020**, *99*, 1349–1356.

19. Bastianoni, S.; Boggia, A.; Castellini, C.; Di Stefano, C.; Niccolucci, V.; Novelli, E.; Paolotti, L.; Pizzigallo, A. Measuring Environmental Sustainability of Intensive Poultry-Rearing System. In *Genetic Engineering, Biofertilisation, Soil Quality and Organic Farming*; Lichtfouse, E., Ed.; Springer: Dordrecht, The Netherlands, 2010; Volume 4, pp. 277–309.
20. Castellini, C.; Bastianoni, S.; Granai, C.; Dal Bosco, A.; Brunetti, M. Sustainability of poultry production using the emergy approach: Comparison of conventional and organic rearing systems. *Agric. Ecosyst. Environ.* **2006**, *114*, 343–350. [CrossRef]
21. Vaarst, M.; Steenfeldt, S.; Horsted, K. Sustainable development perspectives of poultry production. *Worlds Poult. Sci. J.* **2015**, *71*, 609–620. [CrossRef]
22. Chotteau, P.; Beaumond, N.; Deblitz, C.; Hoste, R.; Magdelaine, P.; Mottet, A.; De Roest, K.; Roguet, C.; Sarzeaud, P.; Topliff, M.; et al. The Impact of Increased Operating Costs on Meat Livestock in the EU. Study for the European Parliament. Paris. Institut de l' Elevage. 2009. Available online: http://www.europarl.europa.eu/RegData/etudes/etudes/join/2009/419109/IPOLAGRI_ET(2009)419109_EN.pdf (accessed on 18 March 2019).
23. Van Horne, P.L.M. Competitiveness of the EU poultry Meat Sector, Base Year 2017. In *Wageningen Economic Research Report 2018-116*; Wageningen University & Research: Wageningen, The Netherlands, December 2018. [CrossRef]
24. Leinonen, I.; Kyriazakis, I. How can we improve the environmental sustainability of poultry production? *Proc. Nutr. Soc.* **2016**, *75*, 265–273. [CrossRef]
25. Gerber, P.; Opio, C.; Steinfeld, H. *Poultry Production and the Environment—A Review*; Animal Production and Health Division, FAO: Rome, Italy, 2007; Available online: http://www.fao.org/ag/againfo/home/events/bangkok2007/docs/part2/2_2.pdf (accessed on 11 March 2019).
26. Sell-Kubiak, E.; Wimmers, K.; Reyer, H.; Szwaczkowski, T. Genetic aspects of feed efficiency and reduction of environmental footprint in broilers: A review. *J. Appl. Genet.* **2017**, *58*, 487–498. [CrossRef]
27. Hume, D.A.; Whitelaw, C.B.A.; Archibald, A.L. The future of animal production: Improving productivity and sustainability. *J. Agric. Sci.* **2011**, *149*, 9–16. [CrossRef]
28. De Verdal, H.; Mignon-Grasteau, S.; Bastianelli, D.; Même, N.; Le Bihan-Duval, E.; Narcy, A. Reducing the environmental impact of poultry breeding by genetic selection. *J. Anim. Sci.* **2013**, *91*, 613–622. [CrossRef] [PubMed]
29. Kawsar, M.H.; Chowdhury, S.D.; Raha, S.K.; Hossain, M.M. An analysis of factors affecting the profitability of small-scale broiler farming in Bangladesh. *Worlds Poult. Sci. J.* **2013**, *69*, 676–686. [CrossRef]
30. Martins, J.M.S.; Carvalho, C.M.C.; Litz, F.H.; Silveira, M.M.; Moraes, C.A.; Silva, M.C.A.; Fagundes, N.S.; Fernandes, E.A. Productive and Economic Performance of Broiler Chickens Subjected to Different Nutritional Plans. *Braz. J. Poult. Sci.* **2016**, *18*, 209–216. [CrossRef]
31. Kamran, Z.; Sarwar, M.; Nisa, M.; Nadeem, M.A.; Mahmood, S.; Babar, M.E.; Ahmed, S. Effect of low-protein diets having constant energy-to-protein ratio on performance and carcass characteristics of broiler chickens from one to thirty-five days of age. *Poult. Sci.* **2008**, *87*, 468–474. [CrossRef] [PubMed]
32. Moosavi, M.; Eslami, M.; Chaji, M.; Boujarpour, M. Economic value of diets with different levels of energy and protein with constant ratio on broiler chickens. *J. Anim. Vet. Adv.* **2011**, *10*, 709–711. [CrossRef]
33. Willems, O.W.; Miller, S.P.; Wood, B.J. Assessment of residual body weight gain and residual intake and body weight gain as feed efficiency traits in the Turkey (*Meleagris gallopavo*). *Genet. Sel. Evol.* **2013**, *45*, 26–65. [CrossRef]
34. Lourenço, I.C.; Branco, M.C. Determinants of corporate sustainability performance in emerging markets: The Brazilian case. *J. Clean. Prod.* **2013**, *57*, 134–141. [CrossRef]
35. Reyer, H.; Hawken, R.; Murani, E.; Ponsuksili, S.; Wimmers, K. The genetics of feed conversion efficiency traits in a commercial broiler line. *Sci. Rep.* **2015**, *5*, 16387. [CrossRef]
36. Marcu, A.; Vacaru-Opris, I.; Gabi, D.; Liliana, P.C.; Marcu, A.; Marioara, N.; Ioan, P.; Dorel, D.; Bartolomeu, K.; Cosmin, M. The Influence of Genetics on Economic Efficiency of Broiler Chickens Growth. *Anim. Sci. Biotech.* **2013**, *46*, 339–346.
37. Tolimir, N.; Milošević, N.; Ceranić, C.; Škrbić, Z.; Maslovarić, M. The effect of feeding programmes on broiler performance and production efficiency. In Proceedings of the International Symposium on Animal Science, Belgrade-Zemun, Serbia, 23–25 September 2014; Available online: http://agris.fao.org/agris-search/search.do?recordID=RS2017000916 (accessed on 11 March 2019).
38. Szőllősi, L.; Szűcs, I.; Nábrádi, A. Economic issues of broiler production length. *Econ. Agric.* **2014**, *61*, 633–646. [CrossRef]
39. Tandoğan, M.; Cicek, H. Technical Performance and Cost Analysis of Broiler Production in Turkey. *Braz. J. Poult. Sci.* **2016**, *18*, 169–174. [CrossRef]
40. Abdel-Latif, M.A.; El-Far, A.H.; Elbestawy, A.R.; Ghanem, R.; Mousa, S.A.; Abd El-Hamid, H.S. Exogenous dietary lysozyme improves the growth performance and gut microbiota in broiler chickens targeting the antioxidant and non-specific immunity mRNA expression. *PLoS ONE* **2017**, *12*, e0185153.
41. El-Menawey, M.; Attallah, A.; Abdel-Aziz, Y.; Yossif, M.; Gharib, H. Characterization and constrains of semi-intensive broiler production system in rural sector of Al-Sharkia and El-Qalioba governorates, Egypt. *Egypt. Poult. Sci. J.* **2019**, *39*, 311–326. [CrossRef]
42. Food and Agriculture Organization of the United Nations. Guidelines for Development of a Classification System Related to Farm Typology. 2018. Available online: http://www.fao.org/3/ca6391en/ca6391en.pdf (accessed on 26 March 2019).
43. Neyman, J. On the two different aspects of the representative methods. *J. R. Stat. Soc.* **1934**, *97*, 558–606. [CrossRef]
44. Cochran, W.G. *Sampling Techniques*; John Wiley & Sons: New Delhi, India, 1977.

45. Brannen, J. Mixing Methods: The Entry of Qualitative and Quantitative Approaches into the Research Process. *Int. J. Soc. Res. Methodol.* **2005**, *8*, 173–184. [CrossRef]
46. Daoutopoulos, G.; Kazakopoulos, L.; Koussi, M. *Agricultural Sociology*; Zygos Publications: Thessaloniki, Greece, 2005.
47. Siardos, G. *Social Research Methodology*; Ziti Publications: Thessaloniki, Greece, 2009.
48. Iraizoz, B.; Gorton, M.; Davidova, S. Segmenting farms for analysing agricultural trajectories: A case study of the Navarra region in Spain. *Agric. Syst.* **2007**, *93*, 143–169. [CrossRef]
49. Norušis, M.J. IBM SPSS Statistics 19. In *Statistical Procedures Companion*; Prentice Hall: Upper Saddle River, NJ, USA, 2011.
50. Hair, J.F.; Black, W.C.; Babin, B.J.; Anderson, R.E.; Tatham, L.R. *Multivariate Data Analysis*, 6th ed.; Pearson Prentice Hall: Upper Saddle River, NJ, USA, 2007.
51. Everitt, B.S.; Landau, S.; Leese, M.; Stahl, D. *Cluster Analysis*; John Wiley & Sons: Chichester, UK, 2011.
52. Khan., M.; Afzal, M. Profitability analysis of different farm size of broiler poultry in district dir (lower). *Sarhad J. Agric.* **2018**, *34*, 389–394. [CrossRef]
53. Emaikwu, K.K.; Chikwendu, D.O.; Sania, S. Determinants of flock size in broiler production in Kaduna State of Nigeria. *J. Agric. Ext. Rural Dev.* **2011**, *3*, 202–211.
54. Baba, M.D.; Attahiru, M.; Sanda, A.Y. Estimating Resource Productivity and Efficiency of Smallholder Broiler Production in Bida Metropolis, Niger State, Nigeria. *N. Y. Sci. J.* **2016**, *9*, 24–28.
55. Wang, B.Y.; Chien, L.H.; Roan, S.W. POMA-BROILER: A Computer Simulation Model to Evaluate the Optimal Market Age of Broilers. *J. Anim. Vet. Adv.* **2012**, *11*, 2493–2502.
56. Samarakoon, S.M.R.; Samarasinghe, K. Strategies to improve the cost effectiveness of broiler production. *Trop. Agric. Res.* **2012**, *23*, 338–346. [CrossRef]
57. Goliomytis, M.; Panopoulou, E.; Rogdakis, E. Growth Curves for Body Weight and Major Component Parts, Feed Consumption, and Mortality of Male Broiler Chickens Raised to Maturity. *Poult. Sci.* **2003**, *82*, 1061–1068. [CrossRef] [PubMed]

sustainability

Article

Redefining the Use of Vinification Waste By-Products in Broiler Diets

Elisavet Giamouri [1], Alexandros Mavrommatis [1], Panagiotis E. Simitzis [2], Christina Mitsiopoulou [1], Serkos A. Haroutounian [1], Apostolis Koutinas [3], Athanasios C. Pappas [1] and Eleni Tsiplakou [1,*]

1 Laboratory of Nutritional Physiology and Feeding, Department of Animal Science, School of Animal Biosciences, Agricultural University of Athens, Iera Odos 75, 11855 Athens, Greece
2 Laboratory of Animal Breeding & Husbandry, Department of Animal Science, Agricultural University of Athens, Iera Odos 75, 11855 Athens, Greece
3 Laboratory of Food Process Engineering, Department of Food Science and Human Nutrition, Agricultural University of Athens, Iera Odos 75, 11855 Athens, Greece
* Correspondence: eltsiplakou@aua.gr; Tel.: +30-210-529-4435; Fax: +30-210-529-4413

Abstract: In this study, the use of vinification by-products in broiler diets, as a sustainable and promising way of exploiting them, was examined. In particular, the potential use of ground grape pomace (GGP), wine lees extract (WYC) and grape stem extract (PE) in broiler diets was examined. Growth performance parameters, the weight of selected internal organs, meat quality traits, fatty acid profiles of breast meat and selected haematological parameters were determined. Two hundred and forty one-day-old broilers were assigned to four treatments with four replicate pens of fifteen broilers. There was one control treatment (CON), fed a basal diet, and the GGP, WYC and PE treatments, fed a basal diet supplemented with 25 g/kg GGP, 2 g/kg WYC and 1 g starch including 100 mg pure stem extract/kg PE, respectively. The duration of the experiment was 42 days. The average body weight gain during the starter, grower and finisher stages did not differ among treatments. Similarly, the feed intake, FCR and carcass yield did not show a significant difference. The weight of the internal organs was also similar among treatments. Some positive differences were observed in colour traits of meat and in haematological parameters. In the GGP group, saturated (SFAs) and unsaturated fatty acids (USFAs) were lower and higher, respectively, compared to the CON, WYC and PE groups. Vinification by-products seem to be a promising feed additive in broiler diets providing a sustainable approach to grape waste management without affecting broiler performance.

Keywords: broilers; grape pomace; grape stems; fatty acids; haematological parameters; meat quality; vinification by-products; waste

Citation: Giamouri, E.; Mavrommatis, A.; Simitzis, P.E.; Mitsiopoulou, C.; Haroutounian, S.A.; Koutinas, A.; Pappas, A.C.; Tsiplakou, E. Redefining the Use of Vinification Waste By-Products in Broiler Diets. *Sustainability* **2022**, *14*, 15714. https://doi.org/10.3390/su142315714

Academic Editor: Jacopo Bacenetti

Received: 31 October 2022
Accepted: 22 November 2022
Published: 25 November 2022

1. Introduction

The global population is predicted to increase to 9.7 billion people by 2050 according to The United Nations [1]. As a result, more animal-origin products such as meat is likely to be necessary to cover consumers' high demands. However, the limitation of natural resources is projected to pose an extra obstacle in food production [2]. Meanwhile, the cost of conventional feedstuffs that farmers are facing nowadays is extremely high. Hence, searching for alternative feedstuffs that could be used in animal diets is necessary.

Several metric tons of agro-industrial biomass are wasted every year in the European Union [3]. The waste of this origin could be used in several sectors of the industry, including, but not limited to, animal diets, litter, organic fertilizers or as a source of heating and cooking [4]. Vineyards and wineries create high amounts of by-products since *Vitus vinifera* (the common grape vine) is considered to be the fruit crop with the highest production, exceeding 75 million tons per year, 41% of which is produced in Europe [5]. The majority of grape production is used for winemaking, while the Mediterranean area produces approximately 62% of the wine on a global scale [6]. By-products are produced during the destemming of grapes and during the pressing and extracting procedure. The large quantities of grape waste produced annually pose a challenge for management techniques [7]. The wastes derived from wine making are mainly stems, in particular, the woody part of the grapevine, and the grape pomace, such as the skins, stems and seeds, that compose the solid residue of vinification [8]. Grape pomace accounts for 62% of the total amount of waste produced, while wine lees and stems account for 14% and 12%, respectively [9].

The biodegradation of this kind of waste in open fields, as a means of disposing of them, not only leads to a degradation of the environment and water resources [10], but also causes the waste of bioactive compounds present in these by-products [11]. The incorporation of grape by-products in broiler diets is a more sustainable approach regarding grape waste management compared to the aforementioned disposal methods, and may produce healthy products for human consumption, since winery by-products contain several bioactive compounds providing high added value for incorporation into animal diets [12,13]. A wide range of polyphenols, such as flavonoids, catechins and epicatechins, are dominant in these by-products [14]. Vinification by-products could be added to animal diets in varying levels depending on the antinutritional components present. A high crude fibre content along with antinutritional factors such as tannins present in grape pomace could have a negative effect on broiler performance [15].

This study was part of a project designed to assess the effects of adding vinification by-products (grape pomace, wine lees and stem extract) to broiler diets. Previously, we evaluated how vinification by-products affected the expression of genes involved in the oxidative status and total antioxidant capacity [16], and how the inclusion of grape by-products affected the transcriptional profiling of genes regulating the immune system in selected organs [17]. In the present study, the impact of these by-products in broiler performance, carcass yield, selected haematological parameters, the weight of internal organs, meat quality and breast fatty acid profiles of broilers was evaluated.

2. Materials and Methods

2.1. The Procurement of Grape By-Products

The process of acquiring vinification by-products and the compositional analysis was already presented in a previous article [16].

2.2. Broilers' Trial, Diets and Experimental Procedure

For the experimental trial, two hundred and forty as-hatched, one-day-old, Aviagen Ross 308 broilers were used. Broilers were provided from a commercial hatchery. The experimental trial lasted 42 days and conformed with the guidelines of the European Union Directive EU 63/2010 for the protection of animals used for scientific purposes and the Council of the European Union.

Broilers were randomly assigned to 4 experimental treatments with four replicate floor pens of 15 broilers each. Feeding treatments were, namely, the control, GGP, WYC and PE groups, as previously described [16]. In the control group (CON), broilers were fed a basal diet based on corn and soybean meal; in the GGP group, broilers were fed a basal diet supplemented with 25 g/kg GGP; in the WYC group, broilers were fed a basal diet supplemented with 2 g/kg WYC; and in the PE group, broilers were fed a basal diet supplemented with 1 g starch including 100 mg of pure stem extract/kg (PE). Each pen (2 m^2) was covered with wheat straw litter. The stocking density in each pen did not exceed 33 kg/m^2, according to directive 2007/43/EC. The housing conditions (light and ventilation) were controlled. Every pen was provided with a heating infrared lamp for keeping the broilers warm, while the temperature was set to 32 °C for the first week and was gradually decreased every week. Diets were isoenergetic and isonitrogenous and were formulated according to the Aviagen recommendations for each growth phase, namely, the starter (0–10 days), grower (11–24 days) and finisher (25–42 days) phases. Feed and water were provided ad libitum. The composition of the diets for every growing phase, as well as the determined and calculated analysis of the diets, was described in our previous work [16] (Tables S1 and S2).

2.3. Body Weight and Carcass Evaluation—Sampling

The initial body weight (BW) and body weight at the end of each growing phase, on the 10th, 24th and 42nd days, were recorded. The feed intake was also recorded for every growing phase, and the feed conversion ratio (FCR) was calculated at the end of each growing phase.

On the 42nd day, 32 broilers (8 per treatment and 2 per replicate pen) were randomly selected and sacrificed in order to assess the effect of vinification by-products on carcass yield. Samples of blood and selected internal organs (spleen, liver and bursa of Fabricius) were collected and their weight expressed as the % of the total body weight. Approximately 6 mL of whole blood was immediately transferred to heparin-containing tubes (170 units heparin; BD Vacutainer, Plymouth, UK) and stored in an icebox (Thomas Scientific, Swedesboro, NJ, USA) until its transfer to the Laboratory of Nutritional Physiology and Feeding. Then, the blood samples were centrifuged (SL16R, Thermo Fisher Scientific, Waltham, MA, USA) at 2500 rpm for 15 min at 4 °C to separate the plasma from the cells. The carcasses were kept for 24 h in a fridge at 4 °C and were weighed for the estimation of carcass yield. The pectoralis major breast muscle was removed and used for the measurement of the meat quality traits and the analysis of the fatty acid profiles.

2.4. Determination of Haematological Parameters and Internal Organ Weight

An automatic ABX Pentra 400 analyser (Horiba-ABX, Montpellier, France) was used for the determination of the haematological parameters. In particular, the serum of blood samples from 20 broilers was used to assess aspartate aminotransferase (SGOT-AST) (IU/L), alanine aminotransferase (SGPT-ALT) (IU/L), blood urea nitrogen (BUN) (mg/dL), γ-glutamyltransferase (γ-GT) (IU/L), alkaline phosphatase (SAP) (IU/L), cholesterol (CHOL) (mg/dL), fractions of albumins (ALB) (g/dL), total proteins (CP) (g/dL) and sfairines (SFAIR) (g/dL).

2.5. Meat Quality: pH_{24}, Colour, Shear Force and Cooking Loss

The electrode of a pH meter (HI 99,163 Meat pH Temperature Meter, Hanna Instruments, Nusfalau, Romania) was inserted into the breast muscle for the measurement of the pH 24 h postmortem. The meat colour was determined after remaining in room temperature for 30 min. A Miniscan XE (HunterLab, Reston, USA) was used and set to the L*, a*, b* system (CIE, Commission Internationale de l'Eclairage, 1976), with white and black tiles as standard [18]. For the determination of the cooking loss, meat samples were weighed and placed in plastic bags and were cooked for 30 min at 85 °C in a water bath. Afterwards, the samples were left at room temperature under running tap water and weighed again to calculate the cooking loss (%). The shear force was determined as described by Cason et al. [19]. Every muscle was cut parallel to the muscle fibres in three strips of 1 cm^2 each using a Zwick Testing Machine (Model Z2.5/TN1S; Zwick GmbH & Co, Ulm, Germany) equipped with a shear blade (Warner-Bratzler G146; Instron, Grove City, PA, USA). The peak force measurements were calculated as N/mm^2.

2.6. Fatty Acids in Breast Meat

Breast tissue samples were partially thawed at 4 °C and trimmed to remove any external adipose and connective tissue. The total fatty acids (FAs) were extracted and methylated directly, according to the method of O' Fallon et al. [20]. For the determination of the FA profile, an Agilent 6890 N gas chromatograph equipped with an HP-88 capillary column (60 m × 0.25 mm i.d. with 0.20 μm film thickness, Agilent, Santa Clara, CA, USA) and a flame ionization detector (FID) was used. Each peak was identified and quantified using a 37-component FA methyl ester (FAME) standard mix (Supelco, Sigma-Aldrich, St. Louis, MO, USA).

2.7. Statistical Analyses

The statistical analyses were performed using SPSS IBM software and the results were depicted as means and the standard error of means (SEM). For the broilers' growth performance, the experimental unit consisted of the replicate pen. The dietary effects were monitored using one-way ANOVA, followed by Tukey's test. Statistical significance was set at $p \leq 0.05$. Gender was not included in the statistical model as previously justified.

3. Results

3.1. Growth Performance Parameters and Carcass Yield

Broiler growth performance for every growing period is presented in Table 1. At the end of each of the three growing phases, namely, the starter (0–10 days), grower (11–24 days) and finisher (25–42 days), no differences were measured in body weight gain (ABG). Only the feed intake in the PE group during the grower phase (11–24 days) was higher by approximately 5.45, 7.35 and 8%, compared to the CON, GGP and WYC groups, respectively. Moreover, the BW on the 42nd day numerically increased ($p > 0.05$) in the PE group by approximately 4, 3.9 and 3% in comparison with the CON, GGP and WYC groups, respectively. The FCR was slightly better ($p = 0.049$) in the PE group compared with the CON group during the starter phase. No differences were observed for the carcass yield either.

Table 1. Broiler growth performance on starter, grower and finisher experimental periods among the four dietary treatments.

	Dietary Treatment					
	CON	**GGP**	**WYC**	**PE**	**SEM**	**Significance**
Initial BW (g)	44.08	44.08	44.79	44.50	0.540	0.824
			Days 0–10			
ABG (g)	232.7	238.8	231.4	254.3	8.596	0.247
AFI (g)	288.7	287.5	285.7	292.9	7.895	0.925
BW 10 (g)	276.9	283.0	276.2	298.7	7.245	0.248
FCR	1.24 ^{A}	1.20 ^{AB}	1.24 ^{AB}	1.15 ^{B}	0.023	0.049
Mortality (%)	0	1.67	0	0	0.105	0.426
			Days 11–24			
ABG (g)	952.0	926.8	931.4	991.4	23.569	0.256
AFI (g)	1195 ^{B}	1171 ^{B}	1164 ^{B}	1264 ^{A}	25.896	0.050
BW 24 (g)	1229	1210	1208	1290	27.495	0.215
FCR	1.26	1.26	1.25	1.28	0.008	0.399
Mortality (%)	0	0	3.33	1.67	0.224	0.248
			Days 25–42			
ABG (g)	1747	1772	1797	1813	42.569	0.801
AFI (g)	2722	2737	2749	2857	61.598	0.463
BW 42 (g)	2976	2982	3005	3104	61.812	0.489
FCR	1.56	1.54	1.54	1.58	0.029	0.863
Mortality (%)	0	1.67	1.66	0	0.208	0.588
Carcass yield (%)	75.81	75.93	76.28	77.13	0.450	0.193

BW: body weight; AFI: average feed intake; FCR: feed conversion ratio (g feed/g gain); ABG: average body gain; SEM: pooled standard error of means; CON: control treatment; GGP: 25 g/kg ground grape pomace; PE: 1 g starch including 100 mg pure stem extract/kg; WYC: 2 g/kg wine yeast extract. Means with different capital superscript (A, B) in each row indicate significant difference ($p \leq 0.05$).

3.2. Haematological Parameters and Weight of Internal Organs

As an indicator of broiler health, selected haematological parameters and the weight of specific internal organs were examined. Table 2 presents the analysed data. In particular, a significantly lower value of alanine aminotransferase (SGPT-ALT) was noted in the WYC group with the wine lees extract. The remaining examined blood biochemical parameters were unaffected by the dietary treatments. Sfairines tended to be lower in the WYC treatment compared to the CON group, but no statistical significance was observed. Moreover, no differences were observed in the weight of the internal organs, as indicated in Table 2. The weight of the spleen, liver and bursa of Fabricius did not differ among treatments.

Table 2. Effect of diet on blood serum glutamate oxaloacetate transaminase (SGOT-AST) (IU/L), glutamate pyruvate transaminase (SGPT-ALT) (IU/L), urea nitrogen (BUN) (mg/dL), γ-glutamyl transferase (γ-GT) (IU/L), alkaline phosphatase (IU/L), cholesterol (mg/dL), fractions of albumins (g/dL), CP (g/dL) and sfairines (g/dL).

	Dietary Treatment					
	CON	GGP	WYC	PE	SEM	Significance
SGOT-AST (IU/L)	752	938	871	979	110.45	0.577
SGPT-ALT (IU/L)	14.75^{B}	15.00^{B}	6.50^{A}	16.25^{B}	1.02	0.002
BUN (mg/dL)	1.81	2.27	1.81	2.38	0.18	0.290
γ-GT (IU/L)	16.50	15.75	11.50	14.75	2.75	0.686
SAP (IU/L)	1351	1434	1693	1730	245.78	0.733
CHOL (mg/dL)	123	128.5	129.3	141.8	6.89	0.384
ALB (g/dL)	1.3	1.32	1.32	1.40	0.07	0.714
CP (g/dL)	3.15	2.97	2.72	3.22	0.19	0.225
SFAIR (g/dL)	1.85^{T}	1.65	1.40^{T}	1.82	0.15	0.083
Spleen (% of BW)	0.094^{T}	0.075^{T}	0.081	0.080	0.005	0.095
Liver (% of BW)	1.59	1.49	1.57	1.50	0.050	0.403
Bursa of Fabricius (% of BW)	0.071	0.063	0.065	0.053	0.006	0.192

SEM: pooled standard error of means; CON: control treatment; GGP: 25 g/kg ground grape pomace; PE: 1 g starch including 100 mg pure stem extract/kg; WYC: 2 g/kg wine yeast extract. Means with different capital superscript (A, B, T) in each row indicate significant difference ($p \leq 0.05$).

3.3. Meat Quality Indices

In Table 3, the parameters of meat quality are summarized. The feeding grape by-products affected the colour trait of lightness. A higher value of L* was observed in the groups fed diets containing ground grape pomace (GGP), wine lees extract (WYC) and grape stems extract (PE) compared to the control group, indicating a lighter meat colour. The colour factors a* and b* for the determination of redness and yellowness, respectively, did not differ. Physical traits, such as the pH_{24}, cooking loss and shear force, were similar among treatments.

Table 3. Carcass quality based on selected parameters among the four dietary treatments.

	Dietary treatment					
	CON	GGP	WYC	PE	SEM	Significance
Colour traits						
L*	55.87^{A}	58.73^{B}	59.83^{B}	58.64^{B}	0.876	0.015
a*	7.18	5.95	6.26	6.64	0.305	0.417
b*	18.50	17.13	19.18	17.81	0.715	0.319
Physical traits						
pH_{24}	6.11	6.11	6.05	6.18	0.089	0.594
Cooking loss (%)	11.83	15.34	15.58	15.17	1.451	0.313
Shear force (100 N/mm^2)	14.22	15.92	13.17	14.99	0.915	0.289

L*: lightness; a*: redness; b*: yellowness; SEM: pooled standard error of means; CON: control treatment; GGP: 25 g/kg ground grape pomace; PE: 1 g starch including 100 mg pure stem extract/kg; WYC: 2 g/kg wine yeast extract. Means with different capital superscript (A, B) in each row indicate significant difference ($p \leq 0.05$).

3.4. Fatty Acids in Breast Meat

The percentage of individual fatty acids and of the main classes of fatty acids is presented in Table 4. No significant alterations were observed in the myristic, pentadecanoic, palmitoleic, margaric, stearic, $C_{18:1}$ trans, oleic, α-linolenic, γ-linolenic, eicosadienoic, eicosatrienoic, arachidonic, docosadienoic, eicosapentanoic and docosahexaenoic fatty acids. The cis-vaccenic acid increased significantly in the CON, WYC and PE groups in comparison with the GGP group. Additionally, the linoleic acid had its highest value in the GGP group. Palmitic acid decreased in the breast meat of GGP-fed broilers compared to the other groups. As far as the main classes of fatty acids are concerned, saturated (SFAs) and unsaturated fatty acids (USFAs) were lower and higher, respectively, in the GGP group compared to the CON, WYC and PE groups. The index SFA/USFA was significantly lower in the group fed ground grape pomace. Moreover, the GGP group displayed an increased percentage of polyunsaturated fatty acids and a decreased value of the atherogenic index.

Table 4. The mean individual fatty acids (FAs) (% of total FA) in the breast meat of chickens fed the four diets.

	Dietary Treatment					
Fatty Acids	CON	GGP	WYC	PE	SEM	Significance
Myristic acid ($C_{14:0}$)	0.325	0.311	0.313	0.320	0.010	0.986
Pentadecanoic acid ($C_{15:0}$)	0.230	0.151	0.305	0.248	0.005	0.231
Palmitic acid ($C_{16:0}$)	17.12 ^{B}	15.72 ^{A}	17.27 ^{B}	17.86 ^{B}	0.334	0.012
Palmitoleic acid ($C_{16:1\ n\text{-}7}$)	1.18	0.952	1.08	1.47	0.123	0.250
Margaric acid ($C_{17:0}$)	0.117	0.122	0.104	0.076	0.018	0.394
Stearic acid ($C_{18:0}$)	8.16	7.11	8.48	8.03	0.345	0.379
$C_{18:1\,trans}$	0.02	0.05	0.02	0.00	0.002	0.501
Oleic acid ($C_{18:1\ cis\text{-}9}$)	23.93	23.09	22.40	23.85	0.789	0.680
Cis-vaccenic acid ($C_{18:1\ cis\text{-}11}$)	1.693 ^{B}	1.427 ^{A}	1.811 ^{B}	1.587 ^{AB}	0.098	0.083
Linoleic acid ($C_{18:2\ n\text{-}6\ cis}$)	32.16 ^{B}	37.03 ^{A}	31.28 ^{B}	31.17 ^{B}	1.815	0.013
α-linolenic acid ($C_{18:3\ n\text{-}3}$)	2.835	3.40	2.53	2.78	0.245	0.099
γ-linolenic acid ($C_{18:3\ n\text{-}6}$)	0.20	0.26 ^{T}	0.22	0.26	0.021	0.394
Eicosadienoic acid ($C_{20:2\ n\text{-}6}$)	0.74	0.70	0.88	0.69	0.090	0.547
Eicosatrienoic acid ($C_{20:3\ n\text{-}6}$)	0.702	0.661	0.869	0.805	0.079	0.390
Arachidonic acid ($C_{20:4\ n\text{-}6}$)	9.04	7.70	10.54	9.56	1.456	0.608
Docosadienoic acid ($C_{22:2\ n\text{-}6}$)	0.164	0.137	0.235	0.157	0.035	0.131
Eicosapentanoic acid ($C_{22:5\ n\text{-}6}$)	0.791	0.696	0.940	0.818	0.102	0.581
Docosahexaenoic acid ($C_{22:6\ n\text{-}3}$)	0.532	0.435	0.567	0.465	0.095	0.764
Saturated fatty acids (SFAs)	25.96 ^{B}	23.42 ^{A}	26.48 ^{B}	26.31 ^{B}	0.502	0.024
Unsaturated fatty acids (USFAs)	73.99 ^{B}	76.55 ^{A}	73.39 ^{B}	73.62 ^{B}	0.789	0.022
SFA/UNFA	0.351 ^{B}	0.306 ^{A}	0.362 ^{B}	0.359 ^{B}	0.009	0.028
Monounsaturated fatty acids (MUFA)	26.82	25.52	25.32	26.90	0.986	0.655
Polyunsaturated fatty acids (PUFA)	47.17 ^{A}	51.02 ^{B}	48.07 ^{A}	46.72 ^{A}	0.963	0.013
Atherogenic Index (AI)	0.25 ^{B}	0.22 ^{A}	0.25 ^{B}	0.26 ^{B}	0.007	0.010

Atherogenicity index (AI) was calculated according to the equation (C12:0 + 4 × C14:0 + C16:0)/(PUFA + MUFA); SEM: pooled standard error of means; CON: control treatment; GGP: 25 g/kg ground grape pomace; PE: 1 g starch including 100 mg pure stem extract/kg; WYC: 2 g/kg wine yeast extract. Means with different capital superscript (A, B) in each row indicate significant difference ($p \leq 0.05$).

4. Discussion

An assessment of the vinification by-products as feed additives was carried out in the present study aiming to redefine their use in broiler diets. Ground grape pomace (GGP), wine lees extract (WYC) and grape stem extract (PE) were added to the diet of broilers, and the results revealed that broiler performance, haematological parameters and meat quality were not negatively affected and, in some cases, were improved. The inclusion of 25 g/kg GGP, 2 g/kg WYC and 1 g starch including 100 mg pure stem extract/kg PE seemed to have no negative impact on the average body weight gain of broilers, indicating that the full body weight potential of broilers and carcass yield (%) can be achieved. Moreover, the overall FCR did not differ among treatments, indicating a similar utilization of dietary nutrients. Thus, broilers fed the three different vinification by-products performed well, and no major differences were observed between the control and the experimental groups, indicating that by-products can be used in the poultry industry.

The findings of the present study were in agreement with Kumanda et al. [21], who used red grape pomace in broiler diets at levels of 0, 2.5, 4.5, 5.5 and 7.5%. In their study, they reported no difference in the final body weight of broilers, while the FCR was found to be the lowest in the group with 7.5% of GP. In the same study, the group with the highest inclusion level had the lowest feed intake, and this was attributed to the high level of crude fibre. Many studies, such as those of Singh et al. [22] and Lau and King [23], determined a reduction in the feed intake of broilers as the inclusion level of grape pomace increased. Thus, low levels of grape by-products led to advantageous performances due to their lower fibre content. As a result, it is crucial to identify the optimum tolerance level of these by-products in broiler diets in accordance with antinutritional factors in order to maximize the nutrient utilization and growth performance of broilers.

The values of the assessed haematological parameters were within the normal range, indicating that no negative effect was induced in the broilers' health with the dietary inclusion. In the current study, serum SGPT-ALT significantly decreased in the group fed wine lees extract (WYC) compared to the CON, GGP and PE groups. Despite SGPT-ALT not seemingly being affected by diet, Erinle et al. [24] observed a decrease in this enzyme in broilers fed with 2.5% grape pomace compared to the controls, but with differences that were not statistically significant. SGOT-AST and SGPT-ALT are ideal indicators for the health status and normal function of the liver [25]. The results of the present study indicated that the health of dietary-supplemented broilers was maintained at high levels similar to that of the controls. Ebrahimzadeh et al. [26] did not detect any changes in the concentrations of total protein, glucose and cholesterol in the blood of broilers fed grape pomace. In the present study, the weight of the internal organs was similar for all groups, which suggested no alterations in the organ weight of broilers attributed to the dietary treatments. Similar results were reached by Kumanda et al. [21] and Aditya et al. [27].

The impact of the dietary inclusion of vinification by-products in meat quality characteristics was also evaluated. As far as the colour traits are concerned, the factor L* was higher in the treatment groups in comparison with the CON group. Although Kasapidou et al. [28] did not observe any changes in lightness and yellowness after supplementing diets with 2.5, 5 and 10 g/kg grape pomace, redness seemed to be affected. However, the enhancement of broiler meat lightness when antioxidants such as isoflavones are added to their diets has previously been observed [29]. In a study conducted by Bennato et al. [30], the pH and cooking loss of broiler meat did not change, such as in the present study, while lightness did not change in the aforementioned study in contrast to ours when grape pomace was included in broiler diets. In the same study, a more intense red colour was observed in the experimental groups compared to the control group, and a tendency for more yellow meat was also noted.

Regarding the fatty acid profile of the meat, differences were observed in the main classes of fatty acids. The fatty acid profile of grape pomace was previously illustrated [16] and affected the muscle fatty acid profile in a more preferrable way, as also indicated by the atherogenicity index. A great increase in the percentage of linoleic acid in the meat of the group that was dietary supplemented with ground grape pomace was observed, which was attributed to the fact that this was the major fatty acid present in this by-product [16,31–33]. The percentage of PUFA fatty acids also increased in the grape pomace group. The aforementioned alterations to the fatty acid profile could not be solely attributed to by-products, but also to other dietary ingredients, since oils such as soybean oil are often added to broiler diets in order to balance the energy content, resembling results that were also obtained from the study of Bennato et al. [30]. There is a constant effort to produce animal products that are enriched with polyunsaturated fatty acids and, especially, n-3 fatty acids [34]. The use of vinification by-products may be a feasible strategy to obtain these desired effects.

The utilization of vinification by-products as feed additives was investigated in the present study, aiming to reveal sustainable utilization strategies. By-products used as feed additives could boost the valorisation of waste and promote a circular economy model. The maintenance of broiler performance was the key finding of the present study, along with the improvement of meat PUFA in the case of GGP-fed broilers. However, combining the results of the present study with our previous work, which showed an improved oxidative status as an effect of the dietary supplementation with grape stems (PE) and wine lees extracts (WYC), the potential use of vinification by-products is promising.

Supplementary Materials: The following supporting information can be downloaded at: https://www.mdpi.com/article/10.3390/su142315714/s1, Table S1. composition (%) of the starting (0–10 d), growing (11–24 d) and finishing (25–42 d) phases of the control (CON), ground grape pomace (GGP), wine lees (rich in yeast cell walls) extract (WYC) and grape stems extract (PE) diets; Table S2. composition (%) and calculated analysis of the starting (0–10 d), growing (11–24 d) and finishing (25–42 d) phases of the control (CON), ground grape pomace (GGP), wine lees (rich in yeast cell walls) extract (WYC) and grape stems extract (PE) diets; Table S3. chemical composition (%) and fatty acid profile of ground grape pomace (GGP).

Author Contributions: Conceptualization, A.M.; methodology, A.M., P.E.S., E.G. and C.M., software, A.M.; validation, E.T. and A.C.P.; formal analysis, A.M.; investigation, A.M., E.T. and A.C.P.; resources, E.T.; data curation, A.M.; writing—original draft preparation, E.G., writing—review and editing, A.C.P., E.T., A.M. and P.E.S.; supervision, E.T., S.A.H. and A.K.; project administration, S.A.H. and A.K.; funding acquisition, E.T., S.A.H. and A.K. All authors have read and agreed to the published version of the manuscript.

Funding: This work was supported by the project "INVALOR: Research Infrastructure for Waste Valorization and Sustainable Management" (MIS 5002495), which was implemented under the action "Reinforcement of the Research and Innovation Infrastructure" funded by the Operational Programme "Competitiveness, Entrepreneurship and Innovation" (NSRF 2014–2020) and co-financed by Greece and the European Union (European Regional Development Fund).

Institutional Review Board Statement: The study was conducted according to the guidelines of the European Union Directive on the protection of animals used for scientific purposes (EU 63/2010; Council of the European Union 2010). Considering an experimental design report, the Agricultural University of Athens' Ethical Committee in Research approved the experimental protocol under the no. 38/9-10-2020.

Informed Consent Statement: Not applicable.

Data Availability Statement: All data are presented within the article and Supplementary Materials.

Acknowledgments: We sincerely appreciate the kind assistance of Christos Petropoulos for his contribution in the experimental trial.

Conflicts of Interest: The authors declare no conflict of interest.

References

1. United Nations News. Global Perspective Human Stories. Available online: https://news.un.org/en/story/2022/07/1122272 (accessed on 5 October 2022).
2. Godfray, H.C.J.; Beddington, J.R.; Crute, I.R.; Haddad, L.; Lawrence, D.; Muir, J.F.; Pretty, J.; Robinson, S.; Thomas, S.M.; Toulmin, C. Food security: The challenge of feeding 9 billion people. *Science* **2010**, *327*, 812–818. [CrossRef]
3. FUSIONS. *Estimates of European Food Waste Levels*; IVL Swedish Environmental Research Institute: Stockholm, Sweden, 2020.
4. Ong, K.L.; Kaur, G.; Pensupa, N.; Uisan, K.; Lin, C.S.K. Trends in food waste valorization for the production of chemicals, materials and fuels: Case study South and Southeast Asia. *Bioresour. Technol.* **2018**, *248*, 100–112. [CrossRef]
5. FAO-OIV. Focus on Table and Dried Grapes. 2016. Available online: http://www.fao.org/3/a-i7042e.pdf (accessed on 15 October 2022).
6. FAO. STAT-FAO Statistical Database. 2010. Available online: http://faostat3.fao.org (accessed on 25 August 2021).

7. Skuras, D.; Psaltopoulos, D. A broad overview of the main problems derived from climate change that will affect agricultural production in the Mediterranean area. In *Building Resilience for Adaptation to Climate Change in the Agriculture Sector*; FAO: Rome, Italy, 2012; pp. 217–260.
8. Brenes, A.; Viveros, A.; Chamorro, S.; Arija, I. Use of polyphenol-rich grape by-products in monogastric nutrition. A review. *Anim. Feed Sci. Technol.* **2016**, *211*, 1–17. [CrossRef]
9. Naziri, E.; Nenadis, N.; Mantzouridou, F.T.; Tsimidou, M.Z. Valorization of the major agrifood industrial by-products and waste from Central Macedonia (Greece) for the recovery of compounds for food applications. *Food Res. Int.* **2014**, *65*, 350–358. [CrossRef]
10. Filippi, K.; Georgaka, N.; Alexandri, M.; Papapostolou, H.; Koutinas, A. Valorisation of grape stalks and pomace for the production of bio-based succinic acid by Actinobacillus succinogenes. *Ind. Crops Prod.* **2021**, *168*, 113578. [CrossRef]
11. Ferri, M.; Vannini, M.; Ehrnell, M.; Eliasson, L.; Xanthakis, E.; Monari, S.; Sisti, L.; Marchese, P.; Celli, A.; Tassoni, A. From winery waste to bioactive compounds and new polymeric biocomposites: A contribution to the circular economy concept. *J. Adv. Res.* **2020**, *24*, 1–11. [CrossRef] [PubMed]
12. Georganas, A.; Giamouri, E.; Pappas, A.C.; Papadomichelakis, G.; Galliou, F.; Manios, T.; Tsiplakou, E.; Fegeros, K.; Zervas, G. Bioactive Compounds in Food Waste: A Review on the Transformation of Food Waste to Animal Feed. *Foods* **2020**, *9*, 291. [CrossRef] [PubMed]
13. Leyva-López, N.; Lizárraga-Velázquez, C.E.; Hernández, C.; Sánchez-Gutiérrez, E.Y. Exploitation of Agro-Industrial Waste as Potential Source of Bioactive Compounds for Aquaculture. *Foods* **2020**, *9*, 843. [CrossRef] [PubMed]
14. Dorri, S.; Tabeidian, A.S.; Toghyani, M.; Jahanian, R.; Behnamnejad, F. Effect of different levels of grape pomace on blood serum and biochemical parameters of broiler chicks at 29 and 49 days of age. In Proceedings of the 1st International and 4th National Congress on Recycling of Organic Waste in Agriculture, Isfahan, Iran, 26–27 April 2012.
15. Van Niekerk, R.F.; Mnisi, C.M.; Mlambo, V. Polyethylene glycol inactivates red grape pomace condensed tannins for broiler chickens. *Br. Poult. Sci.* **2020**, *61*, 566–573. [CrossRef]
16. Mavrommatis, A.; Giamouri, E.; Myrtsi, E.D.; Evergetis, E.; Filippi, K.; Papapostolou, H.; Koulocheri, S.D.; Zoidis, E.; Pappas, A.C.; Koutinas, A.; et al. Antioxidant Status of Broiler Chickens Fed Diets Supplemented with Vinification By-Products: A Valorization Approach. *Antioxidants* **2021**, *10*, 1250. [CrossRef]
17. Mavrommatis, A.; Simitzis, P.E.; Kyriakaki, P.; Giamouri, E.; Myrtsi, E.D.; Evergetis, E.; Filippi, K.; Papapostolou, H.; Koulocheri, S.D.; Pappas, A.C.; et al. Immune-Related Gene Expression Profiling of Broiler Chickens Fed Diets Supplemented with Vinification Byproducts: A Valorization Approach II. *Animals* **2021**, *11*, 3038. [CrossRef] [PubMed]
18. Mavrommatis, A.; Zografaki, M.-E.; Marka, S.; Myrtsi, E.D.; Giamouri, E.; Christodoulou, C.; Evergetis, E.; Iliopoulos, V.; Koulocheri, S.D.; Moschopoulou, G.; et al. Effect of a Carotenoid Extract from Citrus reticulata By-Products on the Immune-Oxidative Status of Broilers. *Antioxidants* **2022**, *11*, 144. [CrossRef] [PubMed]
19. Cason, J.A.; Lyon, C.E.; Papa, C.M. Effect of Muscle Opposition during Rigor on Development of Broiler Breast Meat Tenderness. *Poult. Sci.* **1997**, *76*, 785–787. [CrossRef]
20. O'Fallon, J.V.; Busboom, J.R.; Nelson, M.L.; Gaskins, C.T. A direct method for fatty acid methyl ester synthesis: Application to wet meat tissues, oils, and feedstuffs. *J. Anim. Sci.* **2007**, *85*, 1511–1521. [CrossRef] [PubMed]
21. Kumanda, C.; Mlambo, V.; Mnisi, C.M. From landfills to the dinner table: Red grape pomace waste as a nutraceutical for broiler chickens. *Sustainability* **2019**, *11*, 1931. [CrossRef]
22. Singh, A.K.; Berrocoso, J.F.D.; Dersjant-Li, Y.; Awati, A.; Jha, R. Effect of a combination of xylanase, amylase and protease on growth performance of broilers fed low and high fiber diets. *Anim. Feed Sci. Technol.* **2017**, *232*, 16–20. [CrossRef]
23. Lau, D.W.; King, A.J. Pre- and post-mortem use of grape seed extract in dark poultry meat to inhibit development of thiobarbituric acid reactive substances. *J. Agric. Food Chem* **2003**, *51*, 1602–1607. [CrossRef]
24. Erinle, T.J.; Oladokun, S.; MacIsaac, J.; Rathgeber, B.; Adewole, D. Dietary grape pomace–effects on growth performance, intestinal health, blood parameters, and breast muscle myopathies of broiler chickens. *Poult. Sci.* **2022**, *101*, 101519. [CrossRef]
25. Zhang, X. Application of total bile acid, ALT and AST in serum. *Jilin Med. J.* **2011**, *32*, 4840–4841.
26. Ebrahimzadeh, S.K.; Navidshad, B.; Farhoomand, P.; Mirzaei Aghjehgheshlagh, F. Effects of grape pomace and vitamin E on performance, antioxidant status, immune response, gut morphology and histopathological responses in broiler chickens. *S. Afr. J. Anim. Sci.* **2018**, *48*, 324–336. [CrossRef]
27. Aditya, S.; Ohh, S.J.; Ahammed, M.; Lohakare, J. Supplementation of grape pomace (*Vitis vinifera*) in broiler diets and its effect on growth performance, apparent total tract digestibility of nutrients, blood profile, and meat quality. *Anim. Nutr.* **2018**, *4*, 210–214. [CrossRef] [PubMed]
28. Kasapidou, E.; Soddidou, E.N.; Zdragas, A.; Papadaki, C.; Vafeas, G.; Mtlianga, P. Effect of grape pomace supplementation on broiler meat quality characteristics. *Eur. Poult. Sci.* **2016**, *80*, 135–142.
29. Salami, S.A.; Majoka, M.A.; Saha, S.; Garber, A.; Gabarrou, J.-F. Efficacy of Dietary Antioxidants on Broiler Oxidative Stress, Performance and Meat Quality: Science and Market. *Avian Biol. Res.* **2015**, *8*, 65–78. [CrossRef]
30. Bennato, F.; Di Luca, A.; Martino, C.; Ianni, A.; Marone, E.; Grotta, L.; Ramazzotti, S.; Cichelli, A.; Martino, G. Influence of grape pomace intake on nutritional value, lipid oxidation and volatile profile of poultry meat. *Foods* **2020**, *9*, 508. [CrossRef]
31. Ianni, A.; Martino, G. Dietary Grape Pomace Supplementation in Dairy Cows: E_ect on Nutritional Quality of Milk and Its Derived Dairy Products. *Foods* **2020**, *9*, 168. [CrossRef]

32. Manso, T.; Gallardo, B.; Salvá, A.; Guerra-Rivas, C.; Mantecón, A.R.; Lavín, P.; De la Fuente, M.A. Influence of dietary grape pomace combined with linseed oil on fatty acid profile and milk composition. *J. Dairy Sci.* **2016**, *99*, 1111–1120. [CrossRef]
33. Tsiplakou, E.; Zervas, G. The e_ect of dietary inclusion of olive tree leaves and grape marc on the content of conjugated linoleic acid and vaccenic acid in the milk of dairy sheep and goats. *J. Dairy Res.* **2008**, *75*, 270–278. [CrossRef]
34. Bennato, F.; Ianni, A.; Innosa, D.; Grotta, L.; D'Onofrio, A.; Martino, G. Chemical-nutritional characteristics and aromatic profile of milk and related dairy products obtained from goats fed with extruded linseed. *Asian Australas. J. Anim. Sci.* **2020**, *33*, 148. [CrossRef]

sustainability

MDPI

Article

Effect of an Herbal Mixture of Oregano, Garlic, Sage and Rock Samphire Extracts in Combination with Tributyrin on Growth Performance, Intestinal Microbiota and Morphology, and Meat Quality in Broilers

Eleftherios Bonos [1,*], Ioannis Skoufos [1], Ilias Giannenas [2], Erasmia Sidiropoulou [2], Konstantina Fotou [1], Ioanna Stylianaki [2], Olga Tsiftsoglou [3], Diamanto Lazari [3], Brigkita Venardou [1], Dimitrios Galamatis [4], Stylianos Skoufos [1], Evangelia Gouva [1] and Athina Tzora [1]

1 Department of Agriculture, School of Agriculture, University of Ioannina, Kostakioi Artas, 47100 Arta, Greece
2 School of Veterinary Medicine, Faculty of Health Sciences, Aristotle University of Thessaloniki, 54124 Thessaloniki, Greece
3 School of Pharmacy, Faculty of Health Sciences, Aristotle University of Thessaloniki, 54124 Thessaloniki, Greece
4 Department of Animal Science, University of Thessaly, 41335 Larissa, Greece
* Correspondence: ebonos@uoi.gr; Tel.: +30-268-105-0205

Citation: Bonos, E.; Skoufos, I.; Giannenas, I.; Sidiropoulou, E.; Fotou, K.; Stylianaki, I.; Tsiftsoglou, O.; Lazari, D.; Venardou, B.; Galamatis, D.; et al. Effect of an Herbal Mixture of Oregano, Garlic, Sage and Rock Samphire Extracts in Combination with Tributyrin on Growth Performance, Intestinal Microbiota and Morphology, and Meat Quality in Broilers. *Sustainability* **2022**, *14*, 13565. https://doi.org/10.3390/su142013565

Academic Editor: Rajeev Bhat

Received: 4 October 2022
Accepted: 17 October 2022
Published: 20 October 2022

Abstract: The present study investigated the effects of two feed additives, the first containing an herbal mixture of oregano, garlic, sage, and rock samphire extracts and the second containing tributyrin (glyceryl tributyrate) when fed to broiler chickens. A total of 360 one-day-old chicks were randomly allocated to four treatments (6 replicate pens of 15 chicks). One treatment served as the unsupplemented control, whereas the feeds of the other three treatments were supplemented either with the herbal additive (3 g/kg), the tributyrin additive (1 g/kg), or both additives. The duration of the trial was 37 days. Data were collected on growth performance, intestinal microbiota and morphology, and some meat quality parameters. The combined supplementation improved ($p < 0.05$) weigh gain, feed conversion ratio, and the European Efficiency Factor. In the cecum, the combined supplementation lowered ($p < 0.05$) the microbial populations of aerobes, anaerobes, *Escherichia coli*, total *Enterobacteriaceae*, and *Clostridium* spp. compared to the other treatments. Fecal coccidial oocyst counts were also reduced ($p < 0.01$) by the combined supplementation. The herbal mixture supplementation improved ($p < 0.05$) breast and thigh meat resistance to oxidation. In conclusion, the combined dietary supplementation with the examined feed additives could be utilized to improve the performance and intestinal health of broiler chickens.

Keywords: herbs; phytobiotics; organic acids; butyrate; tributyrin; performance; intestinal microbiota; intestinal morphology; meat quality

1. Introduction

Chickens are an essential source of animal protein and meat products that play a vital role in human nutrition and development, especially in high gross domestic product (GDP) countries [1,2]. It is widely acknowledged that the human population is growing fast, particularly in countries where food consumption is expected to be inadequate and disproportional to that growth, and by 2050 it is anticipated to increase to 9.5 billion worldwide [2]. Thus, humanity will have to ensure up to 60% additional food resources compared to the current levels of 8.5 billion tons per year to feed such a population [3]. At the same time, grassland degradation, biodiversity, and greenhouse gas production due to ruminants are particular concerns that have to be addressed to achieve sustainable food production [3]. The broiler meat industry has been expanding rapidly in the last decades, in response to the constant increase of global meat consumption. An estimate growth of

121% in poultry meat production is expected, far more significant to 66% and 43% of beef and pork meat, respectively [1]. This high demand in poultry meat products is also due to their wide acceptance as healthier alternatives characterized by higher protein and lower lipid content compared to other meats and to their cost-effective productivity values in most cultures, religions, and traditions [4].

To achieve these demands of the broiler industry, many feed additives have been employed to improve productivity and other traits, along with improvements in genetics and management [5]. Extensive research has acknowledged the importance of phytobiotics and other feed additives such as organic acids in animal nutrition and sustainability. Currently, the use of dietary growth promoting antibiotics has been banned or placed under major restrictions in most developed countries worldwide, due to their contribution to the development of microbial resistance in humans and animals [6,7]. In chickens a typical example is the presence of the extended-spectrum beta-lactamase (ESBL) bacteria, a major antibiotic–resistance concern for public health [8]. Moreover, consumers are aware of the importance of quality food and prefer natural products, free of any pharmaceutical residues in the food chain. Phytobiotics, as plant derived material, their extracts, and/or their essentials oils, have played an important part in animal nutrition, with numerous studies focusing on their positive effect in feed palatability and intake, anti-inflammatory, antioxidant and/or antimicrobial activity, intestinal function, and growth performance [9,10]. Essential oils derived from oregano (*Origanum* spp.) and sage (*Salvia triloba*) plants, grown abundantly in the Mediterranean area, contain potent antioxidant substances that have the ability to reduce the phenomenon of oxidative stress in tissues and/or serum samples, while improving the physical characteristics of chicken meat [11]. These effects are attributed to their rich content in phenolic compounds [12]. Moreover, the active compounds of garlic (*Allium sativa*) and its derivatives, such as flavonoids and organosulphurs, have shown to possess antioxidant capacity, as they were found to deteriorate lipid oxidation in chickens, along with various therapeutic applications [13,14]. Another plant of interest, rock samphire or sea fennel (*Crithmum maritimum*), is grown in coastal dunes of the Mediterranean basin [15]. Recently, components of its hydro-ethanolic extracts were found to possess antioxidant and antimicrobial properties [16].

Another category of important non-antibiotic feed additives in broiler nutrition are the organic acids (OAs) which represent promising feeding strategy in the industry [17]. The inclusion of OAs in animal diet has been involved in several studies evaluating intestinal tract parameters, immunity, and production performance [18,19]. Some of the most important OAs involved in animal nutrition are butyric, citric, propionic, ascorbic, formic, acetic, benzoic, and fumaric acids [20,21]. Organic acids in their salt or ester forms have a wide application in animal nutrition, as they have the ability to positively alter the immune responses, possess antioxidant and anti-inflammatory capacity, thus maintaining the balance of gastrointestinal homeostasis and the epithelial integrity, and are involved in the energy metabolism of monogastric animals [20,22]. Their mode of action is at least partially linked to the reduction of intestinal pH, which supports the growth of advantageous strains such as lactobacilli while decreasing pathogen populations, such as *Escherichia coli* and/or *Salmonella*, therefore affecting the intestine microbiota [20,23]. Since the health status of farm animals is of high importance and directly correlated with animal welfare traits, performance, and farm profitability, the intestinal inflammation and dysfunction, along with altered gut microbiota, could represent major issues in the poultry industry [22]. Furthermore, few reports have shown a synergistic effect of encapsulated butyric acid, when supplied in combination with other compounds such as oregano and/or attapulgite clay, on broiler performance and intestinal health [24,25].

The aim of the present study was to investigate the potential benefits of the combined use of an herbal feed additive (containing oregano, garlic, sage, and rock samphire) with another feed containing an organic acid (glyceryl tributyrate) on growth performance, intestinal microbiota and morphology, and some meat quality parameters of broiler chickens.

2. Materials and Methods

2.1. Animals, Diets and Experimental Design

The protocol of the experimental project "Innochicken" was co-financed by the European Regional Development Fund (ERDF) under the Operational Program "Epirus 2014–2020", NSRF 2014–2020. Project Code: HΠ1AB-0028192. During the experimental trial, the broiler chickens were managed in compliance with local ethical practices and procedures [26] and following the recommendations for broiler welfare [27].

Three hundred sixty one-day-old male Ross-308 chicks (initial body weight 45.3 ± 0.7 g) were obtained from the PINDOS APSI hatchery and reared in a commercial poultry farm in Gavria, Arta, (latitude 38.617°, longitude 20.767°), Epirus, Greece, during the period of October–December 2019. There were 4 treatment groups, each with 6 replicate pens (length 1 m; width 1.1 m) of 15 chicks. The stocking density was calculated to be 15 birds per m^2 (area of 1.1 m^2 per pen). Throughout the experimental period, commercial breeding and management procedures were applied. Both natural and artificial lighting were used to achieve light of 23 h for the first two days, 16 h from day 3 to day 14, and 21 h from day 15 to slaughter. Ambient temperature and humidity were carefully controlled by a computer monitoring system. All chicks were vaccinated against Newcastle disease, infectious bronchitis, and infectious bursal disease (Gumboro) at the hatchery. *Ad libitum* feeds and drinking water were given to all broilers throughout the trial.

Control treatment (CONTR) chickens were fed commercial typical corn and soybean meal based rations in mash form (Table 1), which did not contain anticoccidials or antibiotics, formulated based on breeder recommendations [28,29]. The diets of the other three treatments were further supplemented with either the tested herbal mixture (3 g/kg; HERB), the tested butyrate additive (1 g/kg; BUTYR), or both the herbal additive (3 g/kg) and the butyrate additive (1 g/kg) together (HERB&BUTYR). The herbal mixture was formulated to provide in the feed: 50 ppm oregano essential oil; 5 ppm garlic essential oil; 1 g/kg dried sage; and 1 g/kg dried rock samphire. The herbal material was supplied by a local farm in Palaiohori, Filiates, Thesprotia, Greece. The butyrate additive contained 100% glyceryl tributyrate ("Butiphorce 1065", NuSana, Land van Cuijk, The Netherlands).

The broiler weights were determined on days 1, 12, 24, and 37. Data on feed consumption and mortality were collected daily. The "European Production Efficiency Factor (EPEF)" was calculated for the overall trial [30], using the formula:

$$\text{EPEF} = [\text{Average daily weight gain (g)} \times \text{Survival rate (\%)}]/[\text{Feed conversion ratio} \times 10]$$

On the last day of the trial (day 37), all broilers were humanely slaughtered under commercial conditions at a nearby abattoir. Four chickens were randomly chosen and processed from each replicate pen (24 per treatment; 96 total). During processing, the gastrointestinal tracts of the selected chickens were carefully removed for further analysis.

2.2. Gastrointestinal Tract Sampling

To collect the gastrointestinal tissues, initially the abdomen of each bird was cleaned with 70% (v/v) ethanol and skin incisions were aseptically performed to access the intestine. Then, the jejunum and the caeca of each chicken were separated and dissected using a sterile scalpel, which was also used to gently scrape off the mucus layer from the intestinal content of each incision site and transfer it to a sterile container and stored.

2.3. Microbiological Analysis

For the bacterial isolation, enumeration, and identification, 1 g of intestinal content from each bird was weighted and homogenized with 9 mL of sterile peptone water solution 0.1%. Miles and Misra plate method (surface drop) was used for bacterial enumeration in which each sample was serially diluted via 10-fold dilutions (from 10^{-1} to 10^{-12}) using standard 96-well microplates, while dilutions were plated onto selective mediums which were incubated properly [31]. Total aerobic and anaerobic bacterial counts were determined using plate count agar medium (PCA) (Oxoid, Basingstoke, UK), while plates were incubated at 30 °C aerobically for 48 h and at 37 °C anaerobically for 48–72 h, respectively. For the isolation and enumeration of *E. coli* and total *Enterobacteriaceae*, MacConkey agar (Merck, Darmstadt, Germany) and Violet Red Bile Glycose (VRBG) agar (Merck) were used, and plates were both incubated aerobically at 37 °C for 24–48 h. De Man, Rogosa, and Sharpe (MRS) agar (Oxoid) was used for the isolation and enumeration of *Lactobacillus* species, while plates were incubated at 37 °C for 48–72 h under anaerobic condition. For the isolation and enumeration of *Clostridium* spp., Tryptose Sulfite Cycloserine (TSC) agar was used (Merck) and plates were incubated at 37 °C for 24–48 h in anaerobic conditions. After incubation time, typical colonies from an appropriate dilution were counted and expressed as log colony-forming units per 1 g wet weight sample (CFU/g). In addition, typical colonies were characterized morphologically by microscopy, gram staining, catalase, and oxidase tests and sub-cultured in order to obtain pure cultures. Lactobacilli were identified using API 50 CHL kits according to the manufacturer's instructions and API LAB Plus software version 3.3.2 (Bio-Merieux, Marcy-l'Étoile, France). Identification of *Clostridium* spp. were performed by VITEK 2 ANC card, while identification of *E. coli* by GN identification card was performed by API LAB Plus software version 3.3.2 (bioMérieux, Marcy l'Etoile, France). ANC and GN identification cards were used in conjunction with VITEK 2 system (bioMérieux, Marcy l'Etoile, France) [32].

Table 1. Ingredients and nutrient content of the diets fed to the control treatment.

Ingredients (%)	**Starter Days 1–14**	**Grower-Finisher Days 15–37**
Maize	49.108	51.339
Wheat	10.000	10.000
Soybean meal (47% crude protein)	33.547	30.715
Soybean oil	2.623	3.769
Salt	0.325	0.315
Sodium carbonate	0.074	0.077
Limestone (Calcium carbonate)	1.512	1.390
Dicalcium phosphate	1.516	1.300
Lysine HCl	0.422	0.327
Methionine DL	0.434	0.374
Threonine	0.189	0.144
Vitamin and mineral premix [1]	0.250	0.250
Total	100.000	100.000
Nutrient content		
Metabolisable energy (Kcal/kg)	3090.0	3180.0
Crude protein (%)	23.50	22.50
Crude fat (%)	5.50	5.80
Total lysine (%)	1.50	1.40
Total methionine (%)	1.40	1.20
Total calcium (%)	1.00	1.00
Available phosphorus (%)	0.60	0.60

[1] Supplying per kg feed: 13,000 IU vitamin A, 4000 IU vitamin D3, 40 mg vitamin E, 9 mg vitamin K, 3 mg thiamine, 7 mg riboflavin, 6 mg pyridoxine, 0.035 mg vitamin B12, 40 mg niacin, 13 mg pantothenic acid, 1.5 mg folic acid, 0.13 mg biotin, 340 mg choline chloride, 55 mg Zn, 155 mg Mn, 20 mg Fe, 12 mg Cu, 0.2 mg Co, 1 mg I, 0.2 mg Se, and phytase 500 FTU.

2.4. Coccidial Oocysts Count

On the 28th and 37th day of the trial, fresh faecal samples were taken from chickens from all experimental pens to determine coccidial oocyst output. The McMaster technique was applied to identify and count coccidian oocysts in these faecal samples [33,34]. The identification of unsporulated oocysts of *Eimeria* spp. were based on their morphological features (oocyst shape, mean oocyst length, and mean oocyst width after microscopic examination).

2.5. Intestinal Morphometric Analysis

To perform the morphometric analysis of the samples from the small intestine, the methodology described by Gava et al. [35] was initially followed. Then, photographic images were taken under light microscopy, using a Nikon microscope coupled with a Microcomp integrated digital imaging analysis system (Nikon Eclipse 200, Tokyo, Japan). Images were analysed with the "Image-Pro Plus" analysis software. Villus height (VH) and crypt depth (CD) were measured as the mean of 10 collected values per sample (Figure 1).

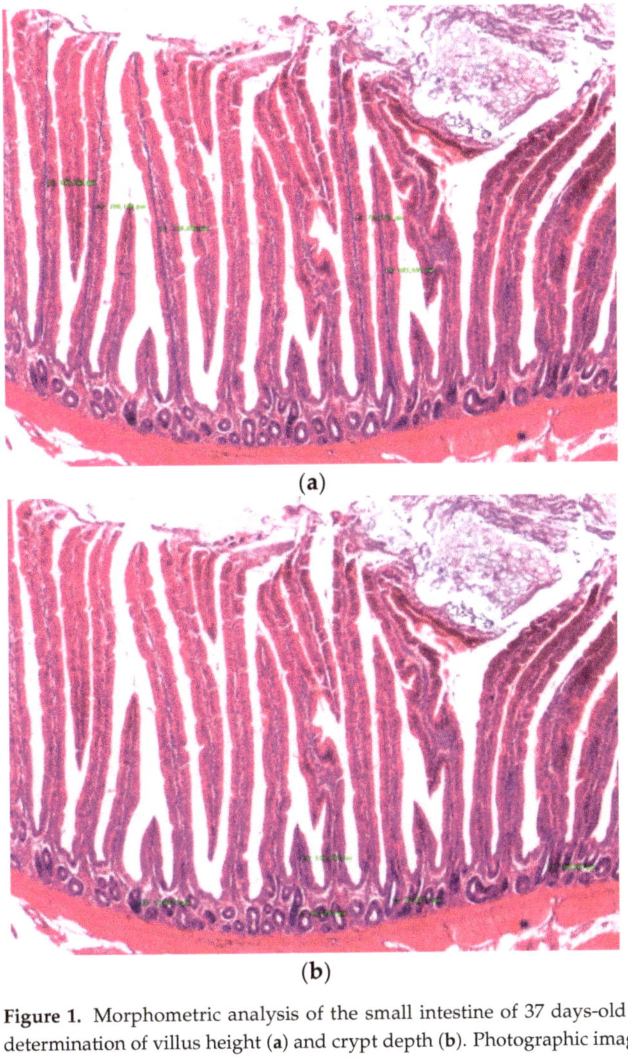

(**a**)

(**b**)

Figure 1. Morphometric analysis of the small intestine of 37 days-old broiler chicken depicting determination of villus height (**a**) and crypt depth (**b**). Photographic images were taken under light microscopy as described by Gava et al. [35].

2.6. Meat Chemical Analysis

The carcasses of the selected birds were initially processed according to the slaughterhouse commercial procedures. Then, breast and thigh meat samples were collected from each carcass; they were carefully skinned, deboned, and grounded using an industrial large meat grinder. Then, sub-samples of 200 g were analyzed for moisture, crude protein, fat, and ash content by near infra-red spectroscopy, using a FoodScanTM Lab (FOSS, Hillerød, Denmark) as described in reference method AOAC 2007.04 [36,37].

2.7. Meat lipid Oxidation Analysis

Breast and thigh meat samples were processed according to Ahn et al. [38] with minor modifications to determine their lipid oxidation status during refrigerated storage, with a UV spectrophotometer (UV 1700 PharmaSpec, Shimadzu, Kyoto, Japan) set at 532 nm. The lipid oxidation of the samples was evaluated as "2-thiobarbituric acid-reactive substances (TBARS)" values, expressed as ng of malondialdehyde (MDA) per g of sample.

2.8. Statistical Analysis

The experimental study design was RCBD (random complete block design). In all measurements, the replication (pen) was considered as the experimental unit. Experimental data were evaluated using the one-way ANOVA (general linear model) and the Kruskal–Wallis tests of the SPSS statistical package (version 20.0) [39]. Microbiology data were log-transformed (Log10) prior to analysis and data homogeneity was examined with the Levene's test. The threshold for significance was set at 5% ($p \leq 0.05$).

3. Results

The effects of the dietary supplementation on broiler chicken performance parameters are presented in Table 2. The combined supplementation of the herbal and the organic acid additives (HERB&BUTYR) significantly increased ($p < 0.05$) the final body weight, compared to the unsupplemented control (CONTR), at the end of the trial (day 37). The feed intake was found to be significantly lower ($p < 0.001$) for the BUTYR treatment compared to the CONTR and HERB treatments during the period of 13–24 days; however, this effect was not noted for the other periods or the overall trial. In addition, overall feed conversion ratio (days 1–37) was significantly lower ($p < 0.05$) and the European Efficiency Factor was significantly improved ($p < 0.05$) for the HERB&BUTYR treatment compared to the unsupplemented CONTR treatment.

Table 2. Effect of combined herbal and organic acid dietary supplementation on the broiler chicken performance parameters.

Live Body Weight on Day (g)	CONTR	HERB	BUTYR	HERB&BUTYR	SEM	p
1	45.0	45.2	45.7	45.4	0.13	0.254
12	303.6	314.5	344.0	310.1	5.54	0.185
24	923.1	1016.6	982.2	960.1	13.90	0.160
37	1685.9 ^{a}	1833.9 ^{ab}	1794.3 ^{ab}	1846.9 ^{b}	19.07	0.023
Feed intake during period (g)						
1–12 days	322.7	323.8	324.4	324.5	0.00	1.000
13–24 days	1007.9 ^{b}	1004.1 ^{b}	968.3 ^{a}	985.7 ^{ab}	4.15	<0.001
25–37 days	1224.6	1318.9	1295.8	1275.5	24.11	0.628
1–37 days	2574.8	2662.6	2603.9	2620.4	25.49	0.675
Feed conversion ratio during period (g feed/g weight gain)						
1–12 days	1.249	1.204	1.113	1.228	0.020	0.332
13–24 days	1.630	1.439	1.529	1.535	0.027	0.133
25–37 days	1.608	1.614	1.602	1.464	0.036	0.423
1–37 days	1.569 ^{b}	1.489 ^{ab}	1.494 ^{ab}	1.456 ^{a}	0.013	0.043
European efficiency factor						
Day 37	270.2 ^{a}	321.7 ^{ab}	315.1 ^{ab}	324.4 ^{b}	7.62	0.026

CONTR: control non-supplemented treatment; HERB: feed supplemented with herbal mixture at 3 g/kg; BUTYR: feed supplemented with glyceryl tributyrate at 1 g/kg; HERB&BUTYR: feed supplemented with herbal mixture at 3 g/kg and glyceryl tributyrate at 1 g/kg. SEM: standard error of the means. ^{a,b}: values in the same row without superscripts in common differ significantly ($p \leq 0.05$).

Table 3 shows the effects of the dietary supplementation on the intestinal microbiota. In the jejunum, *E. coli* and total *Enterobacteriaceae* were significantly increased ($p \leq 0.05$) in the BUTYR treatment compared to the CONTR; *Clostridium* spp. was significantly decreased ($p \leq 0.001$) in the three supplemented treatments compared to the CONTR; and total aerobes and anaerobes were not significantly ($p > 0.05$) affected. In the cecum, total aerobes were significantly decreased ($p \leq 0.01$) in the HERB&BUTYR treatment compared to the HERB treatment; total anaerobes were significantly decreased ($p \leq 0.01$) in the HERB and HERB&BUTYR treatments compared to the CONTR; and *E. coli*, total *Enterobacteriaceae*, and *Clostridium* spp. were significantly decreased ($p \leq 0.001$; $p \leq 0.01$; and $p \leq 0.001$, respectively) in the HERB&BUTYR treatment compared to the other three treatments. In addition, oocysts analysis from the broiler feces on day 28 showed that the HERB&BUTYR treatment had significantly lower ($p < 0.001$) counts compared to all other treatments, while also HERB treatment had lower counts compared to the CONTR. In addition, the oocyst analysis on day 37 showed that the HERB treatments and HERB&BUTYR treatments had significantly lower ($p < 0.01$) counts compared to the CONTR.

Table 3. Effect of combined herbal and organic acid dietary supplementation on the broiler chicken intestinal microbiota.

Jejunum Microbiota (Log CFU/g Digesta)	CONTR	HERB	BUTYR	HERB&BUTYR	SEM	*p*
Total aerobic counts	6.263	6.405	6.604	6.193	0.142	0.748
Total anaerobic counts	8.143	7.851	8.361	8.209	0.116	0.479
Escherichia coli	3.530 ^{a}	5.030 ^{ab}	5.217 ^{b}	3.941 ^{ab}	0.191	0.029
Total *Enterobacteriaceae*	3.829 ^{a}	5.341 ^{ab}	5.419 ^{b}	4.511 ^{ab}	0.172	0.017
Lactobacillus spp.	7.576	7.562	7.873	7.694	0.154	0.883
Clostridium spp.	2.882 ^{b}	2.555 ^{a}	2.465 ^{a}	2.303 ^{a}	0.034	<0.001
Cecum microbiota (Log CFU/g digesta)						
Total aerobic counts	7.751 ^{ab}	8.239 ^{b}	7.748 ^{ab}	7.029 ^{a}	0.093	0.002
Total anaerobic counts	8.824 ^{b}	8.216 ^{a}	8.692 ^{ab}	8.242 ^{a}	0.063	0.010
Escherichia coli	7.440 ^{b}	7.172 ^{b}	7.427 ^{b}	6.176 ^{a}	0.105	0.001
Total *Enterobacteriaceae*	7.564 ^{b}	7.289 ^{b}	7.660 ^{b}	6.531 ^{a}	0.096	0.002
Lactobacillus spp.	8.251	8.271	7.918	8.201	0.119	0.706
Clostridium spp.	4.869 ^{b}	4.816 ^{b}	5.371 ^{b}	3.765 ^{a}	0.122	0.001
Oocyst counts in feces (Log/g)						
Day 28	3.989 ^{c}	3.672 ^{b}	3.835 ^{bc}	3.304 ^{a}	0.037	<0.001
Day 37	4.007 ^{b}	3.475 ^{a}	3.645 ^{ab}	3.388 ^{a}	0.053	0.003

CONTR: control non-supplemented treatment; HERB: feed supplemented with herbal mixture at 3 g/kg; BUTYR: feed supplemented with glyceryl tributyrate at 1 g/kg; HERB&BUTYR: feed supplemented with herbal mixture at 3 g/kg and glyceryl tributyrate at 1 g/kg. SEM: standard error of the means. ^{a,b}: values in the same row without superscripts in common differ significantly ($p \leq 0.05$).

The results of the broiler chicken jejunum morphology analysis are given in Table 4. No significant effects ($p > 0.05$) were identified in jejunum villus height, jejunum crypt depth, or the villus height to crypt depth ratio.

Table 4. Effect of combined herbal and organic acid dietary supplementation on the broiler chicken jejunum morphology.

Jejunum Morphology	CONTR	HERB	BUTYR	HERB&BUTYR	SEM	*p*
Jejunum villus height (μm)	764.6	754.1	710.3	697.5	14.02	0.269
Jejunum crypt depth (μm)	196.4	197.4	203.1	183.0	8.31	0.889
Jejunum Villus height/Crypt depth	3.976	3.931	3.535	3.819	0.1578	0.810

CONTR: control non-supplemented treatment; HERB: feed supplemented with herbal mixture at 3 g/kg; BUTYR: feed supplemented with glyceryl tributyrate at 1 g/kg; HERB&BUTYR: feed supplemented with herbal mixture at 3 g/kg and glyceryl tributyrate at 1 g/kg. SEM: standard error of the means.

The effects of the dietary supplementation on the chemical composition and oxidative stability of the meat are presented in Table 5. Regarding the breast meat, a significantly ($p \leq 0.05$) higher amount of fat was found in the HERB&BUTYR treatment, compared to the unsupplemented CONTR. Moreover, lipid oxidation (expressed as ng MDA/g meat) was found to be significantly lower ($p \leq 0.05$) in the breast meat of the HERB treatment compared to the BUTYR treatment. Regarding the thigh meat, no significant effects ($p > 0.05$) were identified in the chemical composition. However, the lipid oxidation of the thigh meat was significantly lower ($p \leq 0.001$) in the HERB treatment compared to all other treatments, as well as lower in the CONTR and HERB&BUTYR treatments, compared to the BUTYR treatment.

Table 5. Effect of combined herbal and organic acid dietary supplementation on the broiler chicken meat chemical composition and oxidative stability.

Breast Meat Chemical Composition	CONTR	HERB	BUTYR	HERB&BUTYR	SEM	p
Moisture %	74.96	74.78	74.36	75.01	0.122	0.271
Crude protein %	23.84	22.34	22.55	22.24	0.300	0.257
Crude fat %	1.57 ^{a}	1.94 ^{ab}	2.03 ^{ab}	2.40 ^{b}	0.082	0.028
Ash %	0.76	0.76	0.74	0.74	0.024	0.527
Breast meat lipid oxidation						
Day 1 of storage (ng MDA/g)	28.8 ^{ab}	20.4 ^{a}	44.6 ^{b}	21.9 ^{ab}	1.668	0.020
Thigh meat chemical composition						
Moisture %	73.65	72.44	72.09	71.94	0.321	0.303
Crude protein %	19.87	18.78	19.66	18.50	0.247	0.207
Crude fat %	6.24	8.08	7.77	8.70	0.389	0.240
Ash %	0.64	0.71	0.60	0.70	0.022	0.250
Thigh meat lipid oxidation						
Day 1 of storage (ng MDA/g)	21.8 ^{b}	14.2 ^{a}	28.5 ^{c}	24.3 ^{b}	0.685	<0.001

CONTR: control non-supplemented treatment; HERB: feed supplemented with herbal mixture at 3 g/kg; BUTYR: feed supplemented with glyceryl tributyrate at 1 g/kg; HERB&BUTYR: feed supplemented with herbal mixture at 3 g/kg and glyceryl tributyrate at 1 g/kg. SEM: standard error of the means. ^{a,b}: values in the same row without superscripts in common differ significantly ($p \leq 0.05$).

4. Discussion

Feed additives have been widely employed in broiler nutrition to increase performance characteristics, improve farm animal welfare, and ultimately achieve sustainability in addition to high productivity [20,40]. Herbal plant and organic acid-based feed additives have been evaluated and considered among the most promising supplements for poultry and thus have been screened in numerous studies on broiler performance, sometimes with conflicting results [41,42].

The European Production Index related to the outcome of both performance and health. One way to improve overall productivity is the dietary use of functional feed additives that support weight gain by supporting the digestive and immune systems. Additives with beneficial properties include aromatic plants and their extracts that have long been used as alternatives to antibiotic growth promoters, mainly due to their valuable constituents such as polyphenols [43]. The oregano essential oil used in our study was found to be rich in carvacrol (75.06%) and thymol (6.56%), with both constituents playing a significant role on its antioxidant capacity [44]. Oregano as a feed supplement has been found to possess antimicrobial applications against poultry and human pathogens [45]. The principal phytochemicals of garlic that display antioxidant and antibacterial activity are oil-soluble organosulfur compounds such as allicin and/or allyl sulfides [46,47]. The most prominent constituents of garlic essential oil were diallyl sulfides (diallyl disulfide and trisulfide). In addition, garlic is known for its effects on the immune response and the regulation of the gastrointestinal microbiome. Garlic polysaccharides have been found to have the capacity to reduce the expression of inflammatory factors and improve the colon tissue integrity and microbiota [48]. Sage plants are also rich in polyphenols and flavonoids including catechin, rutin, caffeic, and rosmarinic acids, while their volatile oils contain mainly monoterpenes and sesquiterpenes such as carnosic acid [49]. However, the concentration of those constituents may differ to a great extent depending on the *Salvia* species. Aqueous sage extract has shown significant inhibitory activity against different bacterial species such as *Bacillus mycoides*, *Bacillus subtilis*, *Enterobacter cloacae*, and *Proteus* spp. [50], whereas the hydroalcoholic extract has demonstrated strong inhibitory effect on *Streptococcus mutans*, *Lactobacillus rhamnosus*, and *Actinomyces viscosus* [51]. Thus, sage extract has been suggested as a valid alternative source to traditional antibiotics [52]. Rock samphire is rich in polyphenols such as chlorogenic acid and flavonoids including rutin, cirsiliol, and quercetrin [16]. Rock samphire essential oil components have been shown to inhibit the growth of food-borne bacteria, including *E. coli*, *Staphylococcus aureus*, *Bacillus subtilis*, *Pseudomonas aeruginosa*, *Listeria innocua*, and others [53].

Tributyrin is a triglyceride ester of butyric acid and glycerol [54]. It is naturally found in butter and margarine. Butyric acid is also naturally produced by fermenting bacteria in the caecum [55]. It has been suggested that luminal butyric acid is linked to the endocrine regulation of the digestive tract affecting enteroendocrine cells and overall nutrient digestibility [56]. When used as feed additive, the activity of butyric acid in its free form is limited to the crop, proventriculus, and gizzard of the broilers. However, protected forms of butyric acid such as its esters are hydrolyzed by the pancreatic lipase to glycerol tributyrate and dibutyrate that are effective in the small intestine [54,57]. Dietary butyric acid and its esters have been reported to limit the growth of pathogens [58], stimulate feed intake, and improve digestion by stimulating pancreatic secretion [59].

The potential benefits of dietary herbal extracts and organic acids on broiler performance have been widely examined. Nevertheless, published data on the combined supplementation of these two additives are still quite scarce. Studies evaluating combinations of plant essential oils with organic acids in poultry nutrition showed that these combinations increased growth rates and/or feed utilization [19,24,60–63], which is in agreement with the results of our study. It should be noted that such positive effects are not always reported [64–66]. This inconsistency can be explained in part by the wide variety of the tested herbal or organic acid material, as well as the different inclusion rates. In addition, the active substances of plant essential oils and dissociated organic acids are quickly absorbed in the crop and stomach of the birds, thus exerting their direct effects in the anterior part of the gastrointestinal system [67]. Additionally, some phytobiotics increase feed palatability and stimulate the production of digestive enzymes [61,68]. Most dietary organic acids are weak acids that only partially dissociate in aquatic solution and the percentage of dissociation depends on the pH of the gastrointestinal tract [69]. For these reasons, protective technologies, such as encapsulation of the active ingredients, are under examination [17,60]. These methods lower the rate of absorption or degradation of the active ingredients in the anterior part of the gastrointestinal tract and allow them to reach and exert their effects on the latter parts of the intestine such as the ileum and the ceca of the birds, directly affecting the intestinal microbiota as well as the intestinal epithelium.

An important indicator of intestinal health is the integrity of the intestinal epithelium. The villus size, absorptive area, and thickness of the mucus layer regulate the transport of molecules from the intestinal lumen through the epithelium into the bloodstream [70]. Intestinal disease severely limits the villus length and uniformity because bacterial toxins rapidly damage the membranes of the epithelial cells [71,72]. The replacement of the intestinal epithelium poses an important energy cost for the animal. Decreased crypt depth is linked to lower tissue turnover and reduced secretion [71,72]. It has been reported that dietary plant essential oils can increase villus height and cell proliferation in chickens [24,73,74]. Moreover, dietary organic acids can improve villus height in the small intestine, directly by causing enterocyte hyperplasia or limiting apoptosis [63,75] or indirectly through the modification of the microbiota balance [19,76]. In our study, no significant differences were noted on villus height and crypt depth between the supplemented treatment and the unsupplemented control, which could reflect the overall high health status of both the supplemented and unsupplemented treatment groups.

In the last two decades, a large effort has been made to examine natural feed ingredients as potential alternatives to antimicrobial and anticoccidial drugs in meat type poultry [7,77,78]. Recent information on necrotic enteritis and coccidiosis in European countries elucidate their regular occurrence accompanied by major economic losses, especially after the ban of antibiotic growth promoters [79]. Although these drugs are highly effective in preventing or treating clinical disease in broilers, their overuse has caused a global alarm due to the adverse effects of their residues on the environment [12,80]. Many antimicrobials are poorly absorbed or metabolized in the gastrointestinal tract of the birds, resulting in a large amount being excreted in the feces. Subsequently, animal waste is used as fertilizer or reaches surface water, unbalancing the microbiota of natural ecosystems and increasing antibiotic resistance in animal and human pathogens [7,81]. In our study, the chicken control diet did not contain any antimicrobial or anticoccidial drugs. The combined dietary supplementation with the herbal mixture and the organic acid influenced mainly the cecum microbiota, lowering some microbial population such as *E. coli*, *Enterobacteriaceae*, and *Clostridium* spp. Fecal oocyst counts were also lowered in broilers receiving the combined supplementation. It should be noted that the tested broilers in our trial were not exposed to experimental *Eimeria* infection, but natural exposure was expected as they were reared in a commercial setting. Coccidiosis is caused by *Eimeria* species parasites and today it constitutes one of the biggest challenges for the broiler industry due to the ubiquity of these parasites and the increasing occurrence of anticoccidial resistance [82]. Nowadays, restrictions on the use of anticoccidial drugs increase worldwide and anticoccidial vaccination is still not widely available and affordable [83]. Extracts from medicinal aromatic plants are potential alternatives in the effort to control coccidiosis [84–86].

In the present study, the effect of the dietary supplements on the breast and thigh meat chemical composition was additionally evaluated. The breast meat analysis showed that the combined supplementation significantly increased fat percentage, whereas the same effect was not confirmed statistically in the thigh meat. This increase of fat content in the meat could be attributed to the improved gut health and, thus, better feed digestion that increased the metabolisable energy that remained available to the broilers for deposition in the muscle tissue. It is known that dietary composition can regulate lipid metabolism and carcass fat content, affecting key enzymes associated with lipid anabolism; however, the underlying factors have not been fully elucidated [87]. Furthermore, in the present study, resistance to oxidation was mainly affected by the herbal supplementation, whereas the organic acid supplementation seemed to trigger oxidation of lipids in the meat. It is well established that dietary herbal extracts can positively affect meat oxidative stability, influencing various mechanisms of action, i.e., improving the activity of antioxidant enzymes that counteract inflammatory agents such as reactive oxygen species, as well as protecting cells membranes from oxidation, atrophy, or breaks [88,89]. It has been reported that essential oils from *Labiatae* family plants such as oregano, thyme, rosemary, and others can protect refrigerated poultry products by lowering meat oxidation mainly due to the effect of their polyphenol compounds or a sparing effect on tocopherols [90–92]. In addition, a herbal mixture including thyme, rosemary, garlic, and others protected broiler meat from oxidative rancidity during storage [93].

5. Conclusions

Based on the results of this study, the combined dietary supplementation of two feed additives, the first containing a mixture of oregano, garlic, sage, and rock samphire extracts and the second containing glyceryl tributyrate, can improve growth performance parameters of broiler chickens. Moreover, this supplementation can potentially modify beneficially the intestinal microbiota of the broilers and lower the overall oocyst count in the feces. In addition, the combined dietary supplementation seemed to increase breast meat fat content, whereas the herbal feed additive mainly improved the resistance to oxidation of the meat. Additional in vivo studies could potentially determine the optimum combinations of herbal and organic acid feed additives, to achieve sustainable and profitable broiler farming without the use of anticoccidial or ionophores.

Author Contributions: Conceptualization, E.B., I.S. (Ioannis Skoufos), A.T. and I.G.; methodology, E.B., I.S. (Ioannis Skoufos) and I.G.; software, E.B.; validation, I.S. (Ioannis Skoufos); formal analysis, E.B., I.S. (Ioannis Skoufos), I.G., E.S., K.F., I.S. (Ioanna Stylianaki), O.T., D.L., B.V., D.G., S.S. and E.G.; investigation, E.B., I.S. (Ioannis Skoufos), I.G., E.S., K.F., I.S. (Ioanna Stylianaki), O.T., D.L., B.V., D.G., S.S. and E.G.; resources, I.S. (Ioannis Skoufos) and I.G.; data curation, E.B.; writing—original draft preparation, E.B., I.S. (Ioannis Skoufos), I.G. and A.T.; writing—review and editing, E.S., K.F., I.S. (Ioanna Stylianaki), O.T., D.L., B.V., D.G., S.S. and E.G.; visualization, E.B. and I.S. (Ioannis Skoufos); supervision, I.S. (Ioannis Skoufos) and I.G.; project administration, I.S. (Ioannis Skoufos), I.G. and A.T.; funding acquisition, I.S. (Ioannis Skoufos). All authors have read and agreed to the published version of the manuscript.

Funding: This research has been co-financed by Greece and the European Union (European Regional Development Fund) in the context "Research—Create—Innovate" within the Operational Program "Epirus 2014–2020" of the NSRF 2014–2020. Project Code: HΠ1AB-00142. Acronym "InnoChicken".

Institutional Review Board Statement: The research project was approved by the Institutional Review Board of the University of Ioannina (Project code: 61480).

Informed Consent Statement: Not applicable.

Data Availability Statement: The data are contained within the article.

Acknowledgments: The authors wish to thank Michas Poultry Farm, and especially Konstantinos Michas, Apostolos Michas, Pavlos Michas, and Konstantinos Gartzionis for their technical support regarding the experimental trial.

Conflicts of Interest: The authors declare no conflict of interest.

References

1. Mottet, A.; Tempio, G. Global poultry production: Current state and future outlook and challenges. *World's Poult. Sci. J.* **2017**, *73*, 245–256. [CrossRef]
2. Alexandratos, N.; Bruinsma, J. *Word Agriculture towards 2013/2050: The 2012 Revision*; FAO: Rome, Italy, 2012; pp. 1–147.
3. Michalk, D.L.; Kemp, D.R.; Badgery, W.B.; Wu, J.; Zhang, Y.; Thomassin, P.J. Sustainability and future food security—A global perspective for livestock production. *Land Degrad. Dev.* **2018**, *30*, 561–573. [CrossRef]
4. Wahyono, N.D.; Utami, M.M.D. A review of the poultry meat production industry for food safety in Indonesia. *J. Phys. Conf. Ser.* **2018**, *953*, 12125. [CrossRef]
5. Hussein, M.A.; Rehan, I.F.; Rehan, A.F.; Eleiwa, N.Z.; Abdel-Rahman, M.A.; Fahmy, S.G.; Ahmed, A.S.; Youssef, M.; Diab, H.M.; Batiha, G.E.; et al. Egg yolk IgY: A novel trend of feed additives to limit drugs and to improve poultry meat quality. *Front. Vet. Sci.* **2020**, *7*, 350. [CrossRef]
6. Christaki, E.; Bonos, E.; Giannenas, I.; Florou-Paneri, P. Aromatic plants as a source of bioactive compounds. *Agriculture* **2012**, *2*, 228–243. [CrossRef]
7. Wen, R.; Li, C.J.; Zhao, M.J.; Wang, H.; Tang, Y. Withdrawal of antibiotic growth promoters in China and its impact on the foodborne pathogen *Campylobacter coli* of swine origin. *Front. Microbiol.* **2022**, *13*, 1004725. [CrossRef]
8. Salium, E.M.; Vahjen, W.; Zentek, J. Types and prevalence of extended-spectrum beta-lactamase producing *Enterobacteriaceae* in poultry. *Anim. Health Res. Rev.* **2017**, *18*, 46–57. [CrossRef]
9. Gheisar, M.M.; Kim, I.H. Phytobiotics in poultry and swine nutrition—A review. *Ital. J. Anim. Sci.* **2017**, *17*, 92–99. [CrossRef]
10. Giannenas, I.; Bonos, E.; Christaki, E.; Florou-Paneri, P. Oregano: A feed additive with functional properties. In *Therapeutic Foods*; Holban, A.M., Grumezescu, A.M., Eds.; Handbook of Food Engineering; Elsevier Academic Press: London, UK, 2018; Volume 8, pp. 179–208. [CrossRef]
11. Franciosini, M.P.; Casagrande-Proietti, P.; Forte, C.; Beghelli, D.; Acuti, G.; Zanichelli, D.; dal Bosco, A.; Castellini, C.; Trabalza-Marinucci, M. Effects of oregano (*Origanum vulgare* L.) and rosemary (*Rosmarinus officinalis* L.) aqueous extracts on broiler performance, immune function and intestinal microbial population. *J. Appl. Anim. Res.* **2016**, *44*, 474–479. [CrossRef]
12. Hockenhull, J.; Turner, A.; Reyher, K.; Barrett, D.; Jones, L.; Hinchliffe, S.; Buller, H. Antimicrobial use in food-producing animals: A rapid evidence assessment of stakeholder practices and beliefs. *Vet. Rec.* **2017**, *181*, 510. [CrossRef]
13. Putnik, P.; Gabric, D.; Roohinejad, S.; Barba, F.J.; Granato, D.; Mallikarjunan, K.; Lorenzo, J.M.; Kovacevic, D.B. An overview of organosulfur compounds from *Allium* spp.: From processing and preservation to evaluation of their bioavailability, antimicrobial, and anti-inflammatory properties. *Food Chem. Toxicol.* **2019**, *276*, 680–691. [CrossRef]

14. Sheoran, N.; Kumar, R.; Kumar, A.A.; Batra, K.; Sihag, S.; Maan, S.; Maan, N.S. Nutrigenomic evaluation of garlic (*Allium sativum*) and holy basil (*Ocimum sanctum*) leaf powder supplementation on growth performance and immune characteristics in broilers. *Vet. World* **2017**, *10*, 121–129. [CrossRef]
15. Meot-Duros, L.; Magne, C. Antioxidant activity and phenol content of *Crithmum maritimum* L. leaves. *Plant Physiol. Biochem.* **2009**, *47*, 37–41. [CrossRef]
16. Souid, A.; Della Croce, C.M.; Frassinetti, S.; Gabriele, M.; Pozzo, L.; Ciardi, M.; Abdelly, C.; Hamed, K.B.; Magne, C.; Longo, V. Nutraceutical potential of leaf hydro-ethanolic extract of the edible halophyte *Crithmum maritimum* L. *Molecules* **2021**, *26*, 5380. [CrossRef]
17. Yang, X.; Liu, Y.; Yan, F.; Yang, C.; Yang, X. Effects of encapsulated organic acids and essential oils on intestinal barrier, microbial count, and bacterial metabolites in broiler chickens. *Poult. Sci.* **2019**, *98*, 2858–2865. [CrossRef]
18. Rathnayake, D.; Mun, H.S.; Dilawar, M.A.; Baek, K.S.; Yang, C.J. Time for a paradigm shift in animal nutrition metabolic pathway: Dietary inclusion of organic acids on the production parameters, nutrient digestibility, and meat quality traits of swine and broilers. *Life* **2021**, *11*, 476. [CrossRef]
19. Stefanello, C.; Rosa, D.P.; Dalmoro, Y.K.; Segatto, A.L.; Vieira, M.S.; Moraes, M.L.; Santin, E. Protected blend of organic acids and essential oils improves growth performance, nutrient digestibility, and intestinal health of broiler chickens undergoing an intestinal challenge. *Front. Vet. Sci.* **2020**, *6*, 461. [CrossRef]
20. Pearlin, B.V.; Muthuvel, S.; Govidasamy, P.; Villavan, M.; Alagawany, M.; Farag, M.R.; Dhama, K.; Gopi, M. Role of acidifiers in livestock nutrition and health: A review. *J. Anim. Physiol. Anim. Nutr.* **2020**, *104*, 558–569. [CrossRef]
21. Bonos, E.; Christaki, E.; Abrahim, A.; Soultos, N.; Florou-Paneri, P. The influence of mannan oligosaccharides, acidifiers and their combination on caecal microflora of Japanese quail (*Coturnix japonica*). *Anaerobe* **2011**, *17*, 436–439. [CrossRef]
22. Melaku, M.; Zhong, R.; Han, H.; Wan, F.; Yi, B.; Zhang, H. Butyric and citric acids and their salts in poultry nutrition: Effects on gut health and intestinal microbiota. *Int. J. Mol. Sci.* **2021**, *22*, 10392. [CrossRef]
23. Chowdhury, R.; Islam, K.M.S.; Khan, M.J.; Karim, M.R.; Haque, M.N.; Khatun, M.; Pesti, G.M. Effect of citric acid, avilamycin, and their combination on the performance, tibia ash, and immune status of broilers. *Poult. Sci.* **2009**, *88*, 1616–1622. [CrossRef]
24. Tzora, A.; Giannenas, I.; Karamoutsios, A.; Papaioannou, N.; Papanastasiou, D.; Bonos, E.; Skoufos, S.; Bartzanas, T.; Skoufos, I. Effects of oregano, attapulgite, benzoic acid and their blend on chicken performance, intestinal microbiology and intestinal morphology. *J. Poult. Sci.* **2017**, *54*, 218–227. [CrossRef]
25. Bozkurt, M.; Kucukyilmaz, K.; Catli, A.U.; Cinar, M.; Cabuk, M.; Alcicek, A. Effects of administering an essential oil mixture and an organic acid blend separately and combined to diets on broiler performance. *Archiv. Geflug.* **2012**, *76*, 81–87.
26. PD. *Presidential Degree 56/2013 on Harmonization of the Directive 2010/63/EU, on the Protection of Animals Used for Scientific Purposes*; Greek Goverment: Athens, Greece, 2013.
27. NRC. *Guide for the Care and Use of Laboratory Animals*, 8th ed.; National Academy Press: Washington, DC, USA, 2011.
28. Premier Nutrition. *Premier Atlas 2014. Ingredients Matrix*; Premier Nutrition: Brereton, UK, 2014.
29. Aviagen. *ROSS Nutrition Specifications*; Aviagen: Huntsville, AL, USA, 2019.
30. KPI Library. European Production Efficiency Factor (EPEF). Available online: https://kpilibrary.com/kpis/european-production-efficiency-factor-epef-2 (accessed on 3 March 2021).
31. Card, R.M.; Cawthraw, S.A.; Nunez-Garcia, J.; Ellis, R.J.; Kay, G.; Pallen, M.J.; Woodward, M.J.; Anjum, M.F. An in vitro chicken gut model demonstrates transfer of a multidrug resistance plasmid from *Salmonella* to commensal *Escherichia coli*. *mBio* **2017**, *8*, e00777-17. [CrossRef] [PubMed]
32. Bolder, N.M. Microbial challenges of poultry meat production. *World's Poult. Sci. J.* **2007**, *63*, 401–411. [CrossRef]
33. Melkamu, S.; Chanie, M.; Asrat, M. Studies on coccidia in experimental infection with *Eimeria* spp. in Rose-Cobb broiler chicken. *J. Anim. Sci.* **2017**, *7*, 115–122. [CrossRef]
34. Amer, M.M.; Awaad, M.H.H.; El-Khateeb, R.M.; Abu-Elezz, N.M.T.; Sherein-Said, A.; Gheta, M.M.; Kutkat, M.A. Isolation and identification of *Eimeria* from field coccidiosis in chickens. *J. Am. Sci.* **2010**, *6*, 1107–1114. [CrossRef]
35. Gava, M.S.; Moraes, L.B.; Carvalho, D.; Chitolina, G.Z.; Fallavena, L.C.B.; Moraes, H.L.S.; Herpich, J.; Salle, C.T.P. Determining the best sectioning methods and intestinal segments for morphometric analysis in broilers. *Braz. J. Poult. Sci.* **2015**, *71*, 145–150. [CrossRef]
36. Anderson, S. Determination of fat, moisture, and protein in meat and meat products by using the FOSS FoodScan near-infrared spectrophotometer with FOSS artificial neural network calibration model and associated database: Collaborative study. *J. AOAC Int.* **2007**, *90*, 1073–1083. [CrossRef]
37. AOAC. *Official Methods of Analysis*, 18th ed.; Association of Analytical Chemists, AOAC International: Arlington, VA, USA, 2007.
38. Ahn, D.U.; Olson, D.G.; Jo, C.; Love, J.; Jin, S.K. Volatiles production and lipid oxidation on irradiated cooked sausage as related to packaging and storage. *J. Food Sci.* **1999**, *64*, 226–229. [CrossRef]
39. SPSS. *SPSS Statistics for Windows, Release 20.0*; IBM: Armonk, NY, USA, 2018.
40. Jeni, R.E.I.; Dittoe, D.K.; Olson, E.G.; Jeferson, L.; Seidel, D.S.; Ricke, S.C.; Callaway, T.R. An overview of health challenges in alternative poultry production systems. *Poult. Sci.* **2021**, *100*, 101173. [CrossRef]
41. Tzora, A.; Bonos, E.; Giannenas, I.; Karamoutsios, A.; Fotou, K.; Sidiropoulou, E.; Tsiftsoglou, O.S.; Lazari, D.; Stylianaki, I.; Grigoriadou, K.; et al. Effects of dietary essential oils in combination with butyrate on growth performance, intestinal microflora, blood parameters and meat composition of broiler chickens. *Eur. Poult. Sci.* **2021**, *85*. [CrossRef]

42. Abbasi, M.A.; Ghazanfari, S.; Sharifi, S.D.; Gavlighi, H.A. Influence of dietary plant fats and antioxidant supplementations on performance, apparent metabolizable energy and protein digestibility, lipid oxidation and fatty acid composition of meat in broiler chicken. *Vet. Med. Sci.* **2020**, *6*, 54–68. [CrossRef]
43. Giannenas, I.; Florou-Paneri, P.; Papazahariadou, M.; Botsoglou, N.A.; Christaki, E.; Spais, A.B. Effect of diet supplementation with ground oregano on performance of broiler chickens challenged with *Eimeria tenella*. *Archiv. Geflug.* **2004**, *68*, 247–252.
44. Amer, S.A.; Metwally, A.E.; Ahmed, S.A.A. The influence of dietary supplementation of cinnamaldehyde and thymol on the growth performance, immunity and antioxidant status of monosex Nile tilapia fingerlings (*Oreochromis niloticus*). *Egypt. J. Aquat. Res.* **2018**, *44*, 251–256. [CrossRef]
45. Bauer, B.W.; Radovanovic, A.; Willson, N.L.; Bajagai, Y.S.; Thu, T.; Van, H.; Moore, R.J.; Stanley, D. Oregano: A potential prophylactic treatment for the intestinal microbiota. *Heliyon* **2019**, *5*, e02625. [CrossRef]
46. Bhatwalkar, S.B.; Mondal, R.; Krishna, S.B.N.; Adam, J.K.; Govender, P.; Anupam, R. Antibacterial properties of organosulfur compounds of garlic (*Allium sativum*). *Front. Microbiol.* **2021**, *12*, 613077. [CrossRef]
47. Pourali, M.; Kermanshahi, H.; Golian, A.; Ramzi, G.R.; Soukhtanloo, M. Antioxidant and anticoccidial effects of garlic powder and sulfur amino acids on *Eimeria*-infected and uninfected broiler chickens. *Iran. J. Vet. Res.* **2014**, *15*, 227–232.
48. Shao, X.; Sun, C.; Tang, X.; Zhang, X.; Han, D.; Liang, S.; Qu, R.; Hui, X.; Shan, Y.; Hu, L.; et al. Anti-inflammatory and intestinal microbiota modulation properties of jinxiang garlic (*Allium sativum* L.) polysaccharides toward dextran sodium sulfate-induced colitis. *J. Agric. Food Chem.* **2020**, *68*, 12295–12309. [CrossRef]
49. Lopresti, A.L. Salvia (sage): A review of its potential cognitive-enhancing and protective effects. *Drugs R D* **2017**, *17*, 53–64. [CrossRef] [PubMed]
50. Stanojevic, D.; Comic, L.; Stefanovic, O.; Solujic-Sukdolak, S. In vitro synergistic antibacterial activity of *Salvia Officinalis* L., and some preservatives. *Arc. Biol. Sci.* **2010**, *62*, 175–183. [CrossRef]
51. Kermanshah, H.; Hashemi Kamangar, S.; Arami, S.; Mirsalehian, A.; Kamalinejad, M.; Karimi, M.A.; Jabalamoli, F. In vitro evaluation of antibacterial activity of hydroalcoholic extract of *Salvia officinalis* and *Pimpinella anisum* against cariogenic bacteria. *J. Dental. Med.* **2009**, *22*, 149–154.
52. Hamidpour, M.; Hamdipour, R.; Hamidpour, S.; Shahlari, M. Chemistry, pharmacology, and medicinal property of sage (salvia) to prevent and cure illnesses such as obesity, diabetes, depression, dementia, lupus, autism, heart disease, and cancer. *J. Tradit. Complement. Med.* **2014**, *4*, 82–88. [CrossRef] [PubMed]
53. Houta, O.; Akrout, A.; Najja, H.; Neffati, M.; Amri, H. Chemical composition, antioxidant and antimicrobial activities of essential oil from *Crithmum maritimum* cultivated in Tunisia. *J. Essent. Oil Bear. Plants* **2015**, *18*, 1459–1466. [CrossRef]
54. Lum, J.; Sygall, R.; Ros Felip, J.M. Comparison of tributyrin and coated sodium butyrate as sources of butyric acid for improvement of growth performance in Ross 308 broilers. *Int. J. Poult. Sci.* **2018**, *17*, 290–297. [CrossRef]
55. Zhao, L.Q.; Liu, S.B.; Zhang, Z.; Zhang, J.; Jin, X.; Zhang, J.; Jiang, W.; Li, H.; Lin, H. Low and high concentrations of butyrate regulate fat accumulation in chicken adipocytes via different mechanisms. *Adipocyte* **2020**, *9*, 120–131. [CrossRef]
56. Vindrine, K.; Ye, J.; Martin, R.J.; McCutcheon, K.L.; Raggio, A.M.; Pelkman, C.; Durham, H.A.; Zhou, J.; Senerivathne, R.N.; Williams, C.; et al. Resistant starch from high amylose maize (HAMRS2) and dietary butyrate reduce abdominal fat by a different apparent mechanism. *Obesity* **2014**, *22*, 344–348. [CrossRef]
57. Moquet, P.C.A.; Salami, S.A.; Onrust, L.; Hendriks, W.H.; Kwakkel, R.P. Butyrate presence in distinct gastrointestinal tract segments modifies differentially digestive processes and amino acid bioavailability in young broiler chickens. *Poult. Sci.* **2018**, *98*, 167–176. [CrossRef]
58. Yang, X.; Yin, F.; Yang, Y.; Lepp, D.; Yu, H.; Ruan, Z.; Yang, C.; Yin, Y.; Hou, Y.; Leeson, S.; et al. Dietary butyrate glycerides modulate intestinal microbiota composition and serum metabolites in broilers. *Sci. Rep.* **2018**, *8*, 4940. [CrossRef]
59. Brenes, A.; Roura, E. Essential oils in poultry nutrition: Main effects and modes of action. *Anim. Feed Sci. Technol.* **2010**, *158*, 1–14. [CrossRef]
60. Liu, Y.; Yang, X.; Xin, H.; Chen, S.; Yang, C.; Duan, Y.; Yang, X. Effects of a protected inclusion of organic acids and essential oils as antibiotic growth promoter alternative on growth performance, intestinal morphology and gut microflora in broilers. *Anim. Sci. J.* **2017**, *88*, 1414–1424. [CrossRef]
61. Basmacioglu-Malayoglu, M.H.; Baysal, S.; Misirtioglou, Z.; Polat, M.; Yilmaz, H.; Turan, N. Effects of oregano essential oil with or without feed enzymes on growth performance, digestive enzyme, nutrient digestibility, lipid metabolism and immune response of broilers fed on wheat-soybean meal diets. *Br. Poult. Sci.* **2010**, *51*, 67–80. [CrossRef]
62. Zhang, K.Y.; Yan, F.; Keen, C.A.; Waldroup, P.W. Evaluation of microencapsulated essential oils and organic acids in diets for broiler chickens. *Int. J. Poult. Sci.* **2005**, *4*, 612–619. [CrossRef]
63. Zeeshan, M.; Zaneb, H.; Masood, S.; Ashraf, S.; Khan, I.; Rehman, H.F.U.; Din, S.; Hayat, K. Morphological modulation of broiler organs in response to an organic acid–phytogen composite in healthy broilers. *Agriculture* **2022**, *12*, 791. [CrossRef]
64. Yang, X.; Xin, H.; Yang, C.; Yang, X. Impact of essential oils and organic acids on the growth performance, digestive functions and immunity of broiler chickens. *Anim. Nutr.* **2018**, *4*, 388–393. [CrossRef]
65. Gao, Y.Y.; Zhang, X.L.; Xu, L.H.; Peng, H.; Wang, C.K.; Bi, Y.Z. Encapsulated blends of essential oils and organic acids improved performance, intestinal morphology, cecal microflora, and jejunal enzyme activity of broilers. *Czech J. Anim. Sci.* **2019**, *64*, 189–198. [CrossRef]

66. Yakhkeshi, S.; Rahimi, S.; Gharib Naseri, K. The effects of comparison of herbal extracts, antibiotics, probiotics and organic acid on serum lipids, immune response, GIT microbial population, intestinal morphology and performance of broilers. *J. Med. Plants* **2011**, *10*, 80–95.
67. Gheisar, M.M.; Hosseindoust, A.; Kim, I.H. Evaluating the effect of microencapsulated blends of organic acids and essential oils in broiler chickens diet. *J. Appl. Poult. Res.* **2015**, *24*, 511–519. [CrossRef]
68. Jang, I.S.; Ko, Y.H.; Kang, S.Y.; Lee, C.Y. Effect of a commercial essential oil on growth performance, digestive enzyme activity and intestinal microflora population in broiler chickens. *Anim. Feed Sci. Technol.* **2007**, *134*, 304–315. [CrossRef]
69. Partanen, K.; Jalava, T.; Valaja, J. Effects of a dietary organic acid mixture and of dietary fibre levels on ileal and faecal nutrient apparent digestibility, bacterial nitrogen flow, microbial metabolite concentrations and rate of passage in the digestive tract of pigs. *Animal* **2007**, *1*, 389–401. [CrossRef]
70. Hollemans, M.S.; Van Baal, J.; De Vries Reilingh, G.; Kemp, B.; Lammers, A.; De Vries, S. Intestinal epithelium integrity after delayed onset of nutrition in broiler chickens. *Poult. Sci.* **2020**, *99*, 6818–6827. [CrossRef] [PubMed]
71. Bonis, V.; Rossell, C.; Gehart, H. The intestinal epithelium—Fluid fate and rigid structure from crypt bottom to villus tip. *Front. Cell Dev. Biol.* **2021**, *9*, 661931. [CrossRef] [PubMed]
72. Sen, S.; Ingale, S.L.; Kim, Y.W.; Kim, J.S.; Lohakare, J.D.; Kim, E.K.; Kim, H.S.; Ryu, M.H.; Kwon, I.K.; Chae, B.J. Effect of supplementation of *Bacillus subtilis* LS 1–2 to broiler diet on growth performance, nutrient retention, caecal microbiology and small intestinal morphology. *Res. Vet. Sci.* **2011**, *93*, 264–268. [CrossRef] [PubMed]
73. Reisinger, N.; Steiner, T.; Nitsch, S.; Schatzmayr, G.; Applegate, T.J. Effects of a blend of essential oils on broiler performance and intestinal morphology during coccidial vaccine exposure. *J. Appl. Poult. Res.* **2011**, *20*, 272–283. [CrossRef]
74. Giannenas, I.; Bonos, E.; Filliousis, G.; Stylianaki, I.; Kumar, P.; Lazari, D.; Christaki, E.; Florou-Paneri, P. Effect of a polyherbal or an arsenic-containing feed additive on growth performance of broiler chickens, intestinal microbiota, intestinal morphology and lipid oxidation of breast and thigh meat. *J. Appl. Poult. Res.* **2019**, *28*, 164–175. [CrossRef]
75. Gadde, U.; Kim, W.; Oh, S.; Lillehoj, H.S. Alternatives to antibiotics for maximizing growth performance and feed efficiency in poultry: A review. *Anim. Health Res. Rev.* **2017**, *18*, 26–45. [CrossRef]
76. Ahsan, H.; Cengiz, O.; Raza, I.; Kuter, E.; Chacher, M.F.A.; Iqbal, Z.; Umar, S.; Cakir, S. Sodium butyrate in chicken nutrition: The dynamics of performance, gut microbiota, gut morphology, and immunity. *World's Poult. Sci. J.* **2016**, *72*, 265–275. [CrossRef]
77. Suresh, G.; Das, R.K.; Kaur Brar, S.; Rouissi, T.; Avalos Ramirez, A.; Chorfi, Y.; Godbout, S. Alternatives to antibiotics in poultry feed: Molecular perspectives. *Crit. Rev. Microbiol.* **2018**, *44*, 318–355. [CrossRef]
78. Skoufos, I.; Giannenas, I.; Tontis, D.; Bartzanas, T.; Kittas, C.; Panagakis, P.; Tzora, A. Effects of oregano essential oil and attapulgite on growth performance, intestinal microbiota and morphometry in broilers. *S. Afr. J. Anim. Sci.* **2016**, *46*, 77–88. [CrossRef]
79. Glorieux, M.; Newman, L.J.; Wang, Y.T.; De Herdt, P.; Hautekeur, J.; De Gussem, M.; Christiaens, I.; Verbeke, J. Sustainable coccidiosis control implications based on susceptibility of European *Eimeria* field isolates to narasin + nicarbazin from farms using anticoccidial medication or coccidial vaccines. *J. Appl. Poult. Res.* **2022**, *31*, 100263. [CrossRef]
80. Hao, H.; Cheng, G.; Iqbal, Z.; Ai, X.; Hussain, H.I.; Huang, L.; Dai, M.; Wang, Y.Z.; Liu, Z.; Yuan, Z. Benefits and risks of antimicrobial use in food-producing animals. *Front. Microbiol.* **2014**, *5*, 288. [CrossRef]
81. Millet, S.; Maertens, L. The European ban on antibiotic growth promoters in animal feed: From challenges to opportunities. *Vet. J.* **2011**, *187*, 143–144. [CrossRef]
82. Jin, J.Z. Phytogenic feed additives: Nutritional functions and mechanism of action in monogastric animals. *Chin. J. Anim. Nutr.* **2010**, *22*, 1154–1164.
83. Pastor-Fernandez, I.; Kim, S.J.; Billington, K.; Bumstead, J.; Marugan-Hernandez, V.; Kuster, T.; Ferguson, D.J.P.; Vervelde, L.; Blake, D.P.; Tomley, F.M. Development of cross-protective *Eimeria*-vectored vaccines based on apical membrane antigens. *Int. J. Parasitol.* **2018**, *48*, 505–518. [CrossRef]
84. Bozkurt, M.; Ege, G.; Aysul, N.; Aksit, H.; Tuzun, A.E.; Kucukyilmaz, K.; Borum, A.E.; Uygun, M.; Aksit, D.; Aypak, S.; et al. Effect of anticoccidial monensin with oregano essential oil on broilers experimentally challenged with mixed *Eimeria* spp. *Poult. Sci.* **2016**, *95*, 1858–1868. [CrossRef]
85. Jitviriyanon, S.; Phanthong, P.; Lomarat, P.; Bunyapraphatsara, N.; Porntrakulpipat, S.; Paraksa, N. In vitro study of anticoccidial activity of essential oils from indigenous plants against *Eimeria tenella*. *Vet. Parasitol.* **2016**, *228*, 96–102. [CrossRef]
86. Dkhil, M.A.; Thagfan, F.A.; Hassan, A.M.; Al-Shaebi, E.M.; Abdel-Gaber, R.; Al-Quraishy, S. Anthelmintic, anticoccidial and antioxidant activity of Salvadora persica root extracts. *Saudi. J. Biol. Sci.* **2019**, *26*, 1223–1226. [CrossRef]
87. Fouad, A.M.; El-Senousey, H.K. Nutritional factors affecting abdominal fat deposition in poultry: A review. *Asian-Austral. J. Anim. Sci.* **2014**, *27*, 1057–1068. [CrossRef]
88. Moretti, S.; Mrakic-Sposta, S.; Roncoroni, L.; Vezzoli, A.; Dellanoce, C.; Monguzzi, E.; Branchi, F.; Ferretti, F.; Lombardo, V.; Duoneda, L.; et al. Oxidative stress as a biomarker for monitoring treated celiac disease. *Clin. Transl. Gastroenterol.* **2018**, *9*, e157. [CrossRef]
89. Giannenas, I.; Karamoutsios, A.; Bartzanas, T.; Tsinas, A.; Tzora, A.; Skoufos, I. Oregano and sage essential oils improve antioxidant status of raw and cooked breast and thigh chicken meat. *Agro Food Ind. Hi-Tec.* **2016**, *27*, 10–14.
90. Barbarestani, S.Y.; Jazi, V.; Mohebodini, H.; Ashayerizadeh, A.; Shabani, A.; Toghyani, M. Effects of dietary lavender essential oil on growth performance, intestinal function, and antioxidant status of broiler chickens. *Livest. Sci.* **2020**, *233*, 103958. [CrossRef]

91. Giannenas, I.; Florou-Paneri, P.; Botsoglou, N.A.; Christaki, E.; Spais, A.B. Effect of supplementing feed with oregano and/or a-tocopheryl acetate on growth of broiler chickens and oxidative stability of meat. *J. Anim. Feed Sci.* **2005**, *14*, 521–535. [CrossRef]
92. Surai, P.F. Polyphenol compounds in the chicken/animal diet: From the past to the future. *J. Anim. Physiol. Anim. Nutr.* **2014**, *98*, 19–31. [CrossRef]
93. Ashour, E.A.; Abd El-Hack, M.E.; Swelum, A.A.; Osman, A.O.; Taha, A.E.; Alhimaidi, A.R.; Ismail, I.E. Does the dietary graded levels of herbal mixture powder impact growth, carcass traits, blood indices and meat quality of the broilers? *Ital. J. Anim. Sci.* **2020**, *19*, 1228–1237. [CrossRef]

sustainability

MDPI

Article

An Evaluation of the Uses of Different Environmental Enrichments on a Broiler Farm with the Help of Real-Time Monitoring via a Farmer-Assistant System

Fabian Spieß [1], Bernd Reckels [1], Christian Sürie [2], Monika Auerbach [3], Silke Rautenschlein [3], Ottmar Distl [4], Joerg Hartung [5], Jan Berend Lingens [1,*,†] and Christian Visscher [1,†]

1 Institute for Animal Nutrition, University of Veterinary Medicine Hannover, Foundation, 30173 Hannover, Germany
2 Farm for Education and Research Ruthe, University of Veterinary Medicine Hannover, Foundation, 31157 Sarstedt, Germany
3 Clinic for Poultry, University of Veterinary Medicine Hannover, Foundation, 30559 Hannover, Germany
4 Institute for Animal Breeding and Genetics, University of Veterinary Medicine Hannover, Foundation, 30559 Hannover, Germany
5 Institute of Animal Hygiene, Animal Welfare and Farm Animal Behaviour, University of Veterinary Medicine Hannover, Foundation, 30625 Hannover, Germany
* Correspondence: jan.berend.lingens@tiho-hannover.de
† These authors contributed equally to this work.

Citation: Spieß, F.; Reckels, B.; Sürie, C.; Auerbach, M.; Rautenschlein, S.; Distl, O.; Hartung, J.; Lingens, J.B.; Visscher, C. An Evaluation of the Uses of Different Environmental Enrichments on a Broiler Farm with the Help of Real-Time Monitoring via a Farmer-Assistant System. *Sustainability* **2022**, *14*, 13015. https://doi.org/10.3390/su142013015

Academic Editors: George K. Symeon and Vassilios Dotas

Received: 26 August 2022
Accepted: 8 October 2022
Published: 11 October 2022

Abstract: Modern broilers are usually raised in barren environments in large enclosed halls. Various environmental enrichment elements such as perches, elevated platforms, and similar structures were proposed for these barns with the aim of improving the welfare and well-being of the birds. This study compares and evaluates three different types of environmental enrichment. In 2 identical barns, 8100 Ross 308 broilers were housed divided between a control group (CG) and a trial group (TG). In the TG, three types of environmental enrichment (perches, elevated platforms, and a combined structure) were used. A real-time monitoring device (FAS = farmer-assistance system) suspended from the ceiling was used in combination with single photographs to count the number of birds on the enrichment elements. In addition, the body weights of individual birds and their foot pad dermatitis (FPD) scores were collected at days 14, 21 and 28 in both barns. No differences in these parameters were seen between TG and CG. Birds showed highest preference for the elevated platforms (average 31.93 kg/m^2), followed by the combined structure (average 21.36 kg/m^2) and the perches (0.35 kg/m^2). Overall, this study shows that Ross 308 broiler birds significantly prefer elevated platforms over combined structures or simple perches.

Keywords: broiler; enrichment; animal welfare; performance; real-time monitoring

1. Introduction

Broiler meat is the fastest growing sector of food derived from animals [1]. The birds are usually kept indoors in large barns offering only a few structuring elements above ground level for the animals [2,3]. This barren environment can lead to boredom, inactivity, distress, and behavioural- and even health-related disorders [4,5]. In order to avoid or mitigate these negative consequences of such a barren environment, a variety of environmental enrichment elements were tested. The provision of elements such as perches, elevated platforms, and similar structures was postulated for broiler production to improve the welfare and quality of life of these broilers [6]. Providing environmental enrichment can target problems with inactivity by offering incentives, for example, to climb elevated places, while also allowing the birds more flexibility in exhibiting a wider range of specific behaviours [7].

In particular, concerns regarding animal welfare have recently arisen [8]. This development has also brought new challenges: To improve animal welfare, the problems relating to the current forms of husbandry must first be addressed in order to then find, test, and evaluate solutions and improvements. One problem is the structure of the barns, or rather, the lack of structure [6]. On most commercial broiler farms, birds are raised on a flat concrete floor covered with wood shavings [2], and the housing offers no structural elements above the ground level other than the feeding and drinking lines [3]. In combination with high stocking densities, this lack of structuring elements can lead to common production diseases such as foot pad dermatitis (FPD) [9], deep skin dermatitis, and sudden cardiac death [10]. These diseases, also termed cumulative disorders or technopathies, can heavily impair the welfare of the animals [11].

A promising approach is environmental enrichment elements that allow the animals to exhibit more natural behaviours. The term environmental enrichment has been defined by Newberry as "an improvement in the biological functioning of captive animals resulting from modifications to their environment" [12].

In recent years, many different types of environmental enrichment have been tested [13–16]. Most approaches deal with enabling natural behaviours such as pecking, hooting, or seeking elevated perching positions. Others dealt with, for example, the type of feeding [17] or more intensive human care [18]. One important aspect of the natural behaviour of broilers is seeking elevated positions to rest. Day-old broilers develop the urge to seek hiding places and elevated positions as they grow older as it is part of their species-specific behaviour [19]. As a kind of protection mechanism against potential predators, chickens took up elevated positions in the wild [20,21]. This could be the reason why birds have been shown to be eager to take up elevated seating positions and even make an effort to climb up to them [21–23]. In this way, the provision of environmental enrichment elements can help to encourage natural behaviour in birds while also giving them a more secure feeling [24].

Several options for enrichment in the form of elevated structures that offer seating possibility like straw bales [25], perches, and small elevated floors [23,26] have been the focus of research in recent years, for example the elevated plastic platforms 30 cm above the ground accessible via ramps with an angle of 15° used by Kaukonen et al. [27], the galvanized steel pipes 15 cm above the ground used by Aksit et al. [28], or the wooden beams 10 cm above the ground used by Ventura et al. [26]. This is driven by the awareness in society of the importance of animal welfare. Animal numbers per m^2 and stocking densities in general [8] should be reduced and the animals should be offered more opportunities to perform their natural behaviours. Therefore, it is necessary to investigate and evaluate environmental enrichment elements which are offered by industry in regard to broiler health and welfare [4,14].

When testing and evaluating the use of the structures by the birds, digitalisation and modern monitoring devices can be of value especially when assessing the elements in commercial farms [29]. A considerable advantage digitalisation offers is "real-time" monitoring. Data and pictures of animals can be collected even when no farmer is present [30]. The continuous collection of data offers the possibilities to have a look at how the birds interact with the environmental enrichment over the course of 24 h a day. The air conditions in the barn have a significant influence on broiler health [31] and welfare. Knowing the air conditions in the barn together with continuous monitoring of the birds can possibly help to detect healt and welfare problems earlier than usually recognized by the farmer or when they happen at nighttime.

The aim of this field study was to investigate and evaluate the use of elevated platforms, perches and a combination of both which are offered as typical enrichment elements for broiler houses. Research regarding these points is important as there is a lot of potential for environmental enrichment to improve animal health and welfare. The acceptance of the structures by the broilers and their influence on broiler behaviour were analysed by personal observations. In addition, photos of all elements and the birds on them were regularly taken and evaluated to be able to describe how the birds react to the different types of enrichment and to see whether it has an influence on the distribution of the birds on the floor in the barn. The results were compared with the control group without environmental enrichment elements.

2. Materials and Methods

2.1. Animals and Diets

2.1.1. Animals

In this trial, for each of the 8 rounds, 8100 broilers were housed per group. They were randomly distributed between the control group and trial group as hatched. Over the course of 9 fattening rounds, this resulted in a total of 72,900 birds. The genetic strain used was Ross 308.

2.1.2. Diets

The birds were fed pellets ad libitum in four phases (Table 1). The feed was provided by a commercial feed supplier (MEGA Tierernährung GmbH & Co. KG, Visbek, Germany). From the day of arrival to d 7, the birds were fed a starter diet which was afterwards replaced by a grower one diet. At d 20, the birds were fed a grower two diet. Ultimately, at d 29, the birds were fed a finisher diet up until 12 h before departure to slaughter at d 33.

Table 1. Composition of the commercial diets used for control and trial groups.

Ingredients (in %)	Starter	Grower One	Grower Two	Finisher
Crude protein	21.60	19.00	19.00	19.50
Ether extract	5.40	4.70	4.70	7.80
Crude fibre	2.50	3.50	3.20	3.20
Crude ash	5.50	5.40	5.10	4.80
Calcium	0.90	0.75	0.70	0.65
Phosphorus	0.85	0.55	0.50	0.40
Sodium	0.16	0.16	0.15	0.14
Lysine	1.35	1.12	1.12	1.14
Methionine	0.80	0.28	0.54	0.28

2.2. Experimental Design and Housing

The experimental and the control barns had identical sizes, equipment, and management. The rectangular floor space was 16 m by 30 m. The length of the fattening period was 33 days.

The barns were equipped with three conventional feeding lines (Big Dutchman International GmbH, Vechta, Germany) and four conventional water lines (LUBING Maschinenfabrik Ludwig Bening GmbH & Co. KG, Barnstorf, Germany) with drinking nipples (LUBING Maschinenfabrik Ludwig Bening GmbH & Co. KG, Barnstorf, Germany), as shown in Figures 1 and 2. Both barns were evenly littered with wood shavings (GOLDSPAN®, Goldspan GmbH and Co. KG, Goldenstedt, Germany). Figure 2 in addition shows the enrichment elements.

At d 0, the light programme was 24 h of light. At d 1, the light was turned off for 4 h between 23:00 and 03:00. At d 2, the 6-h dark period occurred between 22:00 and 04:00. Starting at d 3, the light was turned off for 8 h from 21:00 to 05:00. From d 21 onwards, the dark period was shortened to 6 h between 22:00 and 04:00.

At the arrival of the birds at d 1, the air in the barn was heated up to 33.5 °C with a gas air-heating system. The air temperature was then continuously lowered until it reached 23 °C at d 33. To control the negative pressure ventilation system and the air quality, the air was measured with temperature and humidity sensors.

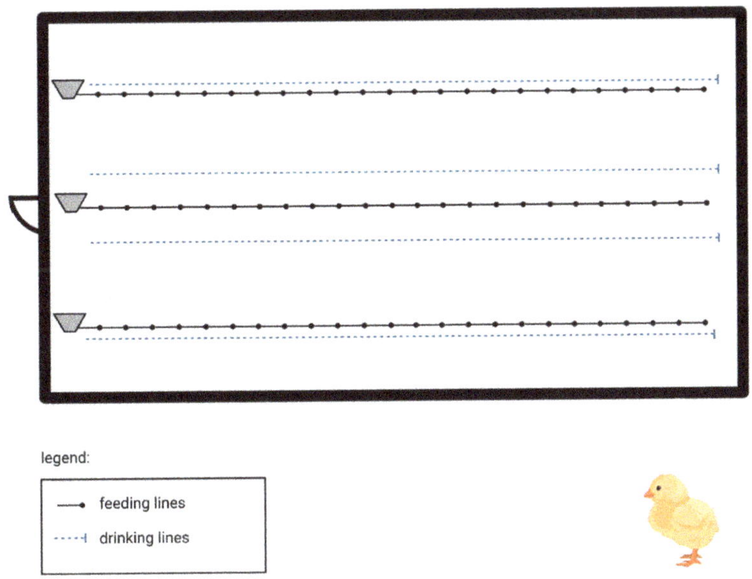

Figure 1. Schematic drawing of the control groups (figure was created with Biorender.com (accessed on 7 October 2022)).

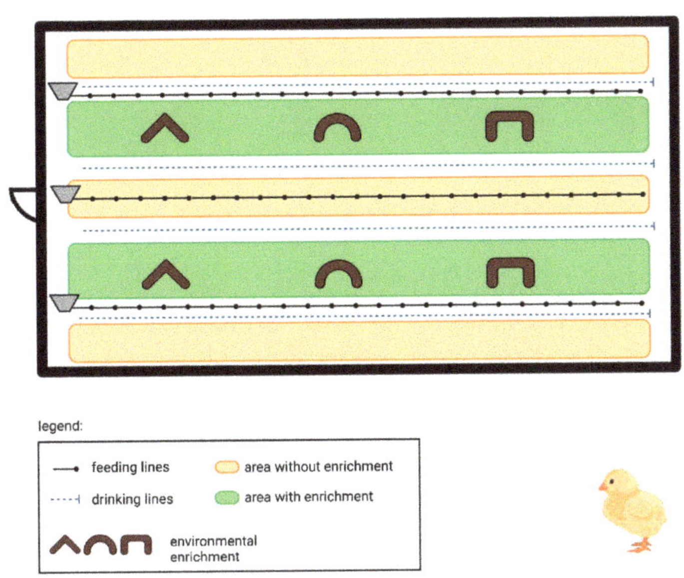

Figure 2. Schematic drawing of the trial groups (figure was created with https://biorender.com/ (accessed on 7 October 2022)).

At d 12, the birds were vaccinated against Newcastle disease; at d 18, against Gumboro; and at d 20 against infectious bronchitis with virus strain Ma5 with conventional vaccines in the recommended dose via the drinking water.

2.3. *Experimental Treats*

2.3.1. Environmental Enrichment

During the trial, three different elements of environmental enrichment were used (Figures 3–5): The first element was perches. The enrichment used for evaluating perches was the so-called "A-Reuter", which was originally made for keeping broiler parent stock. The construction had a length of 5.60 m and was 1.40 m wide. The A-Reuter consisted of 5 metal perches that were equal in length and had a diameter of 1.90 cm each. All perches were mounted on a frame made of the same material.

Figure 3. Photograph of the perch variant of environmental enrichment, "A-Reuter".

Figure 4. Photograph of the elevated platform variant of environmental enrichment, the "Plateau".

Figure 5. Photograph of the combined variant of environmental enrichment: the "Hybrid".

The second element was elevated platforms. The enrichment used as an elevated platform was the so-called Plateau. This consisted of two carriage axles with two tyres each. A rectangular framework was mounted on top of the axles. This framework held the grids, each of which measured 1.20 m long and 0.75 m wide, making the whole Plateau 2.40 m long. On both of the long sides, a ramp consisting of one of those grids was mounted to allow the birds access to the top.

The third category is a combination of perches and elevated plains. For the combination we used the Hybrid (Hölscher + Leuschner GmbH & Co. Kg, Emsbüren, Germany). It is composed of three perches, which are attached at right angles to two elevated plains. Both elevated planes are accessible via two ramps each. The elevated plains are 248 mm wide and 1012 mm long. In-between the planes there are two perches of 1.5 m length. Two of those are mounted underneath the elevated planes. This leaves a space of 992 mm between the inner sides of the elevated planes. Four elements of the Hybrid additionally had a third perch above the planes which is mounted on a rectangular frame. Each perch is 52 mm wide and 78 mm high, with a rounded top part for the birds to sit.

All three environmental enrichments described above, A-Reuter, Plateau, and Hybrid, were used in each of the experimental runs with enrichment. In order to exclude any influence of their position in the barn, their positions were changed from trial to trial in rotation in a clockwise direction. This meant that an environmental enrichment element that was placed at position one, which is located at the front of the trial barn, at the outset of the trial was placed in the following round of trial at position two located in the middle of the trial barn and thereafter at position three at the back of the trial barn.

2.3.2. Farmer-Assistant System

The Farmer-Assistant System (FAS) is a livestock robot that consists of an upper and a lower box connected by a telescopic arm. The upper part contains the battery, the motor, the drive wheels, and the upper camera. This camera provides an overview of the barn. The robot runs on a railway that is located underneath the roof of the barn. The railway enables the FAS to run the sensor box at a height of 70 cm above the broilers without disturbing them, while constantly monitoring the climatic conditions and the flock. The sensors for this are located in the lower box. They continuously measure air temperature, relative humidity of the air, wind speed, carbon dioxide (CO_2), ammonia (NH_3), light, and noise. The lower box also contains a bottom camera and two side cameras in order to be able to observe the broilers more precisely. The collected data are presented to the farmer in a daily report every morning. A cloud-based system stores the raw data for all parameters in a raw form but also in figures mapped over a virtual version of the barn floor. The parameter-free space is calculated through an algorithm. The top camera takes an image, and the artificial intelligence of the robot detects the birds present in that image. The amount of free space is then calculated from the area of the floor minus the space that birds take up on the given image.

2.4. *Measurements*

2.4.1. Growth Performance and Slaughter Data

First, 50 birds were randomly selected, and their weight was measured at d 14, 21, and 28 of life over the course of all rounds of trial (Figure 6). A hanging scale (VEIT Electronics s.r.o., Moravany, the Czech Republic) was used to record the birds' weight. The total weight and the number of birds were measured at the slaughterhouse and the average body weight was then calculated therefrom. The slaughterhouse used a camera-based system common in Germany and according to the common FPD scoring (QS Qualität und Sicherheit GmbH, Bonn, Germany) at slaughter. The scores of this scoring system are defined as 0, 1, 2a and 2b [32] and are afterwards calculated into a number for footpad points.

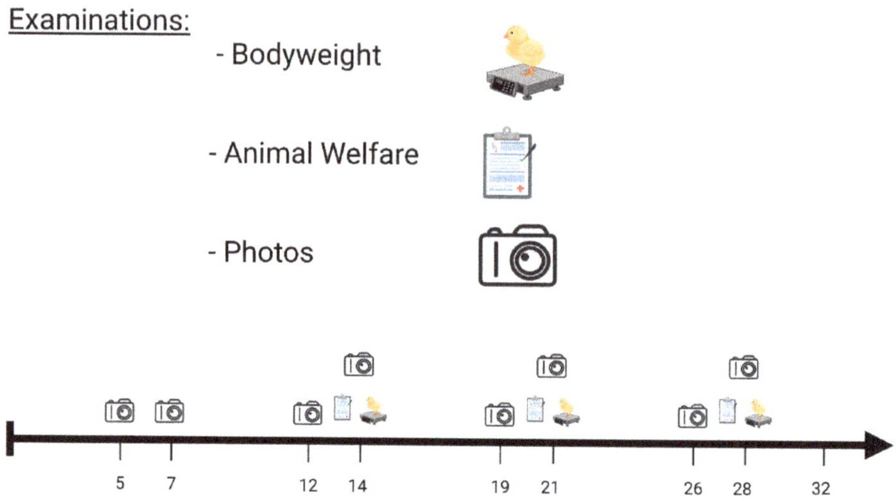

Figure 6. Timeline of examinations made and photographs taken during one trial round (figure was created with https://biorender.com/ (accessed on 7 October 2022)).

2.4.2. Photographs and Evaluation

As described in Figure 6, photographs were taken at d 5, 7, 12, 14, 19, 21, 26, and 28. Each type of environmental enrichment was photographed separately over the course of three trial rounds. Under the roof of the barn, the photos were taken from a windowed corridor so that the animals would not be startled or otherwise affected by the picture taking. To ensure that each image showed the same section of the image, markings were placed on the windows for the individual enrichments to indicate the points for the images. As the positions of the environmental enrichments were rotated each run, the alignment of the individual elements in the barn had to be checked once at the beginning of each run to match the markers. The number of birds per square metre was calculated by dividing the total number of birds on top of one element of environmental enrichment by the size of the element. The size of each element was calculated by multiplying the length of it with its width. The kg per square metre was then calculated by multiplying that value with the average weight of birds on the actual day. Afterwards the difference from the average weight in the barn was calculated by subtracting the value for kg per square metre from the average weight in the barn.

2.4.3. Feed Conversion Ratio

By dividing the feed intake (kg) by the total BW (kg) of all birds, the feed conversation ratio (FCR) could be calculated. This was calculated for all fattening rounds in the control group and trial group. To have a slightly more precise value, it is possible to calculate the corrected FCR, which also takes bird losses into consideration. At first, the cumulative feed intake of the dead animals needs to be calculated. This is the sum of the daily feed intakes of each animal up to the day of departure. Then the corrected feed intake in kg needs to be calculated. The corrected feed intake is obtained by subtracting the cumulative feed intake of the dead animals from the total feed intake. Then the corrected feed intake is divided by the total body weight gain in order to obtain the corrected FCR.

2.4.4. Foot Pad Dermatitis

As an indicator of animal welfare and to control whether the environmental enrichment has an influence on the foot pads, the FPD scores were taken in all of the nine fattening runs. For this purpose, 50 animals were randomly selected from the flock on each of the examination days. The FPD scores of these animals were then recorded, looking at the central plantar area on both feet of these 50 broilers at d 14, 21, and 28 (Figure 6). A seven-point scale according to Mayne et al. [33] was used to evaluate the FPD scores. As described in Figure 7, the first score on this scale is 0, which means the feet show no external signs of FPD. Score 3 marks the point where the central part of the footpad is swollen, red, and hard and where the first necrotic areas are visible. In increasing order, a larger size of the necrotic areas is then described until ultimately Score 7 describes a foot pad where half of the central plantar area is covered in necrotic scales.

Score	Description of footpad
0	No external signs of FPD are visible. Skin of the footpad and digital pads appears normal, no redness, swelling or necrosis is evident. The skin of the footpad feels soft to the touch.
1	Slight swelling and/or redness of the skin of the footpad.
2	The pad feels harder and denser than a non-affected foot. The central part of the pad is raised with swelling and redness and the reticulate scales may be separated. The digital pads may show similar signs.
3	The central and digital footpads are enlarged and swollen with red areas, and as the skin has become compacted, the footpad is hard. The reticulate scales have become enlarged and separated, and small black necrotic areas may occur.
4	Marked swelling and redness around the margins of lesions occur. Reticulate scales die and turn black, forming scale-shaped necrotic areas. The scales around the outside of the black areas may have turned white. The area of necrosis is less than one-eighth of the total area of the footpad.
5	Swelling and redness are evident in the central and digital footpads. The total footpad size is enlarged. Reticulate scales are pronounced, increased in number, and separated from each other. The amount of necrosis extends to a quarter of the footpad. Small necrotic areas may also appear on the digital pads.
6	As score 5, but with half the footpad covered by necrotic cells. The digital pads may have up to half of one pad covered with necrotic cells.
7	A footpad with over half of the footpad covered in necrotic scales.

Figure 7. Foot pad scoring according to Mayne et al. (2007) [33].

2.5. Statistical Analysis

Data analysis was performed using the SAS statistical software package, version 7.1 (SAS Institute, Cary, NC, USA). First a Shapiro–Wilk test for normal distribution was performed. All measurement data were analysed descriptively according to sample size, means, confidence interval, standard deviation, minimum and maximum. For data not normally distributed, a Kruskal–Wallis test was performed, followed by a Wilcoxon two-sample test. Normally distributed data were checked for significant differences with the Ryan–Einot–Gabriel–Welsch-test (one-way ANOVA). The analysis of the values for foot pad scoring was performed using a Kruskal–Wallis test. All statements of statistical significance were based on $p < 0.05$.

3. Results

3.1. Growth Performance and Slaughter Data

3.1.1. Growth Performance

Table 2 displays and statistically compares the average BW for d 14, 21, and 28. The BWs were collected from 50 individual broilers from both control and trial.

Table 2. Average body weight (g) ± standard deviation of individually weighed birds from day 14 to 28 of life in both the control (CG) and trial group (TG).

Day of Life	n	CG	TG	*p*-Value
14	50	524.23 ± 60.94	525.41 ± 52.23	0.7543
21	50	1045.04 ± 124.86	1034.96 ± 128.41	0.2328
28	50	1675.44 ± 197.54	1699.90 ± 186.33	0.0563

As shown in Table 2, no significant differences in average BW were seen between the control and the trial group at d 14, 21, and 28 of life.

3.1.2. Footpad Scores

Table 3 shows the average foot pad disease scores at d 14, 21 and 28 for both trial group and control group.

Table 3. Foot pad disease score ± standard deviation, in accordance with Mayne, scored for both feet of 50 birds per day in the control (CG) and the trial groups (TG).

		CG	TG	*p*-Value
Day of Life	**n**	**FPD Score**	**FPD Score**	
14	450	$0.88^{b} \pm 1.15$	$0.65^{a} \pm 0.99$	0.0019
21	450	$1.64^{b} \pm 1.72$	$1.12^{a} \pm 1.48$	<0.0001
28	450	$1.79^{a} \pm 1.94$	$1.75^{a} \pm 2.13$	0.1836

a, b Means in a row with different superscripts differ significantly ($p < 0.05$).

Table 3 shows significant differences for day 14 and 21 between control and trial group, when the scores in the trial group were significantly lower.

3.1.3. Slaughter Data

The slaughter data were reported after each single round of trial directly from the slaughterhouse. Table 4 displays the average BW and the foot pad scores scored at the slaughterhouse with standard deviation for the control and the trial group at d 33.

Table 4. Slaughter data regarding average body weight (g) ± standard deviation per bird and foot pad points ± standard deviation according to QS Qualität und Sicherheit GmbH, Bonn, Germany, for the control (CG) and trial group (TG) over nine trial rounds.

	CG n = 67,780	TG n = 69,609	*p*-Value
Bodyweight (g)	2059.89 ± 66.66.	2048.44 ± 48.70	0.6830
Foot pad score	12.00 ± 12.47	15.55 ± 10.85	0.5281

As Table 4 displays there were no significant differences between the control and trial group.

3.1.4. Feed Conversion Ratio

The FCR and the corrected FCR are displayed and analysed for the control and trial groups in Table 5.

Table 5. Feed conversion ratio (kg feed/kg body weight gained) and corrected feed conversion ratio (corr. FCR) ± standard deviation in the control (CG) and trial group (TG) over nine trial rounds.

	N	CG	TG	*p*-Value
FCR	9	1.43 ± 0.02	1.41 ± 0.02	0.2060
corr. FCR	9	1.42 ± 0.03	1.41 ± 0.02	0.3279

Table 5 displays no significant differences between the control group and the trial group regarding FCR and corrected FCR.

3.2. Photo Evaluation

3.2.1. Birds on Environmental Enrichment

Table 6 shows the total number of birds observed at the time of each taken photograph for each different type of environmental enrichment.

Table 6. Average total number of birds on the different types of environmental enrichment over the course of three rounds of trial ± standard deviation in the trial groups (TG) at the time of taking photographs.

	Week 1	Week 2	Week 3	Week 4	*p*-Value
"A-Reuter"	1.00 ^{a} ± 0.63	1.83 ^{a} ± 1.47	1.33 ^{a} ± 0.82	2.17 ^{a} ± 0.75	0.2000
"Plateau"	67.50 ^{c} ± 10.86	65.83 ^{c} ± 8.66	68.17 ^{c} ± 5.23	68.67 ^{c} ± 8.19	0.9427
"Hybrid"	49.00 ^{bB} ± 9.78	47.58 ^{bB} ± 5.85	40.92 ^{bA} ± 3.15	38.33 ^{bA} ± 4.46	0.0002
p-Value	<0.0001	<0.0001	<0.0001	<0.0001	

a, b, c Means in a row with different superscripts differ significantly ($p < 0.05$). A, B Means in a column with different superscripts differ significantly ($p < 0.05$).

Table 6 shows significant differences between the three types of environmental enrichment over the course of all four weeks, with the A-Reuter having the lowest number and the Plateau having the highest. However, there were no significant differences within the individual enrichment elements A-Reuter and the Plateau over the course of the four weeks. Regarding the Hybrid, there were significant differences between weeks one and two compared with weeks three and four where the number of birds was significantly lower.

As Table 7 indicates, slight differences were apparent when considering the number of birds per square metre on the different types of environmental enrichment.

Table 7. Average number of birds per square meter on the different types of environmental enrichment over the course of three trial rounds ± standard deviation in the trial groups (TG) at the time of taking photographs.

	Week 1	Week 2	Week 3	Week 4	*p*-Value
"A-Reuter"	0.10 ^{a} ± 0.07	0.18 ^{a} ± 0.17	0.13 ^{a} ± 0.07	0.21 ^{a} ± 0.07	0.2885
"Plateau"	18.68 ^{b} ± 3.06	18.28 ^{c} ± 2.43	18.90 ^{c} ± 1.45	19.01 ^{c} ± 2.25	0.9518
"Hybrid"	16.24 ^{bB} ± 3.27	15.84 ^{bB} ± 1.92	13.54 ^{bA} ± 1.06	12.69 ^{bA} ± 1.42	0.0002
p-Value	<0.0001	<0.0001	<0.0001	<0.0001	

a, b, c Means in a row with different superscripts differ significantly ($p < 0.05$). A, B Means in a column with different superscripts differ significantly ($p < 0.05$).

Table 7 shows no significant differences between the Plateau and the Hybrid for week one. In weeks two to four, the same significant differences between all types of environmental enrichment are seen as in Table 5.

3.2.2. Kilogramme per Square Metre

In Table 8, the respective kg per area for each type of enrichment are displayed.

Table 8. Average mass in kg per square meter of each environmental enrichment ± standard deviation in the trial group.

	Week 1	Week 2	Week 3	Week 4	p-Value
"A-Reuter"	0.02 ^{aA} ± 0.01	0.10 ^{aA} ± 0.09	0.14 ^{aB} ± 0.08	0.35 ^{aC} ± 0.12	<0.0001
"Plateau"	3.70 ^{cA} ± 0.62	9.77 ^{cB} ± 1.26	19.76 ^{cC} ± 1.51	31.93 ^{cD} ± 3.74	<0.0001
"Hybrid"	3.09 ^{bA} ± 0.62	8.45 ^{bB} ± 1.04	14.29 ^{bC} ± 1.12	21.36 ^{bD} ± 2.40	<0.0001
p-Value	<0.0001	<0.0001	<0.0001	<0.0001	

a, b, c Means in a row with different superscripts differ significantly ($p < 0.05$). A, B, C, D Means in a column with different superscripts differ significantly ($p < 0.05$).

As shown in Table 8, significant differences between the Plateau and the Hybrid were apparent over the weeks, with the Plateau having the highest values overall. The A-Reuter had significantly lower values compared with the Hybrid over the course of all weeks.

Table 9 indicates the differences between the displayed values in Table 8 and the average kg/m^2 in the trial group.

Table 9. Difference in mass per kg on each type of enrichment to the average kg per square metre of the barn ± standard deviation of the trial group (TG).

	Week 1	Week 2	Week 3	Week 4	p-Value
"A-Reuter"	−3.25 ^{aD} ± 0.02	−8.87 ^{aC} ± 0.06	−17.59 ^{aB} ± 0.08	−27.88 ^{aA} ± 0.12	<0.0001
"Plateau"	0.45 ^{c} ± 0.62	0.79 ^{c} ± 1.26	2.03 ^{c} ± 1.51	3.70 ^{c} ± 3.74	0.0637
"Hybrid"	−0.16 ^{bC} ±0.62	−0.53 ^{bC} ± 1.04	−3.44 ^{bB} ± 1.11	−6.87 ^{bA} ± 2.39	<0.0001
p-Value	<0.0001	<0.0001	<0.0001	<0.0001	

a, b, c Means in a row with different superscripts differ significantly ($p < 0.05$). A, B, C, D Means in a column with different superscripts differ significantly ($p < 0.05$).

As shown in Table 9, only the mass per square metre for the Plateau was higher than the average over the course of all four weeks. The Plateau was also significantly higher than the Hybrid in all week. The values for A-Reuter were significantly the lowest.

3.3. Free Space

Table 10 shows the weekly average of free space in the barn for nine fattening rounds.

Table 10. Average value of free space ± standard deviation for each week over the course of nine trial rounds.

	Week 1	Week 2	Week 3	Week 4	Week 5
Free space	81.84 ± 5.53	74.56 ± 3.72	62.70 ± 4.37	56.91 ± 4.62	52.38 ± 7.61

As seen in Table 10, the average free space decreased continuously from week one to week five.

Table 11 compares the free space in different zones of the barn over the course of five weeks. For each zone, the weekly average values during the trial with and without enrichment are displayed.

Table 11. Average values of free space ± standard deviation in the different zones of the trial group in the trial rounds with environmental enrichment and without environmental enrichment over the course of five weeks.

		Week 1	Week 2	Week 3	Week 4	Week 5
Zone 1	enriched	79.48 ± 5.51	74.91 ± 2.63	62.63 ± 4.25	58.06 ± 6.44	52.70 ± 9.33
	non-enriched	82.81 ± 5.51	77.95 ± 5.89	65.04 ± 2.72	60.65 ± 1.65	59.20 ± 2.84
Zone 2	enriched	79.71 ± 5.69	69.48 ^{a} ± 3.25	55.74 ^{a} ± 5.05	51.51 ^{a} ± 6.63	45.00 ± 9.14
	non-enriched	83.13 ± 8.59	77.42 ^{b} ± 5.95	65.25 ^{b} ± 5.05	57.69 ^{b} ± 4.99	57.64 ± 5.25

Table 11. *Cont.*

		Week 1	Week 2	Week 3	Week 4	Week 5
Zone 3	enriched	81.82 ± 5.94	73.60 ± 3.50	61.58 ± 4.52	55.33 ± 5.72	46.80 ± 8.42
	non-enriched	80.81 ± 7.02	75.33 ± 3.72	62.56 ± 2.81	52.70 ± 2.18	55.05 ± 9.71
Zone 4	enriched	80.16 ± 5.72	72.67 ^{a} ± 3.62	59.91 ^{a} ± 6.15	54.69 ± 6.27	48.32 ± 11.43
	non-enriched	82.90 ± 7.58	80.00 ^{b} ± 5.52	73.46 ^{b} ± 4.61	58.15 ± 4.54	56.77 ± 9.83
Zone 5	enriched	81.80 ± 5.08	75.12 ± 3.18	63.41 ± 4.46	57.82 ± 4.98	51.59 ± 7.42
	non-enriched	83.83 ± 6.03	77.83 ± 2.87	67.50 ± 1.80	58.65 ± 1.60	59.23 ± 6.31

a, b Means in column in one zone with different superscripts differ significantly ($p < 0.05$).

As Table 11 shows, there were significant differences in Zone 2 and Zone 4 between the enriched and non-enriched trial rounds. The non-enriched trial rounds showed significantly higher values of free space compared to the enriched trial rounds.

4. Discussion

4.1. Influence of the Design and Material on the Usage of the Environmental Enrichment Elements

The design and material of the environmental enrichment elements might have influenced the usage of the different types of them. Lower usage might have been caused by the fact that the industrially manufactured perches were not perfectly fitted for the birds. The diameter of the perches were too large and the small birds were not able to grip the perches optimally with their feet (In our study the diameter was 1.90 cm. Only later, at higher age the birds were able to jump and hold on the perches). Therefore, we found the highest frequency of use with 2.17 as an average in week four.

It is also possible that the birds were not used to the perches in young age and did not explore them sufficiently. Also the size of the perches could have resulted in uncomfortable seating positions for the birds, which makes it less attractive for them to sit there. This contrasts with findings of Bailie and O'Connell who used similar perches in previous studies [34].

Another reason for the little use of the perches could be the height above the floor. If the broilers can only reach the perches with difficulty or have to make a great effort to do so, then the incentive of perching might not have been sufficiently strong. Norring et al. observed that broilers mostly used the lowest available perches only which they could reach easily [35]. This could also explain the lowest value with 1.00 in week one when the birds are at their smallest and therefore have the most difficulties in reaching the perches.

Furthermore, the material used for the "A-Reuter" should also be scrutinised. Metal is easy to clean and disinfect, but it is much harder and probably more slippery as well as uncomfortable than natural materials such as wood. Hongchao et al. observed a higher use of wooden perches compared to synthetic materials in their study [36].

4.2. Usage of the Different Environmental Enrichment Elements

All of the three tested types of environmental enrichment elements were used by birds although the frequency and intensity of usage differed among the different types. The least used type was the "A-Reuter" perch in our trials. It was used significantly less by birds than the other types as described in Table 6 (a maximum value of 2.17 compared to 49.00 and 68.67 for "Hybrid" and "Plateau", respectively). These results align with previous studies showing a general low use of perches by broilers [37–39].

The same applies to the use of the combinated element, the "Hybrid", which were used significantly less compared to the platforms ("Plateau") (highest values 49.00 compared to 68.67 for the "Plateau"). This difference in the use of perches compared to platforms agrees with Kaukonnen et al. [27].

In animals per square metre, these figures were ahead of the "A-Reuter" (16.24 vs. 0.21), but far behind the "Plateau" with a value of 19.01. Compared to the perches of the "A-Reuter", the material could have been advantageous, as the perches were made of plastic. This material is softer than metal and thus more comfortable for the animals' feet. The texture of the perch could also be advantageous. The provided version was not completely round but only rounded on the upper side, which can contribute to a better grip on the perch.

The "Plateau" was used the most of all environmental enrichments in our study, with the average number of birds per square meter ranging between 18.28 and 19.01. There were no differences in the later weeks compared to the earlier weeks, although the animals in the later weeks were clearly larger and heavier. Table 7 shows that the highest value with an average of 19.01 birds per square metre on an environmental enrichment element was even found in week four.

The same development can be seen with kg per square metre where the "Plateau" showed significantly higher values in all four weeks than the other elements (highest value of 31.93 compared to 0.35 for the "A-Reuter" and 21.36 for the "Hybrid". Also, in comparison to the overall average in the whole barn, only the "Plateau" was above average, with up to 3.7 kg more per square metre in week four. These results are aligned with those of Malchow et al. who also observed a higher use of elevated structures the higher the age of the birds [37].

In order to gain insight into how the elements were used by the birds, a total capacity was calculated for each type of environmental enrichment (Appendix A). To calculate this capacity, values of 22 cm per bird for perches [38] and 303.3 cm^2 per bird for plain spaces [39] based on the available data from scientific literature were used.

The utilisation in per cent for the A-Reuter was the lowest of all types of enrichment (1.71%) followed by the Hybrid (19.36%). The highest values were seen for the Plateau with more than half of the theoretically available space covered by bird (57.80%).

4.3. Evaluation of Free Space

The free space was compared between enriched and non-enriched trial rounds, and there were significant differences between those (see Table 10). However, the differences were only apparent for weeks two and three in Zone 2 and Zone 4, as those were the zones with the enrichment elements. These results indicate that an enriched area in the barn can be more attractive to birds. Zone 2 with 69.48 and 55.74% of free space and Zone 4 with 72.67 and 59.91% of free space showed the lowest numbers for weeks three and four in trial rounds with enrichment elements. These results are aligned with those of Ventura et al., who also observed higher numbers of birds in areas with the provision of enrichment elements [26]. In non-enriched trial rounds, the values regarding the free space for these zones were as high or even higher than in the rest of the barn which describes lower numbers of birds when environmental enrichment was absent. This could be related to the fact that broilers are eager to climb elevated structures when these are available [26].

These effects disappeared in weeks four and five, which could also be related to the growth of the broilers, which, in the later stages of fattening does not allow them to choose the space as readily as in earlier weeks.

4.4. Influence of Environmental Enrichment on Growth Performance

The provision of environmental enrichment in this trial did not show negative influence on body weight gain or growth performance, although recent studies have shown otherwise [40,41]. Significantly higher values regarding the body weight for birds housed without enrichment was described by de Jong et al. [40], while Nazareno et al. recently recorded increased body weights with the provision of those [41]. This could be related to the easier accessibility of the platforms used by de Jong et al., which were equipped with ramps. This might have increased the frequency of use of that environmental enrichment and therefore resulted in higher body weights for the birds housed without enrichment. In this trial, the results do not align with either of the findings described above. The results of this trial regarding the growth performance align with those of Jacob et al., who also described no influence of environmental enrichment regarding this factor [42]. The comparison of the FCR also showed no significant differences. This is in accordance with findings of de Jong et al. who did not find a difference in FCR between enriched and non-enriched groups in their trials [40].

5. Conclusions

In conclusion, this study shows a significantly higher use of elevated platforms compared with perches for broilers. The provision of these environmental enrichment elements displayed no negative influences on growth performance and also had neither positive nor negative influences on the FPD scores in our trial. There were signs that enriched areas were more attractive to the birds than non-enriched ones. Further research regarding the type, design and construction of environmental enrichment elements for broilers should be carried in larger production units.

Author Contributions: Conceptualisation, J.H., J.B.L. and C.V.; methodology, F.S., J.H., J.B.L. and C.V.; software, C.V.; validation, F.S., J.B.L. and C.V.; formal analysis, F.S., B.R., J.B.L. and C.V.; investigation, F.S., C.S., M.A., S.R., J.B.L. and C.V.; resources, F.S., C.S., J.H., J.B.L. and C.V.; data curation, F.S., B.R., J.B.L. and C.V.; writing—original draft preparation, F.S.; writing—review and editing, F.S., B.R., M.A., J.H., J.B.L. and C.V.; visualisation, F.S., B.R. and J.B.L.; supervision, J.H., J.B.L. and C.V.; project administration, S.R., O.D., J.B.L. and C.V.; funding acquisition, O.D. All authors have read and agreed to the published version of the manuscript.

Funding: This project was financially supported by the Federal Ministry of Food and Agriculture (BMEL) based on a decision of the Parliament of the Federal Republic of Germany, granted by the Federal Office for Agriculture and Food (BLE) (Funding code 2817MDT305). This Open Access publication was funded by the Deutsche Forschungsgemeinschaft (DFG, German Research Foundation) within the programme LE 824/10-1 "Open Access Publication Costs" and the University of Veterinary Medicine Hannover, Foundation.

Institutional Review Board Statement: The study design was approved by the Animal Welfare Officer of the University of Veterinary Medicine Hannover, Foundation, Germany code "AZ: TVO-2021-V-19".

Informed Consent Statement: Not applicable.

Data Availability Statement: The datasets generated for this study are available on request to the corresponding author.

Acknowledgments: We would like to thank Frances Sherwood-Brock for proofreading the manuscript to ensure correct English language usage.

Conflicts of Interest: The authors declare no conflict of interest.

Appendix A

In order to be able to describe the actual utilisation of the individual elevated seating possibilities as environmental enrichment in a somewhat more striking way, the maximum capacity of each variant was first calculated on the basis of defined standard values from the scientific literature, where Brandes et al. described the space requirement on a perch for broilers at d 28 with 22 cm [38] and Spindler et al., described the floor space requirement for a broiler at d 28 with 303.3 cm^2 [39] Afterwards, the value for the actual utilization was calculated by dividing the average number of birds at d28 by the total capacity.

Table A1. Calculation of the utilisation rate for the environmental enrichment elements with values for space requirements resulting from the scientific work of Brandes et al. [38], and Spindler et al. [39].

Variant of Enrichment	A-Reuter	Plateau	Hybrid
Space requirement	22 cm/Bird	303.3 cm^2/Bird	303.3 cm^2/Bird
Total space available	28 m/element	3.6 m^2/element	6 m^2/element
Total capacity for birds	127.27	118.81	198.02
average number of birds at d28	2.17	68.67	38.33
Utilisation (in %)	1.71	57.80	19.36

References

1. FAO. *Food Outlook—Biannual Report on Global Food Markets*; FAO: Rome, Italy, 2020.
2. Adler, C.; Tiemann, I.; Hillemacher, S.; Schmithausen, A.; Müller, U.; Heitmann, S.; Spindler, B.; Kemper, N.; Büscher, W. Effects of a partially perforated flooring system on animal-based welfare indicators in broiler housing. *Poult. Sci.* **2020**, *99*, 3343–3354. [CrossRef] [PubMed]
3. Giersberg, M.; Hartung, J.; Kemper, N.; Spindler, B. Floor space covered by broiler chickens kept at stocking densities according to Council Directive 2007/43/EC. *Vet. Rec.* **2016**, *179*, 124. [CrossRef] [PubMed]
4. Bach, M.H.; Tahamtani, F.M.; Pedersen, I.J.; Riber, A.B. Effects of environmental complexity on behaviour in fast-growing broiler chickens. *Appl. Anim. Behav. Sci.* **2019**, *219*, 104840. [CrossRef]
5. Dawson, L.C.; Widowski, T.M.; Liu, Z.; Edwards, A.M.; Torrey, S. In pursuit of a better broiler: A comparison of the inactivity, behavior, and enrichment use of fast-and slower growing broiler chickens. *Poult. Sci.* **2021**, *100*, 101451. [CrossRef]
6. Bergmann, S.; Schwarzer, A.; Wilutzky, K.; Louton, H.; Bachmeier, J.; Schmidt, P.; Erhard, M.; Rauch, E. Behavior as welfare indicator for the rearing of broilers in an enriched husbandry environment—A field study. *J. Vet. Behav.* **2017**, *19*, 90–101. [CrossRef]
7. Vasdal, G.; Vas, J.; Newberry, R.C.; Moe, R.O. Effects of environmental enrichment on activity and lameness in commercial broiler production. *J. Appl. Anim. Welf. Sci.* **2019**, *22*, 197–205. [CrossRef]
8. De Jong, I.; Berg, C.; Butterworth, A.; Estevéz, I. Scientific report updating the EFSA opinions on the welfare of broilers and broiler breeders. *EFSA Support. Publ.* **2012**, *9*, 295E. [CrossRef]
9. Spindler, B.; Hartung, J. Prevalence of Pododermatitis in broiler chickens kept according to Directive 2007/43/EC stocking densities. In *Animal Hygiene and Sustainable Livestock Production, Proceedings of the XVth International Congress of the International Society for Animal Hygiene, Vienna, Austria, 3–7 July 2011*; Tribun EU: Brno, Czech Republic, 2011; Volume 1, pp. 39–42.
10. Gentle, M.J. Pain issues in poultry. *Appl. Anim. Behav. Sci.* **2011**, *135*, 252–258. [CrossRef]
11. Bergmann, S.; Louton, H.; Westermaier, C.; Wilutzky, K.; Bender, A.; Bachmeier, J.; Erhard, M.; Rauch, E. Field trial on animal-based measures for animal welfare in slow growing broilers reared under an alternative concept suitable for the German market. *Berl. Münch. Tierärztl. Wochenschr* **2016**, *129*, 453–461.
12. Newberry, R.C. Environmental Enrichment—Increasing the Biological Relevance of Captive Environments. *Appl. Anim. Behav. Sci.* **1995**, *44*, 229–243. [CrossRef]
13. Zahoor, M.S.; Ahmad, S.; Usman, M.; Dawood, M.; El-Sabrout, K.; Hashmi, S.G.M.D.; Khan, E.U.; Hussain, M.; Maqsood, M.A.; Latif, H.R.A. Effects of mirror and coloured balls as environmental enrichment tools on performance, welfare and meat quality traits of commercial broiler. *Trop. Anim. Health Prod.* **2022**, *54*, 1–12. [CrossRef] [PubMed]
14. Riber, A.; Van De Weerd, H.; De Jong, I.; Steenfeldt, S. Review of environmental enrichment for broiler chickens. *Poult. Sci.* **2018**, *97*, 378–396. [CrossRef] [PubMed]
15. Blatchford, R.A.; Archer, G.S.; Mench, J.A. Contrast in light intensity, rather than day length, influences the behavior and health of broiler chickens. *Poult. Sci.* **2012**, *91*, 1768–1774. [CrossRef] [PubMed]
16. De Jong, I.; Gunnink, H. Effects of a commercial broiler enrichment programme with or without natural light on behaviour and other welfare indicators. *Animal* **2019**, *13*, 384–391. [CrossRef]
17. Wood, B.; Rufener, C.; Makagon, M.M.; Blatchford, R.A. The Utility of Scatter Feeding as Enrichment: Do Broiler Chickens Engage with Scatter–Fed Items? *Animals* **2021**, *11*, 3478. [CrossRef]
18. Taylor, P.S.; Hemsworth, P.H.; Rault, J.-L. Environmental Complexity: Additional Human Visual Contact Reduced Meat Chickens' Fear of Humans and Physical Items Altered Pecking Behavior. *Animals* **2022**, *12*, 310. [CrossRef]
19. van de Weerd, H.A.; Day, J.E.L. A review of environmental enrichment for pigs housed in intensive housing systems. *Appl. Anim. Behav. Sci.* **2009**, *116*, 1–20. [CrossRef]
20. Newberry, R.C.; Estevez, I.; Keeling, L.J. Group size and perching behaviour in young domestic fowl. *Appl. Anim. Behav. Sci.* **2001**, *73*, 117–129. [CrossRef]
21. Schrader, L.; Muller, B. Night-time roosting in the domestic fowl: The height matters. *Appl. Anim. Behav. Sci.* **2009**, *121*, 179–183. [CrossRef]
22. Olsson, I.A.S.; Keeling, L.J. The push-door for measuring motivation in hens: Laying hens are motivated to perch at night. *Anim. Welf.* **2002**, *11*, 11–19.
23. Donaldson, C.J.; O'Connell, N.E. The influence of access to aerial perches on fearfulness, social behaviour and production parameters in free-range laying hens. *Appl. Anim. Behav. Sci.* **2012**, *142*, 51–60. [CrossRef]
24. Baxter, M.; Bailie, C.L.; O'Connell, N. Play behaviour, fear responses and activity levels in commercial broiler chickens provided with preferred environmental enrichments. *Animal* **2019**, *13*, 171–179. [CrossRef] [PubMed]
25. Baxter, M.; Bailie, C.L.; O'Connell, N.E. Evalution of dustbathing substrate and straw bales as environmental enrichments in commercial broiler housing. *Appl. Anim. Behav. Sci.* **2018**, *200*, 78–85. [CrossRef]
26. Ventura, B.A.; Siewerdt, F.; Estevez, I. Access to barrier perches improves behavior repertoire in broilers. *PLoS ONE* **2012**, *7*, e29826. [CrossRef] [PubMed]
27. Kaukonen, E.; Norring, M.; Valros, A. Using elevated platforms to improve broiler leg health on commercial broiler farms. In Proceedings of the 50th Congress of the International Society of Applied Ethology, Edinburgh, UK, 12–15 July 2016; pp. 12–15.

28. Aksit, M.; Yardim, Z.K.; Yalcin, S. Environmental enrichment influences on broiler performance and meat quality: Effect of light source and providing perches. *Eur. Poult. Sci.* **2017**, *81*, 182.
29. Ben Sassi, N.; Averós, X.; Estevez, I. Technology and poultry welfare. *Animals* **2016**, *6*, 62. [CrossRef]
30. Spieß, F.; Reckels, B.; Abd-El Wahab, A.; Ahmed, M.F.E.; Sürie, C.; Auerbach, M.; Rautenschlein, S.; Distl, O.; Hartung, J.; Visscher, C. The Influence of Different Types of Environmental Enrichment on the Performance and Welfare of Broiler Chickens and the Possibilities of Real-Time Monitoring via a Farmer-Assistant System. *Sustainability* **2022**, *14*, 5727. [CrossRef]
31. Part, C.E.; Edwards, P.; Hajat, S.; Collins, L.M. Prevalence rates of health and welfare conditions in broiler chickens change with weather in a temperate climate. *R. Soc. Open Sci.* **2016**, *3*, 160197. [CrossRef]
32. *Leitfaden Befunddaten in der Geflügelschlachtung*; Version 01.01.2022; QS Fachgesellschaft Geflügel GmbH: Bonn, Germany, 2022.
33. Mayne, R.; Else, R.; Hocking, P. High litter moisture alone is sufficient to cause footpad dermatitis in growing turkeys. *Br. Poult. Sci.* **2007**, *48*, 538–545. [CrossRef]
34. Bailie, C.; O'Connell, N. The influence of providing perches and string on activity levels, fearfulness and leg health in commercial broiler chickens. *Animal* **2015**, *9*, 660–668. [CrossRef]
35. Norring, M.; Kaukonen, E.; Valros, A. The use of perches and platforms by broiler chickens. *Appl. Anim. Behav. Sci.* **2016**, *184*, 91–96. [CrossRef]
36. Hongchao, J.; Jiang, Y.; Song, Z.; Zhao, J.; Wang, X.; Lin, H. Effect of perch type and stocking density on the behaviour and growth of broilers. *Anim. Prod. Sci.* **2013**, *54*, 930–941. [CrossRef]
37. Malchow, J.; Berk, J.; Puppe, B.; Schrader, L. Perches or grids? What do rearing chickens differing in growth performance prefer for roosting? *Poult. Sci.* **2019**, *98*, 29–38. [CrossRef]
38. Brandes, A.G.; Spindler, B.; Giersberg, M.F.; Kemper, N. Feed Space Allowance and Perch Design Criteria for Broiler Breeders Determined by Biometric Data. *Vet. Sci.* **2022**, *9*, 350. [CrossRef] [PubMed]
39. Spindler, B.; Giersberg, M.; Briese, A.; Kemper, N.; Hartung, J. Spatial requirements of poultry assessed by using a colour-contrast method (KobaPlan). *Br. Poult. Sci.* **2016**, *57*, 23–33. [CrossRef] [PubMed]
40. de Jong, I.C.; Blaauw, X.E.; van der Eijk, J.A.; da Silva, C.S.; van Krimpen, M.M.; Molenaar, R.; van den Brand, H. Providing environmental enrichments affects activity and performance, but not leg health in fast-and slower-growing broiler chickens. *Appl. Anim. Behav. Sci.* **2021**, *241*, 105375. [CrossRef]
41. Nazareno, A.C.; da Silva, I.J.; Delgado, E.F.; Machado, M.; Pradella, L.O. Does environmental enrichment improve performance, morphometry, yield and weight of broiler parts at different ages? *Rev. Bras. Eng. Agrícola Ambient.* **2022**, *26*, 292–298. [CrossRef]
42. Jacob, F.; Salgado, D.; Nää, I.; Baracho, M. Effect of Environmental Enrichment on the Body Weight in Broiler Chickens. *Braz. J. Poult. Sci.* **2020**, *22*. [CrossRef]

sustainability

MDPI

Article

The Influence of Different Types of Environmental Enrichment on the Performance and Welfare of Broiler Chickens and the Possibilities of Real-Time Monitoring via a Farmer-Assistant System

Fabian Spieß [1,*], Bernd Reckels [1], Amr Abd-El Wahab [2], Marwa Fawzy Elmetwaly Ahmed [3], Christian Sürie [4], Monika Auerbach [5], Silke Rautenschlein [5], Ottmar Distl [6], Joerg Hartung [7] and Christian Visscher [1]

1 Institute for Animal Nutrition, University of Veterinary Medicine Hanover, Foundation, 30173 Hannover, Germany; bernd.reckels@tiho-hannover.de (B.R.); christian.visscher@tiho-hannover.de (C.V.)
2 Department of Nutrition and Nutritional Deficiency Diseases, Faculty of Veterinary Medicine, Mansoura University, Mansoura 35516, Egypt; amrwahab5@mans.edu.eg
3 Department of Hygiene and Zoonoses, Faculty of Veterinary Medicine, Mansoura University, Mansoura 35516, Egypt; marwafawzy@mans.edu.eg
4 Farm for Education and Research Ruthe, University of Veterinary Medicine Hannover, Foundation, 31157 Ruthe/Sarstedt, Germany; christian.suerie@tiho-hannover.de
5 Clinic for Poultry, University of Veterinary Medicine Hannover, Foundation, 30559 Hannover, Germany; monika.auerbach@tiho-hannover.de (M.A.); silke.rautenschlein@tiho-hannover.de (S.R.)
6 Institute for Animal Breeding and Genetics, University of Veterinary Medicine Hannover, Foundation, 30559 Hannover, Germany; ottmar.distl@tiho-hannover.de
7 Institute of Animal Hygiene, Animal Welfare and Farm Animal Behaviour, University of Veterinary Medicine Hannover, Foundation, 30625 Hannover, Germany; joerg.hartung@tiho-hannover.de
* Correspondence: fabian.spiess@tiho-hannover.de; Tel.: +49-511-856-7736

Citation: Spieß, F.; Reckels, B.; Abd-El Wahab, A.; Ahmed, M.F.E.; Sürie, C.; Auerbach, M.; Rautenschlein, S.; Distl, O.; Hartung, J.; Visscher, C. The Influence of Different Types of Environmental Enrichment on the Performance and Welfare of Broiler Chickens and the Possibilities of Real-Time Monitoring via a Farmer-Assistant System. *Sustainability* **2022**, *14*, 5727. https://doi.org/10.3390/su14095727

Academic Editors: George K. Symeon and Vassilios Dotas

Received: 13 April 2022
Accepted: 5 May 2022
Published: 9 May 2022

Abstract: The aim of this study is to evaluate the influence of environmental enrichment on the growth performance, litter and/or air quality as well as animal welfare indicators of broilers. Control groups (CG) and trial groups (TG) were housed under identical conditions during six fattening runs, with the TG having three types of environmental enrichment and a Farmer-Assistant System (FAS). A representative number of 50 birds were weighed and litter samples were taken at d 14, 21 and 28. Additionally, the same broilers were examined for foot pad dermatitis (FPD) on those days. The average bodyweight of the birds in the CG was significantly lower (1671 g) only at d 28 compared to TG (1704 g); at d 14, d 21 and d 33 at the slaughterhouse, no significant differences were observed. The dry matter content in the litter did not significantly differ between CG and TG. Birds housed in CG had significantly higher FPD scores at d 14 (1.24) and d 21 (2.19) compared to those housed in TG (0.73 and 1.52, respectively). No effects on air quality parameters, such as CO_2 and NH_3, were seen between the groups. Overall, our study shows no negative influences of environmental enrichment on growth performance, litter and air quality.

Keywords: broiler; enrichment; animal welfare; foot pad health; performance; real-time monitoring

1. Introduction

Broiler production has been the fastest growing sector in animal production worldwide over the past decades, driven by the increasing demand for animal-sourced foods by the rapid population growth, [1] and a change in dietary preferences [2,3]. In 2020, the world broiler meat production exceeded 100 million tonnes [4]. Broilers grow fast with a high efficiency compared to other farm animals, because they consume relatively little feed per kg of produced kg of meat compared to, for example, pigs and cattle [5]. Broiler meat is consumed across cultural and religious communities [6]. However, citizens and consumers, particularly in developed countries such as those in Europe, are increasingly concerned about the living conditions of intensely farmed chickens and the health and welfare problems encountered in densely populated broiler houses with flocks of 20,000 birds and more [7].

These houses offer very little structural orientation for the birds [8]. The animals can freely move on a flat, usually concrete floor covered with litter material, such as chopped straw or wood chips, peat or something similar [9]. They are offered feed and water from automatic feeding systems. A ventilation system combined with a heating system provides ambient indoor temperatures for optimal growth rates. In this barren environment [10] with high stocking density, typical production diseases, often addressed as technopathies or cumulative disorders, such as lameness or foot pad dermatitis (FPD) [11] and deep skin dermatitis as well as sudden cardiac death [12], culminate in the last week of the production cycle.

To avoid or ease these problems that impair the welfare of the birds [13] various measures and attempts have been undertaken [14–20]. These problems can seriously affect productivity, with animal losses and degraded meat quality. Initially, research was carried out to mainly address factors of nutrition [14,15], but in recent years, attention has been paid more and more to housing conditions [16,17]. Due to their fast growth combined with low activity, the animals easily develop lameness and other pathologies, such as FPD, particularly in combination with wet litter [18]. Consequently, birds may suffer from pain and their well-being and health status are reduced, which also results in economic losses for the farmers [19,20].

A solution, or at least an improvement, was seen in the so-called "environmental enrichment", which has been defined as "an improvement in the biological functioning of captive animals resulting from modifications to their environment" [21]. It has been shown in previous studies that environmental enrichment can be used to target problems of low complexity of structure and can therefore increase animal welfare [8,22–24]. It is part of the natural behaviour of birds and broiler chickens to be eager to climb and sit on perches or other structures off the ground [25–27]. This is presumably related to the wildlife strategy of their ancestors to avoid predators [26,28]. Elevated structures allow species-specific behaviours and the broilers have the possibility to choose several different seating positions [29]. These elements and activities may tackle the well-known problem in conventional chicken houses, i.e., the chickens are inactive for approximately 80% of their time [30,31] when not eating or drinking. Movement and choosing several different seating positions can help to distribute the load on the foot pads, while also help to increase muscle activity [32] and reduce lameness [24,33] and FPD [34].

Enrichment options, which have been under research in recent years, are higher places such as small elevated perforated or non-perforated floors, perches [22,27] and straw bales [35], which trigger activity and direct pecking behaviour to straw stems. However, not every tested enrichment element suits the purpose perfectly [36]. Some recent studies have shown that straw bales, for example, while being well-accepted for seating broilers, can also lead to an increase in FPD [20,37].

A common difficulty in assessing animal welfare in commercial broiler flocks is their often large numbers of birds [38]. Modern technology can offer new options of real-time monitoring of not only the birds, but also the climate and everything related to the animals' environment. Especially the high data density generated by this monitoring as well as the possible earlier detection of irregularities could contribute to higher animal welfare [39,40]. Nevertheless, it has to be taken into consideration that these systems could also be used for increasing productivity in addition to the main focus of improving animal welfare [39]. The focus of animal welfare evaluations has often been on "outcome" measures [40], but the climate also has a considerable influence on animal welfare as well as animal health [41]. Thus, a more intensive monitoring can also contribute to improve environmental conditions.

The aim of this study is to test, characterise and improve possible options to offer environmental enrichment in broiler houses. The focus in this paper is to evaluate the influence of environmental enrichment on animal welfare and growth performance with the help of an FAS. The large amount of continuously measured climate data were analysed to evaluate the effects of possible higher numbers of birds in enriched areas. By collecting litter samples, the influence of the environmental enrichment on litter quality was evaluated. Moisture was the main focus, as it has an influence on the condition of the foot pads [42]. Furthermore, the foot pads of individual animals were scored in order to investigate the influence of the environmental enrichment and to give an estimate of the impairment of the well-being of the birds resulting from these injuries.

2. Materials and Methods

2.1. Animals and Diets

2.1.1. Animals

In this study, broilers were raised as hatched in a barn on the Farm for Education and Research in Ruthe, University of Veterinary Medicine Hannover, Foundation, Hannover, Germany. Chicks of the same age and genetics (Ross 308) were distributed randomly between two groups (control group = CG and trial group = TG) on the same day and at the same time. The broilers were housed for 6 fattening runs, with each run having a control and trial group consisting of 8100 birds each. With 16,200 birds in each fattening group, a total number of 97,200 birds were housed in this trial. The length of each of the 6 fattening runs was 33 days.

2.1.2. Diets

The birds were fed ad libitum and had free access to water. A commercial pelleted diet (MEGA Tierernährung GmbH & Co. KG, Visbek, Germany), based on wheat and soybean meal, was offered in a three-phase feeding programme (Table 1). The first phase was the "starter diet", which was offered until d 7 of life, and then exchanged for the "grower I diet". After d 20 of life, "grower II" was introduced, which was fed until d 29, and finally the "finisher diet" was fed until d 33.

Table 1. Chemical composition of the commercial diets used for the control and trial groups.

Ingredients (in %)	Starter	Grower I	Grower II	Finisher
Crude protein (CP)	21.6	19	19.0	19.5
Ether extract	5.4	4.7	4.7	7.8
Crude fibre (CF)	2.5	3.5	3.2	3.2
Crude ash	5.5	5.4	5.1	4.8
Calcium	0.9	0.75	0.7	0.65
Phosphorus	0.85	0.55	0.5	0.4
Sodium	0.16	0.16	0.15	0.14
Lysine	1.35	1.12	1.12	1.14
Methionine	0.8	0.28	0.54	0.28

2.2. Experimental Design and Housing

The broilers were raised for 33 days in two separate, but identically designed, broiler houses at the same time under controlled environmental housing conditions. The barn for each group was 30 m long and 16 m wide.

The feed was provided in four conventional feeding lines (Big Dutchman International GmbH, Vechta, Germany) with four conventional water lines (LUBING Maschinenfabrik Ludwig Bening GmbH & Co. KG, Barnstorf, Germany) next to them (Figure 1). The water lines were equipped with drinking nipples (LUBING Maschinenfabrik Ludwig Bening GmbH & Co. KG). Figure 1 shows an overview of the barn and the different areas.

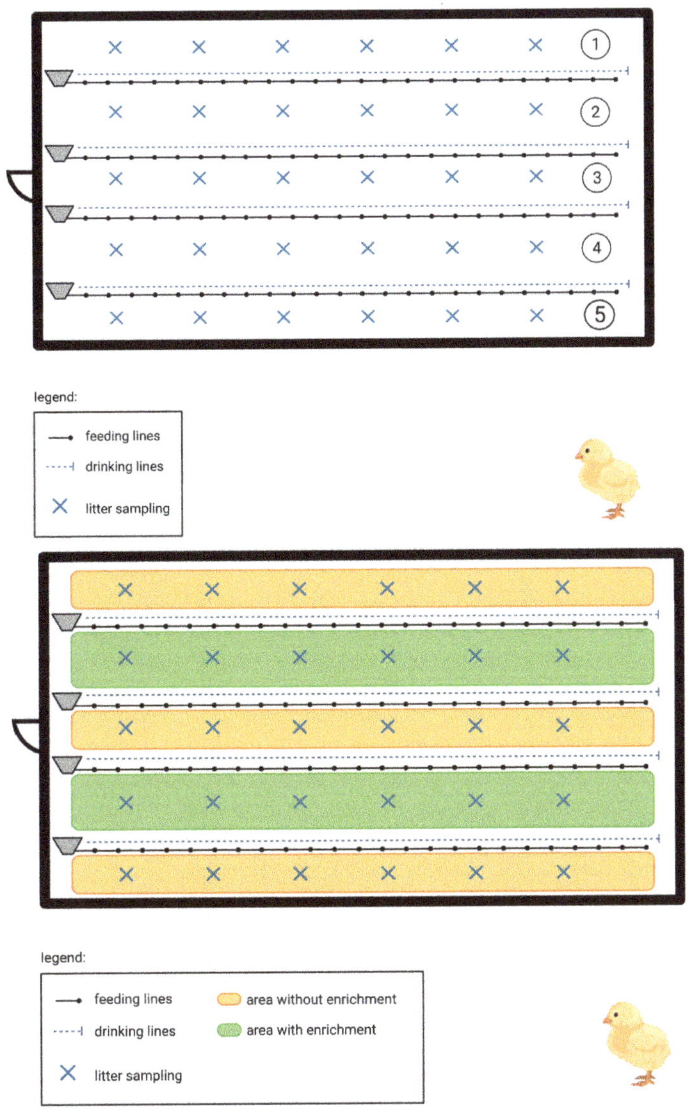

Figure 1. Schematic drawing of the barn for the control (**top**) and trial (**bottom**) groups (the figure was created with Biorender.com).

The chicks were housed on a litter composed of conventional wood shavings (GOLDSPAN®, Goldspan GmbH and Co. KG, Goldenstedt, Germany).

The light programme was 24 h light at d 0. At d 1, the light was turned off from 23:00 to 03:00, and at d 2 from 22:00 to 04:00. From d 3 onwards, the dark period was between 21:00 and 05:00. After d 21, the dark period was shortened to the period from 22:00 until 04:00.

The air temperature at d 1 was 33.5 °C and was successively lowered gradually until d 33 to 23 °C. The barn was heated with a gas air-heating system. The air temperature was measured with temperature and humidity sensors, which were used to control the negative pressure ventilation system.

The birds were vaccinated via the drinking water at d 12 against Newcastle Disease (Poulvac ND Hitchner B1; Zoetis Deutschland GmbH, Berlin, Germany), at d 18 against

Gumboro (AviPro Precise, Elanco Animal Health, Bad Homburg, Germany) and at d 20 against infectious bronchitis with the virus strain Ma5 (Nobilis IB Ma5, Intervet Deutschland GmbH, Unterschleißheim, Germany).

2.3. Experimental Treats

2.3.1. Environmental Enrichment

During the trial, three different types of environmental enrichment were used (Figures 2–4). The first type was the so-called "A-Reuter" (Big Dutchman International GmbH, Vechta, Germany). It consisted of five round perches made of metal, which were mounted on a triangular framework. The dimensions of the perches and, therefore, the whole construction were 5.60 m long by 1.40 m wide.

Figure 2. "A-Reuter".

The second environmental enrichment used was the "Hybrid" (Hölscher + Leuschner GmbH & Co. Kg, Emsbüren, Germany), which was a combination of perches and elevated planes. It consisted of two elevated planes on each side, which were accessible via two ramps, each on either side. Both were 248 mm wide and 1012 mm long. The planes were connected via three perches, each of which was 1.5 m in length. Two perches were mounted under the planes, leaving a space of 992 mm between the inner edges of the planes. The third perch was mounted above the planes and had a triangular framework to hold it. All perches were aligned at right angles towards the planes. The perches were 52 mm wide and 78 mm high, with a rounded top part.

Figure 3. "Hybrid".

The third environmental enrichment was the "Plateau" as an elevated plane variant. It consisted of two grids mounted on a rectangular framework. The framework was supported by two carriage axles with two tyres each. Each grid had a length of 1.20 m and a width of 0.75 m, making the whole enrichment 2.40 m long. On both of the long sides, there was a ramp, also consisting of one of the grids.

Figure 4. Photo of the "Plateau" variant of environmental enrichment. As the photo was taken one day before the birds arrived, the ramps are laying on top of it after disinfection, ready to be attached to the sides with cable ties.

During each trial run, all of the three different environmental enrichments were used and placed in three different positions within the barn: A, B and C. In each of the following trial runs, the positions of the enrichments were changed in a clockwise direction so that A became B, B became C and C became A. This was performed to exclude any negative effects of the positioning.

2.3.2. Farmer-Assistant System

The Farmer-Assistant System (FAS) is a mobile, ceiling-based livestock robot that runs on rails attached to the ceiling of the barn. It consists of top and bottom boxes, which are connected by a telescope arm. The top part contains the battery, the engine, the driving wheels and the top camera, which provides an overview of the stall. The rails allow the robot to permanently circulate through the barn and monitor the broilers continuously. The bottom box is equipped with sensors, which permanently measure air temperature, relative humidity of the air, wind velocity, carbon dioxide (CO_2), ammonia (NH_3), light and noise 70 cm above the barn floor. It does not influence the behaviour of the birds. The robot delivers climate data for each square metre of the barn. The bottom box also contains one bottom and two side cameras for a more detailed monitoring of the broilers. All data are stored in a protected, cloud-based system mapped across the barn floor and presented to the farmers regularly in a daily report.

2.4. *Measurements*

2.4.1. Growth Performance and Slaughter Data

Individual bodyweight (BW) of 50 randomly selected birds per barn (for six runs) was measured at d 14, 21 and 28 of life (Figure 5). Hanging scales (Veit electronics s.r.o., Brno, Czech Republic) were used to record the birds' weight. At the slaughterhouse, the number of delivered birds was noted as well as their BW. After subtracting the discarded birds, an average slaughter weight was calculated. The slaughterhouse also scored the foot pads via a camera-based system according to the common FPD scoring (QS Qualität und Sicherheit GmbH, Bonn, Germany) at slaughter with scores 0, 1, 2 a and 2 b [43].

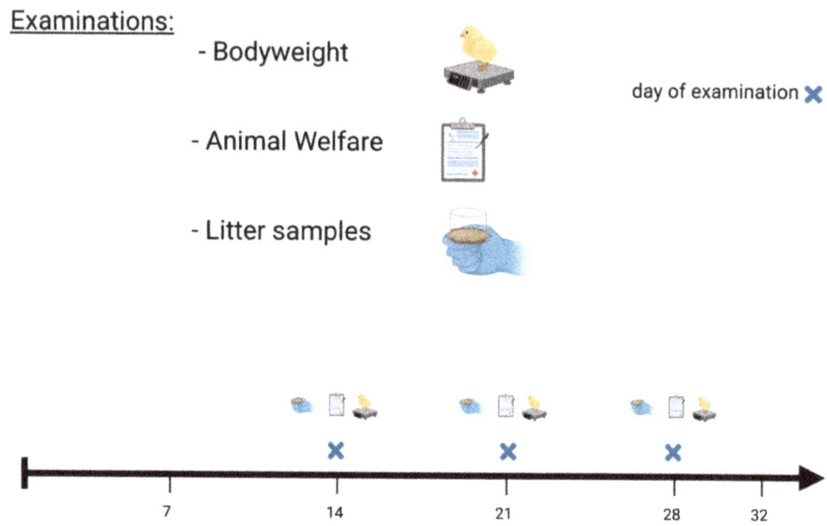

Figure 5. Timeline of the examinations in the trial and control groups.

2.4.2. Feed Conversion Ratio

The feed conversion ratio (FCR) was calculated by dividing the feed intake (kg) by the total BW (kg) of all birds for both the control and trial groups. The corrected FCR was calculated by using the corrected feed intake in kg. The corrected feed intake was obtained by subtracting the cumulative feed intake of the dead animals from the total feed intake. The cumulative feed intake of the dead animals was the sum of the daily feed intake of each animal up to the day of slaughter (d 33).

2.4.3. Litter Dry Matter

Litter samples to measure the dry matter (DM) content were collected at d 14, 21 and 28 of life (Figure 5) from five different rows in each barn (in all six runs). The rows were defined in-between the outside walls and the next drinking line or in the area between feeding and drinking lines (Figure 1). There were six points of sample collection in each row, resulting in thirty samples for each day. The litter was collected by taking all litter down to the ground with one hand at every spot (50 g). All samples were dried at 103 °C for the time needed to reach a constant weight and, afterwards, the DM was measured.

2.4.4. Foot Pad Dermatitis

The external examination of the foot pads (as an indicator of animal welfare) of the birds was performed for 50 randomly selected birds in each barn in all six runs at d 14, 21 and 28 (Figure 5). The foot pads were examined for both feet and the FPD score was recorded looking at the central plantar area. To evaluate the FPD scores, a seven-point scale in accordance with Mayne et al. [44] was used, which is described by Table 2. Regarding this scoring, score 0 refers to no external signs of FPD. Score 3 marks the point where first necrotic areas may occur and where the central part of the foot pad is swollen, red and harder. The other scores relate, in increasing order, to the size of the necrotic areas up until score 7, which describes a foot pad in which half of it is covered in necrotic scales.

Table 2. Footpad scoring in accordance with Mayne et al. (2007).

Score	Description of the Foot Pad
0	No external signs of FPD. Skin of the foot pad and digital pads appears normal, no redness, swelling or necrosis is evident. The skin of the foot pad feels soft to the touch.
1	Slight swelling and/or redness of the skin of the foot pad.
2	The pad feels harder and denser than a non-affected foot. The central part of the pad is raised with swelling and redness and the reticulate scales may be separated. The digital pads may show a similar reaction.
3	The central and digital foot pads are enlarged and swollen with red areas, and, as the skin has become compacted, the foot pad is hard. The reticulate scales have become enlarged and separated, and small black necrotic areas may occur.
4	Marked swelling and redness around the margins of lesions occur. Reticulate scales die and turn black, forming scale-shaped necrotic areas. The scales around the outside of the black areas may have turned white. The area of necrosis is less than one-eighth of the total area of the foot pad.
5	Swelling and redness are evident in the central and digital foot pads. The total foot pad size is enlarged. Reticulate scales are pronounced, increased in number and separated from each other. The amount of necrosis extends to a quarter of the foot pad. Small necrotic areas may also appear on the digital pads.
6	As score 5, but with half of the foot pad covered by necrotic cells. The digital pads may have up to half of one pad covered with necrotic cells.
7	A foot pad with over half of the foot pad covered in necrotic scales.

2.4.5. Carbon Dioxide and Ammonia in the Air

Carbon dioxide (CO_2) and ammonia (NH_3) and the other climate parameters were continuously measured in the air of the barn by the FAS during up to twenty rounds per day on average and for one square metre of the barn floor. Data collection stopped shortly before slaughter, resulting in around 100,000 individual values per run and parameter, in total 2 million pieces of data per day. From these individual values, daily means and weekly means were formed.

2.5. Statistical Analysis

Data analysis was performed using the statistical software package from SAS, Version 7.1 (SAS Inst., Cary, NC, USA). All measured data were analysed descriptively by sample size, mean values, confidence interval, standard deviation, minimum and maximum. The group comparisons as well as the area comparisons were performed by one-way analysis of variance (ANOVA) for independent samples. In general, the Ryan–Einot–Gabriel–Welsch multiple-range test (REGWQ) was used for multiple pairwise means comparisons between the groups. All statements of statistical significance were based on $p < 0.05$.

3. Results

3.1. Growth Performance and Slaughter Data

3.1.1. Growth Performance

In Table 3, the average bodyweight of 50 individual birds of both the control group and trial group for d 14, 21 and 28 are displayed and statistically compared.

As shown in Table 3, there were no significant differences in BW between the groups at d 14 and 21 of life. At d 28, the birds in the CG had a significantly lower BW than those in the TG (1671 g vs. 1703 g).

Table 3. Average bodyweight (g) ± standard deviation of individually weighed birds from day 14 to 28 of life in both the control (CG) and trial groups (TG).

Day of Life	n	CG	TG
14	50	521.17 ^{a} ± 61.19	526.48 ^{a} ± 49.88
21	50	1029.82 ^{a} ± 127.76	1012.02 ^{a} ± 121.75
28	50	1671.09 ^{b} ± 191.70	1703.54 ^{a} ± 182.31

^{a,b} Means in a row with different superscripts differ significantly ($p < 0.05$).

3.1.2. Slaughter Data

The slaughter data were reported from the slaughterhouse. Table 4 shows the average bodyweight and foot pad score with standard deviation for both the control and trial groups at d 33.

Table 4. Slaughter data regarding average bodyweight (g) ± standard deviation per bird and foot pad points ± standard deviation, according to QS GmbH Germany, for the control (CG) and trial groups (TG) over six trials.

	CG (n = 54,276)	TG (n = 54,607)
Bodyweight (g)	2071.71 ± 72.03	2067.29 ± 34.60
Foot pad score	14.71 ± 12.99	17.00 ± 11.80

The results in Table 4 show that there were no significant differences between the control and trial groups regarding bodyweight and foot pad score, which shows that both bodyweight gain and foot health were not negatively affected by the provided environmental enrichment.

3.1.3. Feed Conversion Ratio

In Table 5, both the FCR and corrected FCR are displayed and analysed for the control and trial groups.

Table 5. Feed conversion ratio (kg feed/kg bodyweight gained) and corrected feed conversion ratio ± standard deviation in the control (CG) and trial groups (TG) over six trial runs.

	N	CG	TG
FCR	6	1.39 ± 0.02	1.38 ± 0.02
corr. FCR	6	1.39 ± 0.05	1.38 ± 0.04

As shown in Table 5, both the FCR and corrected FCR had no significant differences between both groups. Neither the control nor the trial groups showed any variances over the six trial runs.

3.2. Litter Quality and FPD Scoring

3.2.1. Litter Quality

Table 6 shows that, over the course of six trial runs, there were no significances differences either for d 14, 21 or 28 between the control and trial groups regarding the average dry matter (g/kg) in the litter.

Table 6. Average dry matter (g/kg) of the litter ± standard deviation in the control (CG) and trial groups (TG) on the examination days.

Day	CG (n = 130)	TG (n = 180)
14	737.89 ± 74.63	735.21 ± 55.76
21	741.35 ± 79.36	737.48 ± 88.26
28	699.55 ± 120.96	701.37 ± 111.22

For a more detailed analysis of the influence of the enrichments, the individual areas were further subdivided. At first, the trial group was separated into the enriched (green area in Figure 1) and non-enriched (yellow area in Figure 1) area and the DM was analysed.

As Table 7 shows, the DM was significantly higher in the enriched areas of the trial group compared to the non-enriched areas.

Table 7. Dry matter (g/kg) ± standard deviation of the litter samples taken in the trial group divided by the enriched and non-enriched areas in the barn on the examination days.

Day	Enriched Area	Non-Enriched Area
14	745.38 ^{a} ± 53.94	728.44 ^{b} ± 56.16
21	745.54 ^{a} ± 92.68	732.10 ^{a} ± 85.20
28	713.31 ^{a} ± 86.77	693.41 ^{a} ± 124.65

^{a,b} Means in a row with different superscripts differ significantly ($p < 0.05$).

In order to be able to compare the individual enrichments, the litter samples were also examined individually in relation to the range of each environmental enrichment. This was performed by comparing the samples from the green areas (Figure 5).

Table 8 shows no significant differences between all three types of enrichment. On all days, the dry matter in the litter was the same for all types of environmental enrichments. Therefore, the individual variants did not have a different influence on the litter quality.

Table 8. Dry matter (g/kg) ± standard deviation of the litter in the trial group regarding the different types of environmental enrichment used in that area.

Day	n (per Enrichment) = 24	Trial
14	Enrichment 1	730.62 ± 52.44
	Enrichment 2	747.04 ± 62.86
	Enrichment 3	758.45 ± 51.21
21	Enrichment 1	748.58 ± 88.81
	Enrichment 2	754.66 ± 90.96
	Enrichment 3	733.37 ± 100.51
28	Enrichment 1	728.0.8 ± 96.58
	Enrichment 2	707.08 ± 93.90
	Enrichment 3	704.75 ± 68.92

3.2.2. FPD Scoring

Table 9 shows the FPD scores at d 14, 21 and 28 in comparison for the control and trial groups with regard to the average values on examination days with standard deviation.

As shown in Table 9, at days 14 and 21, the CG showed significantly higher FPD scores than the TG. At day 28, the control group reached the values of the trial group and no differences were seen.

Table 9. Foot pad disease score ± standard deviation, in accordance with Mayne, scored for both feet of 50 birds per day in the control (CG) and the trial groups (TG).

		CG	TG
Day of Life	**n**	**FPD Score**	**FPD Score**
14	300	$1.24^{a} \pm 1.23$	$0.73^{b} \pm 0.87$
21	300	$2.19^{a} \pm 1.76$	$1.52^{b} \pm 1.61$
28	300	$2.45^{a} \pm 1.93$	$2.38^{a} \pm 2.22$

a,b Means in a row with different superscripts differ significantly ($p < 0.05$).

3.3. *Air Quality*

Table 10 shows the weekly averages of CO_2 and NH_3 in the TG. It displays significant differences for CO_2 between the weeks, except for weeks four and five. For NH_3, there are significant differences between the weeks, except between weeks 1 and 2 and weeks 3 and 4.

Table 10. Average values (ppm) ± standard deviation over periods of one week each for carbon dioxide (CO_2) and ammonia (NH_3) in the air of the trial group.

Week	CO_2	NH_3
1	$2934.21^{A} \pm 606.16$	$0.10^{C} \pm 0.11$
2	$2493.35^{B} \pm 395.68$	$0.77^{C} \pm 0.39$
3	$2209.08^{AB} \pm 683.97$	$2.70^{B} \pm 1.64$
4	$2036.97^{C} \pm 458.46$	$2.67^{B} \pm 1.45$
5	$1909.58^{C} \pm 215.48$	$5.55^{A} \pm 1.00$

A–C Means in a column with different superscripts differ significantly ($p < 0.05$).

In a following step, the areas inside the trial group were divided into enriched and non-enriched areas (green areas in Figure 1) to have a closer look at the influence of the environmental enrichment on the air quality.

Table 11 compares the CO_2 concentrations in the enriched and non-enriched areas in the trial group. Over the course of five weeks, these were compared within the respective group as well as with each other.

Table 11. Average values (ppm) ± standard deviation over periods of one week each for carbon dioxide (CO_2) in the air in the enriched and non-enriched areas of the trial group.

	Enriched Areas	Non-Enriched Areas
Week	CO_2	CO_2
1	$2891.77^{A} \pm 644.66$	$2962.50^{A} \pm 596.46$
2	$2489.57^{AB} \pm 412.09$	$2495.87^{B} \pm 396.47$
3	$2190.66^{BC} \pm 713.40$	$2221.36^{BC} \pm 684.34$
4	$2021.35^{BC} \pm 471.65$	$2047.38^{C} \pm 462.96$
5	$1886.84^{C} \pm 225.49$	$1924.75^{C} \pm 215.15$

A–C Means in a column with different superscripts differ significantly ($p < 0.05$).

As Table 11 shows, there were no significant differences between the enriched and the non-enriched areas in the trial group over the course of all five weeks. It also shows the significances between the weeks for both the enriched and non-enriched areas. Carbon dioxide was significantly higher in the beginning before continuing to decline in both areas.

Table 12 depicts the same comparison as Table 11, but for NH_3 instead of CO_2.

Table 12. Average values (ppm) ± standard deviation over periods of one week each for ammonia (NH_3) in the air in the enriched and non-enriched areas of the trial group.

	Enriched Areas	Non-Enriched Areas
Week	NH_3	NH_3
1	$0.10^{C} \pm 0.12$	$0.10^{C} \pm 0.11$
2	$0.77^{C} \pm 0.41$	$0.77^{C} \pm 0.39$
3	$2.69^{B} \pm 1.69$	$2.71^{B} \pm 1.66$
4	$2.62^{B} \pm 1.45$	$2.70^{B} \pm 1.50$
5	$5.49^{A} \pm 0.98$	$5.59^{A} \pm 1.05$

$^{A–C}$ Means in a column with different superscripts differ significantly ($p < 0.05$).

Table 12 displays no significant differences between both areas regarding ammonia in the air for all five weeks, but it shows significances for each area regarding the weeks. In both the enriched and non-enriched areas, ammonia was significantly lower in week one and continued to increase until week 5. The exception to this was week 4, when the ammonia in the air was slightly lower than in the previous week for both the enriched and non-enriched areas.

4. Discussion

4.1. Influence of Environmental Enrichment on Growth Performance

The comparison of bodyweights from the control and trial groups in this study did not show any significantly lower values for the trial group. The FCR also showed no significant differences between the control and trial groups. This indicates that the presence of environmental enrichment and the assumed higher energy consumption, which could result from the increased physical activity stimulated by the environmental enrichment, had no negative influence on weight development. The FCR also showed no significant differences between the control group and trial group. This indicates that the presence of environmental enrichment and the assumed higher energy consumption, which could result from the increased physical activity stimulated by the environmental enrichment, had no negative influence on weight development. Recent studies have shown different influences of environmental enrichment on the growth performance of broilers. While de Jong et al. recorded significantly higher bodyweights for birds housed without enrichment after d 17 [45], Jacob et al. did not find any differences in bodyweight [46]. More recently, Nazareno et al. described an increased bodyweight due to environmental enrichment [47].

This is contrary to results that showed it is possible that environmental enrichment increases the activity of the birds and that this could have a negative effect on bodyweight development [48]. The results of our study, however, had no such observable effect. There were suggestions that increased activity could have a positive effect on leg health [49], muscle growth [32] and weight gain [50]. Regarding leg health, de Jong et al. found no such effects on leg health [45]. It has also been described in the literature that increased activity through environmental enrichment can also lead to an increased amount of exploratory behaviour and comfort behaviour [51], which is, therefore, an indication of improved animal welfare. However, the temporary, slightly positive effect on bodyweight in this trial was only seen for a few days at the very end of the fattening period, and, with a longer fattening period, the results could be different again.

When taking a look at the FCR, the statistical analysis also shows no differences regarding both groups, which also gives a hint towards environmental enrichment not having any influence on the performance of the birds. This is in accordance with de Jong et al., who described that the FCR did not differ between the enriched and non-enriched housed groups of the same strain [45].

4.2. Influence of Environmental Enrichment on FPD Scoring and Litter Quality

The examination of foot pads in both groups on the farm showed significantly lower FPD scores for the trial group at days 14 and 21. This could be related to the accessibility of more seating positions offered by the environmental enrichment or maybe by the increased activity, which could also lead to improved leg health [16]. A main influence of wet litter on the development of FPD is described in the literature [52], although there are a lot of other factors influencing this as well [53]. The results obtained in the current study reveal that the environmental enrichment had no negative influence on the litter quality. A recent study found that environmental enrichment in form of elevated platforms and straw bales has also no effect on the litter quality, as humidity in the litter differed from 25.1 $\pm$ 5.1% to 48.1 $\pm$ 7.7% with enrichment and from 19.4 $\pm$ 3.4% to 45.1 $\pm$ 9.9% without it (p = 0.12) [54]. Nevertheless, another recent study elucidated that the significant effect of the enriched environment on the litter quality depended on the litter collection point; the litter taken from below the elevated platforms had a significantly higher moisture content than the same area in the control compartments (p= 0.013). However, no significant effects occurred in the litter taken around the feeding troughs and water dispensers in both compartments [55]. Other studies that analysed litter moisture in enriched environments found no significant changes between the enriched and control groups during the production cycle, with 24.6 $\pm$ 4.6% with the enriched environment and 27.9 $\pm$ 6.3% without it [56,57].

In our study, the analysis of the litter samples, however, showed no significance in the DM content of litter sampling between the groups and, therefore, no indication that the differences in FPD scoring were related to wet litter in this trial, although the DM in the litter of our trial never reached the critical value of 65% or less [58]. For the majority of the examination days, the DM content was even closer to what is described as the ideal moisture content of 20–25% [59]. To take a closer look at how the environmental enrichment might have contributed to the lower FPD scores by having a positive influence on the litter quality, the trial group was then divided into enriched and non-enriched areas. The results of this analysis show that the DM content of the litter was significantly higher in the enriched areas at day 14. In combination with the lower FPD scores at days 14 and 21, this could be an indication of a positive effect on the litter quality, as the FPD score was closely related to the litter moisture [58]. In the following weeks, the DM contents of both areas had no significant differences, although the DM contents in the enriched areas were slightly higher. This could explain why the FPD scores became similar at the end of the fattening process.

The results in this trial also show that the type of environmental enrichment does not have a significant influence on the effect on the litter, as there were no significant differences between the three types of environmental enrichment used, although it is described in the literature that perches are used less frequently than elevated platforms [49]. It seems that the possibility of taking a different seating position and being able to sit on grounds other than the litter is enough to improve litter quality and possibly also FPD. In order to evaluate the influences of single types of environmental enrichment more precisely and to examine which type is most suitable for broilers, it could be helpful to carry out trials with only one type of enrichment. Nevertheless, it is necessary to continuously work on the further development of environmental enrichment in order to constantly improve in this area [16,60].

Regarding foot pad scores at the slaughterhouse, there were no significant differences between both groups. Thus, even if it has been discussed that the provision of environmental enrichment could have either a negative [20,61] or a positive influence [62,63] on foot pad health, these previous results are inconclusive. The different types of scoring at the slaughterhouse also play a role, but studies have shown that FPD scoring at the slaughterhouse is well suited to mirror the foot pad health on the farm [64].

4.3. Influence of Environmental Enrichment on Air Quality

A high concentration of ammonia on poultry farms is a potentially dangerous situation both for chickens and farm workers [65,66]. Generally, increasing ammonia volatilisation in the farm air is associated with litter characteristics, such as DM [67]. The current study showed a typical significant increase in the aerial NH_3 level towards the end of the fattening period (5.49 and 5.59 ppm in the control and enriched groups, respectively), but still lower than the maximum values of 20 ppm prescribed by German law [68]. Values of 25 ppm and more have been described to have negative effects on the growth performance of broilers [69]. The results of this study are aligned with the results of Adler et al. [70], which showed that NH_3 concentrations constantly increase towards the end of the fattening cycle, with significantly higher values than at the beginning ($p < 0.01$). The results of our study showed similar significant differences with 0.10 ppm in week 1 and 5.55 ppm in week 5. In contrast to our study, Yang et al. [57] concluded that, under experimental conditions, the NH_3 concentrations in the broiler rooms with an elevated perching platform were 27% lower than those in the rooms without an elevated perching platform. Moreover, Almeida et al. [71] elucidated that NH_3 concentration increased during the production cycle, reaching its highest value at d 42 (25 ppm) in broilers in the control group, while the NH_3 level in the perforated plastic floor reached 2 ppm at d 42.

Concerning the CO_2 concentrations, the CO_2 values obtained in the present study decreased during the broilers' growth, from 2892 to 1887 ppm for the control group and from 2963 to 1925 ppm for the enriched group (Table 11). However, in all cases, these levels were still below the 3000 ppm standard for the protection of broilers established in the European Directive 2007/43/CE [10]. The higher CO_2 concentrations at the beginning can be explained by the fact that gas heating was used in this trial, which leads to higher CO_2 concentrations in the air [72]. The continuous decrease in concentration could then be associated with the steadily lower requirement for temperature and the associated lower heating output. This agrees with the results of Knížatová et al. who described that an increased ventilation rate at the end of the production cycle causes a decrease in CO_2 concentration in the air (inverse relation between CO_2 level and ventilation rate) [73].

5. Conclusions

Overall, the results of our study show no negative influences of the used environmental enrichment installations on the growth performance and FPD scores of Ross broilers during the 33-day fattening periods. There was no negative influence of the different installations on litter and air quality. However, it can be assumed that the provision of enrichment tools, such as perches and elevated platforms, and the birds becoming used to them from an early stage offers more opportunities for the broilers to express better their natural behavioural traits, which can improve individual well-being. Our study did not display any negative influence of the enrichment elements and the mobile-ceiling-based robot on bodyweight gain and animal health. The real-time monitoring and the wealth of data provided by the robot offer vast opportunities to closely monitor broiler flocks and individual birds and give a detailed mapping of air quality and indoor climate conditions in animal barns.

Author Contributions: Conceptualisation, J.H. and C.V.; methodology, F.S., J.H. and C.V.; software, C.V.; validation, F.S., A.A.-E.W., M.F.E.A. and C.V.; formal analysis, F.S., B.R. and C.V.; investigation, F.S., A.A.-E.W., M.F.E.A., C.S., M.A., S.R. and C.V.; resources, F.S., C.S., J.H. and C.V.; data curation, F.S., B.R. and C.V.; writing—original draft preparation, F.S.; writing—review and editing, B.R., A.A.-E.W., M.F.E.A., J.H. and C.V.; visualisation, F.S. and B.R.; supervision, J.H. and C.V.; project administration, S.R., O.D. and C.V.; funding acquisition, O.D. All authors have read and agreed to the published version of the manuscript.

Funding: This project was financially supported by the Federal Ministry of Food and Agriculture (BMEL) based on a decision of the Parliament of the Federal Republic of Germany, granted by the Federal Office for Agriculture and Food (BLE) (Funding code 2817MDT305). This Open Access publication was funded by the Deutsche Forschungsgemeinschaft (DFG, German Research Foundation) within the programme LE 824/10-1 "Open Access Publication Costs" and University of Veterinary Medicine Hannover, Foundation.

Institutional Review Board Statement: The study design was approved by the Animal Welfare Officer of the University of Veterinary Medicine Hannover, Foundation, Germany code "AZ: TVO-2021-V-19".

Informed Consent Statement: Not applicable.

Data Availability Statement: The data presented in this study are available in this manuscript.

Acknowledgments: We would like to thank Frances Sherwood-Brock for proofreading the manuscript to ensure correct English language usage.

Conflicts of Interest: The authors declare no conflict of interest.

References

1. Delgado, C.L. Rising consumption of meat and milk in developing countries has created a new food revolution. *J. Nutr.* **2003**, *133* (Suppl. 2), 3907S–3910S. [CrossRef] [PubMed]
2. Smil, V. Worldwide transformation of diets, burdens of meat production and opportunities for novel food proteins. *Enzym. Microb. Technol.* **2002**, *30*, 305–311. [CrossRef]
3. Vranken, L.; Avermaete, T.; Petalios, D.; Mathijs, E. Curbing global meat consumption: Emerging evidence of a second nutrition transition. *Environ. Sci. Policy* **2014**, *39*, 95–106. [CrossRef]
4. FAO. *Food Outlook: Biannual Report on Global Food Markets*; FAO: Rome, Italy, 2020.
5. Gaudaré, U.; Pellerin, S.; Benoit, M.; Durand, G.; Dumont, B.; Barbieri, P.; Nesme, T. Comparing productivity and feed-use efficiency between organic and conventional livestock animals. *Environ. Res. Lett.* **2021**, *16*, 024012. [CrossRef]
6. Henchion, M.; McCarthy, M.; Resconi, V.C.; Troy, D. Meat consumption: Trends and quality matters. *Meat Sci.* **2014**, *98*, 561–568. [CrossRef] [PubMed]
7. De Jong, I.; Berg, C.; Butterworth, A.; Estevéz, I. Scientific report updating the EFSA opinions on the welfare of broilers and broiler breeders. *EFSA Support. Publ.* **2012**, *9*, 295E. [CrossRef]
8. Bergmann, S.; Schwarzer, A.; Wilutzky, K.; Louton, H.; Bachmeier, J.; Schmidt, P.; Erhard, M.; Rauch, E. Behavior as welfare indicator for the rearing of broilers in an enriched husbandry environment—A field study. *J. Vet. Behav.* **2017**, *19*, 90–101. [CrossRef]
9. Adler, C.; Tiemann, I.; Hillemacher, S.; Schmithausen, A.; Müller, U.; Heitmann, S.; Spindler, B.; Kemper, N.; Büscher, W. Effects of a partially perforated flooring system on animal-based welfare indicators in broiler housing. *Poult. Sci.* **2020**, *99*, 3343–3354. [CrossRef]
10. Giersberg, M.; Hartung, J.; Kemper, N.; Spindler, B. Floor space covered by broiler chickens kept at stocking densities according to Council Directive 2007/43/EC. *Vet. Rec.* **2016**, *179*, 124. [CrossRef]
11. Spindler, B.; Hartung, J. Prevalence of Pododermatitis in broiler chickens kept according to Directive 2007/43/EC stocking densities. In *Animal Hygiene and Sustainable Livestock Production, Proceedings of the XVth International Congress of the International Society for Animal Hygiene, Vienna, Austria, 3–7 July 2011*; Tribun EU: Brno, Czech Republic, 2011; Volume 1.
12. Gentle, M.J. Pain issues in poultry. *Appl. Anim. Behav. Sci.* **2011**, *135*, 252–258. [CrossRef]
13. Bergmann, S.; Louton, H.; Westermaier, C.; Wilutzky, K.; Bender, A.; Bachmeier, J.; Erhard, M.; Rauch, E. Field trial on animal-based measures for animal welfare in slow growing broilers reared under an alternative concept suitable for the German market. *Berl. Münch. Tierärztl. Wochenschr.* **2016**, *129*, 453–461.
14. Oviedo-Rondon, E.O.; Hume, M.E.; Hernandez, C.; Clemente-Hernandez, S. Intestinal microbial ecology of broilers vaccinated and challenged with mixed Eimeria species, and supplemented with essential oil blends. *Poult. Sci.* **2006**, *85*, 854–860. [CrossRef] [PubMed]
15. Waldenstedt, L. Nutritional factors of importance for optimal leg health in broilers: A review. *Anim. Feed Sci. Technol.* **2006**, *126*, 291–307. [CrossRef]
16. Pedersen, I.; Forkman, B. Improving leg health in broiler chickens: A systematic review of the effect of environmental enrichment. *Anim. Welf.* **2019**, *28*, 215–230. [CrossRef]
17. Sans, E.C.O.; Federici, J.F.; Dahlke, F.; Molento, C.F.M. Evaluation of Free-Range Broilers Using the Welfare Quality (R) Protocol. *Braz. J. Poult. Sci.* **2014**, *16*, 297–305. [CrossRef]
18. Bradshaw, R.H.; Kirkden, R.D.; Broom, D.M. A review of the aetiology and pathology of leg weakness in broilers in relation to welfare. *Avian Poult. Biol. Rev.* **2002**, *13*, 45–103. [CrossRef]
19. Mayne, R.K.; Powell, F.; Else, R.W.; Kaiser, P.; Hocking, P.M. Foot pad dermatitis in growing turkeys is associated with cytokine and cellular changes indicative of an inflammatory immune response. *Avian Pathol.* **2007**, *36*, 453–459. [CrossRef]
20. Tahamtani, F.M.; Pedersen, I.J.; Riber, A.B. Effects of environmental complexity on welfare indicators of fast-growing broiler chickens. *Poult. Sci.* **2020**, *99*, 21–29. [CrossRef]

21. Newberry, R.C. Environmental Enrichment—Increasing the Biological Relevance of Captive Environments. *Appl. Anim. Behav. Sci.* **1995**, *44*, 229–243. [CrossRef]
22. Ventura, B.A.; Siewerdt, F.; Estevez, I. Access to barrier perches improves behavior repertoire in broilers. *PLoS ONE* **2012**, *7*, e29826. [CrossRef]
23. De Jong, I.; Gunnink, H. Effects of a commercial broiler enrichment programme with or without natural light on behaviour and other welfare indicators. *Animal* **2019**, *13*, 384–391. [CrossRef] [PubMed]
24. Blatchford, R.A.; Archer, G.S.; Mench, J.A. Contrast in light intensity, rather than day length, influences the behavior and health of broiler chickens. *Poult. Sci.* **2012**, *91*, 1768–1774. [CrossRef] [PubMed]
25. Olsson, I.A.S.; Keeling, L.J. The push-door for measuring motivation in hens: Laying hens are motivated to perch at night. *Anim. Welf.* **2002**, *11*, 11–19.
26. Schrader, L.; Muller, B. Night-time roosting in the domestic fowl: The height matters. *Appl. Anim. Behav. Sci.* **2009**, *121*, 179–183. [CrossRef]
27. Donaldson, C.J.; O'Connell, N.E. The influence of access to aerial perches on fearfulness, social behaviour and production parameters in free-range laying hens. *Appl. Anim. Behav. Sci.* **2012**, *142*, 51–60. [CrossRef]
28. Newberry, R.C.; Estevez, I.; Keeling, L.J. Group size and perching behaviour in young domestic fowl. *Appl. Anim. Behav. Sci.* **2001**, *73*, 117–129. [CrossRef]
29. van de Weerd, H.A.; Day, J.E.L. A review of environmental enrichment for pigs housed in intensive housing systems. *Appl. Anim. Behav. Sci.* **2009**, *116*, 1–20. [CrossRef]
30. Weeks, C.A.; Danbury, T.D.; Davies, H.C.; Hunt, P.; Kestin, S.C. The behaviour of broiler chickens and its modification by lameness. *Appl. Anim. Behav. Sci.* **2000**, *67*, 111–125. [CrossRef]
31. Zuidhof, M.J.; Schneider, B.L.; Carney, V.L.; Korver, D.R.; Robinson, F.E. Growth, efficiency, and yield of commercial broilers from 1957, 1978, and 2005. *Poult. Sci.* **2014**, *93*, 2970–2982. [CrossRef]
32. Pedersen, I.J.; Tahamtani, F.M.; Forkman, B.; Young, J.F.; Poulsen, H.D.; Riber, A.B. Effects of environmental enrichment on health and bone characteristics of fast growing broiler chickens. *Poult. Sci.* **2020**, *99*, 1946–1955. [CrossRef]
33. Reiter, K.; Bessei, W. Effect of locomotor activity on leg disorder in fattening chicken. *Berl. Munch. Tierarztl. Wochenschr.* **2009**, *122*, 264–270. [PubMed]
34. Ohara, A.; Oyakawa, C.; Yoshihara, Y.; Ninomiya, S.; Sato, S. Effect of Environmental Enrichment on the Behavior and Welfare of Japanese Broilers at a Commercial Farm. *J. Poult. Sci.* **2015**, *52*, 323–330. [CrossRef]
35. Baxter, M.; Bailie, C.L.; O'Connell, N.E. Evalution of dustbathing substrate and straw bales as environmental enrichments in commercial broiler housing. *Appl. Anim. Behav. Sci.* **2018**, *200*, 78–85. [CrossRef]
36. Bach, M.H.; Tahamtani, F.M.; Pedersen, I.J.; Riber, A.B. Effects of environmental complexity on behaviour in fast-growing broiler chickens. *Appl. Anim. Behav. Sci.* **2019**, *219*, 104840. [CrossRef]
37. Bilgili, S.F.; Hess, J.B.; Blake, J.P.; Macklin, K.S.; Saenmahayak, B.; Sibley, J.L. Influence of bedding material on footpad dermatitis in broiler chickens. *J. Appl. Poult. Res.* **2009**, *18*, 583–589. [CrossRef]
38. Ben Sassi, N.; Averós, X.; Estevez, I. Technology and poultry welfare. *Animals* **2016**, *6*, 62. [CrossRef]
39. Rowe, E.; Dawkins, M.S.; Gebhardt-Henrich, S.G. A systematic review of precision livestock farming in the poultry sector: Is technology focussed on improving bird welfare? *Animals* **2019**, *9*, 614. [CrossRef]
40. Main, D.; Mullan, S.; Atkinson, C.; Cooper, M.; Wrathall, J.; Blokhuis, H. Best practice framework for animal welfare certification schemes. *Trends Food Sci. Technol.* **2014**, *37*, 127–136. [CrossRef]
41. Part, C.E.; Edwards, P.; Hajat, S.; Collins, L.M. Prevalence rates of health and welfare conditions in broiler chickens change with weather in a temperate climate. *R. Soc. Open Sci.* **2016**, *3*, 160197. [CrossRef]
42. Chuppava, B.; Visscher, C.; Kamphues, J. Effect of different flooring designs on the performance and foot pad health in broilers and turkeys. *Animals* **2018**, *8*, 70. [CrossRef]
43. QS Fachgesellschaft Geflügel GmbH; Bonn, G. *Leitfaden Befunddaten in der Geflügelschlachtung;* Version 01.01.2022; QS Fachgesellschaft Geflügel GmbH: Bonn, Germany, 2022.
44. Mayne, R.; Else, R.; Hocking, P. High litter moisture alone is sufficient to cause footpad dermatitis in growing turkeys. *Br. Poult. Sci.* **2007**, *48*, 538–545. [CrossRef] [PubMed]
45. de Jong, I.C.; Blaauw, X.E.; van der Eijk, J.A.; das Silva, C.S.; van Krimpen, M.M.; Molenaar, R.; van den Brand, H. Providing environmental enrichments affects activity and performance, but not leg health in fast-and slower-growing broiler chickens. *Appl. Anim. Behav. Sci.* **2021**, *241*, 105375. [CrossRef]
46. Jacob, F.; Salgado, D.; Nää, I.; Baracho, M. Effect of Environmental Enrichment on the Body Weight in Broiler Chickens. *Braz. J. Poult. Sci.* **2020**, *22*. [CrossRef]
47. Nazareno, A.C.; da Silva, I.J.; Delgado, E.F.; Machado, M.; Pradella, L.O. Does environmental enrichment improve performance, morphometry, yield and weight of broiler parts at different ages? *Rev. Bras. Eng. Agrícola Ambient.* **2022**, *26*, 292–298. [CrossRef]
48. Aksit, M.; Yardim, Z.K.; Yalcin, S. Environmental enrichment influences on broiler performance and meat quality: Effect of light source and providing perches. *Eur. Poult. Sci.* **2017**, *81*. [CrossRef]
49. Kaukonen, E.; Norring, M.; Valros, A. Perches and elevated platforms in commercial broiler farms: Use and effect on walking ability, incidence of tibial dyschondroplasia and bone mineral content. *Animal* **2017**, *11*, 864–871. [CrossRef]

50. Meyer, M.; Johnson, A.K.; Bobeck, E.A. A novel environmental enrichment device improved broiler performance without sacrificing bird physiological or environmental quality measures. *Poult. Sci.* **2019**, *98*, 5247–5256. [CrossRef]
51. Vasdal, G.; Vas, J.; Newberry, R.C.; Moe, R.O. Effects of environmental enrichment on activity and lameness in commercial broiler production. *J. Appl. Anim. Welf. Sci.* **2019**, *22*, 197–205. [CrossRef]
52. De Jong, I.C.; Gunnink, H.; van Harn, J. Wet litter not only induces footpad dermatitis but also reduces overall welfare, technical performance, and carcass yield in broiler chickens. *J. Appl. Poult. Res.* **2014**, *23*, 51–58. [CrossRef]
53. Abd El-Wahab, A.; Visscher, C.; Kamphues, J. Impact of different dietary protein sources on performance, litter quality and foot pad dermatitis in broilers. *J. Anim. Feed Sci.* **2018**, *27*, 148–154. [CrossRef]
54. Mocz, F.; Michel, V.; Janvrot, M.; Moysan, J.-P.; Keita, A.; Riber, A.B.; Guinebretière, M. Positive Effects of Elevated Platforms and Straw Bales on the Welfare of Fast-Growing Broiler Chickens Reared at Two Different Stocking Densities. *Animals* **2022**, *12*, 542. [CrossRef] [PubMed]
55. Malchow, J.; Schrader, L. Effects of an Elevated Platform on Welfare Aspects in Male Conventional Broilers and Dual-Purpose Chickens. *Front. Vet. Sci.* **2021**, *8*, 660602. [CrossRef] [PubMed]
56. Bailie, C.L.; Baxter, M.; O'Connell, N.E. Exploring perch provision options for commercial broiler chickens. *Appl. Anim. Behav. Sci.* **2018**, *200*, 114–122. [CrossRef]
57. Yang, X.; Huo, X.; Li, G.; Purswell, J.L.; Tabler, T.; Chesser, D.; Zhao, Y. Application of Elevated Perching Platform and Robotic Vehicle in Broiler Production. In Proceedings of the 2019 ASABE Annual International Meeting, Boston, MA, USA, 7–10 July 2019; American Society of Agricultural and Biological Engineers: St. Joseph, MI, USA, 2019.
58. Abd El-Wahab, A.; Visscher, C.; Beineke, A.; Beyerbach, M.; Kamphues, J. Experimental studies on the effects of different litter moisture contents and exposure time to wet litter on development and severity of foot pad dermatitis in young fattening turkeys. *Arch. Geflugelkd.* **2012**, *76*, 55–62.
59. Kuleile, N.; Metsing, I.; Tjala, C.; Jobo, T.; Phororo, M. The effects of different litter material on broiler performance and feet health. *Online J. Anim. Feed Res.* **2019**, *9*, 206–211. [CrossRef]
60. Riber, A.; Van De Weerd, H.; De Jong, I.; Steenfeldt, S. Review of environmental enrichment for broiler chickens. *Poult. Sci.* **2018**, *97*, 378–396. [CrossRef]
61. Đukić Stojčić, M.; Bjedov, S.; Žikić, D.; Perić, L.; Milošević, N. Effect of straw size and microbial amendment of litter on certain litter quality parameters, ammonia emission, and footpad dermatitis in broilers. *Arch. Anim. Breed.* **2016**, *59*, 131–137. [CrossRef]
62. Marušić, D.; Matković, K.; Matković, S.; Pavičić, Ž.; Ostović, M.; Kabalin, A.E.; Lucić, H. Effect of litter type and perches on footpad dermatitis and hock burn in broilers housed at different stocking densities. *S. Afr. J. Anim. Sci.* **2019**, *49*, 546–554.
63. BenSassi, N.; vas, J.; Vasdal, G.; Averós, X.; Estévez, I.; Newberry, R.C. On-farm broiler chicken welfare assessment using transect sampling reflects environmental inputs and production outcomes. *PLoS ONE* **2019**, *14*, e0214070. [CrossRef]
64. Louton, H.; Erhard, M.; Wirsch, K.; Bergmann, S.; Piller, A.; Schmidt, P.; Rauch, E. Comparison of four assessment methods of foot pad dermatitis and hock burn of broilers. *Berl. Munch. Tierarztl. Wochenschr.* **2020**, *133*, 1–11.
65. Ritz, C.; Fairchild, B.; Lacy, M. Implications of ammonia production and emissions from commercial poultry facilities: A review. *J. Appl. Poult. Res.* **2004**, *13*, 684–692. [CrossRef]
66. Naseem, S.; King, A.J. Ammonia production in poultry houses can affect health of humans, birds, and the environment—techniques for its reduction during poultry production. *Environ. Sci. Pollut. Res.* **2018**, *25*, 15269–15293. [CrossRef] [PubMed]
67. Méda, B.; Hassouna, M.; Aubert, C.; Robin, P.; Dourmad, J.-Y. Influence of rearing conditions and manure management practices on ammonia and greenhouse gas emissions from poultry houses. *World's Poult. Sci. J.* **2011**, *67*, 441–456. [CrossRef]
68. TierSchNutztV. *Verordnung zum Schutz Landwirtschaftlicher Nutztiere und Anderer zur Erzeugung Tierischer Produkte Gehaltener Tiere bei Ihrer Haltung*; Tierschutz-Nutztierhaltungsverordnung in der Fassung der Bekanntmachung vom 22 die Durch Artikel 3 Absatz 2 des Gesetzes vom 30; Bundesministerium der Justiz: Berlin, Germany, 2006.
69. Wang, Y.; Meng, Q.; Guo, Y.; Wang, Y.; Wang, Z.; Yao, Z.; Shan, T. Effect of atmospheric ammonia on growth performance and immunological response of broiler chickens. *J. Anim. Vet. Adv.* **2010**, *9*, 2802–2806.
70. Adler, C.; Schmithausen, A.J.; Trimborn, M.; Heitmann, S.; Spindler, B.; Tiemann, I.; Kemper, N.; Büscher, W. Effects of a Partially Perforated Flooring System on Ammonia Emissions in Broiler Housing—Conflict of Objectives between Animal Welfare and Environment? *Animals* **2021**, *11*, 707. [CrossRef] [PubMed]
71. de Almeida, E.A.; de Souza, L.F.A.; Sant'Anna, A.C.; Bahiense, R.N.; Macari, M.; Furlan, R.L. Poultry rearing on perforated plastic floors and the effect on air quality, growth performance, and carcass injuries—Experiment 1: Thermal comfort. *Poult. Sci.* **2017**, *96*, 3155–3162. [CrossRef] [PubMed]
72. Cândido, M.; Xiong, Y.; Gates, R.; Tinôco, I.; Koelkebeck, K. Effects of carbon dioxide on turkey poult performance and behavior. *Poult. Sci.* **2018**, *97*, 2768–2774. [CrossRef]
73. Knížatová, M.; Mihina, Š.; Brouček, J.; Karandušovská, I.; Sauter, G.; Mačuhová, J. Effect of the age and season of fattening period on carbon dioxide emissions from broiler housing. *Czech J. Anim. Sci.* **2010**, *55*, 436–444. [CrossRef]

sustainability

Article

Production Performance, Egg Quality Characteristics, Fatty Acid Profile and Health Lipid Indices of Produced Eggs, Blood Biochemical Parameters and Welfare Indicators of Laying Hens Fed Dried Olive Pulp

Anna Dedousi [1,*], Maria-Zoi Kritsa [1], Mirjana Đukić Stojčić [2], Themistoklis Sfetsas [3], Antonios Sentas [4] and Evangelia Sossidou [1]

1 Veterinary Research Institute, Hellenic Agricultural Organization DIMITRA, 57001 Thessaloniki, Greece; mzkritsa@gmail.com (M.-Z.K.); sossidou@vri.gr (E.S.)
2 Department of Animal Science, Faculty of Agriculture, University of Novi Sad, 21000 Novi Sad, Serbia; mirjana.djukicstojcic@stocarstvo.edu.rs
3 Research & Development, Quality Control and Testing Services, QLAB Private Company, 57008 Thessaloniki, Greece; t.sfetsas@q-lab.gr
4 Quality Control Department of TSAKIRIS FAMILY SA, TSAKIRIS FAMILY SA, 54500 Thessaloniki, Greece; qualitycontrol@tsakirisfamily.gr
* Correspondence: dedousi@vri.gr; Tel.: +30-2310365381

Citation: Dedousi, A.; Kritsa, M.-Z.; Đukić Stojčić, M.; Sfetsas, T.; Sentas, A.; Sossidou, E. Production Performance, Egg Quality Characteristics, Fatty Acid Profile and Health Lipid Indices of Produced Eggs, Blood Biochemical Parameters and Welfare Indicators of Laying Hens Fed Dried Olive Pulp. *Sustainability* **2022**, *14*, 3157. https://doi.org/10.3390/su14063157

Academic Editors: George K. Symeon and Vassilios Dotas

Received: 3 February 2022
Accepted: 5 March 2022
Published: 8 March 2022

Publisher's Note: MDPI stays neutral with regard to jurisdictional claims in published maps and institutional affiliations.

Abstract: This study aimed to evaluate the long-term dietary effects of dried olive pulp (OP) on production performance, fatty acid profile and health lipid indices and quality characteristics of produced eggs, health and welfare indicators of laying hens. It was carried out in a commercial poultry farm using 300 Isa Brown layers at 23 weeks of age. The hens were randomly and equally divided in six dietary groups CON, OP2, OP3, OP4, OP5 and OP6, according to the inclusion rate of OP in the ration (0%, 2%, 3%, 4%, 5% and 6%, respectively). OP feeding increased the percentage of polyunsaturated fatty acids (PUFA) in eggs, decreased that of saturated fatty acids (SFA) and improved the PUFA to SFA ratio and health lipid indices, as indicated by the decrease of AI and TI and the increase in the h/H ratio of produced eggs, in a dose-dependent way. OP-fed layers presented a lower percentage of broken eggshells compared to controls. No adverse effects on birds' performance, egg quality traits, health and welfare parameters were observed but a positive impact on Keel Bone Damage (KBD) incidence and belly plumage damage was recorded. OP feeding at the rates of 5% and 6% seems to be beneficial in improving egg nutrition quality.

Keywords: olive pulp; layers; performance; egg quality; egg lipid profile; health; welfare

1. Introduction

It is well known that the major cost in poultry farming is feed, representing 70% of the total production costs [1]. In order to reduce this cost, the research has recently focused on exploring and evaluating new sources of raw materials from agricultural and industrial by-products for use as animal feed. Key benefits of this practice include lower dependence of animal production on human consumed seeds and reduced waste management costs [2]. Around 2.1 million tons of olive oil is produced annually in Europe, contributing to around 68% of the world's total production, with Spain, Italy and Greece having the leading role in olive oil production [3]. However, the olive oil industry generates considerable amounts of by-products with a harmful environmental impact [4]. The utilization of olive by-products as feedstuff is a promising strategy of recycling this waste, assisting the transition to an efficient circular waste-based economy [2]. It also perfectly fits with the EU Green Deal Strategy for boosting the efficient use of resources by moving to a clean, circular economy and stopping climate change, reversing biodiversity loss and cutting pollution [5].

Olive pulp (OP) is the residue remaining after olive cake (the raw material resulting from the extraction of olive oil) is dried. It is characterized by a high level of essential fatty acids (73% oleic acid, 13% palmitic acid and 7% linoleic acid) as well as high residual oil [6]. Furthermore, it contains oleuropeoside beneficial compounds such as oleuropein and phenolic compounds such as tyrosol [7]. Extensive investigations confirm that dietary polyphenols are strong antioxidants and can be used in poultry for enhancing health and ameliorating the growth performance and quality of animal products [8]. Olive pulp can therefore provide animals with energy and, in particular, polyunsaturated fatty acids, being also a source of many biologically active ingredients with antioxidants, antifungals, antibacterial and anti-tumoral properties [9–14]. Moreover, it is considered a good source of protein, calcium, copper and cobalt [1]. Its high nutritive value and chemical composition makes OP an interesting and low-cost nutrient for productive animals [15,16].

Olive pulp has been previously used at various inclusion rates in laying hens' diet in a limited number of studies with inconsistent results regarding birds' performances, egg quality and health indices [1,2,17–23]. Most of these feeding trials have been carried out in experimental units, evaluating the dietary effect of OP on hens' productivity for a short period of time (6–12 weeks). However, there is lack of research evidence in the available literature regarding the optimal inclusion rate in hens' diet as well as the dietary impact of OP on the lipid profile of egg as well as on hens' welfare. Therefore, the present study was designed in order to investigate the dietary effect of dried OP on laying hens' performance, egg quality characteristics, fatty acid composition and health lipid indices of produced eggs, blood biochemical parameters and welfare indicators of hens, under field conditions, during an entire production cycle. A complimentary goal of this study was to assess the optimal inclusion rate of OP in birds' feed.

2. Materials and Methods

The experimental protocol of the study and implemented animal care procedures were approved by the Committee for Research Ethics of Hellenic Agricultural Organization-DIMITRA (52216/23 October 2020). The study was conducted in accordance with the Declaration of Helsinki.

2.1. Experimental Design

This study was carried out in a commercial poultry farm in Greece and lasted 45 weeks. In total, 300 Isa Brown layers, 23 weeks of age, with initial body weight (BW) 1.58 ± 0.01 kg, were randomly accommodated in 30 enriched cages (10 hens/cage) that were fully equipped and met the requirements of EU Directive 1999/74/EC [24]. Hens were vaccinated and managed according to the breed standards. The light program provided 14 h light (15 lux) per day (14 h light-10 h dark cycle) and kept constant to the end of the experiment. The average temperature in the laying house during the trial was 26 ± 3 °C. Birds were fed with a basal layer diet, formulated according to breed recommendations in mash form. Feed and water were offered ad libitum. After 2 weeks of adaptation, hens were randomly divided in 6 dietary groups CON, OP2, OP3, OP4, OP5 and OP6, with 50 hens/group, 5 replicates-cages/group, 10 hens/replicate-cage. The CON group was fed the basal layer diet provided during the adaptation period and served as control, while the OP groups were fed the basal diet supplemented with dried olive pulp, at the rates of 2%, 3%, 4%, 5% and 6%, respectively. In OP diets, dried olive pulp replaced mainly maize and a small amount of soya meal of the control diet so as to make all rations isonitrogenous and isocaloric (Table 1). The dried OP used in the experiment was a commercial animal feed supplement in the form of flour (*Sparta* INNOLIVE®, Sparta Life S.A., Sparta, Greece). The nutrient and fatty acid composition of OP used in the feeding trial is presented in Table 2.

Table 1. Formulation and nutrient composition of diets containing olive pulp (OP) compared with the control diet (CON).

	CON	OP2	OP3	OP4	OP5	OP6
Ingredients %						
Maize	54.2	53.2	52.2	51.2	50.2	49.2
Soyameal-48	11	10	10	10	10	10
Limestone	9.6	9.6	9.6	9.6	9.6	9.6
Layer concentrate 25% [1]	25	25	25	25	25	25
Olive pulp	0	2	3	4	5	6
MCP	0.2	0.2	0.2	0.2	0.2	0.2
Calculated analysis						
Crude protein (%)	17.68	17.55	17.57	17.59	17.60	17.62
Crude fiber (%)	5.62	5.72	5.68	5.78	5.68	6.48
Fat (%)	6.33	6.48	6.54	6.53	6.50	6.58
Ash (%)	12.70	12.66	12.60	12.59	12.61	12.73
Metabolizable energy (Kcal/kg)	3100	3100	3100	3100	3100	3100
Lysine (%)	0.80	0.80	0.80	0.80	0.80	0.80
Methionine + Cystine (%)	0.66	0.66	0.66	0.66	0.66	0.66
Ca (%)	4.20	4.20	4.20	4.20	4.20	4.20
Av. P (%)	0.33	0.33	0.33	0.33	0.33	0.33

[1] Layer concentrate 25% is a protein/fat mixture for layers including vitamins, minerals, enzymes, yolk coloring, organic acids and mycotoxin binder (Supplementary Table S1). MCP: Monocalcium Phosphate, Av. P: Available phosphorus.

Table 2. Nutrients and fatty acid composition of olive pulp (OP).

Items	Olive Pulp
Moisture (g/100 g)	8.3
Proteins (g/100 g)	9.5
Fat (g/100 g)	14.5
Crude fiber % (*w*/*w*)	25.3
Carbohydrates (g/100 g)	61.0
Ash (g/100 g)	6.7
Lysine %	0.04
Methionine %	0.03
Threonine %	0.37
Ca % (*w*/*w*)	1.52
Mg % (*w*/*w*)	0.15
P % (*w*/*w*)	0.15
ME (Kj/100 g)	1735
Phenolic compounds (mg/kg dry matter)	2410
Fatty acids (% of total fat)	
Lauric (dodecanoic) acid (C12:0)	0.05
Myristic acid (C14:0)	0.05
Myristoleic acid (C14:1)	11.09
Palmitic acid (C16:0)	0.69
Palmitoleic acid (C16:1)	0.19
Margaric acid (C17:0)	0.09
Stearic acid (C18: 0)	2.71
Oleic acid (C18:1)	70.77
α-Linoleic acid (C18:2)	10.77
Linolenic acid(C18:3)	0.48
Arachidic acid (C20:0)	0.85
Eicosenic acid (Gadoleic, C20: 1)	0.11
cis-11,14,17-Eicosatrienoic acid (C 20:3 ω3)	<0.02 *
cis-8,11,14-Eicosatrienoic acid (C 20:3 ω6)	<0.02 *
Arachidonic acid (C 20:4 ω6)	0.13
EPA (C 20:5 n3)	<0.02 *
Behenic acid (C22:0)	0.12

Table 2. *Cont.*

Items	Olive Pulp
Erucic acid (C22:1)	<0.02 *
Docosapentaenoic acid (DPA n3 (C22:5 n3)	<0.02 *
DHA (C 22:6 n3)	<0.02 *
Lignoceric acid (C24:0)	0.18

* This value is the detection limit of the assay. P: Phosphorus, ME: Metabolizable energy, EPA: Eicosapentaenoic acid, DHA: Docosahexaenoic acid.

2.2. *Hen Productivity and Egg Quality*

Laying hens were individually weighted at the onset (23 weeks of age) and at the end of the experiment (68 weeks of age). From 27 to 68 weeks of age, feed intake was measured weekly per replicate, weighing the amount of feed distributed and that of residual and scattered feed and was calculated as daily feed intake per hen. The number of eggs produced as well as those with dirty eggshells or deficits (broken, cracked or without eggshell) were recorded daily per replicate pen. Individual egg weights were recorded per replicate every week. Mortality rate was recorded daily. The following were calculated per each replicate of each treatment group: Hen Day-Egg Production % (HDEP = total number of eggs produced on a day/number of hens present on that day) × 100; Feed Intake (FI = total FI/number of days of the trial period); egg mass (Egg mass = (HDEP × egg weight)/100) and Feed Conversion Ratio (FCR = g feed/g egg mass).

A total of 90 eggs (15 eggs/group, 3 eggs/replicate) were randomly selected at the end of the experiment, in order to determine some internal (albumen and yolk weights and percentages, Haugh unit, yolk color, albumen and yolk height and pH, yolk diameter and index) and external (shape index, eggshell weight, percentage and thickness) egg quality traits. Egg weight was measured on a digital scale with accuracy to the nearest 0.01 g. The egg yolk, egg white (albumen), and eggshell of the cracked egg were weighed on the same digital scale. Then, the proportions of yolk ((yolk weight/egg weight) × 100), albumen ((albumen weight/egg weight) × 100), and shell ((shell weight/egg weight) × 100) in each egg were calculated. The egg shape index value ((width/height) × 100) was calculated using the height and width values of the egg measured with an electronic caliper. To calculate the egg yolk index ((height/diameter) × 100), the height, width and length of the yolk were measured with a tripod micrometer and an electronic caliper, respectively. Haugh unit values were calculated using the egg weight (g) and albumen height (mm) (albumen height + 7.57 − 1.7 × egg weight $^{0.37}$). Albumen height was measured using a tripod micrometer. Yolk color was determined according to the Roche yolk color fan ranging from pale yellow (1) to deep orange (15). Shell thickness (mm) was measured using a dial gauge micrometer. Finally, albumen and yolk pH were measured with a waterproof pH meter.

2.3. *Determination of Egg Fatty Acid Composition, Phenols Content and Health Lipid Indices*

A total of 180 eggs (30 eggs/group, 6 eggs/replicate) were randomly collected from laying hens of intermediate (50 weeks) age, in order to determine fatty acid (FA) profile, total fat and phenols content. In each dietary treatment, 6 final egg samples were formulated after mixing and homogenizing 5 eggs out of the 30 initially collected eggs. The 6 representative egg samples from each group, were analyzed twice for their total phenolic acids content, fatty acids (FA) profile and, the percentage of saturated (SFA), monounsaturated (MUFA) and polyunsaturated fatty acids (PUFA), in the whole edible parts of eggs (albumen + yolk).

All reagents used were of GC-grade and were purchased from Sigma-Aldrich (Merck KGaA, Darmstadt, Germany). The reagents for acid hydrolysis consisted of petroleum ether, boiling area: 40–60 °C, anhydrate sodium sulphate, auxiliary filtration (Celite 545), hydrochloric acid solution (HCl analytical grade). The reagents for Soxhlet Extraction were iso-octane of chromatographic purity, potassium hydroxide, methanolic solution of potassium hydroxide 2 M, methanol with a water content of not more than 0.5% (m/m), plastic syringes with disk filters (cellulose acetate) of 0.45 μm and glass vials with caps. Samples were homogenized by a knife mill. A corresponding quantity depending on the sample was used (±0.1 mg) and was placed in a 250 mL spherical flask. Generally, 5 to 10 g of sample was used. Then, 100 mL of aqueous solution of HCl 4N was added with 2–3 boiling stones and was boiled gently in a 6-position heating mantle at a vertical reflux condenser for 60 min. Then it was left at room temperature to cool down. Afterwards, the solution was filtered and was rinsed with deionized water until neutral pH. The filter paper was dried on a watch glass, at 55 ± 2 °C for 16–18 h (overnight) and the procedure of soxhlet extraction was followed. In a subsequent step, extracted fatty acids were trans-esterified in a methanol potassium hydroxide solution and the fames samples were analyzed by Gas Chromatography - Flame Ionization Detector (GC-FID) [25].

Fatty acid contents were determined by gas chromatography (GC-FID) using a Shimadzu GC-2010 Plus High-End Gas Chromatograph (Shimadzu Europa GmbH, Duisburg, Germany) (GC), equipped with Flame Ionization Detector (Shimadzu Europa GmbH, Duisburg, Germany) (FID), after lipid extraction by the soxhlet procedure, as mentioned. The column used was a Supelco SP2560 (Merck KGaA, Darmstadt, Germany), 100 m × 0.25 mm × 0.20 μm. Helium was used as a carrier gas at a flow rate of 2 mL/min [26]. Injection volume was 1 μL with split ratio 1:50 and injector temperature at 250 °C. The detector temperature was set at 250 °C. The temperature program applied was: initial oven temperature at 110 °C (7 min), increasing at 3 °C/min to 190 °C (2 min), then in a first step at 0.5 °C/min to 205 °C, in a second at 5 °C/min to 230 °C (5 min) and in a third at 5 °C/min to a final temperature of 240 °C for 5 min. The total run time was: 82.67 min. The results were identified using GC solution software comparing mass spectra with retention time peaks. The FA, SFA, MUFA and PUFA values were expressed as weight percentages (% of total FAs).

The phenolic acid content of eggs was determined by Ultra Violet -Visible (UV-VIS) spectrophotometry method using Folin Ciocalteu reagent 2N (Merck KGaA, Darmstadt, Germany). Initially, the sample was extracted with 70% methanol and then was stirred and filtered with a Whatman No 2 filter. Afterwards, 5 mL Folin Ciocalteu 10% was added to the filtrate solution and was stirred. After 3–8 min had passed since the addition of the 10% Folin Ciocalteu reagent, 4 mL of the 7.5% sodium carbonate was added, and the tubes were stirred again. They were left to rest for 60 min and then their optical density was measured with glass cells 10 mm at 765 nm by UV-VIS method. A standard solution of 1000 ppm Gallic acid was used for the calibration curve in the following concentrations: 10 ppm, 20 ppm, 30 ppm, 40 ppm, 50 ppm. Total phenolic acid was expressed as mg Gallic Acid Equivalent (GAE)/g.

Based on the proportions of particular FAs and their groups, the health quality of egg lipids was assessed by calculating: Atherogenic Index (AI), Thrombogenic Index (TI) and the ratio between hypocholesterolemic (h)/hypercholesterolemic (H) fatty acids (h/H). The following equations were used to calculate these indexes:

Atherogenic Index [27,28]:

$$\begin{aligned} AI &= (4 \times C14{:}0 + C16{:}0 + C18{:}0)/(\Sigma MUFA + \Sigma PUFA\text{-}n\text{-}6 + \Sigma PUFA\text{-}n\text{-}3) \\ AI^{**} &= (4 \times C14{:}0) + C16{:}0/(\Sigma MUFA + \Sigma PUFA\text{-}n\text{-}6 + \Sigma PUFA\text{-}n\text{-}3) \end{aligned} \quad (1)$$

Thrombogenic Index [28]:

$$TI = (C14{:}0 + C16{:}0 + C18{:}0)/(0.5 \times \Sigma MUFA + 0.5 \times \Sigma PUFA\text{-}n\text{-}6 + 3 \times \Sigma PUFA\text{-}n\text{-}3 + \Sigma PUFA\text{-}n\text{-}3/\Sigma PUFA\text{-}n\text{-}6) \quad (2)$$

Ratio between hypocholesterolemic and hypercholesterolemic fatty acids [29]:

$$h/H = C18{:}1n9c + C18{:}2n6c + C18{:}3n3c + C18{:}3n6c + C20{:}2n6 + C20{:}3n6 + C20{:}4n6 + C22{:}6n3/C14{:}0 + C16{:}0 \quad (3)$$

where:

Σ = Summatory,

MUFA = monounsaturated FAs and

PUFA = polyunsaturated FAs

The atherogenic index was calculated by using two equation formulas either by including C18:0 fatty acid to the numerator of equation formula (IA), or not (IA**) since both methods are reported to the literature.

2.4. Blood Biochemical Parameters

At 50 weeks of age, blood samples were collected from 2 randomly selected birds per replicate (a total of 60 blood samples, 10 per group) for the determination of selected biochemical parameters. Approximately 2 mL of blood were taken from the brachial vein of each hen and were collected in plastic vacuum tubes (BD Vacutainer® SST™ II Advance, Becton Dickinson, NJ, USA). After clotting, the serum was separated by centrifugation (3000× *g* for 15 min), transferred into plastic vials and forwarded on ice to an ISO-certified commercial veterinary laboratory for analysis. The serum samples were analyzed for Cholesterol, Triglycerides, Aspartate Aminotransferase (AST), Gamma-Glutamyl Transferase G-GT, Uric acid, Blood Urea Nitrogen (BUN) and Glutamate Dehydrogenase (GLDH) using an automatic biochemical analyzer (Advia® 1800 chemistry analyzer—Siemens Healthineers Headquarters, Erlangen, Germany) and commercially available diagnostic kits.

2.5. Welfare Indicators

In the middle of the production cycle (50 week of age) the hens of each group were individually evaluated for the presence of keel bone damage (KBD), plumage damage, comb abnormalities, skin lesions, claws, foot pad dermatitis and toe damage. Assessment of KBD for all laying hens was performed by the same person according to the palpation technique [30]. Each hen was gently held by one person, and another trained person examined and palpated the keel bone. It was only determined whether KBD was present (fracture, deformation, or both score 1) or not (completely straight and flat keel bone, score 0). Moreover, the % of hens in each group presenting deformation, fractures or both was calculated.

The rest of the welfare parameters were determined according to the Welfare Quality Network (2019) protocol [31]. Briefly, in each hen, the plumage of 3 different body parts (head-neck, back-rump and belly) was given a score on a 3-point scale: 0—All body parts had no or slight wear, (nearly) complete feathering (only single feathers lacking); 1—One or more body parts had moderate wear, i.e., damaged feathers (worn, deformed) or one or more featherless areas <5 cm in diameter at the largest extent; 2—One or more body parts had at least one featherless area ≥5 cm in diameter at the largest extent. The percentage of hens presenting each score was then calculated. In order to achieve a single general plumage score per bird the aforementioned scores of the 3 body parts were combined according to the following classification: 0—All body parts had score '0'; 1—One or more body parts had score '1', but no body part had score '2'; 2—One or more body parts had score '2'. Percentage of birds with scoring categories 0, 1, 2 was recorded.

The hens of each group were individually inspected at the rear end, legs and underneath feathers for the presence of skin lesions (3 or more pecks and/or scratches, large unhealed wounds). An individual score was assessed ranging from 0 to 2 as follow: 0—No lesions, only single (<3) pecks (punctiform damage < 0.5 cm diameter) or scratches; 1—At least one lesion < 2 cm diameter at largest extent or ≥3 pecks or scratches; 2—At least one lesion ≥ 2 cm diameter at largest extent. Both feet of each hen were examined for the presence of foot pad dermatitis (swelling-bubble foot) or toe damage (wounds on one or more toes and/or missing (parts of) one or more toes) and scores were assessed according to the following: (a) Foot pad dermatitis: 0—Feet intact, no or minimal proliferation of epithelium; 1—Necrosis or proliferation of epithelium or chronic bumble foot with no or moderate swelling; 2—Swollen (dorsally visible); (b) toe damage: 0—no toe damage; 1—presence of toe damage. The percentage of birds within each scoring category was recorded. Additionally, the percentage of hens in each group presenting normal or long claws was estimated. Comb individual examination was also performed for the presence of abnormalities (blue or black spots or areas, very pale combs, wounds–not being punctiform pecking wounds–or missing parts)—score 1 or not—score 0.

2.6. Statistical Analysis

The data were analyzed using the statistical software Jeffreys's Amazing Statistics Program JASP (JASP v 0.14. https://jasp-stats.org/download/ (accessed on 20 January 2020)); [32]. The significance of the differences of KBD incidence rate, mortality rate, plumage damage, claw length, comb abnormalities, skin lesions, foot pad dermatitis and toe damage among groups was assessed by Chi-square test. For the analysis of laying hens' performance, egg quality traits, egg yolk fatty acid composition, phenols content, health yolk lipid indices and blood biochemical parameters the normality of the data was tested using the Shapiro–Wilk test and the homogeneity of variance was evaluated with Levene's test. One-way Analysis of Variance (ANOVA)was used to compare the average values of the parameters evaluated among dietary treatments. Post hoc analysis was performed using the Tukey test. In cases where the distribution was not normal the comparisons were made with the non-parametric tests Kruskal–Wallis and Mann–Whitney. All comparisons were made at a significance level of $p \leq 0.05$.

3. Results

3.1. Laying Performance

The incorporation of OP in laying hens' diet had a significant impact ($p < 0.05$) on HDEP %, egg weight, percentage of eggs with broken shell and egg mass (Table 3). However, final BW, percentage of eggs with dirty shells, feed consumption and FCR were not significantly different among dietary treatments ($p > 0.05$) as indicated in Table 3. Hens of group OP4 presented significantly lower HDEP % compared to that of hens belonging to the CON, OP2, OP5 and OP6 groups ($p < 0.05$), whereas OP6 hens had significantly higher HDEP % compared to OP2, OP3 and OP4 hens ($p < 0.05$). The highest egg weight was recorded in the CON group followed by the OP6, OP3, OP2, OP5 and OP4 groups, respectively. The differences in egg weight among dietary treatments were significant ($p < 0.05$) except those observed between the OP5 and OP4 groups. Eggs produced by hens that were fed diets with OP at the rate of 2%, 3%, 5% and 6% had a significantly lower percentage of broken shells compared to those produced by the CON and OP4 hens ($p < 0.05$). The percentage of broken eggshells between the CON and OP4 groups did not differ significantly ($p > 0.05$). Eggs produced by hens of the OP4 and OP5 groups had significantly lower egg mass compared to that recorded in eggs of the CON group ($p < 0.05$). The eggs produced by the CON, OP2, OP3 and OP6 hens presented similar egg mass ($p > 0.05$). Overall mortality rate ranged from 2% (OP6 group) to 12% (OP4 group) and the differences among treatments were not significant ($p > 0.05$).

Table 3. The overall performance of laying hens (27–68 week of age). Data are presented as mean ± SE.

Items	Dietary Treatments					
	CON	OP2	OP3	OP4	OP5	OP6
Final BW (Kg)	2.02 ± 0.04	1.94 ± 0.04	1.95 ± 0.02	1.99 ± 0.04	1.93 ± 0.04	1.91 ± 0.02
HDEP (%)	86.75 ± 0.43 ^{ac}	85.43 ± 0.54 ^{a}	83.35 ± 0.75 ^{ab}	83.88 ± 0.53 ^{b}	85.66 ± 0.60 ^{ac}	86.41 ± 0.59 ^{c}
Egg weight (g)	66.23 ± 0.17 ^{a}	64.19 ± 0.16 ^{b}	64.73 ± 0.15 ^{c}	63.78 ± 0.18 ^{d}	63.90 ± 0.16 ^{d}	65.26 ± 0.16 ^{e}
Eggs with broken shell %	1.57 ± 0.16 ^{a}	0.65 ± 0.09 ^{b}	0.74 ± 0.10 ^{b}	1.32 ± 0.15 ^{a}	0.68 ± 0.11 ^{b}	0.75 ± 0.11 ^{b}
Eggs with dirty eggshells %	0.13 ± 0.05	0.07 ± 0.05	0.19 ± 0.06	0.26 ± 0.10	0.09 ± 0.04	0.13 ± 0.07
Feed consumption (g/h/d)	112.42 ± 4.37	116.7 ± 5.23	103.18 ± 3.10	114.28 ± 4.67	113.12 ± 4.90	102.42 ± 3.09
FCR	2.00 ± 0.08	2.18 ± 0.10	2.00 ± 0.07	2.18 ± 0.09	2.12 ± 0.09	1.88 ± 0.06
Egg mass	56.45 ± 0.75 ^{a}	54.05 ± 1.11 ^{ab}	52.59 ± 1.33 ^{ab}	52.87 ± 0.87 ^{b}	53.69 ± 1.00 ^{b}	54.95 ± 1.02 ^{ab}

$^{a–e}$ Means within a row with different superscripts differ significantly ($p < 0.05$).

3.2. *Egg Quality Traits*

Egg quality traits of 68-week-old laying hens that received the control and diets containing different levels of OP are presented in Table 4. No differences were found in shape index, shell weight, albumen weight, albumen and yolk ratio (%), Haugh units, albumen height and yolk pH ($p > 0.05$). Hens of the OP2 group laid eggs with thinner shells than those produced in the rest of the groups and the differences noticed between the OP2 group and the OP3, OP4 and OP5 groups were significant ($p < 0.05$). The lowest shell ratio (%) was observed in the OP6 group and differed significantly ($p < 0.05$) with the highest that was found in the OP2 group. Eggs from the OP5 group presented significantly lower yolk weight compared to eggs from the OP6, OP3 and OP4 groups ($p < 0.05$). The addition of OP in laying hens' diet increased egg yolk diameter, with significant differences being observed between the CON group with the OP4 and OP5 groups, as well as between the OP4 and OP6 groups ($p < 0.05$). Moreover, eggs from the OP6 group had significantly higher yolk height compared to the OP4 eggs ($p < 0.05$). The aforementioned differences in yolk diameter and height reflected in the egg yolk index. In particular, a significantly increased yolk index was found in the OP6 eggs compared to the OP3 and OP4 eggs ($p < 0.05$). A greater yolk color score was recorded in the OP4 and OP6 groups compared to the rest of the dietary treatments, however significant differences were only found between the OP3 and OP4 groups ($p < 0.05$). The albumen pH of the OP5 and OP6 eggs was significantly higher than that recorded in the eggs of the rest of the groups.

Table 4. Egg quality traits of laying hens at 68 weeks of age fed the control and diets containing olive pulp. Data are presented as mean ± SE.

Hen Age: 68 wk	Dietary Treatments					
Items	**CON**	**OP2**	**OP3**	**OP4**	**OP5**	**OP6**
Shape Index (%)	76.50 ± 0.86	76.79 ± 1.48	77.63 ± 0.85	78.00 ± 0.79	77.97 ± 0.88	76.07 ± 1.50
Shell thickness (mm)	0.38 ± 0.01 ^{ab}	0.36 ± 0.01 ^{a}	0.41 ± 0.01 ^{b}	0.41 ± 0.01 ^{b}	0.41 ± 0.02 ^{b}	0.40 ± 0.02 ^{ab}
Shell weight (g)	9.42 ± 0.27	9.62 ± 0.44	9.59 ± 0.14	9.29 ± 0.32	9.35 ± 0.29	9.14 ± 0.24
Shell ratio (%)	13.78 ± 0.23 ^{ab}	15.10 ± 0.63 ^{a}	14.07 ± 0.22 ^{ab}	14.17 ± 0.36 ^{ab}	14.61 ± 0.31 ^{ab}	13.17 ± 0.33 ^{b}
Albumen weight (g)	42.10 ± 1.25	38.59 ± 1.10	41.47 ± 1.01	38.51 ± 1.43	39.39 ± 1.21	43.05 ± 1.08
Albumen ratio (%)	61.51 ± 0.93	60.41 ± 0.60	60.76 ± 1.27	58.72 ± 1.28	61.45 ± 0.68	61.90 ± 0.50
Yolk weight (g)	15.75 ± 0.53 ^{ab}	15.27 ± 0.30 ^{ab}	16.67 ± 0.34 ^{b}	16.45 ± 0.40 ^{b}	14.83 ± 0.33 ^{a}	16.74 ± 0.29 ^{b}
Yolk ratio (%)	23.05 ± 0.62	23.97 ± 0.34	24.45 ± 0.57	25.23 ± 0.79	23.22 ± 0.55	24.14 ± 0.48
Haugh units	94.84 ± 0.95	95.48 ± 0.93	93.93 ± 0.86	94.81 ± 0.97	96.00 ± 0.59	96.22 ± 1.01
Albumen height (mm)	7.11 ± 0.16	7.09 ± 0.19	6.92 ± 0.18	7.01 ± 0.18	7.19 ± 0.13	7.43 ± 0.18
Yolk diameter (mm)	41.88 ± 0.16 ^{a}	42.94 ± 0.33 ^{abc}	43.11 ± 0.29 ^{abc}	43.31 ± 0.28 ^{b}	43.24 ± 0.40 ^{bc}	42.01 ± 0.26 ^{ac}
Yolk height (mm)	18.43 ± 0.18 ^{ab}	18.47 ± 0.24 ^{ab}	18.26 ± 0.17 ^{ab}	18.06 ± 0.21 ^{a}	18.61 ± 0.24 ^{ab}	19.03 ± 0.14 ^{b}
Yolk index	44.01 ± 0.49 ^{ab}	43.05 ± 0.68 ^{ab}	42.36 ± 0.32 ^{b}	41.72 ± 0.47 ^{b}	43.11 ± 0.85 ^{ab}	45.32 ± 0.44 ^{a}
Yolk color	10.50 ± 0.22 ^{ab}	10.60 ± 0.22 ^{ab}	10.20 ± 0.20 ^{a}	11.10 ± 0.23 ^{b}	10.70 ± 0.15 ^{ab}	11.00 ± 0.21 ^{ab}
Albumen pH	8.55 ± 0.04 ^{b}	8.55 ± 0.04 ^{b}	8.57 ± 0.02 ^{b}	8.62 ± 0.02 ^{b}	8.76 ± 0.03 ^{a}	8.82 ± 0.02 ^{a}
Yolk pH	6.30 ± 0.05	6.18 ± 0.03	6.14 ± 0.06	6.22 ± 0.03	6.23 ± 0.05	6.17 ± 0.02

$^{a–c}$ Means within a row with different superscripts differ significantly ($p < 0.05$).

3.3. *Egg Fatty Acid Composition, Phenols Content and Health Lipid Indices*

Significant differences ($p < 0.05$) were detected among dietary treatments on egg total fat, FA composition, total phenols and health lipid indices that were determined at the whole edible parts (albumen and yolk) of eggs obtained from laying hens of intermediate age (Table 5). Eggs from the OP6 group had significantly ($p < 0.05$) lower total fat content than that recorded in the eggs from the rest of the groups, whereas the highest fat content was found in eggs from group OP3 ($p < 0.05$). Hens from the CON and OP2 groups produced eggs with a similar percentage of total fats ($p > 0.05$). The highest phenol content among treatments was found in the OP4 and OP6 eggs ($p < 0.05$) and the lowest in the OP3 eggs ($p < 0.05$). The total phenol content of eggs produced by the CON, OP2 and OP5 hens was not significantly different ($p > 0.05$).

Table 5. The effect of dietary olive pulp (OP) supplementation on egg total fat, fatty acid (FA) composition (g/100 g FA) total phenols and health lipid indices as evaluated in the whole edible parts (albumen and yolk) of eggs obtained from 50-week-old laying hens. Data are presented as mean ± SE.

Hen Age: 50 wk	Dietary Treatments					
Item	CON	OP2	OP3	OP4	OP5	OP6
% Fat	9.09 ± 0.10 [a]	9.06 ± 0.09 [a]	10.93 ± 0.08 [b]	10.26 ± 0.09 [c]	9.49 ± 0.09 [d]	7.66 ± 0.07 [e]
MUFA % (g/100 g Fat)	47.01 ± 0.40 [ab]	46.61 ± 0.28 [ab]	45.56 ± 0.55 [a]	47.42 ± 0.47 [b]	45.43 ± 0.32 [a]	47.74 ± 0.52 [b]
PUFA % (g/100 g Fat)	7.33 ± 0.08 [a]	9.06 ± 0.08 [b]	11.13 ± 0.13 [c]	10.70 ± 0.13 [c]	22.21 ± 0.49 [d]	20.67 ± 0.16 [e]
SFA % (g/100 g Fat)	45.67 ± 0.37 [a]	44.33 ± 0.36 [ab]	43.31 ± 0.58 [bc]	41.86 ± 0.41 [c]	32.35 ± 0.72 [d]	31.59 ± 0.45 [d]
Total Phenols (mg GAE/g)	36.05 ± 0.30 [a]	36.10 ± 0.35 [a]	31.15 ± 0.23 [b]	39.49 ± 0.41 [c]	35.47 ± 0.24 [a]	39.08 ± 0.32 [c]
PUFA/SFA	0.16 ± 0.00 [a]	0.21 ± 0.00 [ab]	0.26 ± 0.01 [b]	0.26 ± 0.00 [b]	0.69 ± 0.03 [c]	0.66 ± 0.01 [c]
PUFA n6	7.42 ± 0.14 [a]	9.46 ± 0.17 [b]	11.59 ± 0.16 [c]	10.83 ± 0.09 [d]	22.36 ± 0.13 [e]	19.30 ± 0.10 [f]
PUFA n3	ND	ND	ND	ND	0.94 ± 0.01 [a]	0.83 ± 0.01 [b]
PUFA n6/PUFA n3					23.71 ± 0.24 [a]	23.31 ± 0.13 [a]
AI	0.84 ± 0.01 [a]	0.82 ± 0.01 [a]	0.74 ± 0.01 [b]	0.69 ± 0.02 [c]	0.47 ± 0.01 [d]	0.49 ± 0.01 [d]
AI**	0.69 ± 0.01 [a]	0.67 ± 0.01 [a]	0.59 ± 0.01 [b]	0.58 ± 0.02 [b]	0.39 ± 0.00 [c]	0.41 ± 0.00 [c]
TI	1.63 ± 0.01 [a]	1.60 ± 0.02 [a]	1.44 ± 0.02 [b]	1.33 ± 0.04 [b]	0.85 ± 0.01 [c]	0.89 ± 0.01 [c]
h/H	1.40 ± 0.01 [a]	1.43 ± 0.02 [a]	1.66 ± 0.02 [b]	1.65 ± 0.05 [b]	2.52 ± 0.03 [c]	2.39 ± 0.03 [d]
Fatty acids						
Caproic acid (C6:0)	0.028 ± 0.003	0.027 ± 0.002	0.018 ± 0.003	0.023 ± 0.002	ND	ND
Caprylic acid (C8:0)	0.180 ± 0.003 [a]	0.173 ± 0.002 [a]	0.112 ± 0.003 [b]	0.130 ± 0.003 [c]	ND	ND
Myristic acid (C14:0)	0.460 ± 0.003 [a]	0.400 ± 0.003 [b]	0.362 ± 0.003 [c]	0.480 ± 0.003 [d]	0.347 ± 0.003 [e]	0.350 ± 0.003 [ec]
Myristoleic acid (C14:1)	0.058 ± 0.003 [a]	0.058 ± 0.003 [a]	0.038 ± 0.003 [b]	0.088 ± 0.003 [c]	0.067 ± 0.002 [ad]	0.077 ± 0.003 [cd]
Pentadecylic acid (C15:0)	0.070 ± 0.003 [ac]	0.223 ± 0.135 [b]	0.080 ± 0.004 [ab]	0.073 ± 0.003 [a]	0.062 ± 0.003 [c]	0.060 ± 0.004 [c]
Palmitic acid (C16:0)	35.710 ± 0.257 [a]	35.408 ± 0.180 [a]	32.683 ± 0.218 [b]	32.882 ± 0.619 [b]	25.372 ± 0.232 [c]	26.142 ± 0.193 [c]
Palmitoleic acid (C16:1)	3.582 ± 0.124 [a]	3.558 ± 0.101 [a]	2.713 ± 0.152 [b]	4.367 ± 0.118 [c]	3.457 ± 0.095 [a]	4.393 ± 0.159 [c]
Margaric acid (C17:0)	0.250 ± 0.003 [a]	0.238 ± 0.003 [a]	0.280 ± 0.003 [b]	0.190 ± 0.003 [c]	0.173 ± 0.004 [d]	0.142 ± 0.003 [e]
Ginkgolic acid (C17:1)	0.040 ± 0.003 [a]	0.030 ± 0.003 [a]	0.040 ± 0.003 [a]	0.040 ± 0.003 [a]	0.090 ± 0.003 [b]	0.088 ± 0.003 [b]
Stearic acid (C18:0)	8.492 ± 0.153 [a]	8.368 ± 0.149 [a]	8.505 ± 0.111 [a]	6.393 ± 0.134 [b]	5.557 ± 0.176 [c]	5.440 ± 0.164 [c]
Elaidic acid (C18:1n9t)	0.110 ± 0.003 [a]	0.048 ± 0.003 [b]	0.048 ± 0.003 [b]	0.070 ± 0.003 [c]	ND	ND
Oleic (C18:1n9c)	43.258 ± 0.262 [ab]	41.613 ± 0.462 [ac]	43.087 ± 0.334 [abc]	43.965 ± 0.568 [b]	41.555 ± 0.275 [c]	43.155 ± 0.382 [abc]
Linoleic acid (C18:2n6c)	7.380 ± 0.139 [a]	9.403 ± 0.169 [b]	11.430 ± 0.155 [c]	10.573 ± 0.093 [d]	21.437 ± 0.125 [e]	18.598 ± 0.102 [f]

Table 5. *Cont.*

Hen Age: 50 wk	Dietary Treatments					
Item	CON	OP2	OP3	OP4	OP5	OP6
γ-Linolenic (C18:3n6)	ND	ND	ND	ND	0.120 ± 0.003^{b}	0.088 ± 0.003^{c}
α-Linolenic (C18:3n3)	ND	ND	ND	ND	0.872 ± 0.003^{b}	0.778 ± 0.003^{c}
Arachidic acid (C20:0)	0.040 ± 0.003^{a}	0.040 ± 0.003^{a}	0.038 ± 0.003^{a}	0.030 ± 0.003^{ab}	0.020 ± 0.003^{b}	0.022 ± 0.003^{b}
Gondoic acid (C20:1n9)	0.302 ± 0.003^{a}	0.348 ± 0.003^{b}	0.398 ± 0.003^{c}	0.440 ± 0.003^{d}	0.000 ± 0.000	0.000 ± 0.000
Eicosadienoic acid (C20:2n6)	0.040 ± 0.003^{a}	0.058 ± 0.003^{b}	0.092 ± 0.003^{c}	0.098 ± 0.003^{c}	0.198 ± 0.003^{d}	0.172 ± 0.003^{e}
Dihomo-γ-linolenic acid (C20:3n6)	ND	ND	ND	ND	0.097 ± 0.003^{a}	0.078 ± 0.003^{b}
Arachidonic acid (C20:4n6)	ND	ND	0.065 ± 0.013^{a}	0.078 ± 0.003^{a}	0.507 ± 0.003^{b}	0.367 ± 0.003^{c}
Docosadienoic acid (C22:2n6)	ND	ND	ND	ND	ND	ND
Docosahexaenoic acid (DHA) (C22:6n3)	ND	ND	ND	ND	0.072 ± 0.003^{a}	0.050 ± 0.003^{b}

$^{a-f}$ Means within a row with different superscripts differ significantly ($p < 0.05$), ND: Not Detected, AI: Atherogenic Index [27], AI**: Atherogenic Index [28], TI: Thrombogenic Index, h/H: hypocholesterolemic (h)/hypercholesterolemic (H) fatty acids.

The percentage of MUFA, PUFA, PUFA n6, PUFA n3 and SFA of eggs as well as the PUFA/SFA and PUFA n6/PUFA n3 ratios were significantly affected ($p < 0.05$) by dietary treatments (Table 5). In the OP4 and OP6 eggs, MUFA constituted about 47.74% of the total amount of FAs. This percentage was similar with that recorded for the CON and OP2 groups ($p > 0.05$) but it was significantly higher than that observed in eggs of the OP3 and OP5 groups ($p < 0.05$). The corresponding value was similar between eggs from the CON, OP2, OP3 and OP5 groups ($p > 0.05$). Hens that were fed diets with OP produced eggs with a significantly higher amount of PUFA and PUFA n6 compared to CON hens ($p < 0.05$). Among the groups, the highest percentage of PUFA and PUFA n6 was recorded in the OP5 eggs ($p < 0.05$). The OP3 and OP4 eggs presented similar levels of PUFA ($p > 0.05$) however, the PUFA n6 percentage was significantly different among all groups ($p < 0.05$). In the eggs derived from CON, OP2, OP3 and OP4 hens, PUFA n3 was not detected. On the other hand, hens fed diets with 5% OP produced eggs with significantly higher amount of PUFA n3 than that detected in the OP6 eggs ($p < 0.05$). Consequently, the PUFA n6/PUFA n3 ratio was calculated only for the OP5 and OP6 eggs and did not differ between these groups. As the incorporation rate of OP increased in the laying hens' diet, the percentage of SFA in produced eggs decreased (Table 5). The eggs derived from OP5 and OP6 hens displayed the lowest SFA level compared with that found in the eggs produced by hens of the other treatments ($p < 0.05$). A similar percentage of SFA was found between CON and OP2 eggs, between OP2 and OP3 eggs, as well as between OP3 and OP4 eggs ($p > 0.05$). The significant differences of MUFA, PUFA and SFA levels observed among eggs of all investigated groups resulted in relevant differences of the PUFA/SFA ratio. In particular, the OP5 and OP6 eggs presented the highest PUFA/SFA ratio whereas the CON eggs presented the lowest one ($p < 0.05$). The corresponding ratio was 0.26 for the OP3 and OP4 eggs which was significantly higher than that of the CON eggs ($p < 0.05$) but similar to that found in the OP2 eggs ($p > 0.05$).

The Atherogenic Index, calculated either with the inclusion of C18:0 fatty acid to the numerator of equation formula (IA), or without it (IA**), was found significantly reduced in the OP5 and OP6 eggs than that recorded in eggs of all the other treatments ($p < 0.05$). The corresponding value in the OP3 and OP4 eggs was higher than that in the CON and OP2 eggs ($p < 0.05$). The lowest values of TI were recorded in the OP5 and OP6 eggs, followed by that of the OP3 and OP4 eggs whereas the highest corresponding value was found in the CON and OP2 eggs ($p < 0.05$). Among experimental groups, the OP6 eggs presented the highest h/H ratio followed by that of the OP5 eggs ($p < 0.05$). At the OP3 and OP4 eggs the corresponding ratio was similar ($p > 0.05$) but differed significantly with the lowest one observed in CON and OP2 eggs ($p < 0.05$).

All of the individual egg's FAs, except caproic acid (C6:0), were affected by the addition of OP in laying hens' diet. The differences observed among dietary treatments were significant ($p < 0.05$) and are presented in detail in Table 5. Generally, the most abundant fatty acid among SFAs recorded in eggs of all investigated groups was the palmitic acid (C16:0). The lowest concentration of palmitic acid, approximately 25.37%, was found in the OP5 and OP6 eggs and the highest, about 35.7%, in the CON and OP2 eggs ($p < 0.05$). The intermediate level of the corresponding value observed in the OP3 and OP4 eggs (about 32.7%) differed significantly from that found in the eggs of the rest of the treatments ($p < 0.05$). According to the results, oleic acid was the most abundant of the MUFAs recorded in eggs of all the studied groups. The highest levels of oleic acid were recorded in the OP4 eggs and differed significantly from the lowest levels observed in the OP5 and OP2 groups ($p < 0.05$). Data analysis showed that, among PUFAs, the most abundant fatty acid in eggs of all treatments was linoleic acid. The hens that were fed diets with OP produced eggs with significantly higher levels of linoleic acid compared to the controls ($p < 0.05$). Moreover, as the incorporation rate of OP in the birds' feed was increasing, a significant elevation of linoleic acid concentration in produced eggs was recorded ($p < 0.05$).

3.4. Blood Biochemical Constituents

Table 6 shows the effect of experimental diets on blood profile of 50-week-old laying hens. The egg-layers that consumed diets containing OP at the rate of 3%, 5% and 6% had a significantly higher serum uric acid concentration compared to the CON hens ($p < 0.05$). Serum cholesterol, triglycerides and BUN levels, were not affected by the addition of olive pulp in hens' feed ($p > 0.05$). Data analysis revealed that no significant effect ($p > 0.05$) on liver enzymes' concentration levels was recorded among dietary groups.

Table 6. The effect of experimental diets on biochemical parameters of laying hens in the middle of production cycle. Data are presented as mean ± SE.

Items	Dietary Treatments					
Hen Age: 50 wk	CON	OP2	OP3	OP4	OP5	OP6
Cholesterol (mg/dL)	64.30 ± 6.15	73.90 ± 5.92	80.20 ± 5.60	98.20 ± 19.80	89.70 ± 6.82	92.80 ± 8.67
Triglycerides (mg/dL)	459.46 ± 120.02	663.92± 104.76	783.41 ± 113.05	881.30 ± 141.00	894.67 ± 137.12	951.78 ± 158.59
Uric acid (mg/dL)	2.19 ± 0.39 ^{a}	3.30 ± 0.27 ^{ab}	4.16 ± 0.41 ^{b}	2.88 ± 0.38 ^{ab}	4.25 ± 0.36 ^{b}	4.29 ± 0.28 ^{b}
BUN (mg/dL)	13.75 ± 4.52	17.44 ± 2.76	21.29 ± 4.00	26.58 ± 4.46	21.71 ± 4.39	25.19 ± 5.34
AST (IU/L)	203.40 ± 6.25	198.90 ± 4.68	187.20 ± 8.38	192.30 ± 6.11	190.80 ± 6.12	195.00 ± 2.82
G-GT (IU/L)	26.28 ± 2.04	27.28 ± 2.39	28.34 ± 1.69	30.39 ± 2.87	26.71 ± 2.65	27.88 ± 3.32
GLDH (U/L)	9.71 ± 2.04	11.15 ± 2.39	10.72 ± 1.69	7.83 ± 2.87	7.67 ± 2.65	8.25 ± 3.32

^{a,b} Means within a row with different superscripts differ significantly ($p < 0.05$). G-GT: Gamma-Glutamyl Transferase

3.5. Welfare Parameters

Data in Table 7 display the dietary influence of OP on KBD incidence as evaluated in middle-aged laying hens and the percentage of birds in each group with deformation, fractures or both. The lowest KBD incidence rate was recorded in the OP3 hens and was significantly different among all groups ($p < 0.05$). Birds of the OP2 group presented the highest KBD incidence rate and was significantly different ($p < 0.05$) than that found in the CON, OP3 and OP6 hens. The incidence of KBD in the OP4 and OP5 hens was significantly different only in comparison with that recorded in the OP3 hens ($p < 0.05$). From the birds with damaged keels, research findings revealed that 2.04% of hens in the CON and OP6 groups had both deformation and fractures, while in the OP3 group, 2.08% of the hens had only fractures and 10.42% had only deformation. Only deformed keels were recorded in the hens of the groups OP2, OP4 and OP5. The differences in keel bone deformation rate found among the dietary treatments followed the same pattern with those concerning the KBD incidence among experimental groups (Table 7).

Table 7. Percentage of hens in the middle of production cycle observed in the 6 dietary treatments (CON, OP2, OP3, OP4, OP5, OP6) for the absence or presence (SCORE 0, 1) of KBD and % of hens presenting deformation, fractures or both.

KBD	Dietary Treatments					
score	CON	OP2	OP3	OP4	OP5	OP6
0	53.06 ^{a}	29.17 ^{b}	87.50 ^{c}	44.90 ^{ab}	48.00 ^{ab}	59.18 ^{a}
1	46.94 ^{a}	70.83 ^{b}	12.50 ^{c}	55.10 ^{ab}	52.00 ^{ab}	40.82 ^{a}
Keel Bone Damage % of hens	**CON**	**OP2**	**OP3**	**OP4**	**OP5**	**OP6**
Deformation	44.90 ^{a}	70.83 ^{b}	10.42 ^{c}	55.10 ^{ab}	52.00 ^{ab}	38.78 ^{a}
Fractures	0.00	0.00	2.08	0.00	0.00	0.00
Deformation & Fractures	2.04	0.00	0.00	0.00	0.00	2.04

$^{a–c}$ Means within a raw for a particular score with different superscripts differ significantly ($p < 0.05$).

Feather damage evaluation revealed significant differences ($p < 0.05$) among experimental groups either on individual body areas or on total plumage score (Table 8). Overall, the higher percentage of hens with the best plumage condition (score 0) was recorded in the OP5 group while the lowest one was found in the OP2 group. The percentage of birds evaluated with score 0 differed significantly between the OP2 and OP5 groups ($p < 0.05$). The percentage of CON hens with a score 1 was significantly lower ($p < 0.05$) than that recorded for hens of the OP6 group. Finally, a significantly higher percentage of birds with score 2 for total feather damage was observed in the hens of the CON and OP2 groups in comparison to the OP4, OP5 and OP6 hens ($p < 0.05$). During the evaluation for total feather damage, none of the hens of the OP5 and OP6 groups was given a score 2.

Table 8. Percentage of hens in the middle of production cycle observed in the 6 dietary treatments (CON, OP2, OP3, OP4, OP5, OP6) scoring for 0, 1, 2 for feather damage in 3 body regions.

Head-Neck	**Dietary Treatments**					
score	**CON**	**OP2**	**OP3**	**OP4**	**OP5**	**OP6**
0	34.69 ^{a}	14.58 ^{b}	27.08 ^{ab}	30.61 ^{ab}	38.00 ^{a}	22.45 ^{ab}
1	51.02 ^{a}	58.34 ^{ab}	62.50 ^{ab}	69.39 ^{ab}	62.00 ^{ab}	77.55 ^{b}
2	14.29 ^{a}	27.08 ^{a}	10.42 ^{ab}	0.00 ^{b}	0.00 ^{b}	0.00 ^{b}
Back-Rump						
score	**CON**	**OP2**	**OP3**	**OP4**	**OP5**	**OP6**
0	93.88	89.58	100.00	100.00	100.00	100.00
1	6.12	10.42	0.00	0.00	0.00	0.00
Belly						
score	**CON**	**OP2**	**OP3**	**OP4**	**OP5**	**OP6**
0	77.55 ^{a}	87.50 ^{ab}	87.50 ^{ab}	97.96 ^{bc}	100.00 ^{c}	100.00 ^{c}
1	18.37 ^{a}	12.50 ^{a}	10.42 ^{ab}	0.00 ^{b}	0.00 ^{b}	0.00 ^{b}
2	4.08	0.00	2.08	2.04	0.00	0.00
Total plumage						
score	**CON**	**OP2**	**OP3**	**OP4**	**OP5**	**OP6**
0	30.61 ^{ab}	14.58 ^{b}	27.08 ^{ab}	30.61 ^{ab}	38.00 ^{a}	22.45 ^{ab}
1	51.02 ^{a}	58.34 ^{ab}	62.50 ^{ab}	67.35 ^{ab}	62.00 ^{ab}	77.55 ^{b}
2	18.37 ^{a}	27.08 ^{a}	10.42 ^{ab}	2.04 ^{b}	0.00 ^{b}	0.00 ^{b}

$^{a–c}$ Means within a raw for a particular score with different superscripts differ significantly ($p < 0.05$); 0—All body parts have no or slight wear, (nearly) complete feathering (only single feathers lacking); 1—One or more body parts have moderate wear, i.e., damaged feathers (worn, deformed) or one or more featherless areas < 5 cm in diameter at the largest extent; 2—One or more body parts have at least one featherless area $\geq$ 5 cm in diameter at the largest extent.

No significant differences in plumage conditions in the back/rump area were observed among groups ($p > 0.05$). A significantly higher percentage of hens with score 0 in the head/neck area was recorded in the CON and OP5 groups than the OP2 group ($p < 0.05$). Moreover, for the same body area, the highest percentage of hens with score 1 was found in the OP6 group. The differences observed among groups in hens scoring 1 for feather damage in the head/neck area were significant ($p < 0.05$) between the OP6 and CON groups. None of the hens consuming diets with OP at the dose of 4%, 5% and 6% were given a score 2 for feather damage in the head/neck area. On the contrary, similar percentages of the CON and OP2 hens ($p > 0.05$) were evaluated with a score 2 in this body area. The percentages of CON and OP2 hens evaluated with a score 2 for feather damage at the head/neck area were significantly different ($p < 0.05$) from those observed in the OP4, OP5 and OP6 hens (Table 8).

Laying hens that received OP at the rates of 4%, 5% and 6% presented the best plumage condition in the belly area (Table 8). For score 0 in feather damage evaluation, the differences observed among experimental groups were significant ($p < 0.05$) between the CON group with the OP4, OP5 and OP6 groups as well as between the OP2 and OP3 groups with the OP5 and OP6 groups. Significantly higher was the percentage of hens evaluated for belly feather damage with score 1 in the CON and OP2 groups than the 0% observed in the OP4, OP5 and OP6 groups ($p < 0.05$). The percentage of hens evaluated for belly feather damage with score 2 was similar in all groups ($p > 0.05$).

In the present study, no problems with foot pad dermatitis or toe damage were recorded in hens of all dietary groups. The birds of all treatments were evaluated with very good scores for comb abnormalities and skin lesions, indicating no evidence of such welfare issues (Table 9). The percentages of hens in each score category for cob abnormalities and skin lesions did not differ among the groups ($p > 0.05$). The addition of OP in laying hens' diet had a significant impact on birds' claws (Table 9). The highest percentages of hens with long nails were found in the OP4, OP5 and OP6 groups, followed by the OP3 and CON groups. All hens of the OP2 group had normal nails in respect of their length. The differences regarding nail length of hens were similar between the CON and OP2 groups ($p > 0.05$). However, significant differences ($p < 0.05$) in the length of hens' nails were observed between the CON and OP2 groups with the OP3 group, between the CON and OP2 groups with the OP4, OP5 and OP6 groups, as well as between the OP3 group with the OP4, OP5 and OP6 groups.

Table 9. Percentage of hens in the middle of production cycle observed in the 6 dietary treatments (CON, OP2, OP3, OP4, OP5, OP6) scoring for comb abnormalities and skin lesions (0, 1, 2) as well as claw length (Normal/Long).

Comb Abnormalities	Dietary Treatments					
score	CON	OP2	OP3	OP4	OP5	OP6
0	97.96	100.00	100.00	100.00	98.00	100.00
1	2.04	0.00	0.00	0.00	2.00	0.00
2	-	-	-	-	-	-
Skin lesions						
score	CON	OP2	OP3	OP4	OP5	OP6
0	100.00	100.00	100.00	97.96	100.00	100.00
1	0.00	0.00	0.00	2.04	0.00	0.00
2	-	-	-	-	-	-
Claw length						
score	CON	OP2	OP3	OP4	OP5	OP6
Normal	93.88 ^{a}	100.00 ^{a}	77.08 ^{b}	22.45 ^{c}	42.00 ^{c}	30.61 ^{c}
Long	6.12 ^{a}	0.00 ^{a}	22.92 ^{b}	77.55 ^{c}	58.00 ^{c}	69.39 ^{c}

$^{a–c}$ Means within a raw with different superscripts differ significantly ($p < 0.05$).

4. Discussion

A large proportion of consumers in today's health conscious society is seeking for properly balanced diets in order to prevent and minimize adverse health problems [33,34]. At the same time, in the last decade, there has been an increased consumer demand for meat and egg products that focuses on animal welfare during production as well as on product safety and quality [35,36]. As a result of this direction, adequate supplementation of poultry diets with novel and beneficial feed additives or supplements is gaining importance as it significantly improves overall poultry production and performance as well as safeguarding the health of birds [37,38]. Previous studies indicate that manipulating the laying hens' diet, by adding different by-products rich in fatty acids and antioxidants, can alter the FA profile of eggs [39–43]. A review of literature highlights that OP is not just an oil by-product, but a source of functional ingredients exploitable to obtain high value-added foods, thus increasing their shelf-life and/or formulating more nutritious and healthy products [44]. The reuse of waste to recover functional compounds is in line with consumer and society's requirements for high quality, safe, processed foods and the reduction of waste to exert a lower environmental impact [45].

The present study revealed that the addition of dried OP in laying hens' diet increased the percentage of PUFA in produced eggs and decreased that of SFA in a manner proportional to the inclusion rate in hens' diet. On the other hand, the concentration of MUFA remained relatively stable in the eggs produced by OP-fed layers compared to the control.

The reduced SFAs concentration recorded in the eggs produced by OP-fed layers could be mainly attributed to a proportional reduction of palmitic acid, which was found to be the most abundant among the SFAs recorded in the eggs of all dietary groups and, to a lesser extent, to a similar reduction of the stearic and myristic acid concentrations. The increased concentration in PUFAs observed in the eggs laid by hens receiving OP is attributed to a similar proportional elevation of linoleic acid concentration, the most abundant fatty acid among PUFAs found in the eggs of all experimental groups. The FA composition analysis of OP used in this study revealed the presence of linoleic acid in a percentage of 10.77%. It could be supported that the increase in the OP incorporation rate in the hens' diet resulted in a concomitant increase of linoleic acid concentration in produced eggs. Since there is lack of research evidence regarding the effect of OP in the lipid profile of eggs when supplementing laying hens' feed, the comparison of our results with those in literature is not possible. However, similar studies carried out with other farm animals have shown that the incorporation of OP in their diet ameliorates the FA profile by decreasing SFAs and increasing unsaturated fatty acids in the final products such as broiler meat [46,47], pork meat [48], rabbit meat [49], lamb meat [50] and small ruminants' milk [51,52] and cheese [52]. The decreased concentration of SFAs in OP eggs observed in this study could be attributed to the high concentration of phenolic compounds of the OP used in the experiment that act as antioxidants and suppress egg lipid oxidation. The antioxidant effect of polyphenols on the final product when supplemented in animals' diet has been previously documented [53]. In this investigation, the increase of OP incorporation rate in the hens' diet did not result in a similar elevation of egg total phenol content. This is not surprising however, since it has been recently shown that the content of polyphenols in body tissues is not directly related to their dietary levels [53].

In line with our findings, Laudadio et al. [54] noticed that feeding high-polyphenols extra-virgin olive oil to laying hens raised the PUFAs and linoleic acid composition, reduced SFAs in egg yolks and improved the PUFA/SFA ratio. Ratios of PUFA/SFA are commonly used to assess the nutritional value of fat. Dietary ratios of PUFA to SFA above 0.45 are considered safe for human consumption [55] and appropriate in order to protect against the development of ischemic heart disease [56]. The current study revealed that feeding laying hens with dried OP increased the PUFA/SFA ratio in a manner proportional to the inclusion rate in the hens' diet. Furthermore, the optimal ratios from a nutritive point of view were achieved with the higher doses of olive pulp (5% and 6%). The differences of MUFA, PUFA and SFA levels observed among the eggs of all the investigated groups resulted in relevant differences of the PUFA/SFA ratio. Additionally, our results indicated that feeding laying hens with the higher dose of OP (6%) significantly reduced the total fat content of eggs. Similar reduction of egg fat composition has been documented by Abd El-Moneim and E.M. Sabic [57], after the inclusion of 5% and 10% of OP in the diet of Japanese laying quails in a twelve-week feeding trial.

Even though PUFA/SFA is the most commonly used index for evaluating the nutritional value of dietary foods, it is considered too general and unsuitable for assessing the atherogenicity of foods [58] since specific SFA and PUFA have different metabolic effects [56]. Fatty acids can either promote or prevent atherosclerosis and coronary thrombosis, based on their effects on serum cholesterol and low density lipoprotein cholesterol concentrations [54]. For this reason, the index of atherogenicity (AI) was developed [28] and characterizes the atherogenic potential of FA. From the main classes of the SFAs, the C14:0 and C16:0 fatty acids are known to be among the most atherogenic, while C18:0 is thought to be neutral with respect to atherogenicity, but is instead considered to be thrombogenic [54,58]. Unsaturated Fatty Acids (UFAs) are considered to be anti-atherogenic as they inhibit the accumulation of plaque and reduce the levels of phospholipids, cholesterol, and esterified fatty acids [27,59]. Therefore, the consumption of foods or products with a lower AI can reduce the levels of total cholesterol and low-density lipoprotein cholesterol (LDL-C) in human blood plasma [60]. Another important index commonly used in many FA composition studies in order to assess the degree of thrombogenicity is TI [58].

This index characterizes the thrombogenic potential of FAs, indicating the tendency to form clots in blood vessels and provides the contribution of different FAs, which denotes the relationship between the pro-thrombogenic FAs (C12:0, C14:0, and C16:0) and the anti-thrombogenic FAs (MUFAs and the n-3 and n-6 families) [28]. It has been reported that animal products with a low index of thrombogenicity decrease the threat of atrial fibrillation [61]. In brief, both the AI and the TI can be used to assess the potential effects of FA composition on Cardiovascular Health (CVH). A FA composition with a lower AI and TI has a better nutritional quality, and its consumption may reduce the risk of coronary heart disease (CHD), but no organization has yet provided the recommended values for the AI and TI [58].

The results of this study indicated that the addition of OP in laying hens' diet decreased both AI and TI in a proportional manner. At the same time incorporation rates of OP above 3% reduced significantly the h/H ratio of produced eggs, especially at the higher doses (5% and 6%). According to Santos-Silva et al. [62], the higher the ratio between the hypocholesterolemic and the hypercholesterolemic fatty acids, the more the oil or fat is appropriate to human nutrition. These observations demonstrate the health beneficial potential associated with the fat intake from eggs produced from OP-fed layers since they presented higher nutrition quality than the control conventional eggs. It is difficult to compare the values of the AI, TI, and h/H indexes obtained in this research for the eggs laid by hens that were fed OP-supplemented diets to those of other studies because, in the literature, there is a lack of work related to the assessment of health lipid indexes of hens' eggs following OP dietary incorporation. However, similar to our results, Laudadio et al. [54] denoted a significant decrease of AI in the yolk of eggs produced by laying hens fed with extra-virgin olive oil, rich in polyphenols. Previous feeding trials performed in other farm animals have shown that the addition of OP in their diet improves the health lipid indices in final products such ewe milk [63] and rabbit meat [49] by reducing both AI and TI indexes.

Regarding laying hens' performance, our findings indicated that including OP up to 6% in laying hens' diet during a production cycle does not compromise their final body weight and does not affect feed consumption, FCR or the percentage of eggs with dirty eggshells. Results concerning body weight are in line with those obtained by other investigators who evaluated the in feed inclusion of OP in hens' diet, at rates ranging from 1% [23] up to 20% [1,21,22] for a period of 6, 12 and 16 weeks, respectively. In respect of feed consumption, our results confirm those observed in previous feeding trials [2,18–20,23]. However, increased feed intake has been previously demonstrated in laying hens that were fed diets containing 4–20% OP compared to birds fed OP-free diets [1,17,21,22]. In terms of FCR, our results are in agreement with those recorded by other authors [2,18,21]. According to Ghasemi et al. [20], the addition of 20% OP in a corn-based diet of laying hens had no impact on FCR which is consistent with our findings. However, when the same amount of OP was incorporated in a wheat-based diet the FCR was increased. Deterioration of FCR was also observed in layers that were fed diets with 10–20% OP compared to controls [1,17,19,22]. The non-significant differences to the produced number of eggs with dirty shells observed among the dietary groups could be an indirect indicator that OP does not cause diarrhea when added to laying hen's diet at the inclusion levels studied.

The present study demonstrated that hens that were fed diets with OP produced lighter eggs compared to controls. Despite this reduction however, OP eggs did not weight less than 63 g. According to Commission Regulation (EC) No 589/2008 [64], eggs should be graded by weight as follows: XL-very large: 73 g and more; L-large: from 63 g up to 73 g; M-medium: from 53 g up to 63 g; S-small: under 53 g. Taking into consideration the European guidelines, hens of all the experimental groups produced L-large eggs as graded by their weight. A numerical decrease in egg weight was also reported by Rezar et al. [23] in hens that were fed a diet with 1% OP compared to those that received a control diet (62.6 g and 64.3 g, respectively). Contrary to our results, other authors found that the addition of OP at the rate of 4.5% [2], 9% [2,18], 12% and 16% [1] in the layers' diet significantly increased the egg weight of produced eggs. On the other hand, no differences in egg weight were recorded by other authors when OP was included in laying hens' diet at levels from 10% up to 20% [17,20–22].

In the current study, a drop of HEDP% of about 3% was recorded in hens of the OP3 and OP4 groups compared to controls. This is considered as a random finding, since the laying hens that were fed the higher dosage rate of OP presented similar HDEP% with the CON hens. A decrease of egg production (%) has also been observed in previous feeding trials in which OP was incorporated in laying hens ratio at the levels of 9% [18], 12% and 16%, respectively [1]. However, other investigators observed no differences on egg production performance of laying hens that were fed diets with OP compared to the control hens [2,17,19–23]. This trial also revealed that hens of the OP4 and OP5 groups laid eggs with a lower egg mass than that recorded in eggs of the CON group. The differences in egg mass observed between the CON and the OP4 and OP5 groups is attributed to the highest HDEP% of CON hens as well as to the heaviest eggs produced by them, compared to the corresponding values recorded for the other two groups. According to some researchers, supplementing laying hens' diet with OP did not affect the egg mass of produced eggs [2,17–22]. However, Abd-El Galil et al. [1] observed a significant decrease of egg mass by 5.8% in hens that were fed diets with 16% olive cake compared to those that were fed a control diet.

A 15–58% lower percentage of broken eggshells was recorded to the OP-fed groups compared to control hens. This finding is very important from a financial point of view because cracked and broken eggshells are regarded as major source of economic loss for egg producers [65]. These results indicate that OP might increase the eggshell strength. Shell strength is influenced by shell thickness and shell matrix organization [66]. Even though eggshell quality data were similar between the OP and CON eggs, a numerical increase of shell thickness and percentage (%) was observed in the OP groups compared to controls. To our knowledge, there are no research findings regarding the dietary effect of OP on eggshell-breaking strength. However, Zhang and Kim [67] showed that supplementing laying hens' diet with 2% and 5% olive oil increased eggshell breaking strength and the shell thickness of produced eggs in comparison to controls. These researchers supported that olive oil, which is considered to be a good solvent of vitamin D, could improve calcium concentrations in eggshells. Vitamin D is known to be a fat-soluble vitamin and has a direct effect on calcium absorption. On the other hand, it has been previously documented that the addition of organic or inorganic sources of combined Zn, Mn and Cu to hens' feed does not significantly influence the amount of eggshell material deposited during eggshell formation but can enhance some mechanical properties like, improved breaking strength and fracture toughness (resistance to fracture) regardless of the source of the trace elements [68]. Olive pulp is regarded as a good source of minerals like Ca, P, Zn Mn and Cu [69]. Thus, it could be supported that the mineral content of OP fed to laying hens could have a positive impact on the eggshell strength of laid eggs, but further research is needed in order to verify this mechanism of action.

Both internal and external egg quality traits were not adversely affected by the incorporation of OP in laying hens' diet. A minor but significant increase of albumen pH was recorded in eggs produced by hens that were fed with 5% and 6% OP compared to CON eggs. Currently, there are no data regarding the dietary effect of OP in albumen pH so the elevation of the corresponding value observed here cannot be explained. The increase in albumen pH has been associated to a loss of CO_2 via eggshell pores [70]. However, all eggs used in our trial for the evaluation of their quality were one day old and they were collected and analyzed at the same day. In general, eggs recently laid have an initial albumen pH of 7.4 to 8.6 [71]. In this study, albumen pH documented in eggs from all dietary treatments was found slightly higher than the reported ranges. From the rest of the quality parameters evaluated in present trial, an increase in yolk diameter was observed in OP4 and OP5 eggs compared to CON eggs which could be due to the numerically higher yolk ratio (%) documented in OP4 and OP5 eggs. Yolk diameter is an important measurement taken for the calculation of yolk index which provides indication on the freshness of the egg. Despite the recorded increase in yolk diameter observed in OP4 and OP5 eggs, the yolk index of eggs produced by OP-fed layers was similar to that recorded in CON eggs. These findings are in agreement with those obtained by other investigators in relevant feeding trials [17,19,20,22]. However, higher inclusion rates of OP in hens' diet than those used in this study, such as 9% [18] and 10% [21] resulted in an increased yolk index of produced eggs. This improvement of yolk index was attributed to the beneficial effect of the unsaturated fatty acids and polyphenols in the OP used [21].

The results concerning the assessment of dietary effect of OP in egg quality traits of laying hens in available literature are inconsistent. According to some researchers, egg shape index [18–21], Haugh unit (HU) [18,20–22], yolk color [2,20–22], shell thickness [2,18,21], shell weight and ratio % [18,21,22], yolk weight and ratio % [21,22] and albumen weight and ratio % [21,22] were not affected by the incorporation of OP in birds' diet, which is consistent with our findings. On the other hand, Zangeneh and Torki [2] observed that the supplementation of 4.5% and 9% of OP in hens' diet increased the shell weight of produced eggs, whereas hens receiving 9% OP laid eggs with a decreased HU compared to the controls and to those fed diets containing 4.5% OP. Increased shell weight and shell ratio (%) of produced eggs was also reported in the study of Al-Harthi and Attia [22] after the supplementation of hens' diet with 10% OP. Contrary to our results, decreased shell thickness by 7.97%, 5.83% and 10.50% has been documented in eggs produced by laying hens that were fed with higher levels of olive cake, 8%, 12% and 16%, respectively [1]. Moreover, previous feeding trials have shown that when OP is incorporated into layers' diet at even higher inclusion rates (16–20%) the produced eggs presented a lower shape index [17,22], decreased HU, paler yolk color as well as reduced shell weight and shell thickness compared to controls [17,19].

The current study revealed that the addition of OP in hens' diet has no adverse effect on the birds' liver and kidney function, as indicated from the evaluation of the biochemical parameters selected. The higher serum uric acid concentration recorded to the layers that consumed diets containing OP at the rate of 3%, 5% and 6% compared to the CON hens is of no clinical significance since it remained within the reported normal ranges of 2–7 mg/dL for this parameter [72]. In accordance with our results, other authors also did not notice any dietary impact of OP on hens' serum concentration of cholesterol [18], triglycerides [19] or both [2,20]. However, other researchers observed a reduction in the serum concentration of cholesterol [19], triglycerides [18] or both [1]. Consistent with our findings, Al-Harthi et al. [21] observed no negative effect in hens' liver function after the supplementation of their diet with 10% and 20% of olive cake. However, AST serum concentration has been shown to increase when OP was incorporated in hens that were fed diets at higher rates (16–17%) compared to controls and to those receiving lower levels of OP [1,17].

The variability of our results regarding the dietary effect of OP in laying hens' performance, egg quality characteristics and health parameters evaluated with those previously recorded in similar feeding trials, could be possibly attributed to differences in the composition of OP and diets used, the inclusion rate and the hens' age and hybrid.

The present investigation revealed a positive dietary effect of OP on KBD at an inclusion level of 3%, as hens of the OP3 group presented the lowest KBD incidence rate among all experimental groups. This finding is highly valued since KBD is a well-recognized health and welfare issue of the modern poultry sector, with a high prevalence in commercial laying hens globally [73,74], that has been shown to cause stress in birds, reduce their productivity and compromise egg quality, posing financial concerns for producers [75–77]. The bone demineralization process seems to play a key role to the pathogenesis of keel fractures [78]. Recent research evidence indicated that keel bone fractures are associated with differences in the concentrations and activities of bone metabolism-related indexes, as well as bone mineral density in laying hens [79] and thus abnormal bone metabolism could be a causative factor of KBD [80]. The exact mechanism implicated in the recorded decrement of KBD incidence in OP3 hens is currently unknown and needs further investigation since there are not similar studies in the available literature. Possibly, the PUFA and phenolic compounds of OP used in the present trial could play a role. Evidence presented over the past 20 years has shown that long chain polyunsaturated fatty acids (LCPUFAs) are beneficial to bone health [81]. Similar action has been attributed to dietary polyphenols [82]. In particular, the phenolic compounds in olive oil have been shown to possess antioxidant properties in vivo and in vitro and influence bone mineral density by acting as free radicals, preventing oxidation-induced damage to bone cells [83,84]. Moreover, a recent work carried out in ovariectomized rats fed with 100 μL and 200 μL/day of olive oil for a period of 3 months revealed an improvement of their bones' biomechanical parameters [85].

Regarding the rest of the welfare parameters evaluated, it was shown that OP improved belly plumage condition of hens fed with 4–6% incorporation rates compared to controls. The feather loss in this body area has been linked to abrasion in housing equipment like perches [86]. Moreover, those hens presented the lowest incidence of score 2 compared to controls, as indicated from the total plumage evaluation. No welfare issues in respect of comb abnormalities, skin lesions, foot pad dermatitis and toe damage were recorded in the present trial. Finally, it was shown that feeding laying hens with OP at a percentage of 3% or more increases the length of their claws. The results regarding belly feather condition and claw length observed in this study could be attributed to the skin health beneficial effect of the bioactive compounds of OP, such as PUFA and polyphenols. Oils rich in essential fatty acids has been shown to improve skin hydration, have a regenerative effect on the damaged epidermal lipid barrier and regulate skin metabolism [87]. Both omega -3 and omega -6 fatty acids are important cell membrane components, essential for the function of epidermal barrier; they exhibit anti-inflammatory and anti-allergic effects, enhance repair processes and soothe irritation [87]. Plant polyphenols are considered as important substances for skin function, with hydrating, smoothing and softening effects [88–90]. Additionally, they soothe irritation and reduce the redness of skin, accelerating the natural regeneration of the epidermis, stabilizing the capillaries, improving microcirculation and elasticity in the skin and protecting against harmful external factors [87]. However, since no data are currently available regarding the dietary effect of OP in feather condition and claw length of laying hens, the results observed here need further investigation.

5. Conclusions

The current study revealed that feeding laying hens a diet with dried OP increased the percentage of PUFA in eggs, decreased that of SFA and improved the PUFA to SFA ratio in a manner proportional to the inclusion rate in the hens' diet. Moreover, an amelioration of health lipid indices was recorded in a dose-depended way, as indicated by the decrease of AI and TI and the increase in the h/H ratio of produced eggs. These observations demonstrate the potential health benefits associated with the fat intake from eggs produced from OP-fed layers, especially those fed with higher inclusion rates, since they presented higher nutrition quality than the control conventional eggs. Olive pulp-fed layers presented 15–58% lower percentage of broken eggshells compared to controls indicating a potential cost benefit from the use of OP. The addition of OP in laying hens' diet for a long period of time did not adversely affect the birds' performance, internal and external egg quality traits or health and welfare parameters evaluated. A positive impact on KBD incidence and belly plumage damage was shown to be possibly due to the bone and skin health beneficial effects of the OP bioactive compounds like PUFA and polyphenols, however further investigation is necessary to verify the exact mechanism implicated in those results. Finally, from the evaluated dietary levels of OP, 5% and 6% seems to be more advantageous for the consumer in terms of egg nutrition quality so they are highly recommended.

Supplementary Materials: The following supporting information can be downloaded at: https://www.mdpi.com/article/10.3390/su14063157/s1, Table S1: Nutritional analysis of Layer concentrate 25%.

Author Contributions: Conceptualization, A.D. and E.S.; methodology, A.D., M.-Z.K., M.Đ.S. and T.S.; software, M.-Z.K. and A.S.; validation, M.Đ.S., T.S., A.S. and E.S.; formal analysis, A.D.; investigation, M.-Z.K. and A.S.; resources, E.S.; data curation, A.D. and E.S.; writing—original draft preparation, A.D.; writing—review and editing, E.S., M.Đ.S. and T.S.; visualization, T.S. and A.S.; supervision, E.S.; project administration, A.D. and E.S.; funding acquisition, A.D. and E.S. All authors have read and agreed to the published version of the manuscript.

Funding: This research was funded by the Veterinary Research Institute, Hellenic Agricultural Organization DIMITRA, grant number 22.1681.255.

Institutional Review Board Statement: The study was conducted in accordance with the Declaration of Helsinki and approved by the Committee for Research Ethics of Hellenic Agricultural Organization-DIMITRA (52216/23 October 2020).

Data Availability Statement: The data presented in this study are available on request from the corresponding author. The data are not publicly available due to privacy.

Acknowledgments: We sincerely thank the TSAKIRIS FAMILY S.A. who participated in this research by offering the experimental animals and facilities. We are also thankful to SPARTA LIFE S.A. for providing the dried olive pulp used in this trial.

Conflicts of Interest: The authors declare no conflict of interest.

References

1. Abd El-Galil, K.; Morsy, A.S.; Emam, K.R.S.; Hassan, A.M. Physiological Responses and Productive Performance of Laying Hens Fed Olive Cake under South Sinai Conditions. *Egypt. Poult. Sci. J.* **2017**, *37*, 293–304.
2. Zangeneh, S.; Torki, M. Effects of B-Mannanase Supplementing of Olive Pulp-Included Diet on Performance of Laying Hens, Egg Quality Characteristics, Humoral and Cellular Immune Response and Blood Parameters. *Glob. Vet.* **2011**, *7*, 391–398.
3. Kostas, E.T.; Durán-Jiménez, G.; Shepherd, B.J.; Meredith, W.; Stevens, L.A.; Williams, O.S.A.; Lye, G.J.; Robinson, J.P. Microwave Pyrolysis of Olive Pomace for Bio-Oil and Bio-Char Production. *Chem. Eng. J.* **2020**, *387*, 123404. [CrossRef]
4. Salomone, R.; Ioppolo, G. Environmental Impacts of Olive Oil Production: A Life Cycle Assessment Case Study in the Province of Messina (Sicily). *J. Clean. Prod.* **2012**, *28*, 88–100. [CrossRef]
5. European Commission. The European Green Deal. In *Communication from The Commission to The European Parliament, The European Council, The Council, The European Economic and Social Committee and The Committee of The Regions*; European Commission: Brussels, Belgium, 2019. Available online: https://eur-lex.europa.eu/legal-content/EN/TXT/HTML/?uri=CELEX:52019DC0640&from=EN (accessed on 10 January 2022).
6. Ibrahim, N.S.; Sabic, E.M.; Abu-Taleb, A.M. Effect of Inclusion Irradiated Olive Pulp in Laying Quail Diets on Biological Performance. *J. Radiat. Res. Appl. Sci.* **2018**, *11*, 340–346. [CrossRef]
7. Sayehban, P.; Seidavi, A.; Dadashbeiki, M.; Ghorbani, A.; Araújo, W.A.G.; Albino, L.F.T. Effects of Different Levels of Two Types of Olive Pulp with or without Exogenous Enzyme Supplementation on Broiler Performance and Economic Parameters. *Rev. Bras. Ciênc. Avícola* **2016**, *18*, 489–500. [CrossRef]
8. Abdel-Moneim, A.E.; Shehata, A.M.; Alzahrani, S.O.; Shafi, M.E.; Mesalam, N.M.; Taha, A.E.; Swelum, A.A.; Arif, M.; Fayyaz, M.; Abd El-Hack, M.E. The Role of Polyphenols in Poultry Nutrition. *J. Anim. Physiol. Anim. Nutr.* **2020**, *104*, 1851–1866. [CrossRef]
9. Visioli, F.; Poli, A.; Gall, C. Antioxidant and Other Biological Activities of Phenols from Olives and Olive Oil. *Med. Res. Rev.* **2002**, *22*, 65–75. [CrossRef]
10. Omar, S.H. Oleuropein in Olive and Its Pharmacological Effects. *Sci. Pharm.* **2010**, *78*, 133–154. [CrossRef]
11. Barbaro, B.; Toietta, G.; Maggio, R.; Arciello, M.; Tarocchi, M.; Galli, A.; Balsano, C. Effects of the Olive-Derived Polyphenol Oleuropein on Human Health. *Int. J. Mol. Sci.* **2014**, *15*, 18508–18524. [CrossRef]
12. Bulotta, S.; Celano, M.; Lepore, S.M.; Montalcini, T.; Pujia, A.; Russo, D. Beneficial Effects of the Olive Oil Phenolic Components Oleuropein and Hydroxytyrosol: Focus on Protection against Cardiovascular and Metabolic Diseases. *J. Transl. Med.* **2014**, *12*, 219. [CrossRef] [PubMed]
13. Casamenti, F.; Grossi, C.; Rigacci, S.; Pantano, D.; Luccarini, I.; Stefani, M. Oleuropein Aglycone: A Possible Drug against Degenerative Conditions. In Vivo Evidence of Its Effectiveness against Alzheimer's Disease. *J. Alzheimer's Dis.* **2015**, *45*, 679–688. [CrossRef] [PubMed]
14. Duborija-Kovacevic, N.; Shavrina, K. New Findings on the Pharmacodynamic Actions of Olive Oil: Our Contribution to Better Evidence about Its Remedial Properties. *Prog. Nutr.* **2018**, *20*, 30–38. [CrossRef]
15. Ranalli, A.; Pollastri, L.; Contento, S.; Di Loreto, G.; Iannucci, E.; Lucera, L.; Russi, F. Acylglycerol and Fatty Acid Components of Pulp, Seed, and Whole Olive Fruit Oils. Their Use to Characterize Fruit Variety by Chemometrics. *J. Agric. Food Chem.* **2002**, *50*, 3775–3779. [CrossRef]
16. Tufarelli, V.; Introna, M.; Cazzato, E.; Mazzei, D.; Laudadio, V. Suitability of Partly Destoned Exhausted Olive Cake as By-Product Feed Ingredient for Lamb Production. *J. Anim. Sci.* **2013**, *91*, 872–877. [CrossRef]

17. Mohebbifar, A.; Afsari, M.; Torki, M. Egg Quality Characteristics and Productive Performance of Laying Hens Fed Olive Pulp Included Diets Supplemented with Enzyme. *Glob. Vet.* **2011**, *6*, 409–416.
18. Zarei, M.; Ehsani, M.; Torki, M. Productive Performance of Laying Hens Fed Wheat-Based Diets Included Olive Pulp with or without a Commercial Enzyme Product. *Afr. J. Biotechnol.* **2011**, *10*, 4303–4312. [CrossRef]
19. Afsari, M.; Mohebbifar, A.; Torki, M. Effects of Dietary Inclusion of Olive Pulp Supplemented with Probiotics on Productive Performance, Egg Quality and Blood Parameters of Laying Hens. *Annu. Res. Rev. Biol.* **2014**, *4*, 198–211. [CrossRef]
20. Ghasemi, R.; Torki, M.; Ghasemi, H.A.; Zarei, M. Single or Combined Effects of Date Pits and Olive Pulps on Productive Traits, Egg Quality, Serum Lipids and Leucocytes Profiles of Laying Hens. *J. Appl. Anim. Res.* **2014**, *42*, 103–109. [CrossRef]
21. Al-Harthi, M.A. The Effect of Different Dietary Contents of Olive Cake with or without Saccharomyces Cerevisiae on Egg Production and Quality, Inner Organs and Blood Constituents of Commercial Layers. *Eur. Poult. Sci.* **2015**, *79*, 83. [CrossRef]
22. Al-Harthi, M.A.; Attia, Y.A. Effect of Citric Acid on the Utilization of Olive Cake Diets for Laying Hens. *Ital. J. Anim. Sci.* **2015**, *14*, 394–402. [CrossRef]
23. Rezar, V.; Levart, A.; Salobir, J. The Effect of Olive by Products and Their Extracts on Antioxidative Status of Laying Hens and Oxidative Stability of Eggs Enriched with N-3 Fatty Acids. *Poljoprivreda* **2015**, *21*, 216–219. [CrossRef]
24. EUR-Lex. *Council Directive 1999/74/EC of 19 July 1999 Laying down Minimum Standards for the Protection of Laying Hens*; European Union: Maastricht, The Netherlands, 1999; Volume OJ L203, pp. 53–57. Available online: http://data.europa.eu/eli/dir/1999/74/oj/eng (accessed on 12 January 2022).
25. EUR-Lex. *Commission Regulation (EEC) No 2568/91 of 11 July 1991 on the Characteristics of Olive Oil and Olive-Residue Oil and on the Relevant Methods of Analysis*; European Union: Maastricht, The Netherlands, 1991; Volume OJ L248, pp. 1–83. Available online: http://data.europa.eu/eli/reg/1991/2568/oj/eng (accessed on 20 January 2022).
26. International Organization for Standardization. *Animal and Vegetable Fats and Oils—Gas Chromatography of Fatty Acid Methyl Esters—Part 4: Determination by Capillary Gas Chromatography*; International Organization for Standardization: London, UK, 2015.
27. Omri, B.; Chalghoumi, R.; Izzo, L.; Ritieni, A.; Lucarini, M.; Durazzo, A.; Abdouli, H.; Santini, A. Effect of Dietary Incorporation of Linseed Alone or Together with Tomato-Red Pepper Mix on Laying Hens' Egg Yolk Fatty Acids Profile and Health Lipid Indexes. *Nutrients* **2019**, *11*, 813. [CrossRef] [PubMed]
28. Ulbricht, T.L.V.; Southgate, D.A.T. Coronary Heart Disease: Seven Dietary Factors. *Lancet* **1991**, *338*, 985–992. [CrossRef]
29. Ko, E.-Y.; Saini, R.K.; Keum, Y.-S.; An, B.-K. Age of Laying Hens Significantly Influences the Content of Nutritionally Vital Lipophilic Compounds in Eggs. *Foods* **2021**, *10*, 22. [CrossRef] [PubMed]
30. Wilkins, L.J.; Brown, S.N.; Zimmerman, P.H.; Leeb, C.; Nicol, C.J. Investigation of Palpation as a Method for Determining the Prevalence of Keel and Furculum Damage in Laying Hens. *Vet. Rec.* **2004**, *155*, 547–549. [CrossRef]
31. Welfare Quality Network. *Welfare Quality Assessment Protocol for Laying Hens*, 2nd ed.; Welfare Quality Network: Helsinki, Finland, 2019.
32. *JASP*; Version 0.14, Computer Software; JASP Team: Amsterdam, The Netherlands, 2020.
33. Cherian, G. Nutrition and Metabolism in Poultry: Role of Lipids in Early Diet. *J. Anim. Sci. Biotechnol.* **2015**, *6*, 28. [CrossRef]
34. Lee, S.A.; Whenham, N.; Bedford, M.R. Review on Docosahexaenoic Acid in Poultry and Swine Nutrition: Consequence of Enriched Animal Products on Performance and Health Characteristics. *Anim. Nutr.* **2019**, *5*, 11–21. [CrossRef]
35. Sossidou, E.N.; Dal Bosco, A.; Elson, H.A.; Fontes, C.M.G.A. Pasture-Based Systems for Poultry Production: Implications and Perspectives. *Worlds Poult. Sci. J.* **2011**, *67*, 47–58. [CrossRef]
36. Hammershøj, M.; Johansen, N.F. Review: The Effect of Grass and Herbs in Organic Egg Production on Egg Fatty Acid Composition, Egg Yolk Colour and Sensory Properties. *Livest. Sci.* **2016**, *194*, 37–43. [CrossRef]
37. Dhama, K.; Latheef, S.K.; Mani, S.; Samad, H.A.; Karthik, K.; Tiwari, R.; Khan, R.U.; Alagawany, M.; Farag, M.R.; Alam, G.M.; et al. Multiple Beneficial Applications and Modes of Action of Herbs in Poultry Health and Production-A Review. *Int. J. Pharmacol.* **2015**, *11*, 152–176. [CrossRef]
38. Laudadio, V.; Lorusso, V.; Lastella, N.M.B.; Dhama, K.; Karthik, K.; Tiwari, R.; Alam, G.M.; Tufarelli, V. Enhancement of Nutraceutical Value of Table Eggs Through Poultry Feeding Strategies. *Int. J. Pharmacol.* **2015**, *11*, 201–212. [CrossRef]
39. Omidi, M.; Rahimi, S.; Torshizi, M.A.K. Modification of Egg Yolk Fatty Acids Profile by Using Different Oil Sources. *Vet. Res. Forum* **2015**, *6*, 137–141. [PubMed]
40. Świątkiewicz, S.; Arczewska-Włosek, A.; Szczurek, W.; Calik, J.; Bederska-Łojewska, D.; Orczewska-Dudek, S.; Muszyński, S.; Tomaszewska, E.; Józefiak, D. Algal Oil as Source of Polyunsaturated Fatty Acids in Laying Hens Nutrition: Effect on Egg Performance, Egg Quality Indices and Fatty Acid Composition of Egg Yolk Lipids. *Ann. Anim. Sci.* **2020**, *20*, 961–973. [CrossRef]
41. Orczewska-Dudek, S.; Pietras, M.; Puchała, M.; Nowak, J. Camelina Sativa Oil and Camelina Cake as Sources of Polyunsaturated Fatty Acids in the Diets of Laying Hens: Effect on Hen Performance, Fatty Acid Profile of Yolk Lipids, and Egg Sensory Quality. *Ann. Anim. Sci.* **2020**, *20*, 1365–1377. [CrossRef]
42. Batkowska, J.; Drabik, K.; Brodacki, A.; Czech, A.; Adamczuk, A. Fatty Acids Profile, Cholesterol Level and Quality of Table Eggs from Hens Fed with the Addition of Linseed and Soybean Oil. *Food Chem.* **2021**, *334*, 127612. [CrossRef]
43. Vlaicu, P.A.; Panaite, T.D.; Turcu, R.P. Enriching Laying Hens Eggs by Feeding Diets with Diferent Fatty Acid Composition and Antioxidants. *Sci. Rep.* **2021**, *11*, 20707. [CrossRef]
44. Difonzo, G.; Troilo, M.; Squeo, G.; Pasqualone, A.; Caponio, F. Functional Compounds from Olive Pomace to Obtain High-added Value Foods—A Review. *J. Sci. Food Agric.* **2021**, *101*, 15–26. [CrossRef]

45. Berbel, J.; Posadillo, A. Review and Analysis of Alternatives for the Valorisation of Agro-Industrial Olive Oil By-Products. *Sustainability* **2018**, *10*, 237. [CrossRef]
46. Papadomichelakis, G.; Pappas, A.C.; Tsiplakou, E.; Symeon, G.K.; Sotirakoglou, K.; Mpekelis, V.; Fegeros, K.; Zervas, G. Effects of Dietary Dried Olive Pulp Inclusion on Growth Performance and Meat Quality of Broiler Chickens. *Livest. Sci.* **2019**, *221*, 115–122. [CrossRef]
47. Saleh, A.; Alzawqari, M. Effects of Replacing Yellow Corn with Olive Cake Meal on Growth Performance, Plasma Lipid Profile, and Muscle Fatty Acid Content in Broilers. *Animals* **2021**, *11*, 2240. [CrossRef] [PubMed]
48. Joven, M.; Pintos, E.; Latorre, M.A.; Suárez-Belloch, J.; Guada, J.A.; Fondevila, M. Effect of Replacing Barley by Increasing Levels of Olive Cake in the Diet of Finishing Pigs: Growth Performances, Digestibility, Carcass, Meat and Fat Quality. *Anim. Feed Sci. Technol.* **2014**, *197*, 185–193. [CrossRef]
49. Dal Bosco, A.; Mourvaki, E.; Cardinali, R.; Servili, M.; Sebastiani, B.; Ruggeri, S.; Mattioli, S.; Taticchi, A.; Esposto, S.; Castellini, C. Effect of Dietary Supplementation with Olive Pomaces on the Performance and Meat Quality of Growing Rabbits. *Meat Sci.* **2012**, *92*, 783–788. [CrossRef] [PubMed]
50. Luciano, G.; Pauselli, M.; Servili, M.; Mourvaki, E.; Serra, A.; Monahan, F.J.; Lanza, M.; Priolo, A.; Zinnai, A.; Mele, M. Dietary Olive Cake Reduces the Oxidation of Lipids, Including Cholesterol, in Lamb Meat Enriched in Polyunsaturated Fatty Acids. *Meat Sci.* **2013**, *93*, 703–714. [CrossRef] [PubMed]
51. Molina-Alcaide, E.; Yáñez-Ruiz, D.R. Potential Use of Olive By-Products in Ruminant Feeding: A Review. *Anim. Feed Sci. Technol.* **2008**, *147*, 247–264. [CrossRef]
52. Vargas-Bello-Pérez, E.; Vera, R.R.; Aguilar, C.; Lira, R.; Peña, I.; Fernández, J. Feeding Olive Cake to Ewes Improves Fatty Acid Profile of Milk and Cheese. *Anim. Feed Sci. Technol.* **2013**, *184*, 94–99. [CrossRef]
53. Serra, V.; Salvatori, G.; Pastorelli, G. Dietary Polyphenol Supplementation in Food Producing Animals: Effects on the Quality of Derived Products. *Animals* **2021**, *11*, 401. [CrossRef]
54. Laudadio, V.; Ceci, E.; Lastella, N.M.B.; Tufarelli, V. Dietary High-Polyphenols Extra-Virgin Olive Oil Is Effective in Reducing Cholesterol Content in Eggs. *Lipids Health Dis.* **2015**, *14*, 5. [CrossRef]
55. Wołoszyn, J.; Haraf, G.; Okruszek, A.; Wereńska, M.; Goluch, Z.; Teleszko, M. Fatty Acid Profiles and Health Lipid Indices in the Breast Muscles of Local Polish Goose Varieties. *Poult. Sci.* **2020**, *99*, 1216–1224. [CrossRef]
56. Simopoulos, A.P. The Importance of the Ratio of Omega-6/Omega-3 Essential Fatty Acids. *Biomed. Pharmacother.* **2002**, *56*, 365–379. [CrossRef]
57. Abd El-Moneim, A.E.; Sabic, E.M. Beneficial Effect of Feeding Olive Pulp and Aspergillus Awamori on Productive Performance, Egg Quality, Serum/Yolk Cholesterol and Oxidative Status in Laying Japanese Quails. *J. Anim. Feed Sci.* **2019**, *28*, 52–61. [CrossRef]
58. Chen, J.; Liu, H. Nutritional Indices for Assessing Fatty Acids: A Mini-Review. *Int. J. Mol. Sci.* **2020**, *21*, 5695. [CrossRef] [PubMed]
59. Monteiro, M.; Matos, E.; Ramos, R.; Campos, I.; Valente, L.M.P. A Blend of Land Animal Fats Can Replace up to 75% Fish Oil without Affecting Growth and Nutrient Utilization of European Seabass. *Aquaculture* **2018**, *487*, 22–31. [CrossRef]
60. Yurchenko, S.; Sats, A.; Tatar, V.; Kaart, T.; Mootse, H.; Jõudu, I. Fatty Acid Profile of Milk from Saanen and Swedish Landrace Goats. *Food Chem.* **2018**, *254*, 326–332. [CrossRef] [PubMed]
61. Watson, T.; Shantsila, E.; Lip, G.Y. Mechanisms of Thrombogenesis in Atrial Fibrillation: Virchow's Triad Revisited. *Lancet* **2009**, *373*, 155–166. [CrossRef]
62. Santos-Silva, J.; Bessa, R.J.B.; Santos-Silva, F. Effect of Genotype, Feeding System and Slaughter Weight on the Quality of Light Lambs. II. Fatty Acid Composition of Meat. *Livest. Prod. Sci.* **2002**, *77*, 187–194. [CrossRef]
63. Chiofalo, B.; Liotta, L.; Zumbo, A.; Chiofalo, V. Administration of Olive Cake for Ewe Feeding: Effect on Milk Yield and Composition. *Small Rumin. Res.* **2004**, *55*, 169–176. [CrossRef]
64. EUR-Lex. *Commission Regulation (EC) No 589/2008 of 23 June 2008 Laying down Detailed Rules for Implementing Council Regulation (EC) No 1234/2007 as Regards Marketing Standards for Eggs*; European Union: Maastricht, The Netherlands, 2008; Volume OJ L163, pp. 6–23. Available online: http://data.europa.eu/eli/reg/2008/589/oj/eng (accessed on 10 December 2021).
65. Mazzuco, H.; Bertechini, A.G. Critical Points on Egg Production: Causes, Importance and Incidence of Eggshell Breakage and Defects. *Ciênc. E Agrotecnol.* **2014**, *38*, 7–14. [CrossRef]
66. Mertens, K.; Kemps, B.; Perianu, C.; De Baerdemaeker, J.; Decuypere, E.; De Ketelaere, B.; Bain, M. Advances in Egg Defect Detection, Quality Assessment and Automated Sorting and Grading. In *Improving the Safety and Quality of Eggs and Egg Products*; Woodhead Publishing Limited: Sawston, UK, 2011; Volume 1, pp. 209–241. [CrossRef]
67. Zhang, Z.F.; Kim, I.H. Effects of Dietary Olive Oil on Egg Quality, Serum Cholesterol Characteristics, and Yolk Fatty Acid Concentrations in Laying Hens. *J. Appl. Anim. Res.* **2014**, *42*, 233–237. [CrossRef]
68. Mabe, I.; Rapp, C.; Bain, M.M.; Nys, Y. Supplementation of a Corn-Soybean Meal Diet with Manganese, Copper, and Zinc from Organic or Inorganic Sources Improves Eggshell Quality in Aged Laying Hens. *Poult. Sci.* **2003**, *82*, 1903–1913. [CrossRef]
69. Habeeb, A.A.M.; Gad, A.E.; EL-Tarabany, A.A.; Mustafa, M.M.; Atta, M.A.A. Using of Olive Oil By-Products in Farm Animals Feeding. *Int. J. Sci. Res. Sci. Technol.* **2017**, *3*, 57–68.
70. Santos, R.R.; Segura, C.J.; Sarmiento, F.L. Egg Quality during Storage of Eggs from Hens Fed Diets with Crude Palm Oil. *Rev. MVZ Córdoba* **2019**, *24*, 7297–7304. [CrossRef]
71. Heath, J.L. Chemical and Related Osmotic Changes in Egg Albumen During Storage. *Poult. Sci.* **1977**, *56*, 822–828. [CrossRef]

72. Brelaz, K.C.B.T.R.; Cruz, F.G.G.; Rufino, J.P.F.; Brasil, R.J.M.; Silva, A.F.; Santos, A.N.A. Serum Biochemistry Profile of Laying Hens Fed Diets with Fish Waste Oil. *Arq. Bras. Med. Vet. E Zootec.* **2021**, *73*, 223–230. [CrossRef]
73. Harlander-Matauschek, A.; Rodenburg, T.B.; Sandilands, V.; Tobalske, B.W.; Toscano, M.J. Causes of Keel Bone Damage and Their Solutions in Laying Hens. *Worlds Poult. Sci. J.* **2015**, *71*, 461–472. [CrossRef]
74. Dedousi, A.; Đukić Stojčić, M.; Sossidou, E. Effects of Housing Systems on Keel Bone Damage and Egg Quality of Laying hens. *Vet. Res. Forum* **2020**, *11*, 299–304. [CrossRef]
75. Riber, A.B.; Casey-Trott, T.M.; Herskin, M.S. The Influence of Keel Bone Damage on Welfare of Laying Hens. *Front. Vet. Sci.* **2018**, *5*, 6. [CrossRef]
76. Nasr, M.A.F.; Murrell, J.; Nicol, C.J. The Effect of Keel Fractures on Egg Production, Feed and Water Consumption in Individual Laying Hens. *Br. Poult. Sci.* **2013**, *54*, 165–170. [CrossRef]
77. Candelotto, L.; Stratmann, A.; Gebhardt-Henrich, S.G.; Rufener, C.; van de Braak, T.; Toscano, M.J. Susceptibility to Keel Bone Fractures in Laying Hens and the Role of Genetic Variation. *Poult. Sci.* **2017**, *96*, 3517–3528. [CrossRef]
78. Gebhardt-Henrich, S.G.; Pfulg, A.; Fröhlich, E.K.F.; Käppeli, S.; Guggisberg, D.; Liesegang, A.; Stoffel, M.H. Limited Associations between Keel Bone Damage and Bone Properties Measured with Computer Tomography, Three-Point Bending Test, and Analysis of Minerals in Swiss Laying Hens. *Front. Vet. Sci.* **2017**, *4*, 128. [CrossRef]
79. Wei, H.D.; Chen, Y.J.; Zeng, X.Y.; Bi, Y.J.; Wang, Y.N.; Zhao, S.; Li, J.H.; Li, X.; Zhang, R.X.; Bao, J. Keel-Bone Fractures Are Associated with Bone Quality Differences in Laying Hens. *Anim. Welf.* **2021**, *30*, 71–80. [CrossRef]
80. Wei, H.; Chen, Y.; Nian, H.; Wang, J.; Liu, Y.; Wang, J.; Yang, K.; Zhao, Q.; Zhang, R.; Bao, J. Abnormal Bone Metabolism May Be a Primary Causative Factor of Keel Bone Fractures in Laying Hens. *Animals* **2021**, *11*, 3133. [CrossRef] [PubMed]
81. Kruger, M.C.; Coetzee, M.; Haag, M.; Weiler, H. Long-Chain Polyunsaturated Fatty Acids: Selected Mechanisms of Action on Bone. *Prog. Lipid Res.* **2010**, *49*, 438–449. [CrossRef] [PubMed]
82. Rubert, M.; Torrubia, B.; Díaz-Curie, M.; de la Piedra, C. Olive Oil and Bone Health. *Rev. Osteoporos. Metab. Miner.* **2020**, *12*, 107–110. [CrossRef]
83. Cicerale, S.; Lucas, L.J.; Keast, R.S.J. Antimicrobial, Antioxidant and Anti-Inflammatory Phenolic Activities in Extra Virgin Olive Oil. *Curr. Opin. Biotechnol.* **2012**, *23*, 129–135. [CrossRef]
84. Oliveras-López, M.-J.; Molina, J.J.M.; Mir, M.V.; Rey, E.F.; Martín, F.; de la Serrana, H.L.-G. Extra Virgin Olive Oil (EVOO) Consumption and Antioxidant Status in Healthy Institutionalized Elderly Humans. *Arch. Gerontol. Geriatr.* **2013**, *57*, 234–242. [CrossRef]
85. Díaz-Curiel, M.; Torrubia, B.; Martín-Fernández, M.; Rubert, M.; De la Piedra, C. Effects of Virgin Olive Oil on Bone Health in Ovariectomized Rats. *Nutrients* **2020**, *12*, 1270. [CrossRef]
86. Giersberg, M.F.; Spindler, B.; Kemper, N. Assessment of Plumage and Integument Condition in Dual-Purpose Breeds and Conventional Layers. *Animals* **2017**, *7*, 97. [CrossRef]
87. Michalak, M.; Pierzak, M.; Kręcisz, B.; Suliga, E. Bioactive Compounds for Skin Health: A Review. *Nutrients* **2021**, *13*, 203. [CrossRef]
88. Anunciato, T.P.; da Rocha Filho, P.A. Carotenoids and Polyphenols in Nutricosmetics, Nutraceuticals, and Cosmeceuticals: Nutricosmetics. *J. Cosmet. Dermatol.* **2012**, *11*, 51–54. [CrossRef]
89. Nichols, J.A.; Katiyar, S.K. Skin Photoprotection by Natural Polyphenols: Anti-Inflammatory, Antioxidant and DNA Repair Mechanisms. *Arch. Dermatol. Res.* **2010**, *302*, 71–83. [CrossRef] [PubMed]
90. Zillich, O.V.; Schweiggert-Weisz, U.; Eisner, P.; Kerscher, M. Polyphenols as Active Ingredients for Cosmetic Products. *Int. J. Cosmet. Sci.* **2015**, *37*, 455–464. [CrossRef] [PubMed]

sustainability

MDPI

Article

Dietary Orange Pulp and Organic Selenium Effects on Growth Performance, Meat Quality, Fatty Acid Profile, and Oxidative Stability Parameters of Broiler Chickens

Evangelos Zoidis [1,†], Panagiotis Simitzis [2], Dimitrios Kampantais [1], Petros Katsoulas [2], Athanasios C. Pappas [1], George Papadomichelakis [1] and Michael Goliomytis [2,*,†]

1 Laboratory of Nutritional Physiology and Feeding, Department of Animal Science, Agricultural University of Athens, 75 Iera Odos, 11855 Athens, Greece; ezoidis@aua.gr (E.Z.); dim.kampantais@hcmr.gr (D.K.); apappas@aua.gr (A.C.P.); gpapad@aua.gr (G.P.)
2 Laboratory of Animal Breeding and Husbandry, Department of Animal Science, Agricultural University of Athens, 75 Iera Odos, 11855 Athens, Greece; pansimitzis@aua.gr (P.S.); katsoulaspetros@gmail.com (P.K.)
* Correspondence: mgolio@aua.gr; Tel.: +30-2105294446
† These two authors should be considered joint first author.

Abstract: In this study, orange pulp (OP) and/or organic Se were fed to broilers in order to investigate their effects on the performance, behavior, breast meat quality, and oxidative stability. A total of 240 chicks were allocated to four groups: a control group; an OP group, fed with OP at 50 g/kg of diet; a Se group, fed with organic Se at 0.15 ppm; and an OP + Se group, fed with OP and organic Se at 50 g/kg and 0.15 ppm, respectively. The selenium and OP + Se groups showed improved meat oxidative stability during frozen storage from 90 to 210 days ($p < 0.05$), whereas the performance and meat quality were unaffected by the dietary treatments ($p > 0.05$), apart from a reduction in the meat pH and the dressing percentage in the OP-supplemented groups ($p < 0.05$). A synergistic action between OP and Se was observed for the meat oxidative stability. The polyunsaturated fatty acid (FA) and α-linolenic acid (ALA) contents in the breast meat lipid fractions were increased in the OP groups ($p < 0.05$). Dietary intervention did not affect the feeding or drinking behaviors of the broilers ($p > 0.05$). The dietary supplementation of broiler chickens with the citrus industry byproduct orange pulp at 50 g/kg, along with organic Se at 0.15 ppm, beneficially improves the meat oxidative stability and the meat nutritional value, with no negative side effects on the performance or the meat quality.

Keywords: broilers; orange pulp; organic selenium; antioxidant activity; meat quality; behavior; fatty acid profile

Citation: Zoidis, E.; Simitzis, P.; Kampantais, D.; Katsoulas, P.; Pappas, A.C.; Papadomichelakis, G.; Goliomytis, M. Dietary Orange Pulp and Organic Selenium Effects on Growth Performance, Meat Quality, Fatty Acid Profile, and Oxidative Stability Parameters of Broiler Chickens. *Sustainability* **2022**, *14*, 1534. https://doi.org/10.3390/su14031534

Academic Editors: George K. Symeon, Vassilios Dotas and Hossein Azadi

Received: 26 November 2021
Accepted: 26 January 2022
Published: 28 January 2022

1. Introduction

In recent years, a continuous increase in livestock production costs has been observed, due to the increased prices of feedstuffs, such as soybean products and cereal grains. As a result, low-input feeding strategies are a prerequisite for minimizing the animal nutrition expenses that are based on alternative feeding resources, such as shrubs, bushes, novel pastures, and agroindustrial byproducts [1]. At the same time, animal products are, nowadays, intended not only to satisfy nutritional needs, but also to protect human organisms against degenerative diseases that are linked to oxidative stress. The consumption of foods rich in functional compounds (nutraceuticals) can, therefore, reinforce the activity of the endogenous systems against these diseases and lead to an improvement in mental and physical wellbeing [2,3]. Agroindustrial byproducts are considered to be a cheap source of these compounds, and their application in animal nutrition minimizes the environmental impact induced by their disposal (due to their high organic load) and enables the sustainability of high-added value ingredients inside the food chain [4].

Dried citrus pulp is produced after the extraction of the juice from citrus fruits and the drying of the residues. It is a mixture of the peel, the inside portions, and the culled fruits of the citrus family that contain biologically active compounds (i.e., flavonoids), with positive and beneficial effects for the human organism [5]. Flavonoids (hesperidin, naringin, etc.) are found in the pulp, albedo, membranes, and the pith of citrus, and they are a class of secondary plant phenolics with significant antioxidant and chelating properties [6]. Citrus pulp and its flavonoids have already been used in diets intended for broilers, with varying results regarding the growth performance parameters, but positive effects on the meat antioxidant properties. No effect on the final weight, the hot carcass weight, nor the carcass yield was observed after the dietary inclusion of dried *Citrus sinensis* peel at the levels of 1.5 or 3.0% [7]. Simitzis et al. [8] and Goliomytis et al. [9] reached the same conclusions following hesperidin (1.5–3.0 g/kg) and naringin (0.75–1.5 g/kg) dietary supplementation, respectively. On the other hand, the use of dried orange residues at 2% [10], or dried tangerine peel extract at 80–480 mg/kg [11], improved the feed intake and body weight gain. When higher doses of citrus pulp (5 or 10%) were examined, reductions in the daily weight gain and the carcass yields were reported [12]. In general, the inclusion of hesperidin [8,13] or naringin [9] into the diet of broilers improves the meat antioxidant capacity. Improved antioxidant capacity in breast and thigh meat was also observed when broiler chickens were fed with *Citrus junos* byproducts fermented with multistrain probiotics, at 5, 10, and 20 g/kg of the diet [14]. However, no data exist regarding the effects of orange pulp dietary supplementation on the meat oxidative stability of broilers.

Selenium (Se) is an essential trace element with several important biological roles in poultry, such as the regulation of the antioxidant defense system, and improvements in the immunity, normal growth, and body maintenance [15]. Selenium protects the organism from oxidative stress through the amendment of the antioxidant defense system, since it regulates the expressions of various enzymes, such as glutathione peroxidases, deiodinases, thioredoxin reductases, and other selenoproteins [16,17]. Sodium selenite (SS) is an inorganic form of Se, and it is the most common Se source used in animal diets. However, dietary supplementation with Se organic forms in poultry diets, such as Se-enriched yeast (SY) and selenomethionine (SM), has been legally allowed since 2000 [18]. Se organic forms have improved the bioavailability and antioxidant properties [19], and they are less toxic and more environmentally friendly [20] than inorganic forms [21,22]. The majority of studies regarding selenium supplementation in broilers do not report any effects of selenium on the growth performance and feed conversion [23,24], but there are also researchers who found an increase in the live weight due to the inclusion of Se in broilers diets [25,26]. In general, the broiler meat quality and the antioxidant properties are improved after Se dietary supplementation [17,23,27].

However, the combined action of dietary Se, which triggers the biological endogenous antioxidant defense system, along with orange pulp, which is rich in natural antioxidants, has not yet been evaluated in broiler diets, and their synergistic effects have not been examined. Therefore, the aim of the present study is to highlight the effects of orange pulp and/or organic selenium dietary supplementation on the broiler performance, behavioral traits, meat quality, fatty acid (FA) profile, and oxidative stability.

2. Materials and Methods

2.1. Birds and Diets

A total of 240 one-day-old Cobb 500 broiler chickens, obtained from a commercial hatchery, as hatched, were housed in a controlled environment. The birds were reared for 42 days in 16 pens, with 15 birds per pen, with a surface area of 1.5 m^2 each. The environmental conditions and management practices were in accordance with the standard Cobb guidelines. Feed, in mash form, and water were provided ad libitum. The lighting program consisted of 23L:1D upon arrival, and it was decreased to 18L:6D at Day 7, remained constant until Day 38, and thereafter gradually increased to 23L:1D at slaughter. The birds were vaccinated for Marek's disease at the hatchery, and for infectious bursal, infectious bronchitis, and Newcastle disease on the farm via drinking water. The 240 chicks were randomly allocated to 4 treatment groups, with 4 replicate pens each. The treatment groups were offered a starter (at 1 day to 11 days of age), a grower (at 12 to 22 days of age), and a finisher diet (at 23 to 42 days of age), in crumbled form (Table 1). One of the treatment groups was offered the diet with no additive and it served as a control (C), whereas the other three treatment groups were offered finisher diets further supplemented with dried orange pulp, *Citrus sinensis*, at 50 g/kg (OP group), or organic Se (SelSaf®, Lesaffre, Cimetiere Bourg, France) at 0.15 ppm (Se group), or OP and organic Se (OP + Se group), at 50 g/kg and 0.15 ppm, respectively, in a factorial design. The main compounds of the organic Se used were L-selenomethionine and L-selenocysteine. The treatment diets were isocaloric and isonitrogenous.

Table 1. Ingredients and calculated chemical compositions of the diets used.

Ingredients g/kg	Starter, Days 1 to 10	Grower, Days 11 to 22	Control, Finisher, Days 23 to 42	Orange Pulp Diet, Finisher with 50 g/kg Orange Pulp, Days 23 to 42	Orange Pulp
Orange pulp	0	0	0	50	
Maize	200	120	0	200	
Wheat	409.3	521.6	673.2	370	
Soybean meal, 46% CP	283	260	220	244	
Sesame meal	30	30	30	50	
Fish meal 72% CP	12.5	0	0	0	
Soybean oil	18	29	40	50	
Limestone	15	13	13	12	
Monocalcium phosphate	13	10	9	10	
Sodium chloride	2	2	2	2	
Sodium bicarbonate	2	2	2	2	
Lysine	6.4	5	4	3.2	
Methionine	3.5	2.8	2.2	2.2	
Threonine	1.7	1	1	1	
Choline	0.6	0.6	0.6	0.6	
Natugrain® wheat	0.1	0.1	0.1	0.1	
Phytase, Natuphos®	0.1	0.1	0.1	0.1	
Antioxidant (BHA: E320)	0.1	0.1	0.1	0.1	
Coccidiostat, Clinacox®, 0.5%	0.2	0.2	0.2	0.2	
Vitamin and mineral premix †	2.5	2.5	2.5	2.5	
Chemical composition g/kg					
Metabolizable energy (MJ/kg)	12.89	13.23	13.57	13.57	
Crude protein	217.8	202	190.1	189.9	58.3
Fat	55.8	64	71.5	92.5	12
Fiber	30.6	30.3	29.5	33	101
Lysine	13.8	12	10.7	10.8	
Methionine + cystine	10.2	9.2	8.3	8.5	
Calcium	10	8.5	8.5	8.5	
Phosphorus	8	6.9	6.7	6.9	
Ash	56.4	49.7	48.1	48.7	

† The vitamin and mineral premix provided per kg of diet: 13,000 IU of vitamin A (retinyl acetate); 5000 IU of cholecalciferol; 80 mg of vitamin E (DL-α-tocopheryl acetate); 4 mg of menadione; 4.2 mg of thiamine; 8 mg of riboflavin; 6 mg of pyridoxin; 20 μg of cobalamin; 75 mg of nicotinic acid; 18 mg of pantothenic acid; 2 mg of folic acid; 240 μg of biotin,; 10 mg of vitamin C (ascorbic acid); 500 mg of choline chloride; 0.23 mg of Co; 1.2 mg of I; 0.35 mg of Se; 50 mg of Fe; 140 mg of Mn; 25 mg of Cu; and 115 mg of Zn.

The OP included both the peels and seeds that remained after the processing of the orange fruits. An inclusion rate of 50 g/kg for the OP was the maximum one for a balanced broiler finisher diet. An inclusion rate of 0.15 ppm for the Se was selected by taking into consideration that the maximum inclusion rate in broiler diets is legally limited to 0.50 ppm in the European Union, and the mineral and vitamin premix already provided 0.35 ppm of Se. The dried OP contents in the hesperidin and naringin were determined at 8.52 and 0.0223 g/kg OP, respectively [28]. Subsequently, the hesperidin and naringin contents of the experimental finisher diets were estimated at 0.426 and 0.001 g/kg, respectively.

The behavior of the broilers was videotaped daily, from 23 days of age till 42 days of age, using video cameras with infrared lighting (TX-1430OA, Turbo-X, Plaisio Computers, Athens, Greece). The numbers of birds standing over a feeder or a drinker with their heads towards the trough were recorded through a camera that was placed in a fixed position in each pen, by using time-lapse photography, every ten minutes of an hour. The data were then stored in a digital video recorder equipped with a hard disk (TX168, Telexper Inc., Union City, CA, USA).

The feed intake and the body weights of the broilers were recorded weekly, and the feed conversion ratio (g of feed:g of body gain, FCR) was calculated on a pen basis (4 pens per treatment). At Day 42 of the experiment, 8 chickens per treatment group, randomly chosen, were individually weighed, electrically stunned, and slaughtered. The carcass, liver, heart, gizzard, and fat pad weights were recorded. The chicken carcasses were then chilled, at 4 °C for 24 h, for the subsequent breast meat quality assessment, the fatty acid profile determination, and the oxidative stability measurement in the pectoralis major muscles.

The study was conducted in accordance with Directive 2010/63/EU of the European Parliament and of the Council on the protection of animals used for scientific purposes. The protocol was approved by the Research Ethics Committee of the Agricultural University of Athens, with the code number: 32/20052015.

2.2. Meat Quality

At 8 chicken carcasses per treatment group, the meat color was measured (thrice per sample) on the surfaces of the right pectoralis major muscles, after exposure to the air at room temperature for 30 min, with a Miniscan XE (HunterLab, Reston, VA, USA) chromameter, set on the L* (lightness), a* (redness), and b* (yellowness) systems. The chroma value, $C^* = (a^{*2} + b^{*2})^{1/2}$), and the hue angle value, $H^* = \tan^{-1}(b^*/a^*)$, were also calculated. The instrument was calibrated with a white tile and a black tile, using illuminant D65, with 0° viewing.

The pH was measured in the right pectoralis major muscle, 24 h after slaughter (pH24), by the insertion of a pH meter electrode (HI 99,163 Meat pH Temperature Meter, Hanna Instruments, Nusfalau, Romania). The calibration was performed with buffers of pH 4.0 and pH 7.0.

The cooking loss and the shear force values were also determined in the right pectoralis major muscle, which was dissected, weighed, placed into a thin-walled plastic bag, and cooked in a water bath at 80 °C for 30 min. Each sample was then cooled under tap water and equilibrated at room temperature. The muscle was weighed again for the determination of the cooking loss (%). The shear force was evaluated by cutting two 1.9-mm-wide × 10 mm × 10 mm strips from the center of the muscles parallel to the muscle fibers. The samples were then cut perpendicular to the fiber direction using a Zwick Testing Machine Model Z2.5/TN1S (Zwick GmbH and Co., Ulm, Germany), equipped with a Warner–Bratzler shear [29]. The peak force values were obtained in N/cm^2.

2.3. Oxidative Stability

The oxidative stability was assessed on the basis of the malondialdehyde (MDA) content. MDA is a secondary product originating from the hydrolysis of lipid hydroperoxides during lipid oxidation. In the present study, the MDA concentrations were determined in the muscle samples from 8 chickens per treatment. The measurements were implemented after storage in plastic sealed bags at 4 °C, for 1, 3, 6, and 9 days, and at −20 °C, for 90, 120, and 210 days after slaughter, by using the selective third-order derivative spectrophotometric method. One breast fillet sample from the left pectoralis major muscle per storage time was used for the MDA determination. In brief, 2 g of each sample (two samples per chicken) were homogenized (Unidrive × 1000, CAT, M. Zipperer GmbH, Ballrechten-Dottingen, Germany) in the presence of 8 mL of aqueous trichloroacetic acid (TCA) (50 g/L), and 5 mL of butylated hydroxytoluene in hexane (8 g/L), and the mixture was centrifuged for 5 min at 3000× *g*. The top hexane layer was discarded, and a 2.5-mL aliquot from the bottom layer was mixed with 1.5 mL of aqueous 2-thiobarbituric acid (8 g/L), to be further incubated at 70 °C for 30 min. Following incubation, the mixture was cooled under tap water and submitted to third-order derivative (3D) spectrophotometry (Hitachi U3010 Spectrophotometer, Hitachi High-Technologies Corporation, Tokyo, Japan), in the range of 500–550 nm. The concentration of MDA (ng/g wet tissue) in the samples was determined as the height of the third-order derivative peak at 521.5 nm, by referring to the standard calibration curve, prepared using 1,1,3,3-tetraethoxypropane, the malondialdehyde precursor [30].

2.4. Fatty acid Profile

The fatty acid profile was determined in the feed samples of the experimental diets and orange pulp, and on the breast-muscle samples, from 8 chickens per treatment group, that were dissected at slaughter. Any external fat and connective tissue were dissected out of the left pectoralis major muscle samples, which were then blended in a domestic food processor (Multi Izzy C-5160, 600 W, Benroubi & Fils SA, Marousi, Greece) until smooth. The blending was performed in short bursts to ensure the homogeneous distribution of intramuscular fat in the sample. The FAs of the diets and the intramuscular fat were extracted and methylated directly, according to O'Fallon et al. [31]. Duplicate samples of 1 (±0.05) g were hydrolyzed for 1.5 h at 55 °C, in 1 N of potassium hydroxide in methanol, containing a known amount (approximately 0.5 mg) of tridecanoic acid (C13:0) as the internal standard. The potassium hydroxide was then neutralized, and the free FAs were methylated by sulphuric acid catalysis (24 N H_2SO_4) for 1.5 h at 55 °C. Hexane (3 mL) was added to the reaction tube, which was vortex-mixed and centrifuged at 1100× *g*. The supernatant hexane layer, containing the FA methyl esters, was kept at −20 °C, until analyzed by gas chromatography. A temperature-programmed run was followed on a Perkin Elmer Autosystem XL gas chromatograph, equipped with a 30 m × 0.25 mm × 0.25-µm internal diameter HP-Innowax capillary column (Agilent Technologies, J&W GC columns, Santa Clara, CA, USA), and a flame ionization detector (FID). The column temperature was programmed for 1 min at 140 °C, was raised by 2.5 °C/min to 200 °C, then to 230 °C by 1 °C/min, and was held for 1 min, and finally increased to 240 °C by 4 °C/min, and held for 10 min. Helium was the carrier gas, at a constant pressure of 18 psi, and the temperatures of both the injector and FID were set at 250 °C. The fatty acids were identified through comparison with the standards purchased from the Sigma-Aldrich Co. (FAME 37 Component; Sigma-Aldrich Co. Supelco, St. Louis, MO, USA), and quantification was achieved using the internal standard (13:0), added prior to hydrolysis. The total weights of the FAs (mg/100 g) were calculated as the sums of the areas for all the FA peaks, compared to the area for the 0.5-mg internal standard. The individual FAs were expressed as the % by weight of the total FAs [31].

2.5. Statistical Analysis

The data were subjected to an analysis of variance with the MIXED procedure of the SAS software [32], with the dietary treatment with OP and Se, and their interaction as the fixed effects in the factorial design. The mean comparisons were tested at a 0.05 significance level with a Bonferroni adjustment. The synergistic effect of the combined supplementation of OP and Se was tested with pairwise comparisons between the OP + Se group means and each of the OP group and Se group means. When the OP + Se group mean was different from both the OP and Se group means, a synergistic effect was considered significant, in accordance with the highest single-agent reference model, which assumes that a synergistic compound combination should produce additional benefits, on top of what the compounds can achieve alone [33]. The results are presented as least square means ± SEM.

3. Results

Table 2 presents the FA profiles of the experimental diets and orange pulp used in the present experiment. No significant differences were determined among the experimental diets, apart from an increase in n-3 fatty acids, and a slight increase in the monounsaturated fatty acid (MUFA) content, whereas the saturated fatty acids (SFA) were slightly decreased in the OP-supplemented diets.

Table 2. Fatty acid (FA) profiles of orange pulp and experimental diets (% of total FA).

			Diets †		
FA	**C**	**OP**	**Se**	**OP + Se**	**Orange Pulp**
12:0	0.01	0.01	0.01	0.01	0.02
14:0	0.00	0.09	0.12	0.09	0.37
15:0	0.03	0.03	0.04	0.03	0.08
16:0	14.28	12.75	12.71	12.47	17.73
7c16:1	0.04	0.04	0.05	0.04	0.27
9c16:1	0.15	0.14	0.16	0.13	1.70
17:0	0.11	0.11	0.10	0.10	0.12
17:1	0.05	0.06	0.06	0.06	0.13
18:0	4.91	4.49	4.23	4.37	7.81
9c18:1	24.52	26.78	26.55	26.40	26.70
11c18:1	0.00	0.00	0.00	0.00	1.71
18:2n-6	48.02	47.00	48.01	48.08	29.80
18:3n-6	0.02	0.01	0.02	0.01	0.19
18:3n-3	4.49	5.86	4.97	5.66	2.56
20:0	0.52	0.45	0.44	0.44	0.09
20:1n-9	0.31	0.27	0.33	0.27	0.33
20:2	0.06	0.06	0.06	0.06	0.50
20:3n-6	0.03	0.03	0.03	0.03	0.49
20:4n-6	0.01	0.01	0.01	0.01	2.36
20:3n-3	0.01	0.01	0.01	0.01	0.08
20:5n-3	0.05	0.01	0.05	0.01	0.19
22:0	0.38	0.35	0.31	0.34	0.03
22:1	0.08	0.05	0.09	0.07	0.03
22:2	0.60	0.53	0.56	0.53	1.30
22:4n-6	0.00	0.01	0.01	0.01	0.66
22:5n-3	0.01	0.05	0.01	0.00	0.55
22:6n-3	0.08	0.03	0.10	0.03	0.46
SFA ‡	20.2	18.27	17.96	17.85	26.25
MUFA ‡	25.2	27.33	27.23	26.96	30.86
PUFA ‡	53.4	53.65	53.81	54.43	39.16
n-3	4.64	6.01	5.13	5.71	3.85
n-6	48.08	47.05	48.06	48.14	33.51
n-6/n-3	10.37	7.89	9.42	8.44	8.70

† C: diet with no additive; OP: diet supplemented with 50 g orange pulp per kg feed; Se: diet supplemented with 0.15 ppm Se; OP + Se: diet supplemented with 0.15 ppm Se and 50 g orange pulp per kg feed. ‡ SFA: saturated FA; MUFA: monounsaturated FA; PUFA: polyunsaturated FA.

Table 3 presents the results on the OP and/or Se dietary supplementation on the broiler performances and carcass traits. The performances and the internal organ weights of the broilers were not affected by dietary supplementation with OP and/or Se ($p > 0.05$). However, the dressing percentage was lower in the group of birds fed the OP diet, in comparison with groups fed the C and Se diets ($p < 0.05$). At the same time, no differences were found in the percentages of chickens standing at the feeder (Pf) or the drinker (Pw) among the experimental groups ($p > 0.05$) (Table 3).

Table 3. Effects of dietary supplementation with orange pulp (*Citrus sinensis*) and/or Se on broiler performance, feeding and drinking behaviors, and carcass and internal organ weights, at 42 days of age ($n = 8$).

Parameter ‡	Treatment †										Source of Variation		
	OP		Se			OP × Se							
	−	+	−	+	SEM	C	OP	Se	OP + Se	SEM	OP	Se	OP × Se
BW, g	2730	2671	2636	2765	71	2741	2532	2720	2809	101	0.556	0.213	0.150
Cumulative FI, g	4532	4463	4507	4488	92	4607	4407	4457	4519	130	0.603	0.889	0.333
FCR, g/g	1.72	1.73	1.72	1.73	0.02	1.73	1.71	1.72	1.74	0.03	0.965	0.669	0.644
Carcass weight, g	2112	2035	2013	2133	57	2117	1911	2108	2159	81	0.348	0.150	0.122
DP, %	77.3	76.1	76.4	77.1	0.30	77.3 ^{a}	75.5 ^{b}	77.4 ^{a}	76.8 ^{ab}	0.4	0.009	0.093	0.196
Liver, %	1.71	1.70	1.70	1.71	0.05	1.64	1.76	1.77	1.65	0.07	0.812	0.945	0.072
Heart, %	0.40	0.42	0.43	0.40	0.02	0.40	0.46	0.41	0.39	0.02	0.431	0.149	0.089
Gizzard, %	1.28	1.22	1.25	1.26	0.07	1.27	1.23	1.30	1.22	0.09	0.509	0.879	0.814
Fat pad, %	1.28	1.45	1.40	1.33	0.10	1.27	1.52	1.30	1.37	0.14	0.252	0.655	0.517
Pf	27.00	26.89	27.60	26.29	1.25	27.88	27.82	26.74	26.68	1.84	0.953	0.476	0.970
Pw	7.22	7.30	7.58	6.94	0.35	7.39	7.89	7.26	6.83	0.47	0.887	0.217	0.362

$^{a, b}$ means in the OP × Se row sharing no common superscript are statistically different ($p < 0.05$). † C: diet with no additive; OP: diet supplemented with 50 g orange pulp per kg feed; Se: diet supplemented with 0.15 ppm Se; OP + Se: diet supplemented with 0.15 ppm Se and 50 g orange pulp per kg of feed. ‡ BW: body weight; FI: feed intake; FCR: feed conversion ratio; DP: dressing percentage. % of BW; Pf: percentage of chickens standing at the feeder; Pw: percentage of chickens standing at the drinker. n for FI, and FCR = 4 replicate pens.

Among the meat quality trait measurements (color, cooking loss, tenderness, and pH24), it was only pH24 that was altered, and this change was due to OP supplementation (Table 4). The meat from the birds fed with OP (alone or in combination with Se) exhibited lower pH24 values ($p < 0.05$), in comparison with the meat from the controls.

Table 4. Effect of dietary supplementation with orange pulp (*Citrus sinensis*) and/or Se on broiler meat quality at 42 days of age ($n = 8$).

Parameter ‡	Treatment †										Source of Variation		
	OP		Se			OP × Se							
	−	+	−	+	SEM	C	OP	Se	OP + Se	SEM	OP	Se	OP × Se
pH24	5.96	5.83	5.92	5.87	0.02	5.98 ^{a}	5.85 ^{b}	5.93 ^{ab}	5.80 ^{b}	0.03	<0.001	0.128	0.896
L*	56.7	58.6	57.5	57.8	0.77	56.5	58.5	56.9	58.6	1.1	0.095	0.821	0.891
a*	3.50	3.65	3.65	3.50	0.26	3.47	3.82	3.53	3.47	0.36	0.686	0.686	0.579
b*	12.4	13.0	12.6	12.9	0.33	11.1	12.6	12.7	12.4	0.80	0.220	0.531	0.230
H*	74.4	74.4	73.9	74.8	0.91	73.9	73.9	74.8	74.9	1.29	0.973	0.491	0.987
C*	12.9	13.5	13.1	13.3	0.36	12.5	13.7	13.4	13.3	0.51	0.253	0.625	0.228
Cooking loss (%)	14.8	15.7	15.5	15.0	0.6	15.1	15.9	14.4	15.6	0.89	0.283	0.598	0.786
Shear force, N/cm^2	23.6	23.9	23.2	24.3	1.5	24.1	22.4	23.1	25.4	2.1	0.893	0.618	0.350

$^{a, b}$ means in the OP × Se row sharing no common superscript are statistically different ($p < 0.05$). † C: diet with no additive; OP: diet supplemented with 50 g orange pulp per kg feed; Se: diet supplemented with 0.15 ppm Se; OP + Se: diet supplemented with 0.15 ppm Se and 50 g orange pulp per kg feed. ‡ L* = lightness; a* = redness; b* = yellowness; H* = hue angle; C* = chroma.

The meat oxidative stability results are presented in Table 5. Lower MDA values in the treatment groups compared to the controls were detected throughout the entire storage period, which is indicative of a reduced rate of lipid peroxidation in the meat from birds fed with OP and/or Se. However, a significant effect ($p < 0.05$) was determined as the storage period increased. Significantly lower MDA values were detected in the breast meat samples of the OP + Se group, from Day 9 of storage and onwards, of the Se group during long-term frozen storage (90 to 210 days), and of the OP group at 150 days of storage, in comparison with the control group ($p < 0.05$). Nevertheless, the factorial analysis results show that the diets supplemented with OP or Se resulted in reduced MDA percentages in the breast meat stored for more than 90 or 9 days, respectively ($p < 0.05$). A synergistic effect of the combined supplementation of OP and Se was determined during the long-term frozen storage period from 150 to 210 days ($p < 0.05$). In the aforementioned period, the OP + Se group exhibited significantly lower MDA values in comparison with either the OP group or the Se group alone ($p < 0.05$).

Table 5. Effects of dietary supplementation with orange pulp (*Citrus sinensis*) and/or Se on the broiler meat (pectoralis major) oxidative stability during storage (ng MDA/g meat) (n = 8).

Storage Time ‡, Days	Treatment †										Source of Variation		
	OP		Se			OP × Se							
	−	+	−	+	SEM	C	OP	Se	OP + Se	SEM	OP	Se	OP × Se
1	9.43	8.08	9.59	7.93	0.64	10.3	8.83	8.52	7.34	0.90	0.147	0.077	0.855
3	12.3	10.8	12.0	11.1	0.59	12.1	11.9	12.4	9.70	0.83	0.094	0.275	0.147
6	17.8	16.5	17.9	16.4	0.96	18.0	17.8	17.6	15.2	1.35	0.331	0.284	0.413
9	43.2	39.2	46.9	35.5	2.64	48.7 ^{a}	45.1 ^{ab}	37.7 ^{ab}	33.3 ^{b}	3.73	0.293	0.005	0.925
90	16.9	14.0	17.8	13.1	0.73	19.5 ^{a}	16.1 ^{ab}	14.3 ^{b}	11.9 ^{b}	1.03	0.01	<0.001	0.612
150	32.4	26.3	32.8	25.9	1.06	36.4 ^{a}	29.2 ^{b}	28.3 ^{b}	23.5 ^{bA}	1.50	<0.001	<0.001	0.414
210	44.0	37.9	43.5	38.4	1.28	45.9 ^{a}	41.0 ^{ab}	42.0 ^{ab}	34.8 ^{bA}	1.82	0.003	0.01	0.522

$^{a, b}$ means in the OP × Se row sharing no common superscript are statistically different ($p < 0.05$). ^{A} denotes that the OP + Se group mean is different from both the OP and Se group means in pairwise comparisons ($p < 0.05$), which is indicative of a synergistic effect. † C: diet with no additive; OP: diet supplemented with 50 g orange pulp per kg feed; Se: diet supplemented with 0.15 ppm Se; OP + Se: diet supplemented with 0.15 ppm Se and 50 g orange pulp per kg feed. ‡ stored for 1, 3, 6, and 9 days at 4 °C, and for 90, 150, and 210 days at −20 °C.

The FA profiles of the lipid fractions of the breast meat are presented in Table 6. The dietary modification due to OP supplementation decreased the percentage of palmitoleic acid (9C16:1) by 38.4%, when OP alone was fed, and by 31%, when fed in combination with Se, compared to the C group. The results of the factorial analysis show that the OP groups exhibited increased percentages of both n-3 and n-6 FAs (3.57 and 32.14% of total FAs, respectively), in comparison with the C and Se groups (3.11 and 27.71% of total FAs, for n-3 and n-6 FAs, respectively, $p < 0.05$). The alpha-linolenic acid (ALA, 18:3n-3) concentration was also beneficially affected by the diet modification because of the OP incorporation. Higher ALA content values were detected in the OP groups in comparison with the C and Se groups. Significant increases for the ALA content was observed for the OP + Se vs. the C and Se groups, by 40.5 and 44.1%, respectively ($p < 0.05$). Increased polyunsaturated fatty acid (PUFA), n-3, and n-6 FA contents in the breast intramuscular fat were also observed for the treatment groups fed with diets modified by OP supplementation, in comparison with the broiler groups that were not fed with OP ($p < 0.05$). However, the most notable PUFA increase observed was for the OP + Se group, compared to the Se group, by 21.3% ($p < 0.05$).

Table 6. Effects of dietary supplementation with orange pulp (*Citrus sinensis*) and/or Se on broiler breast meat fatty acid (FA) profiles at 42 days of age (% of total FA, n = 8).

Parameter	Treatment †										Source of Variation		
	OP		Se			OP × Se							
	−	+	−	+	SEM	C	OP	Se	OP + Se	SEM	OP	Se	OP × Se
Total FA weight (mg 100 g^{-1} meat)	1352.4	1317.2	1215.2	1454.5	163.8	1226.5	1203.8	1478.4	1430.6	231.6	0.880	0.310	0.957
FA													
12:0	0.022	0.016	0.021	0.018	0.003	0.028	0.014	0.018	0.018	0.005	0.180	0.552	0.152
14:0	0.50	0.43	0.47	0.45	0.020	0.51	0.44	0.49	0.42	0.027	0.014	0.467	0.982
15:0	0.092	0.086	0.090	0.088	0.003	0.091	0.090	0.093	0.083	0.004	0.152	0.532	0.305
16:0	22.70	20.34	21.44	21.60	1.270	21.42	21.45	23.97	19.23	1.800	0.870	0.541	0.602
7c16:1	0.34	0.27	0.30	0.32	0.020	0.32	0.27	0.35	0.28	0.029	0.035	0.497	0.728
9c16:1	2.17	1.52	1.87	1.81	0.120	2.32 ^{a}	1.43 ^{b}	2.02 ^{ab}	1.60 ^{b}	0.170	<0.001	0.727	0.176
17:0	0.14	0.15	0.14	0.14	0.009	0.12	0.15	0.15	0.14	0.013	0.402	0.823	0.167
17:1	0.19	0.20	0.20	0.19	0.017	0.18	0.22	0.20	0.18	0.024	0.700	0.642	0.192
18:0	10.41	10.36	10.56	10.20	1.028	9.54	11.59	11.28	9.12	1.454	0.972	0.805	0.159
9c18:1	23.08	21.67	22.45	22.36	0.965	24.02	20.88	22.15	22.46	1.365	0.308	0.918	0.218
11c18:1	1.87	1.69	1.80	1.76	0.060	1.97	1.63	1.77	1.75	0.085	0.043	0.677	0.072
18:2n-6	22.92	27.85	24.50	25.47	1.140	23.34	25.66	22.50	28.45	1.612	0.016	0.551	0.269
18:3n-6	0.19	0.16	0.16	0.20	0.013	0.18	0.14	0.21	0.19	0.018	0.110	0.039	0.443
18:3n-3	1.56	2.09	1.77	1.88	0.108	1.58 ^{b}	1.96 ^{ab}	1.54 ^{b}	2.22 ^{a}	0.153	0.001	0.465	0.344
20:0	0.09	0.10	0.09	0.10	0.008	0.08	0.10	0.10	0.09	0.011	0.773	0.828	0.146
20:1n-9	0.23	0.23	0.25	0.21	0.014	0.27	0.23	0.20	0.22	0.020	0.737	0.042	0.106
20:2	0.55	0.60	0.57	0.58	0.040	0.58	0.56	0.52	0.65	0.056	0.331	0.785	0.223
20:3n-6	0.60	0.56	0.61	0.56	0.041	0.67	0.55	0.54	0.58	0.059	0.503	0.412	0.207
20:4n-6	3.25	3.55	3.47	3.32	0.251	3.41	3.53	3.08	3.56	0.355	0.410	0.677	0.619

Table 6. *Cont.*

Parameter						Treatment †						Source of Variation	
	OP		Se					OP × Se					
20:3n-3	0.08	0.09	0.08	0.09	0.009	0.08	0.09	0.07	0.10	0.012	0.230	0.741	0.234
20:5n-3	0.19	0.17	0.19	0.17	0.018	0.21	0.16	0.17	0.17	0.025	0.302	0.510	0.378
22:0	0.03	0.03	0.03	0.03	0.004	0.03	0.02	0.03	0.04	0.006	0.547	0.229	0.089
22:1	0.02	0.01	0.02	0.01	0.004	0.02	0.01	0.01	0.01	0.005	0.087	0.359	0.257
22:2	1.91	1.86	1.98	1.79	0.188	1.87	2.10	1.96	1.62	0.266	0.842	0.471	0.293
22:4n-6	0.74	0.81	0.79	0.76	0.058	0.74	0.84	0.74	0.79	0.082	0.390	0.759	0.815
22:5n-3	0.65	0.70	0.68	0.67	0.049	0.66	0.71	0.65	0.69	0.069	0.506	0.796	0.953
22:6n-3	0.63	0.52	0.60	0.55	0.050	0.66	0.54	0.60	0.50	0.071	0.135	0.488	0.933
SFA ‡	33.97	31.50	32.84	32.63	2.251	31.82	33.85	36.12	29.14	3.18	0.443	0.948	0.167
MUFA ‡	27.90	25.58	26.88	26.60	1.080	29.10	24.66	26.71	26.50	1.520	0.138	0.857	0.175
PUFA ‡	33.28	38.17	35.40	36.04	1.249	33.98 ab	36.83 ab	32.58 b	39.51 a	1.767	0.001	0.719	0.258
n-3	3.11	3.57	3.32	3.36	0.152	3.19	3.46	3.04	3.67	0.304	0.043	0.879	0.393
n-6	27.71	32.14	29.53	30.32	1.252	28.35	30.71	27.07	33.56	2.503	0.018	0.660	0.253
n-6/n-3	9.07	9.00	8.88	9.17	0.205	8.90	8.86	9.26	9.13	0.437	0.824	0.307	0.959

a, b means in the OP × Se row sharing no common superscript are statistically different ($p < 0.05$). † C: diet with no additive; OP: diet supplemented with 50 g orange pulp per kg feed; Se: diet supplemented with 0.15 ppm Se; OP + Se: diet supplemented with 0.15 ppm Se and 50 g orange pulp per kg feed. ‡ SFA: saturated FA; MUFA: monounsaturated FA; PUFA: polyunsaturated FA.

4. Discussion

The performance and carcass traits were not affected by OP and/or organic Se dietary supplementation, apart from a reduced carcass dressing percentage (DP) in the group of birds fed with OP. This reduction may be attributed to the increased fiber content, by 10%, of the OP diet, in comparison with the control diet, 33 vs. 29.5 g/kg, respectively. The increased fiber levels in the broiler rations are associated with the increased digestive tract weight [34] and, consequently, with the reduced carcass yield. Our results are in partial agreement with Murao et al. [12], who detected a reduced DP in broilers fed with citrus pulp at the level of 10%, but not at 5%, of the diet. On the other hand, in the aforementioned study, the broiler performance was deteriorated because of the high-fiber content of the citrus pulp used (insoluble and soluble fiber contents more than 30%), whereas, in the present study, the fiber content of the orange pulp was much lower, at 10.1% [28], and the performance traits were unaffected by the OP supplementation. The broiler performance in the present study was also not influenced by the dietary supplementation with organic Se, which is in agreement with a number of studies related to organic selenium supplementation in broiler chickens, at dosages ranging from 0.1 to 0.5 ppm [35–39]. The data concerning the growth performance parameters are also supported by the behavioral recordings, since no significant differences in the percentages of chickens standing at the feeder (Pf) or the drinker (Pw) were observed among the OP and/or organic Se groups.

The cooking loss, shear force, and the color attributes of the broiler meat remained unaffected by OP and/or Se dietary supplementation. However, the pH24 values were lower in the breast muscles from the birds fed with OP, in comparison with the controls. In close agreement with our study, Mourao et al. [12] report that the meat from broilers fed with citrus pulp at 10% exhibited reduced pH24 values, in comparison with the meat from broilers fed with 5% or 0% citrus pulp. On the other hand, it has been reported that ostriches fed with a diet supplemented with citrus pulp at 20% showed increased meat pH values and, consequently, decreased cooking losses, in comparison with the controls [40]. However, in our study, not only was the pH lower in the meat from broilers fed with OP compared to the controls, but this reduction was also not accompanied by an increased cooking loss, which was unaffected by the dietary intervention.

The meat oxidative stability of the broilers fed with OP + Se was improved during storage of more than nine days. Dietary supplementation with OP and/or Se alone also enhanced the meat oxidative stability, though to a lesser extent, in comparison with the supplementation of a combination of OP and Se. The favorable combined effects of OP and Se are attributed to their synergistic effects, which were determined in the present study. The antioxidant properties of added Se and hesperidin, which were present in the added citrus pulp at a rate of 0.426 g/kg, are most likely responsible for the beneficial effects of the experimental diets in the broiler meat. The dietary supplementation of broiler chickens with hesperidin at 0.75 and 1.5 g/kg feed [9], or at 5 mg/kg feed [13], or at 1.5 and 3 g/kg feed [8], improved the meat oxidative stability, without any negative side effects on the performance or the meat quality. However, undetermined substances found in citrus pulp, such as carotenoids or other phenolic compounds, may have also contributed to the improved meat oxidative stability in the OP-supplemented diets. An extra group of birds, offered a diet supplemented with pure hesperidin at 0.426 g/kg, would assist us in answering the question of whether the improved meat oxidative stability determined in the OP group was attributed solely to hesperidin or not. Unfortunately, our resources did not allow for an expansion of the experimental design with the addition of an extra treatment.

Dietary supplementation with organic Se beneficially improves the meat oxidative stability, and, subsequently, the meat shelf life, of broiler chickens, at doses of 0.3–0.6 ppm [41], 0.1–0.4 ppm [16], 0.5 ppm [42], 0.3 ppm [43], or 0.15–3 ppm [44], and of turkeys at the level of 0.3 ppm [45], in agreement with the present study.

The antioxidant activity of the flavonoid, hesperidin, is presumed to be analogous to that of vitamin E, and this is attributed to its scavenging property of lipid peroxyl radicals, which, in turn, terminates the chain reactions of the lipid peroxidation in the cell membrane [46]. On the other hand, Se is strongly associated with the biological endogenous antioxidant defense system because it is an essential constituent for the formation of the antioxidant enzyme, glutathione peroxidase (GSH-Px), which, along with the non-Se-containing enzymes, such as catalase and superoxide dismutase, comprise the primary antioxidant defense system [43].

In the present study, we found the presence of a synergism between the citrus pulp and the Se. The substances under investigation exhibited synergistic action and, therefore, their antioxidant activity was enhanced when they were fed together. The interaction among natural antioxidants, which may result in an enhanced or decreased oxidative stability of the meat, has been reported in a number of studies. A combination of rosemary and green tea extracts, and a combination of green tea and natural tocopherols, exhibited synergistic antioxidant action in chicken patties when fed to broiler chickens at 200 mg/kg [47]. Synergistic antioxidant effects in meat have also been observed between oregano and rosemary essential oils, when fed in broilers at a dose of 300 mg/kg [48], and between oregano oil and α-tocopheryl acetate, when fed to turkeys at 200 mg/kg [49]. On the other hand, a considerable decrease in the antiradical activity values of an ethanol solution containing resveratrol, catechin, and quercetin was observed, in comparison with each of the aforementioned phenol compounds alone, because of an unfavorable interaction among the phenols [50]. However, this decreased antiradical activity of the combination of the antioxidants was measured by the α, α-diphenyl-β-picrylhydrazyl (DPPH) free radical scavenging method, and it has not been tested in vivo.

The synergistic action between Se and natural antioxidants has already been reported. The combined supplementation of Se and vitamin E improved the antioxidant status of the skeletal muscles of heat-stressed broiler chickens, more than Se and vitamin E alone [51]. The dietary supplementation of vitamin E in combination with organic Se has a synergistic effect in minimizing lipid peroxidation, and it improves the antioxidative status in the seminal plasma of cockerels, which may result in an increased spermatozoa count and enhanced motility, and a reduced percentage of dead spermatozoa under heat-stress conditions [52]. It has been proposed that the synergistic effects involve two antioxidants, of which one reacts with the peroxy radical, and is consumed, and the second regenerates the first, effectively sparing [53]. A sparing effect of dietary grape pomace, rich in natural antioxidants, on liver vitamin E has also been observed in broiler chickens [54].

Dietary modification by OP supplementation significantly improved the breast meat FA profiles because of the increases in the health-promoting ALA and PUFA contents. A mean increase in the ALA content by 25.4% was determined between the OP-supplemented groups and the non-OP-supplemented groups. Therefore, the supplementation of broiler diets with OP not only improved the antioxidant properties of the breast meat stored in the refrigerator, but also enhanced its nutritional value through fortification with health-promoting n-3 fatty acids and PUFAs. The observed increase in the n-3 fatty acid content may be attributed to the increased n-3 fatty acid content of the OP-supplemented diets, in comparison with the non-OP-supplemented diets. On the other hand, the increased PUFA content in the breast muscles of broilers fed with OP may possibly be attributed to the protective action of the OP antioxidants on PUFAs from the oxidation breakdown, since no PUFA differences were observed among the experimental diets. In agreement with our study, the PUFAs and the ALA content in broiler meat were increased as a result of the dietary supplementation with citrus pulp at 10% of the feed [12]. A significant increase in the PUFA content has also been reported for the meat of ostriches fed with dried citrus pulp at 20% of the feed [40]. However, in the aforementioned studies, the increase in the PUFA content was accompanied by decreases in both the MUFA and SFA intramuscular fat contents, and no effect of dried citrus pulp on the n-3 fatty acid content in ostrich meat was observed. Species and diet differences may explain the discrepancies found among the published studies.

5. Conclusions

The dietary supplementation of broiler chicks with OP and organic Se, at 50 g/kg and 0.15 ppm, respectively, improved the oxidative stability of breast meat during storage, with the improvement being more noticeable for long-term frozen storage. The observed synergistic action of the OP and organic Se suggests that the combined supplementation of the byproduct of the citrus industry, which is rich in natural antioxidants and Se, supports both the first-line enzymatic and second-line nonenzymatic antioxidant defense systems, with beneficial effects on the product shelf life. Dietary modification by OP supplementation improved the breast meat nutritional value through the fortification with PUFAs and, more importantly, the health-promoting ALA n-3 fatty acid. Nevertheless, the beneficial effects observed in the present study were not accompanied by negative side effects on the performance or meat quality, suggesting that the inclusion of OP into broiler diets, at 50 g/kg of feed, along with organic Se, may improve the meat nutritional value and the shelf life, with benefits for both the farmer and the consumer. More research is required in order to elucidate the underlying mechanisms that explain the synergistic action between OP and organic Se.

Author Contributions: Conceptualization, E.Z., P.S. and M.G.; formal analysis, M.G.; funding acquisition, E.Z. and P.S.; investigation, E.Z., P.S., D.K., P.K., A.C.P., G.P. and M.G.; project administration, E.Z. and M.G.; supervision, E.Z. and M.G.; writing—original draft, P.S. and M.G.; writing—review and editing, E.Z., D.K., P.K., A.C.P., G.P. and M.G. All authors have read and agreed to the published version of the manuscript.

Funding: This work was funded by the project, "Innovation and Entrepreneurship—Exploitation of Research at the Agricultural University of Athens", which was co-financed by the European Union (European Regional Development Fund), and the Greek State, grant number: 770001.

Institutional Review Board Statement: The study was conducted according to the guidelines of Directive 2010/63/EU of the European Parliament, and of the Council on the protection of animals used for scientific purposes. The protocol was approved by the Research Ethics Committee of the Agricultural University of Athens, with the approval number: 32/20052015.

Informed Consent Statement: Not applicable.

Data Availability Statement: The data presented in this study are available upon request from the corresponding author.

Conflicts of Interest: The authors declare no conflict of interest.

References

1. Vasta, V.; Luciano, G. The effects of dietary consumption of plants secondary compounds on small ruminants' products quality. *Small Rumin. Res.* **2011**, *101*, 150–159. [CrossRef]

2. Pisoschi, A.M.; Pop, A. The role of antioxidants in the chemistry of oxidative stress: A review. *Eur. J. Med. Chem.* **2015**, *97*, 55–74. [CrossRef] [PubMed]
3. Embuscado, M.E. Spices and herbs: Natural sources of antioxidants—A mini review. *J. Funct. Foods* **2015**, *18*, 811–819. [CrossRef]
4. Simitzis, P.E.; Deligeorgis, S.G. Agroindustrial By-Products and Animal Products: A Great Alternative for Improving Food-Quality Characteristics and Preserving Human Health. In *Food Quality: Balancing Health and Disease. Handbook of Food Bioengineering*, 1st ed.; Holban, A.M., Grumezescu, A.M., Eds.; Academic Press Elsevier: London, UK, 2018; Volume 13, pp. 253–290.
5. Bampidis, V.A.; Robinson, P.H. Citrus by-products as ruminant feeds: A review. *Anim. Feed Sci. Technol.* **2006**, *128*, 175–217. [CrossRef]
6. Kamboh, A.A.; Arain, M.A.; Mughal, M.J.; Zaman, A.; Arain, Z.M.; Soomro, A.H. Flavonoids: Health promoting phytochemicals for animal production—A review. *J. Anim. Health Prod.* **2015**, *3*, 6–13. [CrossRef]
7. Ebrahimi, A.; Qotbi, A.A.A.; Seidavi, A. The effects of different levels of dried *Citrus sinensis* peel on broiler carcass quality. *Acta Sci. Vet.* **2013**, *41*, 1169.
8. Simitzis, P.E.; Symeon, G.K.; Charismiadou, M.A.; Ayoutanti, A.G.; Deligeorgis, S.G. The effects of dietary hesperidin supplementation on broilers performance and chicken meat characteristics. *Can. J. Anim. Sci.* **2011**, *91*, 275–282. [CrossRef]
9. Goliomytis, M.; Kartsonas, N.; Charismiadou, M.A.; Symeon, G.; Simitzis, P.E.; Deligeorgis, S.G. The influence of naringin or hesperidin dietary supplementation on broiler meat quality and oxidative stability. *PLoS ONE* **2015**, *10*, e0141652. [CrossRef] [PubMed]
10. Abbasi, H.; Seidavi, A.; Liu, W.; Asadpour, L. Investigation on the effect of different levels of dried sweet orange (*Citrus sinensis*) pulp on performance, carcass characteristics and physiological and biochemical parameters in broiler chicken. *Saudi J. Biol. Sci.* **2015**, *22*, 139–146. [CrossRef]
11. Jiang, X.R.; Zhang, H.J.; Wang, J.; Wu, S.G.; Yue, H.Y.; Lü, H.Y.; Cui, H.; Bontempo, V.; Qi, G.H. Effect of dried tangerine peel extract supplementation on the growth performance and antioxidant status of broiler chicks. *Ital. J. Anim. Sci.* **2016**, *15*, 642–648. [CrossRef]
12. Mourão, J.L.; Pinheiro, V.M.; Prates, J.A.M.; Bessa, R.J.B.; Ferreira, L.M.A.; Fontes, C.M.G.A.; Ponte, P.I.P. Effect of dietary dehydrated pasture and citrus pulp on the performance and meat quality of broiler chickens. *Poult. Sci.* **2008**, *87*, 733–743. [CrossRef] [PubMed]
13. Kamboh, A.A.; Zhu, W.Y. Individual and combined effects of genistein and hesperidin supplementation on meat quality in meat-type broiler chickens. *J. Sci. Food Agric.* **2013**, *93*, 3362–3367. [CrossRef] [PubMed]
14. Ahmed, S.T.; Mun, H.S.; Islam, M.M.; Kim, S.S.; Hwang, J.A.; Kim, Y.J.; Yang, C.J. Effects of *Citrus junos* by-products fermented with multistrain probiotics on growth performance, immunity, caecal microbiology and meat oxidative stability in broilers. *Br. Poult. Sci.* **2014**, *55*, 540–547. [CrossRef] [PubMed]
15. Surai, P.F. *Natural Antioxidants in Avian Nutrition and Reproduction*; Nottingham University Press: Nottingham, UK, 2002.
16. Bakhshalinejad, R.; Akbari Moghaddam Kakhki, R.; Zoidis, E. Effects of different dietary sources and levels of selenium supplements on growth performance, antioxidant status and immune parameters in Ross 308 broiler chickens. *Br. Poult. Sci.* **2018**, *59*, 81–91. [CrossRef] [PubMed]
17. Zoidis, E.; Papadomichelakis, G.; Pappas, A.C.; Theodorou, G.; Fegeros, K. Effects of selenium and cadmium on breast muscle fatty-acid composition and gene expression of liver antioxidant proteins in broilers. *Antioxidants* **2019**, *8*, 147. [CrossRef] [PubMed]
18. Food and Drug Administration (FDA). Food and Drug Administration Approves Food Additive Petition for Selenium Yeast. In *Food and Drug Administration Veterinarian Newsletter (July/August)*; Food and Drug Administration (FDA): Washington, DC, USA, 2000.
19. Wang, Y.; Zhan, X.; Yuan, D.; Zhang, X.; Wu, R. Influence of dietary selenomethionine supplementation on performance and selenium status of broiler breeders and their subsequent progeny. *Biol. Trace Elem. Res.* **2011**, *143*, 1497–1507. [CrossRef]
20. Kim, Y.; Mahan, D. Comparative effects of high dietary levels of organic and inorganic selenium on selenium toxicity of growing-finishing pigs. *J. Anim. Sci.* **2001**, *79*, 942–948. [CrossRef]
21. Attia, Y.A.; Abdalah, A.A.; Zeweil, H.S.; Bovera, F.; Tag El-Din, A.A.; Araft, M.A. Effect of inorganic or organic selenium supplementation on productive performance, egg quality and some physiological traits of dual-purpose breeding hens. *Czech J. Anim. Sci.* **2010**, *55*, 505–519. [CrossRef]
22. Hassan, F.; Mobarez, S.; Mohamed, M.; Attia, Y.; Mekawy, A.; Mahrose, K. Zinc and/or Selenium Enriched Spirulina as Antioxidants in Growing Rabbit Diets to Alleviate the Deleterious Impacts of Heat Stress during Summer Season. *Animals* **2021**, *11*, 756. [CrossRef]
23. Cai, S.J.; Wu, C.X.; Gong, L.M.; Song, T.; Wu, H.; Zhang, L.Y. Effects of nano-selenium on performance, meat quality, immune function, oxidation resistance, and tissue selenium content in broilers. *Poult. Sci.* **2012**, *91*, 2532–2539. [CrossRef]
24. Payne, R.; Southern, L. Comparison of inorganic and organic selenium sources for broilers. *Poult. Sci.* **2005**, *84*, 898–902. [CrossRef] [PubMed]
25. Dlouhá, G.; Ševčíková, S.; Dokoupilová, A.; Zita, L.; Heindl, J.; Skřivan, M. Effect of dietary selenium sources on growth performance, breast muscle selenium, glutathione peroxidase activity and oxidative stability in broilers. *Czech J. Anim. Sci.* **2008**, *53*, 265–269. [CrossRef]
26. Ševčíková, S.; Skřivan, M.; Dlouhá, G.; Koucký, M. The effect of selenium source on the performance and meat quality of broiler chickens. *Czech J. Anim. Sci.* **2006**, *51*, 449–457. [CrossRef]

27. Zhou, X.; Wang, Y. Influence of dietary nano elemental selenium on growth performance, tissue selenium distribution, meat quality, and glutathione peroxidase activity in guangxi yellow chicken. *Poult. Sci.* **2011**, *90*, 680–686. [CrossRef]
28. Goliomytis, M.; Kostaki, A.; Avgoulas, G.; Lantzouraki, D.Z.; Siapi, E.; Zoumpoulakis, P.; Simitzis, P.; Deligeorgis, S.G. Dietary supplementation with orange pulp (*Citrus sinensis*) improves egg yolk oxidative stability in laying hens. *Anim. Feed Sci. Technol.* **2018**, *244*, 28–35. [CrossRef]
29. Cason, J.A.; Lyon, C.E.; Papa, C.M. Effect of muscle opposition during rigor on development of broiler breast meat tenderness. *Poult. Sci.* **1997**, *76*, 785–787. [CrossRef]
30. Botsoglou, N.A.; Fletouris, D.J.; Papageorgiou, G.E.; Vassilopoulos, V.N.; Mantis, A.J.; Trakatellis, A.G. A rapid, sensitive and specific thiobarbituric acid method for measuring lipid peroxidation in animal tissues, food and feedstuff samples. *J. Agric. Food Chem.* **1994**, *42*, 1931–1937. [CrossRef]
31. O'Fallon, J.V.; Busboom, J.R.; Nelson, M.L.; Gaskins, C.T. A direct method for fatty acid methylester synthesis: Application to wet meat tissues, oils, and feedstuffs. *J. Anim. Sci.* **2007**, *85*, 1511–1521. [CrossRef]
32. Sas/Stat. *Statistical Analysis Systems, Version 9.1.3*; SAS Institute Inc.: Cary, NC, USA, 2005.
33. Yadav, B.; Wennerberg, K.; Aittokallio, T.; Tang, J. Searching for Drug Synergy in Complex Dose–Response Landscapes Using an Interaction Potency Model. *Comput. Struct. Biotechnol. J.* **2015**, *13*, 504–513. [CrossRef]
34. Tabook, N.M.; Kadim, I.T.; Mahgoub, O.; Al-Marzooqi, W. The effect of date fiber supplemented with an exogenous enzyme on the performance and meat quality of broiler chickens. *Br. Poult. Sci.* **2006**, *47*, 73–82. [CrossRef]
35. Chen, G.; Wu, J.; Li, C. The effect of different selenium levels on production performance and biochemical parameters of broilers. *Ital. J. Anim. Sci.* **2013**, *12*, 486–491. [CrossRef]
36. Rama Rao, S.V.; Prakash, B.; Raju, M.V.L.N.; Panda, A.K.; Poonam, S.; Murthy, O.K. Effect of supplementing organic selenium on performance, carcass traits, oxidative parameters and immune responses in commercial broiler chickens. *Asian-Australas. J. Anim. Sci.* **2013**, *26*, 247–252.
37. Zdunczyk, Z.; Gruzauskas, R.; Semaskaite, A.; Juskiewicz, J.; Raceviciute-Stupeliene, A.; Wroblewska, M. Fatty acid profile of breast muscle of broiler chickens fed diets with different levels of selenium and vitamin E. *Arch. Geflügelk.* **2011**, *75*, 264–267.
38. Heindle, J.; Ledvinka, Z.; Englmaierova, M.; Zita, L.; Tumova, E. The effect of dietary selenium sources and levels on performance, selenium content in muscle and glutathione peroxidase activity in broiler chickens. *Czech J. Anim. Sci.* **2010**, *55*, 572–578. [CrossRef]
39. Perić, L.; Milošević, N.; Žikić, D.; Kanački, Z.; Džinić, N.; Nollet, L.; Spring, P. Effect of selenium sources on performance and meat characteristics of broiler chickens. *J. Appl. Poult. Res.* **2009**, *18*, 403–409. [CrossRef]
40. Lanza, M.; Fasone, V.; Galofaro, V.; Barbagallo, D.; Bella, M.; Pennisi, P. Citrus pulp as an ingredient in ostrich diet: Effects on meat quality. *Meat Sci.* **2004**, *68*, 269–275. [CrossRef]
41. Ibrahim, D.; Kishawy, A.T.Y.; Khater, S.I.; Hamed Arisha, A.; Mohammed, H.A.; Abdelaziz, A.S.; Abd El-Rahman, G.I.; Elabbasy, M.T. Effect of dietary modulation of selenium form and level on performance, tissue retention, quality of frozen stored meat and gene expression of antioxidant status in ross broiler chickens. *Animals* **2019**, *9*, 342. [CrossRef]
42. Korzeniowska, M.; Króliczewska, B.; Kopeć, W. Effect of dietary selenium on protein and lipid oxidation and the antioxidative potential of selected chicken culinary parts during frozen storage. *J. Chem.* **2018**, *2018*, 3492456. [CrossRef]
43. Li, J.; Zhang, L.; Yang, Z.; Zhang, Z.; Jiang, Y.; Gao, F.; Zhou, G. Effects of different selenium sources on growth performance, antioxidant capacity and meat quality of local Chinese Subei chickens. *Biol. Trace Elem. Res.* **2018**, *181*, 340–346. [CrossRef]
44. Pappas, A.C.; Zoidis, E.; Papadomichelakis, G.; Fegeros, K. Supranutritional selenium level affects fatty acid composition and oxidative stability of chicken breast muscle tissue. *J. Anim. Physiol. Anim. Nutr.* **2012**, *96*, 385–394. [CrossRef]
45. Mikulski, D.; Jankowski, J.; Zduńczyk, Z.; Wróblewska, M.; Sartowska, K.; Majewska, T. The effect of selenium source on performance, carcass traits, oxidative status of the organism, and meat quality of turkeys. *J. Anim. Feed Sci.* **2009**, *18*, 518–530. [CrossRef]
46. Van Acker, F.A.A.; Schouten, O.; Haenen, G.R.M.M.; Van Der Vijgh, W.J.F.; Bast, A. Flavonoids can replace a-tocopheryl as an antioxidant. *FEBS Lett.* **2000**, *473*, 145–148. [CrossRef]
47. Smet, K.; Raes, K.; Huyghebaert, G.; Haak, L.; Arnouts, S.; De Smet, S. Lipid and protein oxidation of broiler meat as influenced by dietary natural antioxidant supplementation. *Poult. Sci.* **2008**, *87*, 1682–1688. [CrossRef] [PubMed]
48. Basmacioglu, H.; Tokusoglu, O.; Ergul, M. The effect of oregano and rosemary essential oils or α-tocopheryl acetate on performance and lipid oxidation of meat enriched with n-3 PUFAs in broilers. *S. Afr. J. Anim. Sci.* **2004**, *34*, 197–210.
49. Papageorgiou, G.; Botsoglou, N.; Govaris, A.; Giannenas, I.; Iliadis, S.; Botsoglou, E. Effect of dietary oregano oil and α-tocopheryl acetate supplementation on iron-induced lipid oxidation of turkey breast, thigh, liver and heart tissues. *J. Anim. Physiol. Anim. Nutr.* **2003**, *87*, 324–335. [CrossRef]
50. Pinelo, M.; Manzocco, L.; Nunez, M.J.; Nicoli, M.C. Interaction among phenols in food fortification: Negative synergism on antioxidant capacity. *J. Agric. Food Chem.* **2004**, *52*, 1177–1180. [CrossRef]
51. Ghazi, S.; Habibian, M.; Moeini, M.M.; Abdolmohammadi, A. Effects of dietary selenium, vitamin E, and their combination on growth, serum metabolites and antioxidant defense system in skeletal muscle of broilers under heat stress. *Biol. Trace Elem. Res.* **2012**, *148*, 322–330. [CrossRef]
52. Ebeid, T.A. Vitamin E and organic selenium enhances the antioxidative status and quality of chicken semen under high ambient temperature. *Br. Poult. Sci.* **2012**, *53*, 708–714. [CrossRef]

53. Brewer, M. Natural antioxidants: Sources, compounds, mechanisms of action, and potential applications. *Compr. Rev. Food Sci. Food Saf.* **2011**, *10*, 221–247. [CrossRef]
54. Goñi, I.; Brenes, A.; Centeno, C.; Viveros, A.; Saura-Calixto, F.; Rebolé, A.; Arija, I.; Estevez, R. Effect of dietary grape pomace and vitamin e on growth performance, nutrient digestibility, and susceptibility to meat lipid oxidation in chickens. *Poult. Sci.* **2007**, *86*, 508–516. [CrossRef]

sustainability

Article

Effects of In Ovo Feeding of γ-Aminobutyric Acid on Growth Performances, Plasma Metabolites, and Antioxidant Status in Broilers Exposed to Cyclic Heat Stress

Chris-Major Ncho [1,2,†], Akshat Goel [1,3,†], Chae-Mi Jeong [1,2], Vaishali Gupta [1,2] and Yang-Ho Choi [1,2,3,*]

1 Department of Animal Science, Gyeongsang National University, Jinju 52828, Korea; major159@gnu.ac.kr (C.-M.N.); genesakshat@gnu.ac.kr (A.G.); wjdcoa@gnu.ac.kr (C.-M.J.); vaishali2020@gnu.ac.kr (V.G.)
2 Division of Applied Life Sciences (BK21 Plus Program), Gyeongsang National University, Jinju 52828, Korea
3 Institute of Agriculture and Life Sciences, Gyeongsang National University, Jinju 52828, Korea
* Correspondence: yhchoi@gnu.ac.kr
† These authors contributed equally to this work.

Abstract: γ-aminobutyric acid (GABA) is an amino acid used for mitigating the detrimental effects of heat stress in broilers. In addition, a growing body of literature suggests that the in ovo feeding of various nutrients can enhance the post-hatch thermotolerance of broilers. Therefore, we hypothesized that the supplementation of GABA during incubation might have positive effects in heat-stressed broilers. Chicks hatched from eggs were divided into three groups described as follows: chicks hatched from eggs incubated at normal temperature and then raised under thermoneutral temperature (CON); chicks hatched from eggs incubated at normal temperature but raised under cyclic heat stress (HS) (CON+HS); and chicks hatched from eggs injected with 60 mg of GABA dissolved in 0.6 mL of distilled water but raised under cyclic HS (G10+HS). The HS was applied between 28 and 31 days of age with ambient temperatures raised from 22 ± 1 °C to 33 ± 1 °C for 6 h daily. Compared to the CON group, average daily weight gain was significantly lower in the CON+HS but not in the G10+HS group. Feed intake was significantly decreased in both the CON+HS and G10+HS groups. Compared to the CON group, plasma corticosterone levels were significantly increased in the CON+HS group, but not the G10+HS group. Hepatic mRNA levels of the acetyl-CoA carboxylase gene (ACC) were significantly reduced in the G10+HS group compared to the CON group. In addition, positive Pearson correlation coefficients were found in mRNA levels between fatty acid synthase (FAS) and nicotinamide adenine dinucleotide phosphate oxidase 1 (NOX1) ($r = 0.55$, $p < 0.05$), NOX1 and NOX4 ($r = 0.65$, $p < 0.01$), and catalase (CAT) and superoxide dismutase (SOD) ($r = 0.62$, $p < 0.05$). Taken together, the results suggest that this study can serve as a basis for future work focusing on the in ovo feeding of GABA as a technique to combat heat stress in broilers.

Keywords: in ovo feeding; GABA; heat stress; broilers; antioxidant status

Citation: Ncho, C.-M.; Goel, A.; Jeong, C.-M.; Gupta, V.; Choi, Y.-H. Effects of In Ovo Feeding of γ-Aminobutyric Acid on Growth Performances, Plasma Metabolites, and Antioxidant Status in Broilers Exposed to Cyclic Heat Stress. *Sustainability* **2021**, *13*, 11032. https://doi.org/10.3390/su131911032

Academic Editors: George K. Symeon, Vassilios Dotas and Chunjiang An

Received: 21 August 2021
Accepted: 30 September 2021
Published: 5 October 2021

1. Introduction

Due to global warming, high rearing temperature is becoming one of the most common issues in poultry production [1]. Heat stress (HS) occurs when the ambient temperature exceeds the optimal range recommended for growth. Due to the high selection for fast growth performance, broilers in the last phase of rearing are especially sensitive to HS [2]. Indeed, HS can impair the performance, metabolism, and health of broilers. The effects of HS can include reduced feed intake (FI), body weight gain (BW), and increased water consumption [3]. Excessively high ambient temperature is strongly correlated with high mortality rates [4]. In addition, previous reports have shown that HS induces oxidative damage by increasing the production of reactive oxygen species in cells [5,6].

Several strategies have been adopted to mitigate the deleterious effects of HS in poultry. The most common approaches include improvements in ventilation, the selection of resistant strains, and nutritional management [7]. The latter consists of the dietary supplementation of various types of substances such as γ-aminobutyric acid (GABA). GABA is a four-carbon non-protein amino acid involved in inhibitory synaptic transmission [8]. In addition, GABA has been implicated in several regulatory functions such as memory, blood pressure, and respiration [9]. In broilers, GABA is used as a feed supplement to reduce the adverse effect of high environmental temperature [10]. Previous research demonstrated its effectiveness for increasing overall growth performance, improving nutrient absorption, and reducing oxidative stress in chickens under high temperatures [11,12].

Recently, in ovo feeding has been considered a cost-effective way to alleviate the drawbacks of post-hatch and incubational HS in broilers. Originally, in ovo feeding was developed as a tool for providing a continuous supply of critical nutrients during the first few days of the chick's life [13]. The methodology consists of injecting nutrients into the amnion of eggs so that by consuming the content of the amnion, the embryo can access nutrients before it starts to hatch [14]. The most recent advances involve testing the heat resistance of birds after the in ovo feeding of prebiotics [15] or amino acids [16]. Methionine–cysteine in ovo injection improved embryonic development, antioxidant status, and jejunum histomorphometry of broilers chicks exposed to incubational HS [17]. Furthermore, the damages (high corticosterone and HSP70 mRNA levels) occasioned by incubational HS were alleviated after the in ovo feeding of sulfur-containing amino acids [18]. Interestingly, evidence suggests that broilers' resistance toward post-hatch HS can also be improved by in ovo feeding. For example, the in ovo feeding of L-leucine was proven effective in affording thermotolerance in broilers under acute heat stress, primarily by changing amino acid metabolism up to market age [19]. In addition, in ovo injection of galactooligosaccharides reduced the harmful effects of hyperthermia on feed efficiency during the finisher feeding phase [20]. When injected in ovo, we found that GABA could improve the early heat resistance of broilers by enhancing antioxidant functions and regulating fatty-acid-metabolism-related genes in 10-day-old chicks [21].

There have been no studies addressing the potential long-term effects of in ovo feeding of GABA in broilers. Therefore, the objective of this study was to investigate the effects of in ovo feeding of GABA on the growth performances, organ indexes, blood biochemical parameters, antioxidant status, and gene expression of stress-related genes in broilers subjected to cyclic HS near market age.

2. Materials and Methods

All the experimental procedures for this study were approved by the Institutional Animal Care and Use Committee of Gyeongsang National University (GNU-200916-C0058).

2.1. Incubation and In Ovo Feeding Procedure

Three hundred eggs were obtained from a commercial breeder farm with 47-week-old Arbor Acres hens (Hapcheon, Korea). According to standard incubation conditions, the eggs were set for incubation in an incubator (Rcom Co., Ltd., Gimhae, Korea). Briefly, from embryonic day (ED) 1 to ED 18, eggs were subjected to 37.8 °C and 56% relative humidity (RH), and from ED 18 until hatching, the incubation temperature was maintained at 36.8 °C with 70% RH. On ED 10, unfertilized eggs were removed from the incubator after candling. On ED 17.5, the eggs were distributed into three groups of equal numbers (n = 60/group) with similar weight (66 ± 4 g) using the Solver module of Microsoft Excel (Microsoft Excel 2016; Microsoft Corp., Redmond, WA, USA). In our previous study, we found that there were no significant differences (hatching parameters and biological parameters) between a sham control (distilled water injected) and a non-injected control [21]; therefore, we did not include a sham treatment in this trial.

Moreover, the solution and methodology used for the in ovo feeding procedure were similar to those previously described [21]. The first two groups (total 120 eggs) were considered as control and were not injected. The third group was injected with 0.6 mL of 10% (0.1 g mL^{-1}) GABA (#A2129 Sigma-Aldrich Inc., St. Louis, MO, USA) dissolved in distilled water. For in ovo injection, after a hole was made on the blunt end of each egg using a dental drill (Saeshin, Daegu, Korea), they were subsequently injected using a 1 mL syringe with a 23 G and 1-inch needle (Kovax-Syringe® Korea Vaccine Co. Ltd., Seoul, Korea), followed by being sealed using surgical tape (3M™ Micropore ™, Saint Paul, MN, USA) and returned to the incubator. The injection targeted the amnion. The control eggs were taken out of the incubator for the same amount of time without injection and then returned to the incubator. It should be mentioned that the incubation temperature was reduced to 20 °C due to an expected general blackout for 6 h between EDs 19 and 20 in the animal complex area.

2.2. Feeding Experiment and HS Challenge

After hatching, a total of sixty unsexed one-day-old broilers were raised in battery brooders under a thermally controlled environment at 34 ± 1 °C and 50% RH, and then the temperature was gradually decreased to 22 ± 1 °C on day 28. A commercially available feed, in crumbled form and water, were available ad libitum under continuous lighting. On day 28, the chicks were allocated into three different treatment groups: chicks hatched from non-injected eggs and reared at thermoneutral temperature (CON); chicks hatched from non-injected eggs and exposed to cyclic HS (CON+HS); and chicks hatched from GABA in ovo injected eggs and exposed to cyclic HS (G10+HS). Each treatment had 20 chicks kept in 5 cages with 4 chicks (unsexed) in each cage. They were then subjected to a daily cyclic HS challenge. An overview of the design is presented in Figure 1.

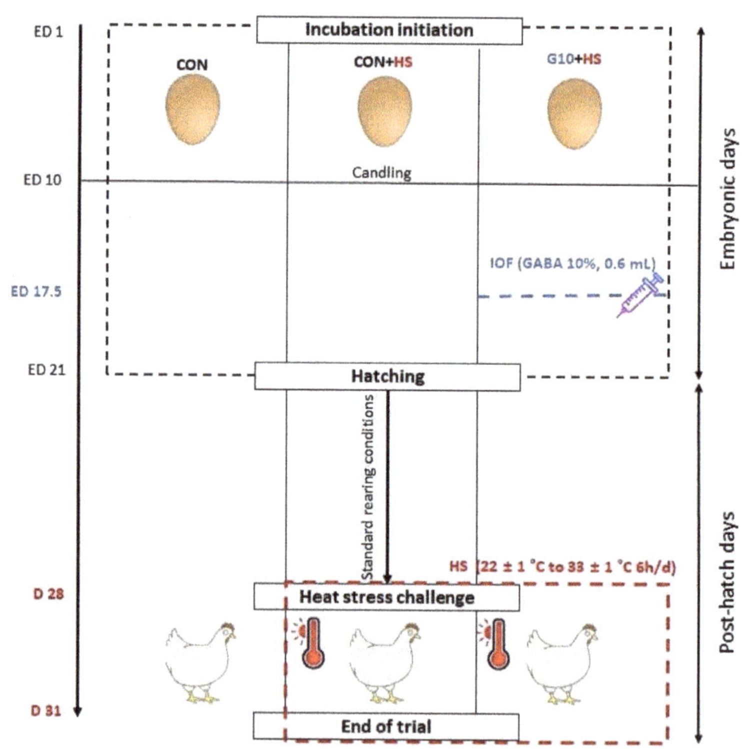

Figure 1. Study design. At 17.5 days of incubation, eggs were injected in ovo with or without 0.6 mL of GABA at 10%. Hatchlings obtained were raised until day 28 and subjected to cyclic heat stress (from 22 °C to 33 °C for 6 h daily) between 28 and 31 days of age or kept at a thermoneutral temperature (22 °C).

As previously executed [22], birds were challenged with a cyclic HS between 28 and 31 days of age with minor modifications. In one of two rooms, the ambient temperature was gradually increased from 22 ± 1 °C to 33 ± 1 °C over 6 h (from 10:00 a.m. to 4:00 p.m.) and then returned to 22 ± 1 °C over 1 h (Figure 2).

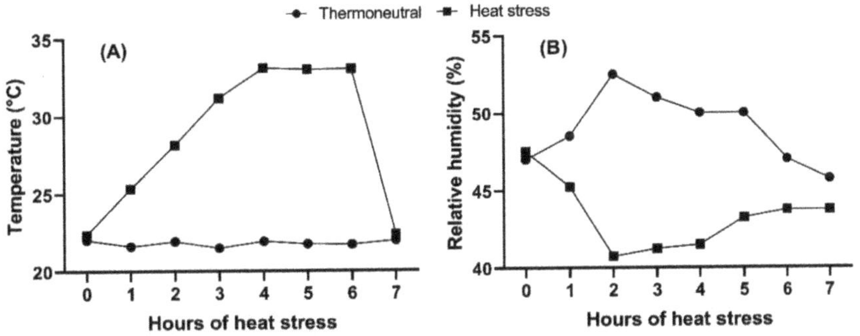

Figure 2. Average temperature (**A**) and relative humidity (**B**) variation in thermoneutral and heat stress environments. At 17.5 days of incubation, eggs were injected in ovo with or without 0.6 mL of GABA at 10%. Hatchlings obtained were raised until day 28 and subjected to cyclic heat stress (from 22 °C to 33 °C for 6 h daily) or kept at a thermoneutral temperature from days 28 to 31.

Chicks under thermoneutral temperature remained at 22 ± 1 °C until day 31. The body weight of individual birds and feed intake of each cage was measured weekly and at the end of the trial (day 31).

2.3. Rectal Temperature, Blood and Tissues Sampling

After the heat exposure on day 31, five birds close to the average body weight in each treatment were selected for blood and tissue sampling. The birds were euthanized with carbon dioxide before sampling. Blood was collected from heart puncture and transferred into heparinized vacuum containers (#367874, BD Co., Ltd., Franklin Lakes, NJ, USA). The blood samples were thereafter centrifuged at 2000× *g* for 10 min at 4 °C, and plasma was collected and then stored at −20 °C for later analysis. Organs were dissected free, and then weighed (the liver, spleen, bursa, heart, proventriculus, and gizzard), or the length was measured (the duodenum, jejunum, ileum, and caecum). Liver samples were snap-frozen in liquid nitrogen and stored at 2212−80 °C for further analysis. Rectal temperatures of birds selected for sampling were recorded before and after the cyclic HS exposure using a digital thermometer (HANNA instruments Inc., Padova, Italy) by inserting the probe 3 cm into the cloaca until the temperature readings stabilized.

2.4. Plasma Corticosterone, Metabolites, and Oxidative Stress Markers

Plasma corticosterone levels were detected by using an ELISA kit (ADI-901-097, ENZO Life Sciences, Inc., Farmingdale, NY, USA) according to the kit's instructions. Absorbance was measured at 405 nm by using a Multiskan FC microplate photometer (ThermoFisher Scientific, Inc., Waltham, MA, USA). Plasma metabolite concentrations were measured according to the manufacturer guide using a VetTest Chemistry Analyzer (IDEXX Laboratories, Inc., Westbrook, ME, USA) with dry-slide technology. Plasma samples were used for 2,2-diphenyl-1-picrylhydrazyl radical scavenging activity assay (DPPH-RSA) and malondialdehyde analysis. DPPH-RSA was based on a method developed previously [23]. Briefly, plasma (20 μL) was diluted in 480 μL of 10 mmol/L sodium–potassium phosphate (pH 7.4) followed by the addition of 500 μL of 0.1 mmol/L (DPPH) free radical. Samples were incubated for 30 min in a dark environment. After incubation, samples were centrifuged briefly for 3 min at 6000× *g* followed by the absorbance reading at 520 nm. The percentage of inhibition was calculated based on the following formula: %

inhibition = [1 − (A1/A0)] × 100, where A0 is the absorbance of the control and A1 is the absorbance of test samples. Malondialdehyde concentrations were analyzed based on a previously developed method [24] with slight modifications. Briefly, 400 µL of plasma were added in tubes containing 400 µL of 40% trichloroacetic acid (TCA). Subsequently, 800 µL of 0.67% thiobarbituric acid (TBA) was added to the mixture, which was then kept in a boiling water bath at 95 °C for 45 min. After cooling on ice, samples were centrifuged at 6000× *g* for 30 s followed by an absorbance reading at 530 nm. The antioxidant balance was then calculated by dividing the DPPH-RSA value by the MDA.

2.5. Real-Time PCR for mRNA Quantification

Total RNA extraction from the liver was performed using Trizol™ reagent (ThermoFisher Scientific, Inc.), following the manufacturer's protocol. The concentrations and purities of the samples were confirmed by reading the optical density of each sample in a Nanodrop (ThermoScientific, Inc.). Subsequently, the reverse transcription reaction was conducted using a PrimeScriptTM first-strand cDNA synthesis kit (Takara, Tokyo, Japan), following the manufacturer's guide. The cDNA synthesized was then used to perform real-time PCR using a StepOnePlus™ system (Applied Biosystems, Inc., Waltham, MA, USA) according to the following protocol: 10 min at 95 °C followed by 40 cycles of 15 s at 95 °C and 1 min at 60 °C. Each reaction well was composed of 20 µL volume containing Power SYBR™ green PCR master mix (ThermoScientific, Inc.), and a 10 pmol concentration of forward and reverse primer specific for each gene and cDNA. Information related to the primers is presented in Table 1. Target gene quantification was normalized to the Ct values of GAPDH. Relative expression was determined using the $2^{-\Delta\Delta ct}$ algorithm.

Table 1. Primer sequences used to evaluate the hepatic gene expression of broilers.

Gene	Sequence	Accession Number	Reference
ACC	F: CACTTCGAGGCGAAAAAC R: GGAGCAAATCCATGACCA	XM_015295697.2	This study
CAT	F: ACCAAGTACTGCAAGGCGAA R: TGAGGGTTCCTCTTCTGGCT	NM_001031215.1	[25]
FAS	F: CAATGGACTTCATGCCTC R: GCTGGGTACTGGAAGACA	NM_205155.3	This study
GAPDH	F: TTGGCATTGTGGAGGGTCTTA R: GTGGACGCTGGGATGATGTT	NM_204305.1	This study
GPx1	F: AACCAATTCGGGCACCAG R: CCGTTCACCTCGCACTTCTC	NM_001277853.2	[26]
HSP70	F: GCTGAACAAGAGCATCAATCCA R: CAGGAGCAGATCTTGCACATTT	AY143693.1	This study
HSP90	F: CCCGAGCAAGCTGGATTCT R: GGTCATCCCTATGCCGGTATC	NM_001109785	This study
NOX1	F: GCGAAGACGTGTTCCTGTAT R: GAACCTGTACCAGATGGACTTC	NM_001101830.1	[27]
NOX4	F: CCTCTGTGCTTGTACTGTGTAG R: GACATTGGAGGGATGGCTTAT	NM_001101829.1	[27]
SOD	F: AGGGGGTCATCCACTTCC R: CCCATTTGTGTTGTCTCCAA	NM_205064.1	[25]

Abbreviations: ACC, acetyl-CoA carboxylase; CAT, catalase; FAS, fatty acid synthase; GAPDH, glyceraldehyde-3-phosphate dehydrogenase; GPx1, glutathione peroxidase 1; HSP70, heat shock protein 70; HSP90, heat shock protein 90; NOX1, nicotinamide adenine dinucleotide phosphate oxidase 1; NOX4, nicotinamide adenine dinucleotide phosphate oxidase 4; SOD, superoxide dismutase.

2.6. Statistical Analysis

A replicate (cage) was considered an experimental unit for all the evaluated response parameters in this trial. A completely randomized design with 5 replicates per treatment was used. Each replicate was constituted of 4 unsexed birds. IBM SPSS Statistics for Windows software (IBM SPSS 25; IBM Corp., Armonk, NY, USA) was used to analyze all data via a one-way ANOVA. When the ANOVA was significant, a Tukey post hoc test was performed to assess the difference between means. As previously executed [28], before statistical analyses, the data in each group were subjected to the ROUT method of GraphPad prism 8 (GraphPad Software, Inc., La Jolla, CA, USA) to detect outliers. All percentage data below 20% and above 80% were arcsine-transformed. The data were tested for the normality of distribution using Shapiro–Wilk tests before applying ANOVA; therefore, the assumption for applying the ROUT method to potentially detect more than one outlier was met. The ROUT test revealed outliers only in the ACC mRNA levels. Results are presented as means ± SEM, and the significance level was set at $p < 0.05$. To detect potential associations between heat shock proteins, antioxidants and-fatty acid-metabolism-related gene expression, Pearson correlations between the relative mRNA levels of the genes studied were calculated to develop a correlation matrix.

3. Results

3.1. Growth Performances

Average daily body weight gain (ADG), average daily feed intake (ADFI), and feed conversion ratio (FCR) are shown in Table 2. In ovo GABA feeding did not affect the initial body weight of chicks. From days 1 to 28, chicks were raised under standard temperatures, and no significant differences in ADG, ADFI, and FCR were found among all treatments. After the cyclic HS, ADG was significantly reduced in the CON+HS compared to the CON group ($p < 0.05$), whereas ADFI was lower in both CON+HS and G10+HS groups ($p = 0.001$). However, there were no significant differences observed in FCR among all groups.

Table 2. Effects of the in ovo injection of γ-aminobutyric acid on growth parameters in broilers subjected to cyclic heat stress.

Parameters	Treatments			
	CON	CON+HS	G10+HS	*p*-Value
Initial body weight (g/bird)	45.1 ± 0.1	44.8 ± 0.2	45.0 ± 0.1	0.342
Bodyweight (g/bird)				
7 days	166 ± 4.3	177 ± 1.7	165 ± 5.2	0.111
21 days	981 ± 43.8	1011 ± 11.9	964 ± 48.4	0.684
28 days	1624 ± 65.0	1613 ± 39.1	1551 ± 82.0	0.697
31 days	1968 ± 77.5	1794 ± 96.7	1795 ± 119.7	0.295
ADG (g/birds)				
0 to 7 days	17.2 ± 0.6	18.8 ± 0.2	17.1 ± 0.7	0.102
8 to 21 days	58.2 ± 3	59.6 ± 0.9	57.0 ± 3.2	0.786
22 to 28 days	91.8 ± 4.5	86.0 ± 4.0	84 ± 2.7	0.496
28 to 31 days	86.0 ± 4.5 ^{a}	45.2 ± 2.4 ^{b}	61.0 ± 12.8 ^{ab}	0.011
ADFI (g/bird)				
0 to 7 days	18.9 ± 0.6	20.2 ± 0.4	18.8 ± 0.8	0.242
8 to 21 days	76.1 ± 3.0	77.4 ± 1.2	74.2 ± 3.6	0.73
22 to 28 days	142 ± 3.0	133 ± 5.6	132 ± 5.9	0.381
28 to 31 days	147 ± 7.8 ^{a}	94.1 ± 5.7 ^{b}	109 ± 9.1 ^{b}	0.001
FCR				
0 to 7 days	1.10 ± 0.02	1.07 ± 0.02	1.10 ± 0.01	0.537
8 to 21 days	1.29 ± 0.02	1.29 ± 0.01	1.31 ± 0.03	0.641
22 to 28 days	1.55 ± 0.05	1.55 ± 0.02	1.59 ± 0.04	0.788
28 to 31 days	1.72 ± 0.1	2.10 ± 0.1	1.97 ± 0.2	0.316

At 17.5 days of incubation, eggs were injected in ovo with or without 0.6 mL of GABA at 10%. Hatchlings obtained were raised until day 28 and subjected to cyclic heat stress (from 22 °C to 33 °C for 6 h daily) or kept at a thermoneutral temperature from days 28 to 31. Data were analyzed via one-way ANOVA (n = 5 cages/treatment). Means in the same row with different superscript letters are significantly different ($p < 0.05$). Abbreviations: ADG, average daily weight gain; ADFI, average daily feed intake; FCR, feed conversion ratio.

3.2. Organ Weight, Length, and Rectal Temperature

Table 3 shows the results of the organ weight in all experimental groups. There were no statistical differences between treatment groups for the proventriculus, gizzard, heart, bursa, liver, and spleen in relative and absolute weights.

Table 3. Effects of the in ovo injection of γ-aminobutyric acid on organ weights in broilers subjected to cyclic heat stress.

Absolute Organs Weight (g)	Treatments			
	CON	CON+HS	G10+HS	*p*-Value
Proventriculus	8.40 ± 1.4	8.70 ± 1.2	8.10 ± 1.1	0.959
Gizzard	15.5 ± 0.9	19.3 ± 2.3	18.1 ± 2.0	0.345
Heart	13.4 ± 0.4	12.1 ± 0.7	11.7 ± 0.6	0.132
Bursa	3.20 ± 0.2	3.30 ± 0.3	3.50 ± 0.5	0.827
Liver	56.7 ± 4.0	59.3 ± 3.2	52.2 ± 2.5	0.339
Spleen	2.10 ± 0.2	2.00 ± 0.1	1.90 ± 0.3	0.788
Relative Organs Weight (%)				
Proventriculus	0.4 ± 0.06	0.47 ± 0.06	0.44 ± 0.06	0.744
Gizzard	0.75 ± 0.02	0.88 ± 0.06	0.82 ± 0.06	0.151
Heart	0.65 ± 0.02	0.65 ± 0.04	0.64 ± 0.02	0.837
Bursa	0.15 ± 0.01	0.18 ± 0.02	0.18 ± 0.01	0.46
Liver	2.73 ± 0.2	3.2 ± 0.2	2.8 ± 0.1	0.099
Spleen	0.1 ± 0.01	0.1 ± 0.01	0.1 ± 0.01	0.896

At 17.5 days of incubation, eggs were injected in ovo with or without 0.6 mL of GABA at 10%. Hatchlings obtained were raised until day 28 and subjected to cyclic heat stress (from 22 °C to 33 °C for 6 h daily) or kept at a thermoneutral temperature from days 28 to 31. Data were analyzed via one-way ANOVA (n = 5 birds/treatment).

The organ lengths of the duodenum, jejunum, ileum, and caecum are given in Table 4. The CON+HS group exhibited a significantly shorter absolute duodenum length compared to the CON ($p < 0.05$). However, there were no significant differences in the absolute length of the jejunum, ileum, and caecum and the relative length of all organs.

Table 4. Effects of the in ovo injection of γ-aminobutyric acid on organ lengths in broilers subjected to cyclic heat stress.

Absolute Organ Length (cm)	Treatments			
	CON	CON+HS	G10+HS	*p*-Value
Duodenum	28.4 ± 0.75 ^{a}	24.3 ± 1.1 ^{b}	26.7 ± 0.86 ^{ab}	0.026
Jejunum	67.4 ± 2.6	59.9 ± 1.5	64.7 ± 4.7	0.286
Ileum	70.0 ± 1.1	59.9 ± 1.4	64.5 ± 4.8	0.095
Caecum	18.4 ± 0.77	15.7 ± 0.82	17 ± 0.53	0.122
Relative Organ Length (cm/100g)				
Duodenum	1.38 ± 0.05	1.31 ± 0.06	1.44 ± 0.03	0.232
Jejunum	3.26 ± 0.10	3.23 ± 0.10	2.92 ± 0.70	0.813
Ileum	3.40 ± 0.12	3.23 ± 0.10	3.49 ± 0.25	0.59
Caecum	0.89 ± 0.06	0.85 ± 0.04	0.92 ± 0.05	0.608

At 17.5 days of incubation, eggs were injected in ovo with or without 0.6 mL of GABA at 10%. Hatchlings obtained were raised until day 28 and subjected to cyclic heat stress (from 22 °C to 33 °C for 6 h daily) or kept at a thermoneutral temperature from days 28 to 31. Data were analyzed via one-way ANOVA (n = 5 birds/treatment). Means in the same row with different superscript letters are significantly different ($p < 0.05$).

Table 5 indicates the results related to rectal temperature. No significant differences in rectal temperature were observed between all treatments under standard housing temperature on day 28. Moreover, on day 31 after cyclic HS, chicks treated with CON+HS and G10+HS had significantly higher rectal temperature values compared to the CON group.

Table 5. Effects of the ovo injection of γ-aminobutyric acid on rectal temperature (°C) in broilers subjected to cyclic heat stress.

Broiler Age	Treatments			
	CON	CON+HS	G10+HS	*p*-Value
Day 28 (before HS)	41.5 ± 0.1	41.6 ± 0.1	41.6 ± 0.1	0.836
Day 31 (after HS)	41.6 ± 0.1 ^{a}	43.2 ± 0.1 ^{b}	43.4 ± 0.1 ^{b}	<0.001

At 17.5 days of incubation, eggs were injected in ovo with or without 0.6 mL of GABA at 10%. Hatchlings obtained were raised until day 28 and subjected to cyclic heat stress (from 22 °C to 33 °C for 6 h daily) or kept at a thermoneutral temperature from days 28 to 31. Data were analyzed via one-way ANOVA (n = 5 birds/treatment). Means in the same row with different superscript letters are significantly different ($p < 0.05$).

3.3. Plasma Corticosterone, Metabolites, and Oxidative Stress Markers

After cyclic HS exposure, the plasma DPPH-RSA of the CON+HS was significantly decreased ($p < 0.05$) compared to the CON group (Table 6). Malondialdehyde levels and the antioxidant balance in the plasma appeared to be similar among all treatment groups.

Table 6. Effects of the ovo injection of γ-aminobutyric acid on plasma oxidative stress markers in broilers subjected to cyclic heat stress.

Parameters	Treatments			
	CON	CON+HS	G10+HS	*p*-Value
DPPH-RSA (%)	49.6 ± 1.0 ^{a}	44.6 ± 1.9 ^{b}	45 ± 0.9 ^{ab}	0.036
MDA (nmol/mL)	0.67 ± 0.1	0.67 ± 0.1	0.61 ± 0.1	0.828
Antioxidant balance (U)	82.2 ± 16.0	69.8 ± 6.7	80.7 ± 13.9	0.762

At 17.5 days of incubation, eggs were injected in ovo with or without 0.6 mL of GABA at 10%. Hatchlings obtained were raised until day 28 and subjected to cyclic heat stress (from 22 °C to 33 °C for 6 h daily) or kept at a thermoneutral temperature from days 28 to 31. Data were analyzed via one-way ANOVA (n = 5 birds/treatment). Means in the same row with different superscript letters are significantly different ($p < 0.05$). Abbreviations: DPPH-RSA, 2,2-diphenyl-1-picrylhydrazyl radical scavenging activity assay; MDA, malondialdehydes.

Table 7 shows the effects of the in ovo feeding of GABA on plasma metabolites and corticosterone levels. No significant effects were detected for triglycerides, cholesterol, and total protein in plasma. However, HS significantly increased plasma glucose levels in both CON+HS and G10+HS groups. In addition, HS significantly increased plasma corticosterone levels in the CON+HS group compared to the CON group. Interestingly, plasma corticosterone levels in the G10+HS group were not statistically different from the CON group.

Table 7. Effects of the in ovo injection of γ-aminobutyric acid on plasma metabolites and corticosterone in broilers subjected to cyclic heat stress.

Parameters	Treatments			
	CON	CON+HS	G10+HS	*p*-Value
Cholesterol (mg/dL)	120 ± 11.4	126 ± 4.0	119 ± 5.4	0.825
Glucose (mg/dL)	245 ± 3.3 ^{a}	277 ± 6.8 ^{b}	280 ± 9.6 ^{b}	0.008
Total protein (g/dL)	3.20 ± 0.14	3.10 ± 0.11	3.10 ± 0.18	0.928
Triglycerides (g/dL)	17.8 ± 3.7	18.2 ± 3.9	18.6 ± 5.9	0.093
Corticosterone (ng/mL)	0.76 ± 0.04 ^{a}	4.22 ± 1.41 ^{b}	1.52 ± 0.24 ^{ab}	0.027

At 17.5 days of incubation, eggs were injected in ovo with or without 0.6 mL of GABA at 10%. Hatchlings obtained were raised until day 28 and subjected to cyclic heat stress (from 22 °C to 33 °C for 6 h daily) or kept at a thermoneutral temperature from days 28 to 31. Data were analyzed via one-way ANOVA (n = 5 birds/treatment). Means in the same row with different superscript letters are significantly different ($p < 0.05$).

3.4. Hepatic mRNA Relative Expression

Figures 3–6 present the effects of the in ovo injection of GABA on hepatic mRNA expression of CAT, GPx1, SOD, ACC, NOX1, NOX4, FAS, HSP70, and HSP90 in heat-exposed chickens.

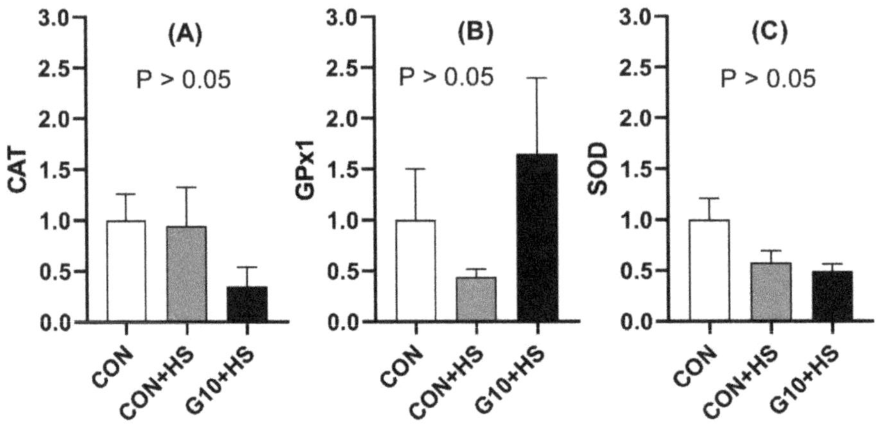

Figure 3. Effects of the in ovo injection of GABA on the relative mRNA expression of hepatic CAT (**A**), GPx1 (**B**), and SOD (**C**) in broilers exposed to cyclic heat stress. At 17.5 days of incubation, eggs were injected in ovo with or without 0.6 mL of GABA at 10%. Hatchlings obtained were raised until day 28 and subjected to cyclic heat stress (from 22 °C to 33 °C for 6 h daily) or kept at a thermoneutral temperature from days 28 to 31. Data were analyzed via one-way ANOVA (n = 5 birds/treatment). Abbreviations: CAT, catalase; GPx1, glutathione peroxidase 1; SOD, superoxide dismutase.

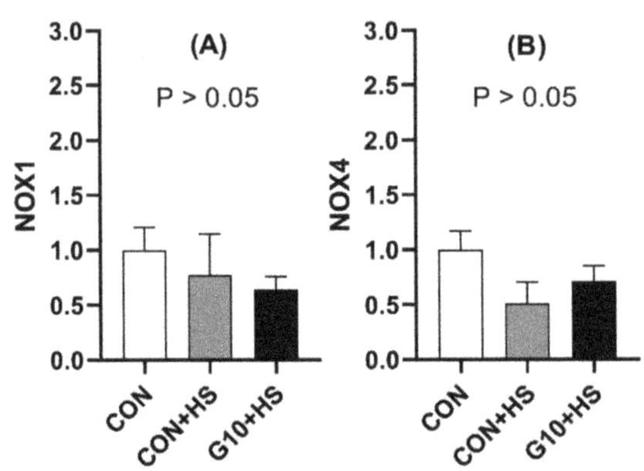

Figure 4. Effects of the in ovo injection of GABA on relative mRNA expression of hepatic NOX1 (**A**) and NOX4 (**B**) in broilers exposed to cyclic heat stress. At 17.5 days of incubation, eggs were injected in ovo with or without 0.6 mL of GABA at 10%. Hatchlings obtained were raised until day 28 and subjected to cyclic heat stress (from 22 °C to 33 °C for 6 h daily) or kept at a thermoneutral temperature from days 28 to 31. Data were analyzed via one-way ANOVA (n = 5 birds/treatment). Abbreviations: NOX1, nicotinamide adenine dinucleotide phosphate oxidase 1; NOX4, nicotinamide adenine dinucleotide phosphate oxidase 4.

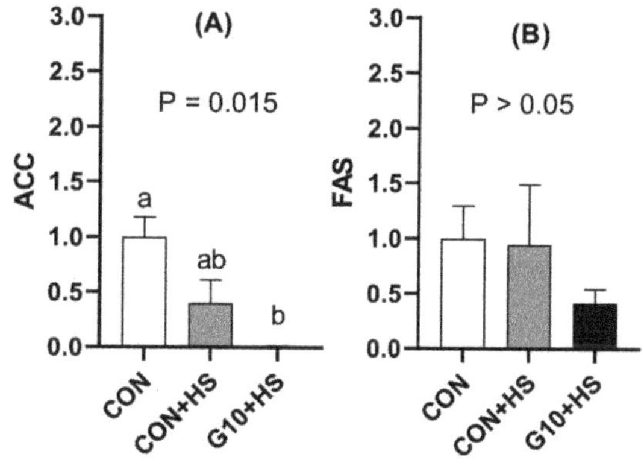

Figure 5. Effects of the in ovo injection of GABA on relative mRNA expression of hepatic ACC (**A**) and FAS (**B**) in broilers exposed to cyclic heat stress. At 17.5 days of incubation, eggs were injected in ovo with or without 0.6 mL of GABA at 10%. Hatchlings obtained were raised until day 28 and subjected to cyclic heat stress (from 22 °C to 33 °C for 6 h daily) or kept at a thermoneutral temperature from days 28 to 31. Abbreviations: ACC, acetyl-CoA carboxylase; FAS, fatty acid synthase. Data were analyzed via one-way ANOVA (n = 5 birds/treatment for the FAS gene). The ROUT test revealed outliers in the data set of ACC (n = 4 birds/treatment).

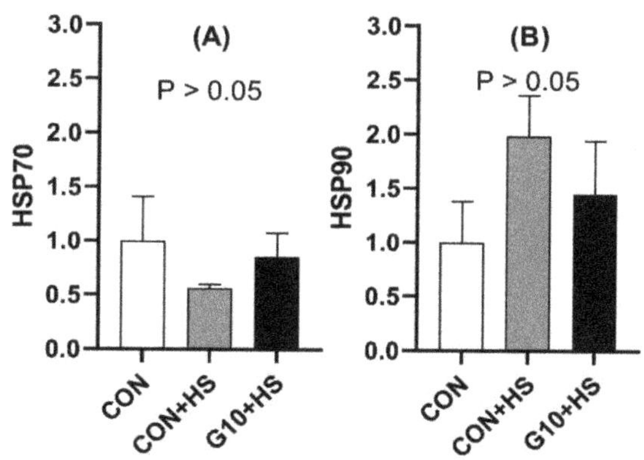

Figure 6. Effects of the in ovo injection of GABA on relative mRNA expression of hepatic HSP70 (**A**) and HSP90 (**B**) in broilers exposed to cyclic heat stress. At 17.5 days of incubation, eggs were injected in ovo with or without 0.6 mL of GABA at 10%. Hatchlings obtained were raised until day 28 and subjected to cyclic heat stress (from 22 °C to 33 °C for 6 h daily) or kept at a thermoneutral temperature from days 28 to 31. Data were analyzed via one-way ANOVA (n = 5 birds/treatment). Abbreviations: HSP70, heat shock protein 70; HSP90, heat shock protein 90.

There were no significant variations in the relative mRNA expression of CAT, SOD, and GPx1 in response to either cyclic HS or the in ovo feeding of GABA (Figure 3).

In addition, there were no differences between treatment groups concerning relative mRNA expression for the NOX family genes (NOX1 and NOX4), the HSP family genes (HSP70 and HSP90), and the FAS gene.

However, the in ovo feeding of GABA downregulated the expression of ACC in heat-stressed broilers compared to the CON group ($p < 0.05$) (Figure 5).

Table 8 shows the Pearson correlation r-values examined between the relative mRNA levels of genes pairwise. Positive correlations were found between FAS and NOX1 (r = 0.55, $p < 0.05$), NOX1 and NOX4 (r = 0.65, $p < 0.01$), and CAT and SOD (r = 0.62, $p < 0.05$).

Table 8. Pearson correlation values of the relative mRNA levels of the genes studied in the liver.

	HSP70	FAS	NOX1	NOX4	CAT	SOD	HSP90	GPx1	ACC
HSP70	1	0.38	0.15	0.25	−0.24	0.16	−0.04	−0.05	0.35
FAS		1	0.55 *	0.30	−0.28	−0.03	−0.17	−0.14	0.13
NOX1			1	0.65 **	−0.26	0.16	0.10	0.02	−0.20
NOX4				1	−0.24	−0.02	−0.22	−0.15	0.20
CAT					1	0.62 *	0.21	0.17	0.47
SOD						1	0.00	0.37	0.42
HSP90							1	−0.19	−0.09
GPx1								1	−0.23
ACC									1

At 17.5 days of incubation, eggs were injected in ovo with or without 0.6 mL of GABA at 10%. Hatchlings obtained were raised until day 28 and subjected to cyclic heat stress (from 22 °C to 33 °C for 6 h daily) or kept at a thermoneutral temperature from days 28 to 31. Data were analyzed via Pearson's correlation test. Abbreviations: ACC, acetyl-CoA carboxylase; CAT, catalase; FAS, fatty acid synthase; GAPDH, glyceraldehyde-3-phosphate dehydrogenase; GPx1, glutathione peroxidase 1; HSP70, heat shock protein 70; HSP90, heat shock protein 90; NOX1, nicotinamide adenine dinucleotide phosphate oxidase 1; NOX4, nicotinamide adenine dinucleotide phosphate oxidase 4; SOD, superoxide dismutase. * Correlation is significant at the 0.05 level. ** Correlation is significant at the 0.01 level.

4. Discussion

In poultry, HS is known for its adverse impacts on health, physiology, and efficiency [29,30]. Chronic HS exposure induces a reduction in broilers' feed intake and body weight while concomitantly increasing FCR [31]. Similarly, in our study, 4 days of cyclic HS led to a significant reduction in ADFI of the birds reared under HS compared to those under normal housing temperature. For regulating their body temperature under HS, broilers reduce their feed consumption to promote heat loss and decrease their metabolic heat production [32]. Therefore, even though reducing FI helps cope with high ambient temperature, it is naturally followed by severe growth impairment [33]. These findings were confirmed in the results of our study. The ADG of heat-stressed birds was statistically lower (by 47.4% for CON+HS vs. 29.1% for G10+HS) compared to the control birds.

Interestingly, the in ovo feeding of GABA could mitigate the weight loss (by 35%) occasioned by cyclic HS. This might be explained by the fact that birds hatched from eggs injected in ovo recorded a higher ADFI (15.8%) than control birds under the same high environmental conditions. Additionally, our results showed that the in ovo feeding of GABA numerically reduced the FCR (15.9%) compared to the control birds under HS. Previous results have already documented the orexigenic effect of dietary GABA supplementation [11,34]. These results might suggest that the mechanism behind in ovo feeding is closely linked to how dietary supplementation works. Indeed, in ovo feeding given during the late embryonic stages targeted the amnion [35]. The embryo, by consuming the content of the amnion, could access the injected nutrients before it started to hatch [14]. GABA's ability to improve the feed intake of HS-exposed birds was explained by the fact that it can lower the expression of anorexigenic neuropeptides and upregulate orexigenic neuropeptides [11].

The organ index is an important parameter reflecting the development status of organs. A recent study [36] reported an increase in the liver index when broilers were reared under chronic HS. There was a weakly significant ($p = 0.099$) tendency to increase the liver index of HS birds in the present study. In poultry, the liver is known as the main organ of fat synthesis. Consequently, higher liver indexes might be correlated with higher fat synthesis during HS [37]. Gastrointestinal length can be an indicator of digestive health. HS significantly decreased the absolute organ length of the duodenum, but its relative length was not affected. Therefore, no conclusive effect of either HS or the in ovo feeding of GABA on duodenum length could be ascertained. Rectal temperature is generally utilized as a marker for HS resistance in broilers [38]. Previous reports have recognized the HS-stimulating effect of rectal temperature in broilers [11,34]. As expected, birds belonging to both groups under cyclic HS had a higher rectal temperature in the current study.

DPPH-RSA has been widely used as a method to evaluate plasma total antioxidant capacity (TAC) [39]. In the present study, the in ovo feeding of GABA resulted in improved TAC in broilers subjected to cyclic HS. Researchers [40] previously reported that GABA supplementation increased TAC in chickens under HS. These findings indicate that similarly to dietary GABA, in ovo GABA injection could instigate the retrieval of antioxidant functions after cyclic HS exposure. In response to cyclic HS, broilers exhibited higher plasma levels of corticosterone. These results of the present study show consistency with those reported by earlier studies [41,42], mentioning that heat-challenged birds tend to have increased plasma corticosterone compared to those at thermoneutral temperature. This might be explained by the fact that heat stress activates the hypothalamic–pituitary–adrenal (HPA) axis, leading to increased corticosterone secretion [43]. Therefore, enhanced corticosterone release under HS may have reduced overall feed intake and body weight gain [44,45], as observed in the current study.

Interestingly, the in ovo feeding of GABA could reduce the plasma corticosterone levels of the birds exposed to HS. Earlier studies have revealed that dietary and GABA-enriched barley bran could lower adrenocorticotropic hormone (ACTH) and corticosterone levels in pigs and rats [46,47]. Indeed, under stressful conditions, the hypothalamus releases the corticotropin-releasing hormone, which stimulates the secretion of ACTH from the anterior pituitary gland [48]. Subsequently, ACTH provokes the synthesis of glucocorticoids such as corticosterone to combat stress [49]. Furthermore, GABA was shown to inhibit ACTH secretion by activating α1- and α2-adrenergic receptors [50]. Thus, we could hypothesize that the potential effect of GABA on corticosterone is linked to its inhibitory effect on ACTH.

In poultry, acetyl-CoA carboxylase (ACC) and fatty acid synthase (FAS) are the key enzymes responsible for fat synthesis. ACC is involved in converting acetyl-CoA into malonyl-CoA and then into palmitate, whereas FAS determines the maximum capacity for fatty acid synthesis in tissues [51]. Our results showed that the hepatic mRNA levels of ACC were downregulated in heat-stressed broilers after the in ovo injection of GABA. In addition, a numerically lower value of relative mRNA levels of FAS was observed in the birds fed GABA in ovo. As previously mentioned, the liver is known to be the main organ of fat synthesis in poultry, and one of the deleterious effects of HS in broilers is lipid accumulation, engendered by the upregulation of FAS and ACC genes [36,52]. Therefore, lower mRNA hepatic levels of FAS and ACC indicate less fatty acid synthesis during HS in birds that were fed GABA in ovo.

Interestingly, we observed a positive correlation between FAS and NOX1 mRNA levels. The NOX gene family, responsible for ROS generation, was affected by heat stress [27], suggesting that the in ovo feeding of GABA may decrease ROS generation during cyclic HS. The mechanism by which GABA potentially reduces ROS generation could be related to its effect on glutathione peroxidase synthesis [53,54]. Significant positive correlations were detected between CAT and SOD mRNA levels. SOD and CAT are among the major antioxidant enzymes [55]. SOD is responsible for the dismutation of superoxide radicals to molecular oxygen and hydrogen peroxide (H_2O_2), whereas CAT is responsible for transforming H_2O_2 into water and oxygen [56]. The actions of both enzymes are synchronized, which may justify the observed positive correlation.

As the Materials and Methods section mentioned, a general blackout for 6 h led to a decrease in the temperature (from 36.8 to 20 °C) during the incubation. However, a potential effect of the incubation temperature drop can be neglected because all the treatment groups were kept in the same incubator. Even though a cold temperature during embryogenesis has been shown to enhance growth performance, such temperature manipulations were lower than 20 °C and repetitive to affect the thermoregulatory mechanism of chicks [57,58]. Furthermore, all the relevant data reported were cross-checked with the literature and showed no discrepancies in the current study. For example, the recorded growth performances of Arbor Acres in this study were concordant to recently published articles [59,60]. In addition, the average BWs obtained after hatch (45.1 g), 7 days (165.7 g), 21 days (981.1 g), and 28 days (1624.1 g) were in line with the Arbor Acres broiler performances (as-hatched performance) objective, which recommends 43 g, 204 g, 978 g, and 1567 g for the same periods, respectively [61]. Similarly, plasma parameters [62] were within the same range of previously reported results.

5. Conclusions

The results from the current study suggested that the in ovo feeding of GABA at 0.1 g. mL^{-1} increased ADG and plasma TAC while reducing plasma corticosterone and ACC mRNA levels in heat-stressed broilers. However, further research is needed to elucidate whether the in ovo feeding of GABA should be considered as a viable technique for enhancing heat resistance in broilers.

Author Contributions: Conceptualization, C.-M.N., A.G. and Y.-H.C.; methodology, C.-M.N., C.-M.J. and V.G.; software, C.-M.N.; validation, A.G. and Y.-H.C.; data curation, C.-M.N., A.G. and V.G.; writing—original draft preparation, C.-M.N.; writing—review and editing, C.-M.N., A.G. and Y.-H.C.; supervision, Y.-H.C. All authors have read and agreed to the published version of the manuscript.

Funding: This work was supported by the Gyeongsang National University Fund for Professors on Sabbatical Leave, 2018.

Institutional Review Board Statement: The study was approved by the Institutional Animal Care and Use Committee of Gyeongsang National University (GNU-200916-C0058) and was conducted according to the guidelines.

Informed Consent Statement: Not applicable.

Data Availability Statement: All the relevant data are available in the paper.

Acknowledgments: The authors are thankful to the Korean Government Scholarship Program (KGSP) and to the Brain Pool Program funded by the Ministry of Science and ICT through the National Research Foundation of Korea.

Conflicts of Interest: The authors declare no conflict of interest.

References

1. Goel, A.; Ncho, C.M.; Choi, Y.H. Regulation of gene expression in chickens by heat stress. *J. Anim. Sci. Biotechnol.* **2021**, *12*, 11. [CrossRef] [PubMed]
2. Habashy, W.S.; Milfort, M.C.; Adomako, K.; Attia, Y.A.; Rekaya, R.; Aggrey, S.E. Effect of heat stress on amino acid digestibility and transporters in meat-type chickens. *Poult. Sci.* **2017**, *96*, 2312–2319. [CrossRef] [PubMed]
3. Donkoh, A. Ambient temperature: A factor affecting performance and physiological response of broiler chickens. *Int. J. Biometeorol.* **1989**, *33*, 259–265. [CrossRef] [PubMed]
4. Warriss, P.D.; Pagazaurtundua, A.; Brown, S.N. Relationship between maximum daily temperature and mortality of broiler chickens during transport and lairage. *Br. Poult. Sci.* **2005**, *46*, 647–651. [CrossRef] [PubMed]
5. Attia, Y.A.; Al-Harthi, M.A.; Elnaggar, A.S. Productive, physiological and Immunological responses of two broiler strains fed different dietary regimens and exposed to heat stress. *Ital. J. Anim. Sci.* **2017**, *17*, 1. [CrossRef]
6. Attia, Y.A.; Hassan, S.S. Broiler tolerance to heat stress at various dietary protein/energy levels. *Eur. Poult Sci* **2017**, *81*.
7. Renaudeau, D.; Collin, A.; Yahav, S.; de Basilio, V.; Gourdine, J.L.; Collier, R.J. Adaptation to hot climate and strategies to alleviate heat stress in livestock production. *Anim. Int. J. Anim. Biosci.* **2012**, *6*, 707–728. [CrossRef]
8. Cobb, S.R.; Buhl, E.H.; Halasy, K.; Paulsen, O.; Somogyi, P. Synchronization of neuronal activity in hippocampus by individual GABAergic interneurons. *Nature* **1995**, *378*, 75–78. [CrossRef]
9. Bongianni, F.; Mutolo, D.; Nardone, F.; Pantaleo, T. GABAergic and glycinergic inhibitory mechanisms in the lamprey respiratory control. *Brain Res.* **2006**, *1090*, 134–145. [CrossRef]
10. Ncho, C.M.; Jeong, C.; Gupta, V.; Goel, A. The effect of gamma-aminobutyric acid supplementation on growth performances, immune responses, and blood parameters of chickens reared under stressful environment: A meta-analysis. *Environ. Sci. Pollut. Res.* **2021**. [CrossRef]
11. Al Wakeel, R.A.; Shukry, M.; Abdel Azeez, A.; Mahmoud, S.; Saad, M.F. Alleviation by gamma amino butyric acid supplementation of chronic heat stress-induced degenerative changes in jejunum in commercial broiler chickens. *Stress* **2017**, *20*, 562–572. [CrossRef]
12. Yang-Ho, C. Effects of γ-aminobutyric acid on mortality in laying hens during summer time. *J. Agric. Life Environ. Sci.* **2019**, *53*, 131–139.
13. Kadam, M.M.; Barekatain, M.R.; Bhanja, S.K.; Iji, P.A. Prospects of in ovo feeding and nutrient supplementation for poultry: The science and commercial applications—A review. *J. Sci. Food Agric.* **2013**, *93*, 3654–3661. [CrossRef]
14. Uni, Z.; Ferket, R.P. Methods for early nutrition and their potential. *World's Poult. Sci. J.* **2019**, *60*, 101–111. [CrossRef]

15. Tavaniello, S.; Slawinska, A.; Prioriello, D.; Petrecca, V.; Bertocchi, M.; Sirri, F.; Salvatori, G.; Maiorano, G. Effect of galactooligosaccharides delivered in ovo on meat quality traits of broiler chickens exposed to heat stress. *Poult. Sci.* **2019**, *99*, 612–619. [CrossRef]
16. Han, G.; Yang, H.; Wang, Y.; Haraguchi, S.; Miyazaki, T.; Bungo, T.; Tashiro, K.; Furuse, M.; Chowdhury, V.S. L-Leucine increases the daily body temperature and affords thermotolerance in broiler chicks. *Asian-Australas. J. Anim. Sci.* **2019**, *32*, 842–848. [CrossRef]
17. Elwan, H.A.M.; Elnesr, S.S.; Xu, Q.; Xie, C.; Dong, X.; Zou, X. Effects of In Ovo Methionine-Cysteine Injection on Embryonic Development, Antioxidant Status, IGF-I and TLR4 Gene Expression, and Jejunum Histomorphometry in Newly Hatched Broiler Chicks Exposed to Heat Stress during Incubation. *Animals* **2019**, *9*, 25. [CrossRef]
18. Elnesr, S.S.; Elwan, H.A.M.; Xu, Q.Q.; Xie, C.; Dong, X.Y.; Zou, X.T. Effects of in ovo injection of sulfur-containing amino acids on heat shock protein 70, corticosterone hormone, antioxidant indices, and lipid profile of newly hatched broiler chicks exposed to heat stress during incubation. *Poult. Sci.* **2019**, *98*, 2290–2298. [CrossRef]
19. Han, G.; Ouchi, Y.; Hirota, T.; Haraguchi, S.; Miyazaki, T.; Arakawa, T.; Masuhara, N.; Mizunoya, W.; Tatsumi, R.; Tashiro, K.; et al. Effects of l-leucine in ovo feeding on thermotolerance, growth and amino acid metabolism under heat stress in broilers. *Anim. Int. J. Anim. Biosci.* **2020**, *14*, 1701–1709. [CrossRef]
20. Slawinska, A.; Zampiga, M.; Sirri, F.; Meluzzi, A.; Bertocchi, M.; Tavaniello, S.; Maiorano, G. Impact of galactooligosaccharides delivered in ovo on mitigating negative effects of heat stress on performance and welfare of broilers. *Poult. Sci.* **2020**, *99*, 407–415. [CrossRef] [PubMed]
21. Ncho, C.M.; Goel, A.; Jeong, C.M.; Youssouf, M.; Choi, Y.H. In ovo injection of gaba can help body weight gain at hatch, increase chick weight to egg weight ratio, and improve broiler heat resistance. *Animals* **2021**, *11*, 1364. [CrossRef] [PubMed]
22. Goel, A.; Kim, B.-J.; Ncho, C.-M.; Jeong, C.-M.; Gupta, V.; Jung, J.-Y.; Ha, S.-Y.; Lee, D.-H.; Yang, J.-K.; Choi, Y.-H. Dietary Supplementation of Shredded, Steam-Exploded Pine Particles Decreases Pathogenic Microbes in the Cecum of Acute Heat-Stressed Broilers. *Animals* **2021**, *11*, 2252. [CrossRef] [PubMed]
23. Gerasopoulos, K.; Stagos, D.; Kokkas, S.; Petrotos, K.; Kantas, D.; Goulas, P.; Kouretas, D. Feed supplemented with byproducts from olive oil mill wastewater processing increases antioxidant capacity in broiler chickens. *Food Chem. Toxicol.* **2015**, *82*, 42–49. [CrossRef]
24. Jyothi, P.; Riyaz, N.; Nandakumar, G.; Binitha, M.P. A study of oxidative stress in paucibacillary and multibacillary leprosy. *Indian J. Dermatol. Venereol. Leprol.* **2008**, *74*, 80. [CrossRef]
25. Akbarian, A.; Michiels, J.; Golian, A.; Buyse, J.; Wang, Y.; De Smet, S. Gene expression of heat shock protein 70 and antioxidant enzymes, oxidative status, and meat oxidative stability of cyclically heat-challenged finishing broilers fed origanum compactum and curcuma xanthorrhiza essential oils. *Poult. Sci.* **2014**, *93*, 1930–1941. [CrossRef]
26. Khan, A.Z.; Khan, I.U.; Khan, S.; Afzal, S.; Hamid, M.; Tariq, M.; Haq, I.U.; Ullah, N.; Khan, M.A.; Bilal, S.; et al. Selenium-enriched probiotics improve hepatic protection by regulating pro-inflammatory cytokines and antioxidant capacity in broilers under heat stress conditions. *J. Adv. Vet. Anim. Res.* **2019**, *6*, 355–361. [CrossRef]
27. Habashy, W.S.; Milfort, M.C.; Rekaya, R.; Aggrey, S.E. Expression of genes that encode cellular oxidant/antioxidant systems are affected by heat stress. *Mol. Biol. Rep.* **2018**, *45*, 389–394. [CrossRef]
28. Bonato, M.; Borges, L.L.; Ingberman, M.; Fávaro, C.; Mesa, D.; Caron, L.F.; Beirão, B.C.B. Effects of yeast cell wall on immunity, microbiota, and intestinal integrity of Salmonella-infected broilers. *J. Appl. Poult. Res.* **2020**, *29*, 545–558. [CrossRef]
29. Nawab, A.; Ibtisham, F.; Li, G.; Kieser, B.; Wu, J.; Liu, W.; Zhao, Y.; Nawab, Y.; Li, K.; Xiao, M.; et al. Heat stress in poultry production: Mitigation strategies to overcome the future challenges facing the global poultry industry. *J. Therm. Biol.* **2018**, *78*, 131–139. [CrossRef]
30. Ncho, C.M.; Gupta, V.; Goel, A. Effect of thermal conditioning on growth performance and thermotolerance in broilers: A systematic review and meta-analysis. *J. Therm. Biol.* **2021**, *98*, 102916. [CrossRef]
31. Sohail, M.U.; Hume, M.E.; Byrd, J.A.; Nisbet, D.J.; Ijaz, A.; Sohail, A.; Shabbir, M.Z.; Rehman, H. Effect of supplementation of prebiotic mannan-oligosaccharides and probiotic mixture on growth performance of broilers subjected to chronic heat stress. *Poult. Sci.* **2012**, *91*, 2235–2240. [CrossRef] [PubMed]
32. Nilsson, J.A.; Molokwu, M.N.; Olsson, O. Body Temperature Regulation in Hot Environments. *PLoS ONE* **2016**, *11*, e0161481. [CrossRef] [PubMed]
33. Bonnet, S.; Geraert, P.A.; Lessire, M.; Carre, B.; Guillaumin, S. Effect of high ambient temperature on feed digestibility in broilers. *Poult. Sci.* **1997**, *76*, 857–863. [CrossRef]
34. El-Naggar, K.; El-Kassas, S.; Abdo, S.E.; Kirrella, A.A.K.; Al Wakeel, R.A. Role of gamma-aminobutyric acid in regulating feed intake in commercial broilers reared under normal and heat stress conditions. *J. Therm. Biol.* **2019**, *84*, 164–175. [CrossRef]
35. Uni, Z.; Ferket, P.R. Enhancement of Development of Oviparous Species by In Ovo Feeding. U.S. Patent 6592878B2, 15 July 2003.
36. Lu, Z.; He, X.; Ma, B.; Zhang, L.; Li, J.; Jiang, Y.; Zhou, G.; Gao, F. Dietary taurine supplementation decreases fat synthesis by suppressing the liver X receptor alpha pathway and alleviates lipid accumulation in the liver of chronic heat-stressed broilers. *J. Sci. Food Agric.* **2019**, *99*, 5631–5637. [CrossRef]
37. Baziz, H.A.; Geraert, P.; Padilha, J.; Guillaumin, S. Chronic heat exposure enhances fat deposition and modifies muscle and fat partition in broiler carcasses. *Poult. Sci.* **1996**, *75*, 505–513. [CrossRef]
38. He, S.; Li, S.; Arowolo, M.A.; Yu, Q.; Chen, F.; Hu, R.; He, J. Effect of resveratrol on growth performance, rectal temperature and serum parameters of yellow-feather broilers under heat stress. *Anim. Sci. J.* **2019**, *90*, 401–411. [CrossRef]
39. Janaszewska, A.; Bartosz, G. Assay of total antioxidant capacity: Comparison of four methods as applied to human blood plasma. *Scand. J. Clin. Lab. Investig.* **2002**, *62*, 231–236. [CrossRef]

40. Chen, Z.; Xie, J.; Hu, M.Y.; Tang, J.; Shao, Z.F.; Li, M.H. Protective effects of γ -aminobutyric acid (GABA) on the small intestinal mucosa in heat-stressed wenchang chicken. *J. Anim. Plant Sci.* **2015**, *25*, 78–87.
41. Ramiah, S.K.; Awad, E.A.; Mookiah, S.; Idrus, Z. Effects of zinc oxide nanoparticles on growth performance and concentrations of malondialdehyde, zinc in tissues, and corticosterone in broiler chickens under heat stress conditions. *Poult. Sci.* **2019**, *98*, 3828–3838. [CrossRef]
42. Zulkifli, I.; Liew, P.; Israf, D.; Omar, A.; Hair-Bejo, M. Effects of early age feed restriction and heat conditioning on heterophil/lymphocyte ratios, heat shock protein 70 expression and body temperature of heat-stressed broiler chickens. *J. Therm. Biol.* **2003**, *28*, 217–222. [CrossRef]
43. Song, Z.; Liu, L.; Sheikhahmadi, A.; Jiao, H.; Lin, H. Effect of heat exposure on gene expression of feed intake regulatory peptides in laying hens. *J. Biomed. Biotechnol.* **2012**, *2012*, 484869. [CrossRef] [PubMed]
44. Lara, L.J.; Rostagno, M.H. Impact of Heat Stress on Poultry Production. *Animals* **2013**, *3*, 356–369. [CrossRef] [PubMed]
45. Quinteiro-Filho, W.; Ribeiro, A.; Ferraz-de-Paula, V.; Pinheiro, M.; Sakai, M.; Sá, L.; Ferreira, A.; Palermo-Neto, J. Heat stress impairs performance parameters, induces intestinal injury, and decreases macrophage activity in broiler chickens. *Poult. Sci.* **2010**, *89*, 1905–1914. [CrossRef] [PubMed]
46. Xu, Q.-l.; Zhang, Q.-l.; Tang, G.-p.; Tang, G.-w. Effect of γ-aminobutyric Acid on Performance and Hormones of Finishing Pig in Heat Stress Environment. *Acta Ecol. Anim. Domastici* **2009**, *6*, 28–32.
47. Vijaya Abinaya, R.; Pichiah, P.B.T.; Sara Thomas, S.; Kim, S.G.; Han, D.W.; Song, Y.S.; Oh, S.H.; Cha, Y.S. γ-amino butyric acid-enriched barley bran lowers adrenocorticotropic hormone and corticosterone levels in immobilized stressed rats. *J. Food Biochem.* **2017**, *41*, e12324. [CrossRef]
48. Aguilera, G. Regulation of pituitary ACTH secretion during chronic stress. *Front. Neuroendocrinol.* **1994**, *15*, 321–350. [CrossRef]
49. Vitale, G.; Salvioli, S.; Franceschi, C. Oxidative stress and the ageing endocrine system. *Nat. Rev. Endocrinol.* **2013**, *9*, 228–240. [CrossRef]
50. Makara, G.; Stark, E. Effect of gamma-aminobutyric acid (GABA) and GABA antagonist drugs on ACTH release. *Neuroendocrinology* **1974**, *16*, 178–190. [CrossRef]
51. Richards, M.P.; Rosebrough, R.W.; Coon, C.N.; McMurtry, J.P. Feed intake regulation for the female broiler breeder: In theory and in practice. *J. Appl. Poult. Res.* **2010**, *19*, 182–193. [CrossRef]
52. Lu, Z.; He, X.F.; Ma, B.B.; Zhang, L.; Li, J.L.; Jiang, Y.; Zhou, G.H.; Gao, F. Increased fat synthesis and limited apolipoprotein B cause lipid accumulation in the liver of broiler chickens exposed to chronic heat stress. *Poult. Sci.* **2019**, *98*, 3695–3704. [CrossRef]
53. Zhu, Y.Z.; Cheng, J.L.; Ren, M.; Yin, L.; Piao, X.S. Effect of gamma-Aminobutyric Acid-producing Lactobacillus Strain on Laying Performance, Egg Quality and Serum Enzyme Activity in Hy-Line Brown Hens under Heat Stress. *Asian-Australas. J. Anim. Sci.* **2015**, *28*, 1006–1013. [CrossRef]
54. Chen, Z.; Tang, J.; Sun, Y.; Xie, J. Protective effect of γ-aminobutyric acid on antioxidation function in intestinal mucosa of Wenchang chicken induced by heat stress. *J. Anim. Plant Sci.* **2013**, *23*, 1634–1641.
55. Surai, P.F.; Kochish, I.I.; Fisinin, V.I.; Kidd, M.T. Antioxidant Defence Systems and Oxidative Stress in Poultry Biology: An Update. *Antioxidants* **2019**, *8*, 235. [CrossRef]
56. Vriend, J.; Reiter, R.J. The Keap1-Nrf2-antioxidant response element pathway: A review of its regulation by melatonin and the proteasome. *Mol. Cell. Endocrinol.* **2015**, *401*, 213–220. [CrossRef]
57. Nyuiadzi, D.; Travel, A.; Méda, B.; Berri, C.; Guilloteau, L.A.; Coustham, V.; Wang, Y.; Tona, J.K.; Collin, A. Effect of low incubation temperature and low ambient temperature until 21 days of age on performance and body temperature in fast-growing chickens. *Poult. Sci.* **2017**, *96*, 4261–4269. [CrossRef]
58. Nyuiadzi, D.; Berri, C.; Dusart, L.; Travel, A.; Méda, B.; Bouvarel, I.; Guilloteau, L.A.; Chartrin, P.; Coustham, V.; Praud, C.; et al. Short cold exposures during incubation and postnatal cold temperature affect performance, breast meat quality, and welfare parameters in broiler chickens. *Poult. Sci.* **2020**, *99*, 857–868. [CrossRef]
59. Wang, G.; Li, X.; Zhou, Y.; Feng, J.; Zhang, M. Effects of Heat Stress on Gut-Microbial Metabolites, Gastrointestinal Peptides, Glycolipid Metabolism, and Performance of Broilers. *Animals* **2021**, *11*, 1286. [CrossRef]
60. Zhang, Q.; Zhang, S.; Cong, G.; Zhang, Y.; Madsen, M.H.; Tan, B.; Shi, S. Effects of Soy Protein Concentrate in Starter Phase Diet on Growth Performance, Blood Biochemical Indices, Carcass Traits, Immune Organ Indices and Meat Quality of Broilers. *Animals* **2021**, *11*, 281. [CrossRef]
61. Aviagen Resource Center. *Arbor Acres Plus Arbor Acres Plus S Broiler Performance Objectives*; Aviagen Resource Center: Tattenhall, UK, 2019; Available online: https://es.aviagen.com/assets/Tech_Center/AA_Broiler/AASF-AAFF-BroilerPO2019-EN.pdf (accessed on 20 August 2021).
62. Liu, Y.; Yang, Y.; Yao, R.; Hu, Y.; Liu, P.; Lian, S.; Lv, H.; Xu, B.; Li, S. Dietary supplementary glutamine and L-carnitine enhanced the anti-cold stress of Arbor Acres broilers. *Arch. Anim. Breed* **2021**, *64*, 231–243. [CrossRef]

sustainability

MDPI

Article

Growth Performance, Gut Health, Welfare and Qualitative Behavior Characteristics of Broilers Fed Diets Supplemented with Dried Common (*Olea europaea*) Olive Pulp

Anna Dedousi [1,*], Charalampos Kotzamanidis [1], Maria-Zoi Kritsa [1], Antiopi Tsoureki [2], Aggeliki Andreadelli [2], Sotiris I. Patsios [3] and Evangelia Sossidou [1]

1 Veterinary Research Institute, Hellenic Agricultural Organization DIMITRA, 57001 Thessaloniki, Greece
2 Institute of Applied Biosciences, Centre for Research and Technology Hellas, 6th km Charilaou-Thermi Road, Thermi, 57001 Thessaloniki, Greece
3 Laboratory of Natural Resources and Renewable Energies, Chemical Process & Energy Resources Institute (CPERI), Centre for Research and Technology-Hellas (CERTH), Thermi, 57001 Thessaloniki, Greece
* Correspondence: dedousi@vri.gr; Tel.: +30-2310365381

Abstract: The present study investigated the dietary impact of dried olive pulp (OP) on growth performance, gut health and some welfare and behavior characteristics of broilers. It was conducted in a commercial poultry farm using 108 13 day-old Ross male broilers. Chickens were equally and randomly assigned to 3 dietary treatments, CON, OP3 and OP6, based on the incorporation rate of OP in the ration (0%, 3%, and 6%, respectively). A beneficial impact on foot pad dermatitis (FPD) and feather cleanliness of OP-fed broilers was recorded. No adverse effects on qualitative behavior characteristics evaluated and on the overall growth performance of chickens were observed. No significant differences in the fecal microbiota population were observed among the groups. Changes of β-diversity in an age-dependent way were only observed. The feces of chickens across all age and dietary groups were mainly dominated by the phylum *Firmicutes* (62.3 to 95.1%), mainly represented by the genus *Lactobacillus* (32.9 to 78.2%), *Proteobacteria* (2.0 to 35.6%), and *Actinobacteria* (1.5 to 11.4%). Supplementing broilers' diets with 3% and 6% OP beneficially affected chickens' health and welfare without compromising their growth performance and gut health.

Keywords: olive pulp; broilers; growth performance; gut health; welfare

Citation: Dedousi, A.; Kotzamanidis, C.; Kritsa, M.-Z.; Tsoureki, A.; Andreadelli, A.; Patsios, S.I.; Sossidou, E. Growth Performance, Gut Health, Welfare and Qualitative Behavior Characteristics of Broilers Fed Diets Supplemented with Dried Common (*Olea europaea*) Olive Pulp. *Sustainability* **2023**, *15*, 501. https://doi.org/10.3390/su15010501

Academic Editors: George K. Symeon and Vassilios Dotas

Received: 6 December 2022
Revised: 22 December 2022
Accepted: 24 December 2022
Published: 28 December 2022

1. Introduction

Poultry gut health has wide implications for birds' systemic health, animal welfare, production efficiency, food safety, and environmental impact [1]. Factors affecting intestinal health are diet, microbiota and environment. [2]. The gut microbiota is a very important organ because it modulates several physiological functions such as nutrition, metabolism, and immunity, thus, maintaining the health of the host. It also affects host behavior and physiology [3,4]. Moreover, modulating microbiota composition, through, for instance, anti- or probiotic treatment, influences anxiety, stress and activity [5,6], as well as the serotonergic and immune systems in rodents [7,8]. Similarly, in poultry, changes in microbiota composition influence fearfulness, memory, as well as serotonergic and immune systems [9–11]. Later studies in laying hens have shown that microbiota transplantation at earlylife affects behavioral responses, serotonin and immune characteristics in chicken lines divergently chosen on feather pecking [12]. Microbiota is defined by host genes and the environment, with nutrition being one of the most significant factors. Alterations in the dietary composition may, thus, cause changes in microbiota [13].

Using antibiotics as growth promoters in farm animals has been totally forbidden by the European Union (EU) since 2006 (EC Regulation No. 1831/2003) [14] due to their possible negative effects on animal health and food safety [15,16]. In a global effort to minimize the use of drugs in poultry and livestock production, phytogenic feed additives abundant in bioactive compounds with anti-microbial, antioxidant and anti-inflammatory properties seem promising alternatives to antibiotics [17,18]. They have positive effects on farm animals since they improve growth parameters by ameliorating diet properties, promoting animals' production performance, and improving food quality [17]. Additionally, plant-derived feed supplements have been recently obtaining interest as a way of maintaining gut health in poultry [19,20]. One such feed additive of plant origin is olive pulp (OP), which is a by-product of the olive oil extraction process. As it is well known, the olive oil industry produces large quantities of by-products that are harmful to the environment, which in recent years have been attempted to be further exploited within the principles of the circular economy [21]. Towards this direction, supplementing animals' diets with olive by-products contributes to the recycling of these wastes, helping the transition to an effective circular waste-based economy [22]. Other advantages of this strategy are the lower dependence of animal production on human-consumed seeds and the decreased waste management costs [22]. Since feed represents 70% of total production costs in poultry farming, new sources of raw materials from agricultural and industrial by-products have been evaluated as feedstuff in order to decrease those costs [18,23].

OP is the remainder after the olive cake is dried. It is rich in essential fatty acids (73% oleic acid, 13% palmitic acid and 7% linoleic acid) and has elevated residual oil [24]. Moreover, it has oleuropeoside advantageous compounds such as oleuropein and phenolics like tyrosol [25]. Previous reports demonstrate that dietary polyphenols are potent antioxidants that can be utilized in poultry for improving health, growth performance and animal product quality. [26]. Thus, OP can supply animals with energy and, specifically, polyunsaturated fatty acids providing them also with various biologically active ingredients with antioxidants, antifungals, and antibacterial and anti-tumoral properties [27–32]. Furthermore, it is regarded as a good source of protein, calcium, copper and cobalt [33]. Its high nutritive value and chemical composition make OP an attractive and low-cost nutrient for farm animals [34,35].

OP has been formerly used at various incorporation rates (2–20%) in broilers' diets in numerous studies with promising though not always consistent findings regarding birds' performances [25,36–45]. In general, doses up to 10% do not seem to adversely affect broiler growth [25,38–40,43], meat quality [41,45] and health parameters [40,42,44]. On the other hand, higher dietary levels of OP have been shown to impair body weight gain, feed conversion ratio and food consumption of broilers [36,43,46]. It has been previously documented that broiler chickens fed dried OP-supplemented diets utilize OP more efficiently when its inclusion rate is increased gradually with birds' age [41].

Current literature indicates that there is a lack of research evidence regarding the optimal inclusion rate of OP in chickens' diets as well as its dietary impact on birds' welfare and behavior. Moreover, the dietary effect of OP on broilers' gut microbiota has been investigated in a limited number of studies [40,47]. On the other hand, feeding broilers with 1500 ppm of a bioactive olive pomace extract from common olive (*Olea europaea*) decreased some of the negative effects that a short-term fasting period caused in birds' intestines [19]. Furthermore, in vitro studies have demonstrated the anticoccidial activity of OP extract against *Eimeria* oocysts in broiler chickens [48]. Taking into consideration all the above, the aim of this investigation was to assess the dietary effect of dried OP on broilers' growth performance and gut microbiota, as well as on some welfare and qualitative behavior characteristics of birds. An additional objective of this study was to determine the optimal inclusion rate of OP in chickens' diets.

2. Materials and Methods

2.1. Animals, Diets and Experimental Design

This trial was conducted in a Greek commercial poultry farm, and the duration of the experiment was 29 days. In total, 108 Ross male broilers, 13-day-olds, with an initial body weight (BW) of 428.01 ± 2.69 g were used in the study. Chickens were randomly accommodated in 9 consecutive floor pens of 1 m^2 (12 birds/pen) in an environmentally controlled poultry house. Each pen was covered with rice husk and it was provided with nipple drinkers and a bell feeder. During the experiment, broilers were allowed free access to fresh water and feed was offered *ad libitum* in mash form. The stoking density in each floorpen (33 kg/m^2) met the requirements of EU Directive 2007/43/EC [49]. The lighting, temperature, and relative humidity were controlled according to the Ross 308 management guidelines [50].

Broilers were equally and randomly assigned in 3 dietary treatments identified CON, OP3, and OP6, based on the inclusion rate of OP in their ration (0%, 3% and 6% respectively) with 36 chickens/group, 3 replicates-floor pens/group. A three-phase feeding program was applied in each dietary treatment, consisting of a grower diet fed from 13 to 20 days of age, a finisher 1 and a finisher 2 diet fed from 21 to 32 days of age and from 33 to 41 days of age, respectively. In total 9 diets were formulated (Table 1). In OP-diets, dried OP substituted mainly wheat and a small quantity of sunflower oil of the CON-diet in order to formulate all ratios at an isonitrogenous and isocaloric basis. The dried OP used in the investigation was a commercial animal feed supplement in the form of flour (Sparta INNOLIVE®, Sparta Life S.A., Sparta, Greece), and its' nutrient and fatty acid composition are described in detail by [51].

Table 1. Formulation and nutrient composition of diets containing Olive Pulp (OP) compared with the Control diet (CON).

	Grower (13–20 Days of Age)			**Finisher 1 (21–32 Days of Age)**			**Finisher 2 (33–41 Days of Age)**		
Items	**CON**	**OP3**	**OP6**	**CON**	**OP3**	**OP6**	**CON**	**OP3**	**OP6**
Ingredients									
Wheat	36.395	33.615	30.955	40.95	38.105	35.35	44.87	42.135	39.28
Corn	25	25	25	25	25	25	25	25	25
Soya meal	31.3	31.4	31.47	26.43	26.58	26.72	22.33	22.43	22.6
Olive pulp	0	3	6	0	3	6	0	3	6
Sunflower oil	4.1	3.9	3.6	4.9	4.7	4.45	5.15	4.9	4.7
MCP *	0.42	0.42	0.43	0.215	0.22	0.22	0.185	0.19	0.195
Premix [1]	0.4	0.4	0.4	0.4	0.4	0.4	0.4	0.4	0.4
Limestone	1.05	0.93	0.8	0.86	0.74	0.61	0.9	0.78	0.66
NaCl	0.28	0.28	0.28	0.28	0.28	0.28	0.28	0.28	0.28
Methionine	0.35	0.34	0.35	0.3	0.3	0.3	0.25	0.25	0.25
Lysine	0.26	0.27	0.27	0.24	0.25	0.25	0.24	0.24	0.24
Threonine	0.15	0.15	0.15	0.135	0.135	0.135	0.11	0.11	0.11
RONOZYME® HiPhos	0.02	0.02	0.02	0.015	0.015	0.015	0.01	0.01	0.01
Mycotoxins Binder	0.2	0.2	0.2	0.2	0.2	0.2	0.2	0.2	0.2
Coccidiostat	0.05	0.05	0.05	0.05	0.05	0.05	0.05	0.05	0.05
Antioxidant	0.015	0.015	0.015	0.015	0.015	0.015	0.015	0.015	0.015
Xylanase	0.01	0.01	0.01	0.01	0.01	0.01	0.01	0.01	0.01
Total	100	100	100	100	100	100	100	100	100
Chemical analysis									
Crude protein (%)	20.774	20.8	20.833	18.985	19.003	19.008	17.49	17.52	17.57
Crude fiber (%)	2.75	3.45	4.15	2.65	3.35	4.05	2.57	3.27	3.91
Fat (%)	5.95	6.15	6.25	6.75	6.95	7.1	7.01	7.15	7.25
Ash (%)	4.76	4.8	4.85	4.12	4.18	4.22	3.93	3.98	4.04
Calculated analysis									
ME (kcal/kg)	2901	2903	2903	2993	2998	2999	3040	3042	3044
Lysine (%)	1.256	1.259	1.254	1.117	1.12	1.118	1.013	1.009	1.075
Methionine (%)	0.652	0.649	0.648	0.579	0.576	0.577	0.511	0.509	0.507
Ca (%)	0.63	0.631	0.63	0.515	0.516	0.514	0.514	0.516	0.517
P (%)	0.462	0.461	0.46	0.398	0.397	0.397	0.377	0.376	0.375
Na (%)	0.12	0.12	0.12	0.12	0.12	0.12	0.12	0.12	0.12

[1] Premix Finisher contains (per kg of product): Vitamin A, 2,500,000 IU; vitamin D3, 1,250,000 IU; vitamin E, 20,000 mg; vitamin K3, 1500 mg; biotin, 35,000 mcg; folic acid, 300 mg; vitamin B1, 1500 mg; vitamin B2, 1500 mg; vitamin B6, 750 mg; vitamin B12, 6000 mcg; niacinamide, 7500 mg; calcium D-pantothenate, 3750 mg; choline chloride, 150,000 mg; carbonate (siderite), 12,500 mg; cooper as copper sulphate pentahydrate, 2500 mg; manganic oxide, 27,500 mg; zinc oxide, 20,000 mg; calcium iodate anhydrous, 300 mg; coated granulated sodium selenite, 75 mg; citric acid, 14 mg; orthophosphoric acid, 3.50 mg; butylhydroxytoluene (BHT), 35 mg; butylated hydroxyanisole (BHA), 8.75 mg; calcium carbonate, 55.90%; calcium, 22.21%; phosphorous, 0.01%. * MCP: Monocalcium Phosphate.

2.2. Production Traits and Fecal Sampling

Broilers' body weight (BW) was determined at the onset of the experiment (13 days of age) and at 20, 27, 34 and 41 days of age. Body weight gain (BWG) and feed consumption (FC) per bird were calculated on weekly intervals as well as for the whole experimental period (13–41 days of age). Based on FC and BWG, the feed conversion ratio (FCR) per bird was calculated both at weekly intervals and during the entire experiment. The mortality rate was recorded daily.

2.3. Welfare and Behavior Indicators

At the age of 34 and 41 days, the chickens of each treatment group were individually evaluated for feather cleanliness, foot pad dermatitis (FPD), and hock burn, as well as for quality behavior traits, according to the Welfare Quality (2009) protocol [52]. First, observations of the quality behavior characteristics (active, fearful, depressed, calm, bored, friendly and feeding behavior) were undertaken in order to avoid confounding data due to handling stress. For the quality behavior traits, individual visual observations were performed (no recording) with a duration of 2 min for each pen. In each group, the number of birds expressing a specific type of behavior was noted, and then it was divided by the total number of live birds in this group and multiplied by 100. All observations were performed at the same time, 9 o'clock in the morning, by the same assessor at each evaluation period (34 and 41 days of age).

Then, each bird was gently held by one person, and its breast was examined for the feather cleanliness assessment by scoring on a 3-point scale: Score 0—completely clean feathers; Score 1—slight feather soiling; Score 2—moderate feather soiling; and Score 3—severe feather soiling. The percentage of chicks presenting each score was then calculated. Both feet of birds were examined for the presence of foot pad dermatitis (swelling-bubble foot) or hock burn, and scores were assessed according to the following: (a) FPD: 0—no evidence of FPD; Score 1 & 2—minimal evidence of FPD, Score 3 & 4—evidence of FPD. (b) Hock burn 0—no evidence of hock burn; Score 1 & 2—minimal evidence of hock burn. Score 3 & 4—evidence of hock burn. The percentage of birds with each scoring category was then estimated.

2.4. Microbiota Analysis

2.4.1. Sample Collection and Processing

At the onset of the experiment (13 days of age), two samples of fresh feces (approximately 1 g each) were randomly collected from 2-floor pens (1 sample per floor pen) for microbiota analysis. At the middle (27 days of age) and at the end of the trial (41 days of age), 9 samples of fresh feces in each time point (1 sample/pen-floor) were collected for microbiota analysis. At each time point, approximately 1 g of fresh feces/pen–on the floor was aseptically gathered by the plastic glove of one use from 5 different points of each pen (4 corners and the center) and was put in sterile plastic tubes. Upon collection, the samples were kept frozen at −20 °C preceding DNA extraction. Total genomic DNA was isolated using the Quick-DNA fecal/soil Microbe Microprep Isolation Kit (Irvine, CA, USA) following the manufacturer's protocol. Afterward, genomic DNA was stored at −20 °C until further analysis.

2.4.2. 16S rRNA Library Construction and Sequencing

DNA concentration of samples was measured using a Qubit 4.0 Fluorometer with the Qubit® dsDNA BR assay kit (Invitrogen, Carlsbad, CA, USA). Bacterial community composition was assessed through sequencing of the V3–V4 hypervariable regions of the prokaryotic 16S ribosomal RNA gene (16S rRNA). Library construction was performed following Illumina's 16S Metagenomic Sequencing Library Preparation (15044223 B) protocol, using 2× KAPA HiFi HotStart ReadyMix Reagent (KAPA BIOSYSTEMS, Woburn, MA, USA). Gene-specific primers described in Klindworth et al. [53] were used for amplification of the V3–V4 hypervariable regions after addition at the 5′ end of an Illumina (Illumina Inc.,

San Diego, CA, USA) overhang adapter nucleotide sequence. The sequences of the primers used are presented in Table S1. PCR amplicon and library purification from primer-dimer species and unincorporated primers was performed using Agencourt AMPure XP magnetic beads (Beckman Coulter-Life Sciences, Indianapolis, IN, USA). Library quantification was conducted through fluorometric quantification with the Qubit® dsDNA BR assay kit, and size evaluation was performed on a Fragment Analyzer system (Agilent Technologies Inc., Santa Clara, CA, USA) with a DNF-915 dsDNA Reagent kit. The libraries' molarity was assessed by a qPCR performed on a RotorGene Q thermocycler (Qiagen, Hilden, Germany) using the KAPA Library Quantification kit for Illumina sequencing platforms (KAPA BIOSYSTEMS, Woburn, MA, USA). Sequencing was performed on a MiSeq platform using the MiSeq® reagent kit v3 (2×300 cycles) (Illumina, San Diego, CA, USA).

2.5. Statistical Analysis

The statistical analysis of the data was performed using Jeffreys's Amazing Statistics Program JASP (v 0.14; JASP Team, 2020) software [54]. The significance of the differences in the percentages of feather cleanliness, foot pad dermatitis (FPD), hock burn and qualitative behavior traits among dietary treatments was assessed by the Chi-square test. For the analysis of broilers' growth performance: BW, BWG, FC and FCR, the normality of the data was tested employing the Shapiro-Wilk test; Homogeneity of variance was evaluated with Levene's test. One-way ANOVA was used to compare the average values of the parameters evaluated among groups. Post hoc analysis was performed using Tukey's test. When the distribution was not normal, the non-parametric tests Kruskal-Wallis and Mann-Whitney were used to make the comparisons. All comparisons were made at a significance level of $p \leq 0.05$.

Bioinformatic Analysis

Bacterial community analysis was performed using Quantitative Insights Into Microbial Ecology2 (Qiime2) (version 2020.8) [55]. Raw reads were imported into QIIME2 and were trimmed for adapters using the cutadapt plug-in [56]. Paired-end trimmed reads were joined using the VSEARCH plug-in [57], and quality filtering was performed with a minimum quality score of 28. Reads passing filters were dereplicated and clustered into Operational Taxonomic Units (OTUs) with 99% sequence similarity using the VSEARCH tool [57] and the open-reference method. Subsequently, chimera filtering was conducted with the VSEARCH plug-in [57], and sequence taxonomy was assigned by aligning the sequences against the SILVA 132 reference database [58] at 99% sequence identity using the BLAST plug-in [59]. Archaeal, mitochondrial, chloroplastic and unassigned sequences were removed from the data.

The resulting OTU table and biom file were imported in R (version 4.0.3) [60] for further processing. OTU counts and taxonomic assignment were merged into a phyloseq and an ampvis2 object and analyzed using the phyloseq [61] and amvis2 [62] R packages, respectively. Results visualization was performed by combining functions provided by the ggplot2 R package [63]. All bar plots were normalized to 100% as abundance estimations within each sample, so percentages do not reflect the true biomass fraction of each sample.

The α-diversity was calculated with the phyloseq package using the Observed, Chao1, and ACE (abundance-based coverage estimator) indices for richness estimation as well as the Shannon, Simpson, Inverse Simpson, and Fisher indices for evenness estimation. Kruskal-Wallis non-parametric test was used to determine significant differences in the α-diversity and the relative abundance across treatments.

To assess the similarity of microbiome structure in samples, the β-diversity was calculated using the Bray-Curtis index. Principal coordinate analysis (PCoA) based on the Bray-Curtis distance matrix was performed; results visualization was performed by combining functions provided by the ggplot2 and MicrobiotaProcess R packages [64]. ANOSIM was performed using the vegan package [65].

For investigating the differences in taxonomic abundances at phylum, family, genus, and species level between CON and dietary treatment groups and between the different dietary treatments, differential abundance analysis was conducted with the DESeq2 (v) R package [66]. Prior to the analysis, OTUs were agglomerated at the species level. The Wald significance test was used for *p*-value calculation. Log2(fold-change) values were shrunk with the 'apeglm' method [67]. The thresholds used for defining differentially abundant taxa were p value < 0.01 and | log2(fold-change) | > 0.5.

2.6. Availability of Data and Materials

Raw fastq files are available through the NCBI Sequence Read Archive under the BioProject ID PRJNA885374.

3. Results

3.1. Growth Performance

The incorporation of OP in broilers' diet had a significant impact ($p < 0.05$) on their BW, BWG, FC and FCR (Table 2). However, at 41 days of age, birds of all groups presented similar BW ($p > 0.05$), while BWG, FC and FCR for the whole experimental period were not significantly different among dietary treatments ($p > 0.05$), as indicated in Table 2. OP-fed broilers presented significantly lower BW at 20 days of age compared to controls ($p < 0.05$). At 27 days of age, control birds and those that received 6% OP with their diet had significantly higher BW compared to OP3 chicks ($p < 0.05$). A similar pattern regarding BW was recorded at day 34 of age; however, the observed differences were significant only between CON and OP3 groups ($p < 0.05$). During the grower phase (13–20 days of age), CON chicks presented significantly higher BWG compared to OP birds ($p < 0.05$), whereas FC and FCR did not differ significantly among groups ($p > 0.05$). From 21 to 27 days of age, similar BWG and FCR were recorded among dietary treatments ($p > 0.05$), while CON birds presented significantly lower FC compared to OP chicks ($p < 0.05$). From 28 to 34 days of age, FCR was higher for OP broilers compared to controls, with significant differences being observed between CON and OP6 groups ($p < 0.05$). At this period, BWG and FC were not significantly different among groups. From 35 to 41 days of age, broilers from all groups presented similar BWG ($p > 0.05$). However, OP birds consumed significantly less feed compared to controls ($p < 0.05$). The observed differences in FC during this period resulted in significant differences in the FCR among all dietary treatments ($p < 0.05$). In particular, the best FCR was recorded in the OP3 group, followed by OP6 and CON groups ($p < 0.05$). During the feeding trial, no deaths were observed among dietary treatments.

3.2. Welfare and Behavior Indicators

Data in Table 3 display the dietary influence of OP on broilers' feather cleanliness as evaluated at the age of 34 and 41 days of their life. At 34 days of age, the percentage of CON chickens with moderate feather soiling (score 2) was significantly higher compared to OP birds ($p < 0.05$). At 41 days of age, the highest percentage of broilers with completely clean feathers (score 0) was recorded in the OP group and the lowest one in the CON group ($p < 0.05$). Moreover, at this time point, a lower percentage of chickens with moderate feather soiling (score 2) was found in OP groups compared to controls, and the observed differences were found significant between CON and OP6 groups ($p < 0.05$).

A significantly higher percentage of broilers with no evidence of FPD (score 0) was observed in OP-fed broilers compared to controls ($p < 0.05$) on both days 34 and 41 of their age (Table 3). Similar results were found for the percentages of broilers evaluated with scores 1 and 2 (indicating minimal evidence of FPD), with the observed differences among groups presented in detail in Table 3. Finally, at day 41 of age, the percentage of CON chickens that were evaluated with a score of 3 for FPD was higher than that recorded in OP birds, with significant differences being noticed between CON and OP3 groups ($p < 0.05$).

Table 2. Growth performance of broilers fed the control and diets containing different levels of OP. Data are presented as mean ± SE.

	CON	OP3	OP6
Body weight (g)			
Day 13	428.61 ± 2.34	426.39 ± 2.87	429.03 ± 2.88
Day 20	873.33 ± 8.18 [a]	809.86 ± 9.70 [b]	812.22 ± 12.45 [b]
Day 27	1346.67 ± 17.52 [a]	1276.11 ± 19.00 [b]	1350.71 ± 19.51 [a]
Day 34	1943.75 ± 33.79 [a]	1791.94 ± 32.70 [b]	1879.00 ± 30.56 [ab]
Day 41	2604.44 ± 37.06	2492.78 ± 38.79	2531.32 ± 45.67
Body weight gained (g)			
13–20 d	444.72 ±7.51 [a]	383.47 ± 5.47 [b]	383.19 ± 18.51 [b]
21–27 d	473.33 ±18.28	466.25 ± 18.19	537.17 ± 18.34
28–34 d	597.08 ± 24.81	515.83 ± 35.63	530.34 ± 37.26
35–41 d	660.69 ± 30.21	700.83 ± 19.14	649.41 ± 34.54
Total period (13–41 d)	543.96 ± 28.25	516.60 ± 36.37	525.03 ± 31.00
Feed consumption (g)			
13–20 d	754.58 ± 31.28	825.00 ± 58.21	845.28 ± 18.55
21–27 d	1062.50 ± 25.09 [a]	1462.36 ± 86.81 [b]	1472.16 ± 102.20 [b]
28–34 d	1274.03 ± 11.40	1483.47 ± 55.25	1567.12 ± 122.41
35–41 d	2155.69 ± 37.29 [a]	1501.53 ± 66.77 [b]	1627.97 ± 67.16 [b]
Total period (13–41 d)	1311.70 ± 157.57	1318.09 ± 90.68	1378.13 ± 101.30
FCR			
13–20 d	1.70 ± 0.05	2.16 ± 0.18	2.22 ± 0.11
21–27 d	2.25 ± 0.05	3.16 ± 0.31	2.75 ± 0.25
28–34 d	2.14 ± 0.09 [a]	2.91 ± 0.28 [ab]	2.96 ± 0.11 [b]
35–41 d	3.27 ± 0.09 [a]	2.14 ± 0.09 [b]	2.51 ± 0.03 [c]
Total period (13–41 d)	2.34 ± 0.18	2.59 ± 0.17	2.61 ± 0.11

[a,b,c] Means within a raw at a particular age with different superscripts differ significantly ($p < 0.05$).

Table 3. Percentage of broilers observed in the 3 dietary treatments (CON, OP3, OP6) scoring for welfare parameters (feather cleanliness, foot pad dermatitis, hock burn) at the age of 34 and 41 days of their life.

	Day 34			Day 41		
Score	**CON**	**OP3**	**OP6**	**CON**	**OP3**	**OP6**
			Feather cleanliness [1]			
0	36.11	61.11	51.43	13.89 [a]	47.22 [b]	50.00 [b]
1	44.44	38.89	48.57	47.22	38.89	44.12
2	19.44 [a]	0.00 [b]	0.00 [b]	36.11 [a]	13.89 [ab]	5.88 [b]
3	-	-	-	2.78	0	0
			Foot pad dermatitis [2]			
0	47.22 [a]	91.67 [b]	82.86 [b]	41.67 [a]	86.11 [b]	79.41 [b]
1	27.78 [a]	5.56 [b]	2.86 [b]	13.89	11.11	8.82
2	22.22 [a]	2.78 [b]	8.57 [ab]	25.00 [a]	2,78 [b]	2.94 [b]
3	2.78	0.00	5.71	16.67 [a]	0 [b]	2.94 [ab]
4	-	-	-	2.78	0	5.88
			Hock burn [3]			
0	97.22	94.44	100	94.44	91.67	97.06
1	2.78	2.78	0	5.56	2.78	2.94
2	-	-	-	0	2.78	0
3	-	-	-	-	-	-
4	0	2.78	0	0	2.78	0

[a,b] Means within a raw at a particular age (Day 34, Day 41) for each score category 0-4 with different superscripts differ significantly ($p < 0.05$). [1] Feather cleanliness: Score 0: indicates completely clean feathers; Score 1: indicates slight feather soiling; Score 2: indicates moderate feather soiling and Score 3: indicates severe feather soiling; [2] Footpad dermatitis: Score 0: indicates no evidence of FPD, Score 1 & 2: indicate minimal evidence of FPD, Score 3 & 4: indicate evidence of FPD; [3] Hock burn: Score 0: indicates no evidence of Hock burn, Score 1 & 2: indicate minimal evidence of Hock burn, Score 3 & 4: indicate evidence of Hock burn.

In the present study, the broilers of all experimental groups were evaluated with very good scores for hock burn, indicating no evidence of such welfare issue (Table 3). In addition, the incorporation of OP in broilers' diet had no significant effect ($p > 0.05$) on hock burn (Table 3) or the qualitative behavior characteristics evaluated (Table 4).

Table 4. Percentage of broilers observed in the 3 dietary treatments (CON, OP3, OP6) scoring for qualitative behavior characteristics (active, fearful, depressed, calm, bored, friendly and feeding behavior at the age of 34 and 41 days of their life.

Quality Behavior Traits	Day 34			Day 41		
	CON	OP3	OP6	CON	OP3	OP6
Active	19.44	30.56	22.22	16.67	25	30.55
Fearful	0	0	0	0	0	0
Depressed	0	0	0	0	0	0
Calm	63.89	55.55	61.11	72.22	52.78	52.78
Bored	0	0	2.78	0	0	0
Friendly	0	2.78	0	0	2.78	0
Feeding	16.67	11.11	13.89	11.11	19.44	16.67

No significant differences were detected among groups ($p > 0.05$).

3.3. Dietary Effects of OP on Fecal Microbiome

3.3.1. Microbiome Diversity Measures

To evaluate the fecal microbial diversity in response to OP feeding, α-diversity was compared between the control and the OP-fed groups. For the statistical tests, samples were grouped into three categories named (a) based on diet, (b) based on sampling time and (c) based on both diet and sampling time. Calculation of richness and evenness indices showed that no statistically significant differences were found among all groups over time, indicating that OP does not modulate the richness and biodiversity of fecal microbial communities (Figure 1). However, though not statistically significant, a numerical increase of indices for richness estimation (Observed, Chao1, and ACE) in OP6 related to the OP3 dietary group by 27 and 41 days of age was observed, indicating a direct relationship between the OP dose and fecal microbial diversity.

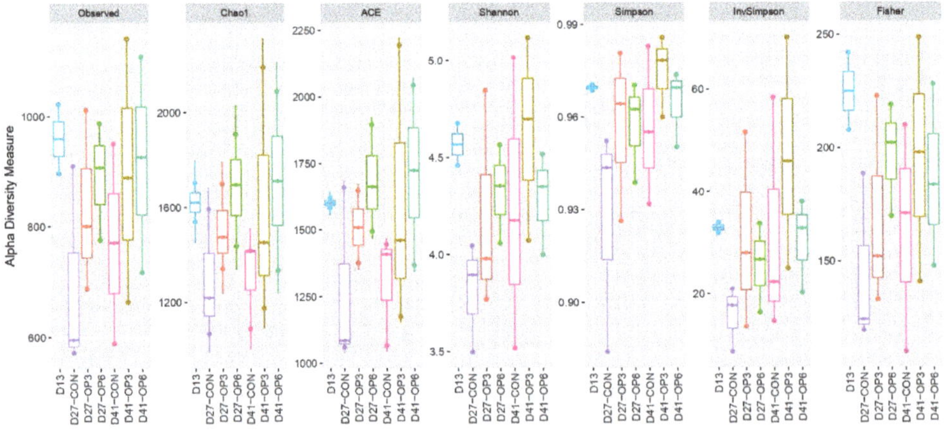

Figure 1. Alpha-diversity boxplots across dietary treatment groups (CON, OP3, OP6) of day 13 (D13; onset of the experiment), 27 (D27), and 41 (D41) chickens. Richness was estimated using the Observed, Chao1, and ACE indices. Evenness was estimated using the Shannon, Simpson, Inverse Simpson, and Fisher indices. No significant differences were found between any of the treatment groups using the Kruskal-Wallis non-parametric test.

The β-diversity analysis was performed to estimate the difference or the similarity in the microbiome composition among groups. Pairwise ANOSIM revealed that the microbial composition was significantly affected by age (p = 0.0027, R = 0.3268; Table S2); PCoA visualization of the β-diversity among samples collected at the onset of the experiment (13 days of age), at 27 days of age, and at the end of the study (41 days of age) revealed a trend of clustering of microbial communities based on sampling time (Figure 2A). On the other hand, pairwise ANOSIM did not show significant differences in microbial composition among samples from different dietary treatments (p = 0.4947, R = −0.01152; Table S3); PCoA plots showed that there was no clear clustering of the samples based on the different dietary treatments (Figure 2B).

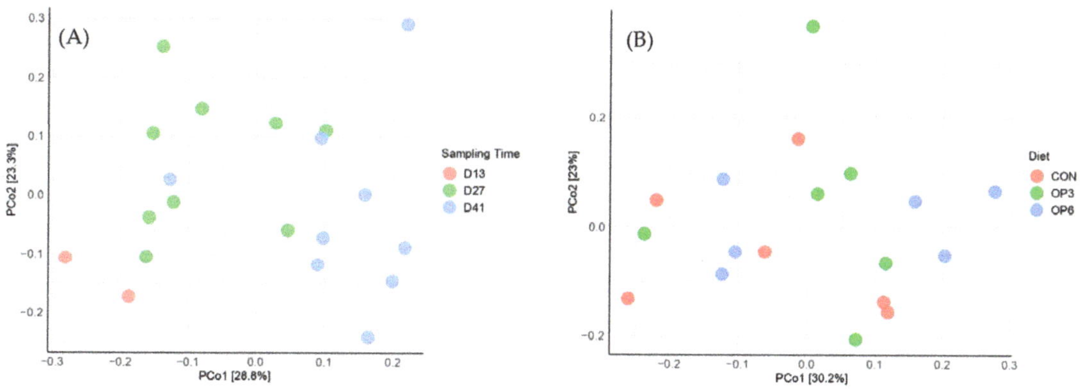

Figure 2. Principal coordinate analysis (PCoA) based on the Bray-Curtis distance matrix of chicken fecal microbiota by age; on day 13 (onset of the experiment), 27 and 41 chickens (**A**), and by dietary treatment; chickens fed with OP3% (OP3), OP6% (OP6), and control group (CON) (**B**). Statistical significance determined by ANOSIM is indicated in each plot.

3.3.2. Microbial Community Composition

To obtain further insights into the impact of OP feeding on the fecal microbiome, we assessed the relative abundance of bacterial taxa in samples collected at the onset of the experiment (13 days of age), at 27 days of age, and at the end of the study (41 days of age). Calculation of the relative abundance at different levels from phylum to genus and multiple test comparisons revealed that the predominant taxa were similar between the different dietary treatments at 27 and 41 days of age as well as between the OP groups and the control group at each time point. *Firmicutes* (62.3 to 95.1%), *Proteobacteria* (2.0 to 35.6%), and *Actinobacteria* (1.5 to 11.4%) formed the vast majority of microbiota at the phylum level across all age and dietary groups (Figure 3A,C). At the genus level, *Lactobacillus* (32.9 to 78.2%) was the dominant genera in the feces of chicken in all OP and CON groups, followed by *Streptococcus* (0.7 to 22.7%) and *Staphylococcus* (0.9 to 15.3%) (Figure 3B,D). Though not statistically significant, a numerical increase and decrease in shifts in the main genera were observed in the OP6 and OP3 dietary groups compared to CON, at 41 days of age; Interestingly, an increase in *Staphylococcus* (0.9 to 14.5%) in OP dietary treatments compared to CON has been revealed.

To identify differential taxa between CON and OP dietary treatment groups an abundant differential analysis has been performed. The staphylolococcaceal family was significantly more present in the OP3 dietary group than in the CON group, at 41 days of age (Figure 4A, Table S4). At the same time point, testing between OP6 dietary treatment group and CON group showed that four families (*Lactobacillaceae*, *Staphylococcaceae*, *Corynebacteriaceae*, and *Peptostreptococcaceae*) were enriched in OP6 group (Figure 4B, Table S5).

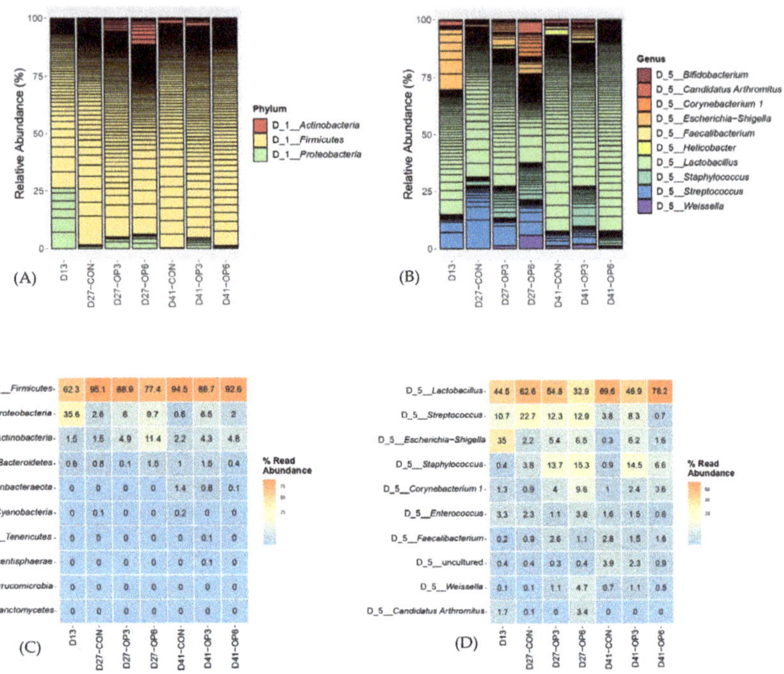

Figure 3. Microbial community composition. Bar charts showing the relative abundance of the most predominant microbial phyla (**A**) and genera (**B**), in fecal samples, in each dietary treatment group and time point. Heatmap analysis of the relative abundance of the top microbial phyla (**C**) and genera (**D**) in the feces of day 13, 27 and 41 chickens, in each dietary treatment group.

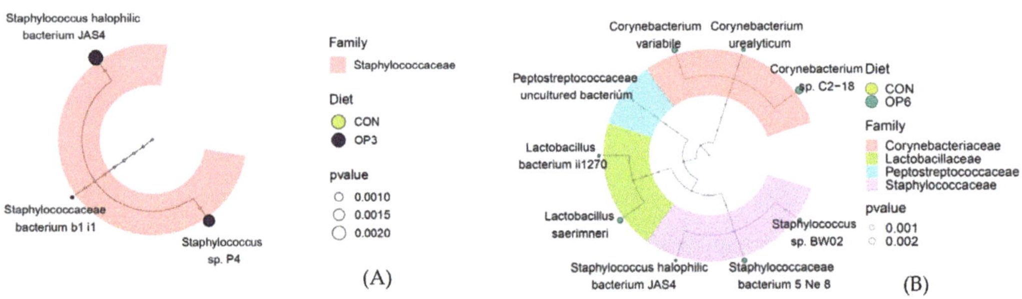

Figure 4. The phylogenetic tree of the species was found to be differentially abundant between the groups CON and 3% (**A**), and 6% (**B**) OP dietary treatments, at 41-day fecal samples. The differentially abundant species' names are shown. The final node's size is proportional to the *p*-value for the corresponding species. The color of the node denotes the group in which the differentially abundant species are more abundant. Clades with different colors correspond to different families.

4. Discussion

This is the first report that investigates the impact of OP on welfare and behavior traits of broilers. The studied inclusion rates were chosen according to the results of a previous study of ours carried out in laying hens [51]. In our former report, a positive dietary effect of dried OP on welfare parameters of hens was found in incorporation rates ranging from 3% to 6%. Thus, we decided to perform this trial using the lowest (3%) and the highest dose rates (6%) based on the beneficial impact on hens' welfare characteristics.

Moreover, the chickens used in the present trial were selected from a commercial flock that was introduced in poultry house at the age of 1 day-old. A 12-days adaptation period was used before starting the experiment, in order to provide chickens the appropriate time to adjust to the new poultry house environment and also reduce stress factors related to their transport from the hatchery to the poultry farm. In addition, this period was also required for the stabilization of their body weight.

The present study revealed that the addition of 3% and 6% OP in growers' diet retarded BWG of chickens during this period and resulted in their lower BW at day 20 compared to controls. However, during the growing phase, the chickens of all groups consumed similar amount of feed. This finding indicates the compromised nutrient utilization of feed in OP fed growers in comparison to control birds due to the high fiber content of OP, and consequently to the higher fiber content of OP diets compared to CON diet, confirming previous reports [36,43]. It seems that the immature digestive system of young broilers, which still develops at the age of 28 days of age [68,69] is not capable of digesting the high fiber content of OP diets. Jiménez-Moreno et al. [70] indicated that moderate amounts of structural insoluble fiber (2.5%) rather than (5%) are required to improve gizzard development, gastrointestinal function and nutrient digestibility in young broilers 21 days of age. Commercial broiler diets are typically formulated to contain a maximum of 2–3% crude fiber [71]. In the present study, CON diets were within crude fiber content recommendations in all three feeding phases, whereas OP diets exceeded them, as indicated by the chemical composition of experimental ratios. In line with our findings, Papadomichelakis et al. [41] also observed a negative impact in broilers growth performance when they consumed diets containing 5% of dried OP compared to controls during the grower phase. Thus, in order to provide young broilers the time required for their digestive system to adapt to the introduction of fibrous OP to their diets, these authors suggested the gradual increase of OP in birds fed with age.

It has been previously documented that broiler chickens undergoing compensatory growth also exhibit greater than normal feed intake relative to body weight and some associated digestive adaptation [72]. The results of this trial concerning FC during the period from 21 to 27 days of life are in agreement with the aforementioned authors. More specifically, in order to cover their needs for maintenance and growth, OP-fed broilers consumed significantly greater amounts of feed compared to CON birds during 21 to 27 days of life. The increased FC observed in OP-fed broilers resulted in similar BWG among chickens of all groups during 21 to 27 days of life. However, despite the similarities in BWG during this time period, only the chicks that consumed diets with 6% OP managed to achieve similar BW with control chicks. Increased feed intake in broilers fed finisher diets with 5% OP have also been recorded in previous reports [45]. During the last two weeks of the finishing period, considerable differences of FCR were recorded among groups. At first, a deterioration of FCR in OP-fed broilers compared to controls was recorded from 28 to 34 day of life followed by the opposite result the next period (35–41 d). The observed differences in feed efficiency were the net result of the numerical or significant differences in FC seen among groups during each time period.

Despite the observed differences in productive traits of broilers as recorded at weekly intervals among groups, overall growth performance of chickens as indicated by final BW at 41 day of life, BWG, FC and FCR during the whole experimental period (13–41 d) was not affected by the addition of OP in birds' diet. These data imply the capability of broilers to adapt to and efficiently utilize diets containing OP at levels of 3% and 6% gradually with age without compromising their growth. The mechanisms implicated in this adaptation could be a combination of factors acting individually or synergistic to a more functional digestive tract. Both the type of diet (higher fiber content in OP ratios compared to control) and the age of broilers with continue developing gastrointestinal system play a key role. Inclusion of insoluble dietary fiber in broiler diets has been shown to regulate intestinal morphology, gut microbiota, nutrient absorption, digestive organ development, and growth performance [73]. However, limited factors such as digestive enzymes' secretion and activities, as well as the surface area for absorption, prevent young broilers with an immature gastrointestinal tract to properly digest and absorb nutrients [74]. As the chickens grow, these issues are improving, resulting in the amelioration of nutrient utilization. Additionally, the role that the microbiome plays, or the absence of it, in the development of young chicks, must be taken into account. The gastrointestinal tract of the hatchling is sterile [75] but is quickly colonized by microbiome through the feed and environment. It is worth mentioning that a stable microbiome, with high-species diversity and an even distribution of predominant species, is established by the third week of life [76,77].

The results of this investigation concerning overall growth performance are in agreement with those observed by other researchers who evaluated the in feed inclusion of OP in broilers' diet, at various incorporation rates such as 5% and 10% [25,38,39,46], 2%, 4%, 6% and 8% [40], 5% [42,43], 2.5%, 5% and 7.5% [36] and 6% [37]. On the other hand, lower BW at slaughter, increased feed intake and FCR in broilers fed grower and finisher ratios containing 2.5% and 5% olive paste flour respectively compared to controls, was reported by Fotou et al. [45]. Moreover, decreased FC [37,43,46], deterioration of FCR [37,41–43], decreased BWG [36,37] and final BW [36,37,43] of broilers fed diets supplemented with OP at rates of 8% [41,42], 3% and 9% [37], 10% [36,43], 15% [43] and 20% [46] compared to control chicks have been previously documented. The deterioration of OP fed broilers' productive traits in former reports has been attributed to the high crude fiber content of OP, and consequently OP diets that have been shown to negatively affect nutrient utilization [36,43,45]. Finally, Saleh et al. [44] observed higher BWG in broilers with 35 days of age that consumed diets containing 4% OP compared to control chickens.

The lack of consistency of our results considering the dietary effect of OP in broiler growth performance with those formerly reported in similar studies could be due to differences in the composition of OP and diets used, the dose rate of OP, as well as chickens age and hybrid. The crude fiber content of olive by- products used is of great importance. Similar to other agro-industrial by-products, dried olive residues are characterized by a high variability in the chemical composition, due to different oil extraction methods and olive varieties, or the following processing out such as drying or destoning [78]. Destoning, has been shown to reduce the crude fiber content of dried OP thus allowing higher dietary dried OP incorporation rates without negative impact on growth performance of broiler chickens [25].

The incorporation of OP in broilers' diet at both studied levels (3% and 6%) had a positive effect on both FPD and the cleanliness of the birds'feathers as evaluated at day 34 and 41 of their life. This finding is very important for chickens health and welfare but also from the farmers' financial point of view. It is well known that FPD is a very common and well recognized problem in broiler industry [79], that negatively affects birds productivity and welfare [80] especially when lesions are severe and painful and it has been associated with reduced mobility, lameness and consequently with behavioral restrictions of birds [81,82]. It has also been shown that FPD is highly correlated to systemic bacterial infections since pathogens can invade to the chickens through damaged epithelium on the foot pads causing bumblefoot [83]. Furthermore, financial losses due to FPD are mainly attributed to slaughterhouse condemnation of carcasses with contact dermatitis lesions [79]. Chicken legs are a highly profitable by-product for the industry, and poor footpad conditions due to FPD downgrade the product quality, resulting in condemnations and downgrading and, consequently, in loss of income [81]. On the other hand, plumage cleanliness is important for thermoregulation and when the feathers are wet or soiled by bedding material they may lose their protective properties, having negative effects on the welfare of birds [84]. Dirty feathers can provide information regarding chickens living conditions and feather cleanliness estimation is a good indicator for management quality and litter humidity [85].

FPD is a multifactorial problem with litter quality, nutrition and gut health being some of the factors implicated in its' incidence [79–81]. The most important risk factor for the development of FPD is considered the litter condition [86]. The litter moisture and ammonia concentration from accumulated fecal material can burn and weaken the dermis of the footpad [87], with an increased severity of FPD resulting from the prolonged exposure of feet to wet litter. Moisture causes the outer layer of the dermis to soften, posing a risk of microbial contamination, leading to necrosis [88]. In a number of studies, dirty feathers and FPD are highly correlated; possibly due to a common cause which is litter humidity [84,89,90]. Moreover, birds with FPD prefer to spend more time sitting due to pain, thus soiling their feathers [81]. The results of the present study confirm the findings of the aforementioned authors.

The precise mechanism implicated in the positive nutritional effect of OP in both FPD and feather cleanliness of broilers observed in this study is presently unknown and requires further investigation due to the lack of similar studies in available literature. As mentioned above, nutrition and diet composition are considered major factors in the onset of FPD because they have a direct effect on feces moisture and eventually litter quality as well as on gut health. Our findings could be partially attributed to the higher fiber content of OP diets compared to the control diet which might have ameliorated either litter quality or gut health of birds or both. It has been previously documented that dietary fiber intake can directly influence wet litter, depending on the type, source, level and chemical composition of the fiber, as well as the diet composition [91]. Dietary non-starch polysaccharides (NSP), specifically insoluble NSP, have been shown to provoke beneficial effects on gut health, litter quality and nutrient utilization, by increasing crop and gizzard activity, stimulating digestive enzyme production and enhancing bacterial fermentation in the hind gut [91]. The presence of NSP in the cell wall of OP has been documented by several authors [25,38,39]. A number of studies have reported that adding lignocellulose-based products like OP [46] in poultry diets positively affects fecal consistency and litter quality [92–95]. In part, this could be due to the presence of fiber that increases digesta retention time and water holding capacity, leading to increased water absorption in the gastrointestinal tract and thus decreased moisture in the excreta. Furthermore, short chain fatty acids (SCFAs), the final products of fermentation of dietary fiber by the intestinal microbiota, have been shown to enhance the absorption of water [96,97]. Moreover, previous reports demonstrated that SCFAs, such as propionate and acetate, present toxic effects on some pathogenic bacteria [98,99].

The higher percentage of birds with no evidence of FPD and completely clean feathers recorded in OP groups compared to CON could also be due to the advantageous for the skin health effect of OP bioactive compounds like polyunsaturated fatty acids (PUFA) and polyphenols. Former reports demonstrate that oils with high content of essential fatty acids ameliorate skin hydration, regenerate the damaged epidermal lipid barrier and modulate skin metabolism [100]. Additionally, plant polyphenols are regarded as important substances for skin function, with hydrating, smoothing and softening actions [101–103]. Since this is the first study investigating the dietary effect of OP in FPD and feather cleanliness of broilers, comparison of our findings cannot be made. However, the results of a recent similar trial carried out in laying hens confirm the dietary impact of OP in keratin components of the skin [51]. In particular, Dedousi et al. [51] observed improved belly feather condition and longer claws in laying hens fed diets with 3–6% incorporation rates of OP compared to the CON diet.

No welfare issues in respect of hock burns were recorded in the present study. Even though nutrition plays an important role in the incidence of leg problems in broilers and consequently to possible behavior changes, the incorporation of OP in broilers' diet did not affect the qualitative behavior characteristics evaluated. These results could be attributed to the minimal evidence of FPD observed in birds of all groups. It has been previously shown that severe lesions of FPD have been associated with hock burns [89,104] and behavioral restrictions of chickens [81,89] and turkeys [105] such as reduced activity due to pain.

To improve our understanding of the mechanisms involved in the positive dietary effect of OP in both FPD and feather cleanliness of broilers observed in this study, we examined the impact of OP in modulating fecal microbiota. Fecal samples were utilized in this study due to concerns that cloacal swabs may not collect enough material for sufficient analysis and the need to not sacrifice birds. Estimation of the α and β-diversity in response to OP feeding showed that there was no significant effect of dietary treatments on fecal microbiota diversity; only changes of β-diversity on an age dependent way were observed. Although in a limited number, previous studies in broilers also revealed that OP did not affect the diversity of the chicken ceacal microbiota. According to them, no significant differences were observed in cecum microbiota of broilers fed 2–8% OP [40] or 750 ppm of a bioactive olive pomace extract from common olive (*Olea europaea*) [47] compared to those fed the control diets.

The feces of chickens across all age and dietary groups were mainly dominated by the phylum *Firmicutes* (62.3 to 95.1%) mainly represented by the genus *Lactobacillus* (32.9 to 78.2%), *Proteobacteria* (2.0 to 35.6%), and *Actinobacteria* (1.5 to 11.4%) (Figure 3A–D). These results are in agreement with previous studies which reported the same dominant phyla in the cecal [106] and fecal microbiota [107] in chickens. Although it has been shown that bioactive compounds of olive pomace such as oleuropein and hydroxytyrosol [108] as well as the use of fiber as a tool in order to encourage extended digesta passage rate in the fore gut can help to promote beneficial microbiota [91], in our study we found that the incorporation of OP in broilers' diet at both studied levels (3% and 6%) did not modify the relative abundance of taxa in feces.

Strikingly, though not statistically significant, at the genus level a numerical increase in *Staphylococcus* (0.9 to 14.5%) in OP dietary treatments compared to CON was revealed. Differential abundant analysis also confirmed differentially enrichment with *Staphylococcaceae* of OP3 and OP6 microbiota compared to CON. *Lactobacillaceae* was also one of the differentially abundant taxa which were identified to be enriched in 41 days of age samples from birds treated with OP6, compared to CON. The latter is in agreement with previous studies reporting *Lactobacillus* was identified at increased counts in ileal samples of birds fed with olive leaves and pomace extracts compared to that of the CON [47,108,109]. Differentially enrichment of *Lactobacillaceae* in OP6 dietary treatment could be attributed to their ability to metabolize phenolic compounds contained in OP thus favoring their growth [110]. Lactobacilli have a principal role in shaping the immune system repertoire by improving antibody-mediated immune responses and modulating cytokine expression [111]. On the other hand, staphylococci constitute a common part of chicken intestinal microbiota acting as symbiotic or pathogenic [112]. Similar to our findings, *Staphylococcaceae* have also been recorded in broiler fecal microbiota samples in former reports [113].

Overall, comparing the OP treated groups with their respective control groups, our results suggest that OP treatments did not significantly affect the fecal microbiota population and consequently positive dietary effects of OP on both FPD and feather cleanliness of broilers are not ascribed to the modulation of gut microbiota. They could however be attributed to the skin health beneficial [101–103] and immunomodulatory effects of the bioactive compounds of OP [47]. These results are in agreement with previous studies which have been proposed that positive effects of OP in pigs and chickens are related to its anti-inflammatory properties rather than to alterations in gut microbial ecology and function [47,114]. However, it should be noted that our results should be evaluated with caution; it has been proposed that fecal samples from chickens can be used to detect changes in the intestinal microbial community but the two microbiota differ significantly in diversity and the fecal microbiota does not accurately represents the intestinal one [107,115].

5. Conclusions

The current study revealed that supplementing broilers' diet with dried OP positively affected both FPD and feather cleanliness of birds. These observations demonstrate the potential health and welfare benefits for OP fed chickens and also indicate a possible cost –benefit relation for the producers. The non-significant differences observed among dietary treatments regarding fecal microbial population indicate that the positive dietary effects of OP on both FPD and feather cleanliness could possibly be related to the higher fiber content of OP diets compared to control diet which might have ameliorated litter quality rather than gut health of birds. They could also be associated with the skin health beneficial effect and the anti-inflammatory properties of OP bioactive compounds like PUFA and polyphenols, however further investigation is necessary to verify the exact mechanism implicated in those results. The addition of OP in broilers' diet at both studied levels (3% and 6%) did not adversely affect the qualitative behavior characteristics evaluated as well as the overall growth performance of chicks. These data imply the capability of broilers to adapt to and efficiently utilize diets containing OP at both levels of 3% and 6% gradually with age without compromising their growth.

Taking into account the encouraging findings of our research on the parameters studied, the nutritional effect of OP on the quality and organoleptic characteristics of broilers' meat could be explored in the future. A more detailed investigation of the intestinal microbiome could also be conducted to assess the potential probiotic action of some of the bioactive substances of the olive paste.

Supplementary Materials: The following supporting information can be downloaded at: https://www.mdpi.com/article/10.3390/su15010501/s1, Table S1: Sequences of primers for NGS libraries.; Table S2: Pairwise ANOSIM analysis of the chicken fecal microbiota comparing samples from different age groups.; Table S3: Pairwise ANOSIM analysis of the chicken fecal microbiota comparing samples from different dietary treatments.; Table S4: Differential abundance analyses of 41-day fecal samples, comparing OP3 to the untreated group CON.; Table S5: Differential abundance analyses of 41-day fecal samples, comparing OP6 to the untreated group CON.

Author Contributions: Conceptualization, A.D. and E.S.; methodology, A.D., C.K., A.T. and A.A.; software, A.D., M.-Z.K., S.I.P. and C.K; validation, A.D., C.K. and E.S.; formal analysis, A.D., C.K., A.T. and A.A.; investigation, A.D., M.-Z.K. and C.K.; resources, S.I.P. and E.S.; data curation, A.D. and E.S.; writing—Original draft preparation, A.D. and C.K.; writing—Review and editing, E.S. and S.I.P.; visualization, M.-Z.K., A.T. and A.A.; supervision, E.S.; project administration, A.D., S.I.P. and E.S.; funding acquisition, E.S. All authors have read and agreed to the published version of the manuscript.

Funding: This research was funded by the Veterinary Research Institute, Hellenic Agricultural Organization DIMITRA, grant number 22.1681.255.

Institutional Review Board Statement: The study was conducted in accordance with the Declaration of Helsinki and approved by the Committee for Research Ethics of Hellenic Agricultural Organization DIMITRA (22072/26 April 2021).The Research Ethics Committee of Hellenic Agricultural Organization-DIMITRA has approved the experimental protocol and implemented animal care procedures of the investigation (22072/26 April 2021).

Informed Consent Statement: Not applicable.

Data Availability Statement: The data presented in this study are available on request from the corresponding author. The data are not publicly available due to privacy.

Acknowledgments: We sincerely thank DIMITRIADIS CHRISTOS S.A. poultry farm that valuably contributed in this research by offering the experimental animals and facilities. We are also thankful to SPARTA LIFE S.A. for providing the dried OP used in this trial.

Conflicts of Interest: The authors declare no conflict of interest.

References

1. Diaz Carrasco, J.M.; Casanova, N.A.; Fernández Miyakawa, M.E. Microbiota, Gut Health and Chicken Productivity: What Is the Connection? *Microorganisms* **2019**, *7*, 374. [CrossRef] [PubMed]
2. Ducatelle, R.; Goossens, E.; De Meyer, F.; Eeckhaut, V.; Antonissen, G.; Haesebrouck, F.; Van Immerseel, F. Biomarkers for monitoring intestinal health in poultry: Present status and future perspectives. *Vet. Res.* **2018**, *49*, 43. [CrossRef] [PubMed]
3. O'Mahony, S.M.; Clarke, G.; Borre, Y.E.; Dinan, T.G.; Cryan, J.F. Serotonin, tryptophan metabolism and the brain-gut-microbiome axis. *Behav. Brain Res.* **2015**, *277*, 32–48. [CrossRef] [PubMed]
4. Foster, J.A.; Rinaman, L.; Cryan, J.F. Stress & the gut-brain axis: Regulation by the microbiome. *Neurobiol. Stress* **2017**, *7*, 124–136. [CrossRef] [PubMed]

5. Hoban, A.E.; Moloney, R.D.; Golubeva, A.V.; McVey Neufeld, K.A.; O'Sullivan, O.; Patterson, E.; Stanton, C.; Dinan, T.G.; Clarke, G.; Cryan, J.F. Behavioural and neurochemical consequences of chronic gut microbiota depletion during adulthood in the rat. *Neuroscience* **2016**, *339*, 463–477. [CrossRef] [PubMed]
6. Ceylani, T.; Jakubowska-Doğru, E.; Gurbanov, R.; Teker, H.T.; Gozen, A.G. The effects of repeated antibiotic administration to juvenile BALB/c mice on the microbiota status and animal behavior at the adult age. *Heliyon* **2018**, *4*, e00644. [CrossRef] [PubMed]
7. Cao, Y.-N.; Feng, L.-J.; Liu, Y.-Y.; Jiang, K.; Zhang, M.-J.; Gu, Y.-X.; Wang, B.-M.; Gao, J.; Wang, Z.-L.; Wang, Y.-M. Effect of *Lactobacillus rhamnosus* GG supernatant on serotonin transporter expression in rats with post-infectious irritable bowel syndrome. *WJG* **2018**, *24*, 338–350. [CrossRef]
8. Yousefi, B.; Eslami, M.; Ghasemian, A.; Kokhaei, P.; Salek Farrokhi, A.; Darabi, N. Probiotics importance and their immunomodulatory properties. *J. Cell. Physiol.* **2019**, *234*, 8008–8018. [CrossRef]
9. Cox, C.M.; Dalloul, R.A. Immunomodulatory role of probiotics in poultry and potential *in ovo* application. *Benef. Microbes* **2015**, *6*, 45–52. [CrossRef]
10. Parois, S.; Calandreau, L.; Kraimi, N.; Gabriel, I.; Leterrier, C. The influence of a probiotic supplementation on memory in quail suggests a role of gut microbiota on cognitive abilities in birds. *Behav. Brain Res.* **2017**, *331*, 47–53. [CrossRef]
11. Yan, F.F.; Wang, W.C.; Cheng, H.W. Bacillus subtilis based probiotic improved bone mass and altered brain serotoninergic and dopaminergic systems in broiler chickens. *J. Funct. Foods* **2018**, *49*, 501–509. [CrossRef]
12. van der Eijk, J.A.J.; de Vries, H.; Kjaer, J.B.; Naguib, M.; Kemp, B.; Smidt, H.; Rodenburg, T.B.; Lammers, A. Differences in gut microbiota composition of laying hen lines divergently selected on feather pecking. *Poult. Sci.* **2019**, *98*, 7009–7021. [CrossRef] [PubMed]
13. Wang, H.; Liang, S.; Li, X.; Yang, X.; Long, F.; Yang, X. Effects of encapsulated essential oils and organic acids on laying performance, egg quality, intestinal morphology, barrier function, and microflora count of hens during the early laying period. *Poult. Sci.* **2019**, *98*, 6751–6760. [CrossRef] [PubMed]
14. EUR-lex. Regulation (EC) No 1831/2003 of the European Parliament and of the Council of 22 September 2003 on Additives for Use in Animal Nutrition. 2003. Available online: https://eur-lex.europa.eu/legal-content/EN/TXT/?uri=CELEX%3A32003R1831 (accessed on 12 January 2022).
15. The Community Summary Report on Trends and Sources of Zoonoses and Zoonotic Agents in the European Union in 2007. *EFSA J.* **2009**, *7*, 223r. [CrossRef]
16. Fernando, U.; Biswas, D.; Allan, B.; Willson, P.; Potter, A.A. Influence of Campylobacter jejuni fliA, rpoN and flgK genes on colonization of the chicken gut. *Int. J. Food Microbiol.* **2007**, *118*, 194–200. [CrossRef] [PubMed]
17. Papatsiros, V.G.; Katsoulos, P.D.; Koutoulis, K.C.; Karatzia, M.; Dedousi, A.; Christodoulopoulos, G. Alternatives to antibiotics for farm animals. *CABI Rev.* **2013**, *8*, 32. [CrossRef]
18. Lillehoj, H.; Liu, Y.; Calsamiglia, S.; Fernandez-Miyakawa, M.E.; Chi, F.; Cravens, R.L.; Oh, S.; Gay, C.G. Phytochemicals as antibiotic alternatives to promote growth and enhance host health. *Vet. Res.* **2018**, *49*, 76. [CrossRef]
19. Herrero-Encinas, J.; Blanch, M.; Pastor, J.J.; Menoyo, D. Diet Supplementation with a Bioactive Pomace Extract from Olea europaea Partially Mitigates Negative Effects on Gut Health Arising from a Short-Term Fasting Period in Broiler Chickens. *Animals* **2020**, *10*, 349. [CrossRef]
20. Rebollada-Merino, A.; Bárcena, C.; Ugarte-Ruiz, M.; Porras, N.; Mayoral-Alegre, F.J.; Tomé-Sánchez, I.; Domínguez, L.; Rodríguez-Bertos, A. Effects on Intestinal Mucosal Morphology, Productive Parameters and Microbiota Composition after Supplementation with Fermented Defatted Alperujo (FDA) in Laying Hens. *Antibiotics* **2019**, *8*, 215. [CrossRef]
21. Salomone, R.; Ioppolo, G. Environmental impacts of olive oil production: A Life Cycle Assessment case study in the province of Messina (Sicily). *J. Clean. Prod.* **2012**, *28*, 88–100. [CrossRef]
22. Zangeneh, S.; Torki, M. Effects of B-Mannanase Supplementing of Olive Pulp-Included Diet on Performance of Laying Hens, Egg Quality Characteristics, Humoral and Cellular Immune Response and Blood Parameters. *Glob. Vet.* **2011**, *7*, 391–398.
23. Upadhaya, S.D.; Kim, I.H. Efficacy of Phytogenic Feed Additive on Performance, Production and Health Status of Monogastric Animals—A Review. *Ann. Anim. Sci.* **2017**, *17*, 929–948. [CrossRef]
24. Ibrahim, N.S.; Sabic, E.M.; Abu-Taleb, A.M. Effect of inclusion irradiated olive pulp in laying quail diets on biological performance. *J. Radiat. Res. Appl. Sci.* **2018**, *11*, 340–346. [CrossRef]
25. Sayehban, P.; Seidavi, A.; Dadashbeiki, M.; Ghorbani, A.; Araújo, W.A.G.; Albino, L.F.T. Effects of Different Levels of Two Types of Olive Pulp with or without Exogenous Enzyme Supplementation on Broiler Performance and Economic Parameters. *Rev. Bras. Cienc. Avic.* **2016**, *18*, 489–500. [CrossRef]
26. Abdel-Moneim, A.E.; Shehata, A.M.; Alzahrani, S.O.; Shafi, M.E.; Mesalam, N.M.; Taha, A.E.; Swelum, A.A.; Arif, M.; Fayyaz, M.; Abd El-Hack, M.E. The role of polyphenols in poultry nutrition. *J. Anim. Physiol. Anim. Nutr.* **2020**, *104*, 1851–1866. [CrossRef] [PubMed]
27. Visioli, F.; Poli, A.; Gall, C. Antioxidant and other biological activities of phenols from olives and olive oil. *Med. Res. Rev.* **2002**, *22*, 65–75. [CrossRef] [PubMed]
28. Omar, S.H. Oleuropein in Olive and its Pharmacological Effects. *Sci. Pharm.* **2010**, *78*, 133–154. [CrossRef]
29. Barbaro, B.; Toietta, G.; Maggio, R.; Arciello, M.; Tarocchi, M.; Galli, A.; Balsano, C. Effects of the Olive-Derived Polyphenol Oleuropein on Human Health. *IJMS* **2014**, *15*, 18508–18524. [CrossRef]

30. Bulotta, S.; Celano, M.; Lepore, S.M.; Montalcini, T.; Pujia, A.; Russo, D. Beneficial effects of the olive oil phenolic components oleuropein and hydroxytyrosol: Focus on protection against cardiovascular and metabolic diseases. *J. Transl. Med.* **2014**, *12*, 219. [CrossRef]
31. Casamenti, F.; Grossi, C.; Rigacci, S.; Pantano, D.; Luccarini, I.; Stefani, M. Oleuropein Aglycone: A Possible Drug against Degenerative Conditions. *Vivo Evid. Eff. Against Alzheimer's Dis.* **2015**, *45*, 679–688. [CrossRef]
32. Duborija-Kovacevic, N.; Shavrina, K. New findings on the pharmacodynamic actions of olive oil: Our contribution to better evidence about its remedial properties. *Prog. Nutr.* **2018**, *20*, 30–38. [CrossRef]
33. Abd El-Galil, K.; Morsy, A.S.; Emam, K.R.S.; Hassan, A.M. Physiological responses and productive performance of laying hens fed olive cake under South Sinai conditions. *Egypt. Poult. Sci. J.* **2017**, *37*, 293–304.
34. Ranalli, A.; Pollastri, L.; Contento, S.; Di Loreto, G.; Iannucci, E.; Lucera, L.; Russi, F. Acylglycerol and Fatty Acid Components of Pulp, Seed, and Whole Olive Fruit Oils. Their Use to Characterize Fruit Variety by Chemometrics. *J. Agric. Food Chem.* **2002**, *50*, 3775–3779. [CrossRef]
35. Tufarelli, V.; Introna, M.; Cazzato, E.; Mazzei, D.; Laudadio, V. Suitability of partly destoned exhausted olive cake as by-product feed ingredient for lamb production. *J. Anim. Sci.* **2013**, *91*, 872–877. [CrossRef]
36. Rabayaa, E.; Omar, J.M.A.; Othman, R.A. Utilization of Olive Pulp in Broiler Rations. *An-Najah Univ. J. Res.* **2001**, *15*, 133–144.
37. Mohamed, M.E.; Ali, A.M.; Shetaewi, M.M.; Said, K.I.; Abdel-Ghaffar, M.A. Utilization of Olive Pomace Meal and Orange By-Product Meal In Broiler Rations. *Sinai J. Appl. Sci.* **2013**, *2*, 113–128. [CrossRef]
38. Al-Harthi, M.A. The efficacy of using olive cake as a by-product in broiler feeding with or without yeast. *Ital. J. Anim. Sci.* **2016**, *15*, 512–520. [CrossRef]
39. Al-Harthi, M. The Effect of Olive Cake, with or Without Enzymes Supplementation, on Growth Performance, Carcass Characteristics, Lymphoid Organs and Lipid Metabolism of Broiler Chickens. *Rev. Bras. Cienc. Avic.* **2017**, *19*, 83–90. [CrossRef]
40. Sateri, S.; Seidavi, A.; Bouyeh, M.; Neumann, P.; Kutzler, M.; Laudadio, V.; Loperfido, F.; Tufarelli, V. Effect of olive meal and supplemental enzymes on performance traits, blood biochemistry, humoral immunity response and caecal microbiota of broilers. *SA J. An. Sci.* **2017**, *47*, 804. [CrossRef]
41. Papadomichelakis, G.; Pappas, A.C.; Tsiplakou, E.; Symeon, G.K.; Sotirakoglou, K.; Mpekelis, V.; Fegeros, K.; Zervas, G. Effects of dietary dried olive pulp inclusion on growth performance and meat quality of broiler chickens. *Livest. Sci.* **2019**, *221*, 115–122. [CrossRef]
42. Pappas, A.C.; Tsiplakou, E.; Papadomichelakis, G.; Mitsiopoulou, C.; Sotirakoglou, K.; Mpekelis, V.; Haroutounian, S.A.; Fegeros, K.; Zervas, G. Effects of olive pulp addition to broiler diets on performance, selected biochemical parameters and antioxidant enzymes. *J. Hellenic Vet. Med. Soc.* **2019**, *70*, 1687–1696. [CrossRef]
43. ELbaz, A.M.; Thabet, H.; Gad, G. Productive and Physiological Performance of Broilers Fed Diets Containing Different Levels of Olive Pulp. *J. Anim. Poult. Prod.* **2020**, *11*, 435–439. [CrossRef]
44. Saleh, A.A.; Paray, B.A.; Dawood, M.A.O. Olive Cake Meal and Bacillus licheniformis Impacted the Growth Performance, Muscle Fatty Acid Content, and Health Status of Broiler Chickens. *Animals* **2020**, *10*, 695. [CrossRef]
45. Fotou, E.; Moulasioti, V.; Angelis, I.; Tsiouris, V.; Patsias, A.; Tellis, C.; Moussis, V.; Boti, M.; Tsoukatos, D. Influence of dietary olive paste flour on the performance and oxidative stress in chickens raised in field conditions. *J. Hellenic Vet. Med. Soc.* **2021**, *72*, 3239–3248. [CrossRef]
46. Saleh, A.; Alzawqari, M. Effects of Replacing Yellow Corn with Olive Cake Meal on Growth Performance, Plasma Lipid Profile, and Muscle Fatty Acid Content in Broilers. *Animals* **2021**, *11*, 2240. [CrossRef]
47. Herrero-Encinas, J.; Blanch, M.; Pastor, J.J.; Mereu, A.; Ipharraguerre, I.R.; Menoyo, D. Effects of a bioactive olive pomace extract from Olea europaea on growth performance, gut function, and intestinal microbiota in broiler chickens. *Poult. Sci.* **2020**, *99*, 2–10. [CrossRef]
48. Debbou-Iouknane, N.; Nerín, C.; Amrane, M.; Ghemghar, M.; Madani, K.; Ayad, A. In Vitro Anticoccidial Activity of Olive Pulp (*Olea europaea* L. var. Chemlal) Extract Against Eimeria Oocysts in Broiler Chickens. *Acta Parasit.* **2019**, *64*, 887–897. [CrossRef]
49. EUR-lex. COUNCIL DIRECTIVE 2007/43/EC of 28 June 2007 Laying down Minimum Rules for the Protection of Chickens Kept for Meat Production. 2007. Available online: https://eur-lex.europa.eu/legal-content/EN/TXT/HTML/?uri=CELEX:32007L0043&from=EN (accessed on 12 April 2022).
50. Aviagen, W. Ross 308: Broiler's Management and Nutrition Specification. ROSS An Aviagen Brand 2018. Available online: https://en.aviagen.com/assets/Tech_Center/Ross_Broiler/Ross-BroilerHandbook2018-EN.pdf (accessed on 20 April 2022).
51. Dedousi, A.; Kritsa, M.-Z.; Đukić Stojčić, M.; Sfetsas, T.; Sentas, A.; Sossidou, E. Production Performance, Egg Quality Characteristics, Fatty Acid Profile and Health Lipid Indices of Produced Eggs, Blood Biochemical Parameters and Welfare Indicators of Laying Hens Fed Dried Olive Pulp. *Sustainability* **2022**, *14*, 3157. [CrossRef]
52. *Welfare Quality®. Welfare Quality®Assessment Protocol for Poultry (Broilers, Laying Hens)*; Welfare Quality®Consortium: Lelystad, The Netherlands, 2009.
53. Klindworth, A.; Pruesse, E.; Schweer, T.; Peplies, J.; Quast, C.; Horn, M.; Glöckner, F.O. Evaluation of general 16S ribosomal RNA gene PCR primers for classical and next-generation sequencing-based diversity studies. *Nucleic Acids Res.* **2013**, *41*, e1. [CrossRef]
54. JASP Team. JASP (Version 0.14) [Computer Software]. Available online: https://jasp-stats.org/previous-versions/ (accessed on 21 December 2020).

55. Bolyen, E.; Rideout, J.R.; Dillon, M.R.; Bokulich, N.A.; Abnet, C.C.; Al-Ghalith, G.A.; Alexander, H.; Alm, E.J.; Arumugam, M.; Asnicar, F.; et al. Reproducible, interactive, scalable and extensible microbiome data science using QIIME 2. *Nat. Biotechnol.* **2019**, *37*, 852–857. [CrossRef]
56. Martin, M. Cutadapt removes adapter sequences from high-throughput sequencing reads. *EMBnet J.* **2011**, *17*, 10. [CrossRef]
57. Rognes, T.; Flouri, T.; Nichols, B.; Quince, C.; Mahé, F. VSEARCH: A versatile open source tool for metagenomics. *PeerJ* **2016**, *4*, e2584. [CrossRef] [PubMed]
58. Quast, C.; Pruesse, E.; Yilmaz, P.; Gerken, J.; Schweer, T.; Yarza, P.; Peplies, J.; Glöckner, F.O. The SILVA ribosomal RNA gene database project: Improved data processing and web-based tools. *Nucleic Acids Res.* **2012**, *41*, D590–D596. [CrossRef] [PubMed]
59. Camacho, C.; Coulouris, G.; Avagyan, V.; Ma, N.; Papadopoulos, J.; Bealer, K.; Madden, T.L. BLAST+: Architecture and applications. *BMC Bioinform.* **2009**, *10*, 421. [CrossRef]
60. R Core Team. 2020. Available online: http://www.r-project.org/index.html (accessed on 15 December 2020).
61. McMurdie, P.J.; Holmes, S. phyloseq: An R Package for Reproducible Interactive Analysis and Graphics of Microbiome Census Data. *PLoS ONE* **2013**, *8*, e61217. [CrossRef]
62. Andersen, K.S.; Kirkegaard, R.H.; Karst, S.M.; Albertsen, M. Ampvis2: An R Package to Analyse and Visualise 16S RRNA Amplicon Data. *bioRxiv* **2018**. [CrossRef]
63. Wickham, H. *ggplot2: Elegant Graphics for Data Analysis*; Springer: Cham, Switzerland, 2016; ISBN 978-3-319-24275-0.
64. MicrobiotaProcess: An R Package for Analysis, Visualization and Biomarker Discovery of Microbiome. Available online: https://github.com/YuLab-SMU/MicrobiotaProcess (accessed on 22 April 2022).
65. Oksanen, J.; Blanchet, F.G.; Friendly, M.; Kindt, R.; Legendre, P.; McGlinn, D.; Minchin, P.R.; O'Hara, R.B.; Simpson, G.L.; Solymos, P.; et al. Vegan: Community Ecology Package. Available online: https://CRAN.R-project.org/package=vegan (accessed on 23 March 2022).
66. Love, M.I.; Huber, W.; Anders, S. Moderated estimation of fold change and dispersion for RNA-seq data with DESeq2. *Genome Biol.* **2014**, *15*, 550. [CrossRef]
67. Zhu, A.; Ibrahim, J.G.; Love, M.I. Heavy-tailed prior distributions for sequence count data: Removing the noise and preserving large differences. *Bioinformatics* **2019**, *35*, 2084–2092. [CrossRef]
68. Sklan, D.; Noy, Y. Functional development and intestinal absorption in the young poult. *Br. Poult. Sci.* **2003**, *44*, 651–658. [CrossRef]
69. Thomas, D.V.; Ravindran, V.; Ravindran, G. Nutrient digestibility and energy utilisation of diets based on wheat, sorghum or maize by the newly hatched broiler chick. *Br. Poult. Sci.* **2008**, *49*, 429–435. [CrossRef]
70. Jiménez-Moreno, E.; González-Alvarado, J.M.; de Coca-Sinova, A.; Lázaro, R.P.; Cámara, L.; Mateos, G.G. Insoluble fiber sources in mash or pellets diets for young broilers. 2. Effects on gastrointestinal tract development and nutrient digestibility. *Poult. Sci.* **2019**, *98*, 2531–2547. [CrossRef]
71. Choct, M. Fibre—Chemistry and Functions in Poultry Nutrition. 2015. Available online: https://www.wpsa-aeca.es/aeca_imgs_docs/16478_fibra_mingan.pdf (accessed on 22 July 2022).
72. Zubair, A.K.; Leeson, S. Compensatory growth in the broiler chicken: A review. *World's Poult. Sci. J.* **1996**, *52*, 189–201. [CrossRef]
73. Tejeda, O.J.; Kim, W.K. Role of Dietary Fiber in Poultry Nutrition. *Animals* **2021**, *11*, 461. [CrossRef] [PubMed]
74. Ravindran, V.; Abdollahi, M.R. Nutrition and Digestive Physiology of the Broiler Chick: State of the Art and Outlook. *Animals* **2021**, *11*, 2795. [CrossRef] [PubMed]
75. Harrow, S.A.; Ravindran, V.; Butler, R.C.; Marshall, J.W.; Tannock, G.W. Real-Time Quantitative PCR Measurement of Ileal *Lactobacillus salivarius* Populations from Broiler Chickens To Determine the Influence of Farming Practices. *Appl. Environ. Microbiol.* **2007**, *73*, 7123–7127. [CrossRef] [PubMed]
76. McBurney, W.; Tannock, G.; Ravindran, V. A culture-independent approach to the analysis of the gut microflora of broilers. *Proc. Aust. Poult. Sci. Symp.* **2003**, *15*, 131–134.
77. Rinttilä, T.; Apajalahti, J. Intestinal microbiota and metabolites—Implications for broiler chicken health and performance. *J. Appl. Poult. Res.* **2013**, *22*, 647–658. [CrossRef]
78. Joven, M.; Pintos, E.; Latorre, M.A.; Suárez-Belloch, J.; Guada, J.A.; Fondevila, M. Effect of replacing barley by increasing levels of olive cake in the diet of finishing pigs: Growth performances, digestibility, carcass, meat and fat quality. *Anim. Feed. Sci. Technol.* **2014**, *197*, 185–193. [CrossRef]
79. Amer, M.M. REVIEW: Footpad dermatitis (FPD) in chickens. *Korean J. Food Health Converg.* **2020**, *6*, 11–16. [CrossRef]
80. Swiatkiewicz, S.; Arczewska-Wlosek, A.; Jozefiak, D. The nutrition of poultry as a factor affecting litter quality and foot pad dermatitis—An updated review. *J. Anim. Physiol. Anim. Nutr.* **2017**, *101*, e14–e20. [CrossRef]
81. Freeman, N.; Tuyttens, F.A.M.; Johnson, A.; Marshall, V.; Garmyn, A.; Jacobs, L. Remedying Contact Dermatitis in Broiler Chickens with Novel Flooring Treatments. *Animals* **2020**, *10*, 1761. [CrossRef]
82. Choi, J.; Kim, W.K. Dietary Application of Tannins as a Potential Mitigation Strategy for Current Challenges in Poultry Production: A Review. *Animals* **2020**, *10*, 2389. [CrossRef] [PubMed]
83. Heidemann Olsen, R.; Christensen, H.; Kabell, S.; Bisgaard, M. Characterization of prevalent bacterial pathogens associated with pododermatitis in table egg layers. *Avian Pathol.* **2018**, *47*, 281–285. [CrossRef] [PubMed]
84. Saraiva, S.; Saraiva, C.; Stilwell, G. Feather conditions and clinical scores as indicators of broilers welfare at the slaughterhouse. *Res. Vet. Sci.* **2016**, *107*, 75–79. [CrossRef] [PubMed]

85. Arnould, C.; Butterworth, A.; Knierim, U. Standardisation of clinical scoring in poultry. In *Assessment of Animal Welfare Measures for Layers and Broilers*; Forkman, B., Keeling, L., Eds.; School of City and Regional Planning, Cardiff University: Cardiff, UK, 2009; pp. 7–30. ISBN 978-1-902647-79-1.
86. Grimes, J.L.; Smith, J.; Williams, C.M. Some alternative litter materials used for growing broilers and turkeys. *World's Poult. Sci. J.* **2002**, *58*, 515–526. [CrossRef]
87. Kaukonen, E.; Norring, M.; Valros, A. Effect of litter quality on foot pad dermatitis, hock burns and breast blisters in broiler breeders during the production period. *Avian Pathol.* **2016**, *45*, 667–673. [CrossRef] [PubMed]
88. Mayne, R.K. A review of the aetiology and possible causative factors of foot pad dermatitis in growing turkeys and broilers. *World's Poult. Sci. J.* **2005**, *61*, 256–267. [CrossRef]
89. de Jong, I.C.; Gunnink, H.; van Harn, J. Wet litter not only induces footpad dermatitis but also reduces overall welfare, technical performance, and carcass yield in broiler chickens. *J. Appl. Poult. Res.* **2014**, *23*, 51–58. [CrossRef]
90. Göransson, L.; Yngvesson, J.; Gunnarsson, S. Bird Health, Housing and Management Routines on Swedish Organic Broiler Chicken Farms. *Animals* **2020**, *10*, 2098. [CrossRef]
91. Kheravii, S.K.; Morgan, N.K.; Swick, R.A.; Choct, M.; Wu, S.-B. Roles of dietary fibre and ingredient particle size in broiler nutrition. *World's Poult. Sci. J.* **2018**, *74*, 301–316. [CrossRef]
92. Farran, M.T.; Pietsch, M.; Chabrillat, T. Effect of lignocellulose on the litter quality and the ready to cook carcass yield of male broilers. In Proceedings of the Actes des 10èmes Journées de la Recherche Avicole et Palmipèdes à Foie Gras, La Rochelle, France, 26–28 March 2013; pp. 917–921.
93. Bogusławska-Tryk, M.; Szymeczko, R.; Piotrowska, A.; Burlikowska, K.; Śliżewska, K. Ileal and Cecal Microbial Population and Short-Chain Fatty Acid Profile in Broiler Chickens Fed Diets Supplemented with Lignocellulose. *Pak. Vet. J.* **2015**, *35*, 212–216.
94. Milošević, N.; Đukić Stojčić, M.; Perić, L.; Vukić-Vranješ, M. Effect of lignocellulose on egg production and egg quality. In Proceedings of the 20th European Symposium on Poultry Nutrition, Prague, Czech Republic, 24–27 August 2015; p. P-046.
95. Kheravii, S.K.; Swick, R.A.; Choct, M.; Wu, S.-B. Coarse particle inclusion and lignocellulose-rich fiber addition in feed benefit performance and health of broiler chickens. *Poult. Sci.* **2017**, *96*, 3272–3281. [CrossRef] [PubMed]
96. Musch, M.W.; Bookstein, C.; Xie, Y.; Sellin, J.H.; Chang, E.B. SCFA increase intestinal Na absorption by induction of NHE3 in rat colon and human intestinal C2/bbe cells. *Am. J. Physiol. Gastrointest. Liver Physiol.* **2001**, *280*, G687–G693. [CrossRef] [PubMed]
97. Gupta, N.; Martin, P.M.; Prasad, P.D.; Ganapathy, V. SLC5A8 (SMCT1)-mediated transport of butyrate forms the basis for the tumor suppressive function of the transporter. *Life Sci.* **2006**, *78*, 2419–2425. [CrossRef] [PubMed]
98. Cherrington, C.A.; Hinton, M.; Mead, G.C.; Chopra, I. Organic Acids: Chemistry, Antibacterial Activity and Practical Applications. In *Advances in Microbial Physiology*; Elsevier: Amsterdam, The Netherlands, 1991; Volume 32, pp. 87–108. ISBN 978-0-12-027732-2.
99. van der Wielen, P.W.J.J.; Biesterveld, S.; Notermans, S.; Hofstra, H.; Urlings, B.A.P.; van Knapen, F. Role of Volatile Fatty Acids in Development of the Cecal Microflora in Broiler Chickens during Growth. *Appl. Environ. Microbiol.* **2000**, *66*, 2536–2540. [CrossRef]
100. Michalak, M.; Pierzak, M.; Kręcisz, B.; Suliga, E. Bioactive Compounds for Skin Health: A Review. *Nutrients* **2021**, *13*, 203. [CrossRef]
101. Anunciato, T.P.; da Rocha Filho, P.A. Carotenoids and polyphenols in nutricosmetics, nutraceuticals, and cosmeceuticals: Nutricosmetics. *J. Cosmet. Dermatol.* **2012**, *11*, 51–54. [CrossRef]
102. Nichols, J.A.; Katiyar, S.K. Skin photoprotection by natural polyphenols: Anti-inflammatory, antioxidant and DNA repair mechanisms. *Arch. Dermatol. Res.* **2010**, *302*, 71–83. [CrossRef]
103. Zillich, O.V.; Schweiggert-Weisz, U.; Eisner, P.; Kerscher, M. Polyphenols as active ingredients for cosmetic products. *Int. J. Cosmet. Sci.* **2015**, *37*, 455–464. [CrossRef]
104. Haslam, S.M.; Knowles, T.G.; Brown, S.N.; Wilkins, L.J.; Kestin, S.C.; Warriss, P.D.; Nicol, C.J. Factors affecting the prevalence of foot pad dermatitis, hock burn and breast burn in broiler chicken. *Br. Poult. Sci.* **2007**, *48*, 264–275. [CrossRef]
105. Hocking, P.M.; Wu, K. Traditional and commercial turkeys show similar susceptibility to foot pad dermatitis and behavioural evidence of pain. *Br. Poult. Sci.* **2013**, *54*, 281–288. [CrossRef] [PubMed]
106. Videnska, P.; Sedlar, K.; Lukac, M.; Faldynova, M.; Gerzova, L.; Cejkova, D.; Sisak, F.; Rychlik, I. Succession and Replacement of Bacterial Populations in the Caecum of Egg Laying Hens over Their Whole Life. *PLoS ONE* **2014**, *9*, e115142. [CrossRef] [PubMed]
107. Yan, W.; Sun, C.; Yuan, J.; Yang, N. Gut metagenomic analysis reveals prominent roles of Lactobacillus and cecal microbiota in chicken feed efficiency. *Sci. Rep.* **2017**, *7*, 45308. [CrossRef] [PubMed]
108. Sarıca, S.; Ürkmez, D. The use of grape seed-, olive leaf- and pomegranate peel-extracts as alternative natural antimicrobial feed additives in broiler diets. *Eur. Poult. Sci.* **2016**, *80*, 121.
109. Amini, Z.; Parsaei, S.; Houshmand, M.; Naghiha, R. Effect of Olive leaf powder on performance and ileal bacteria count of broilers. *Vet. Res. Forum.* **2019**, *10*, 255–259. [CrossRef]
110. García-Ruiz, A.; Bartolomé, B.; Martínez-Rodríguez, A.J.; Pueyo, E.; Martín-Álvarez, P.J.; Moreno-Arribas, M.V. Potential of phenolic compounds for controlling lactic acid bacteria growth in wine. *Food Control* **2008**, *19*, 835–841. [CrossRef]
111. Alizadeh, M.; Shojadoost, B.; Astill, J.; Taha-Abdelaziz, K.; Karimi, S.H.; Bavananthasivam, J.; Kulkarni, R.R.; Sharif, S. Effects of in ovo Inoculation of Multi-Strain Lactobacilli on Cytokine Gene Expression and Antibody-Mediated Immune Responses in Chickens. *Front. Vet. Sci.* **2020**, *7*, 105. [CrossRef]
112. Syed, M.A.; Ullah, H.; Tabassum, S.; Fatima, B.; Woodley, T.A.; Ramadan, H.; Jackson, C.R. Staphylococci in poultry intestines: A comparison between farmed and household chickens. *Poult. Sci.* **2020**, *99*, 4549–4557. [CrossRef]

113. Videnska, P.; Rahman, M.M.; Faldynova, M.; Babak, V.; Matulova, M.E.; Prukner-Radovcic, E.; Krizek, I.; Smole-Mozina, S.; Kovac, J.; Szmolka, A.; et al. Characterization of Egg Laying Hen and Broiler Fecal Microbiota in Poultry Farms in Croatia, Czech Republic, Hungary and Slovenia. *PLoS ONE* **2014**, *9*, e110076. [CrossRef]
114. Liehr, M.; Mereu, A.; Pastor, J.J.; Quintela, J.C.; Staats, S.; Rimbach, G.; Ipharraguerre, I.R. Olive oil bioactives protect pigs against experimentally-induced chronic inflammation independently of alterations in gut microbiota. *PLoS ONE* **2017**, *12*, e0174239. [CrossRef]
115. Stanley, D.; Geier, M.S.; Chen, H.; Hughes, R.J.; Moore, R.J. Comparison of fecal and cecal microbiotas reveals qualitative similarities but quantitative differences. *BMC Microbiol.* **2015**, *15*, 51. [CrossRef] [PubMed]

sustainability

MDPI

Article

Growth Performance, Meat Quality, Welfare and Behavior Indicators of Broilers Fed Diets Supplemented with *Yarrowia lipolytica* Yeast

Anna Dedousi [1,*], Sotiris I. Patsios [2], Maria-Zoi Kritsa [1], Konstantinos N. Kontogiannopoulos [3], Maria Ioannidou [1], Antonios Zdragas [1] and Evangelia N. Sossidou [1]

1 Veterinary Research Institute, Hellenic Agricultural Organization DIMITRA, GR 57001 Thessaloniki, Greece
2 Laboratory of Natural Resources and Renewable Energies, Chemical Process & Energy Resources Institute (CPERI), Centre for Research and Technology-Hellas (CERTH), Thermi, GR 57001 Thessaloniki, Greece
3 Ecoresources P.C., 3 Kolchidos Str., GR 55131 Thessaloniki, Greece
* Correspondence: dedousi@vri.gr; Tel.: +30-2310365381

Abstract: This study investigated the dietary impact of dried *Yarrowia lipolytica* yeast (YLP) on the growth performance, meat quality, welfare and behavior indicators of broilers. It was performed in a commercial poultry farm using 108 13 day-old Ross 308 male broilers. The chicks were randomly and equally divided into three dietary groups CON, YLP3 and YLP5, according to the incorporation rate of YLP in the feed (0%, 3%, and 5%, respectively). A positive effect on foot pad dermatitis (FPD) of YLP-fed broilers was observed without any adverse effects on welfare, behavior, meat quality and the overall growth performance of the broilers. YLP significantly decreased the malondialdehyde (MDA) values in breast and thigh meat. YLP3 birds presented a superior nutrient quality of breast meat, as indicated by the increased concentration of monounsaturated fatty acids (MUFAs) and polyunsaturated fatty acids (PUFAs), decreased levels of saturated fatty acids (SFAs), a better PUFA/SFA ratio and improved health lipid indices. A significant elevation of n-3 PUFAs was observed in the thigh meat of YLP-fed groups, compared to the CON groups. A positive effect on the overall sensory acceptance of thigh meat was detected in the YLP5 group. YLP feeding, at the rate of 3%, seems to be beneficial for improving the meat nutrition quality.

Keywords: *Yarrowia lipolytica*; broilers; sustainable protein feed supplement; growth performance; welfare; meat quality

Citation: Dedousi, A.; Patsios, S.I.; Kritsa, M.-Z.; Kontogiannopoulos, K.N.; Ioannidou, M.; Zdragas, A.; Sossidou, E.N. Growth Performance, Meat Quality, Welfare and Behavior Indicators of Broilers Fed Diets Supplemented with *Yarrowia lipolytica* Yeast. *Sustainability* **2023**, *15*, 1924. https://doi.org/10.3390/su15031924

Academic Editors: George K. Symeon and Vassilios Dotas

Received: 9 December 2022
Revised: 14 January 2023
Accepted: 17 January 2023
Published: 19 January 2023

1. Introduction

Proteins are essential for a balanced diet, and given the significant rise of the global population in the near future, much larger quantities of food-grade protein will be needed [1]. Foods of animal origin are needed for global food security, since they comprise 25% of the protein intake worldwide [1]. Protein feed stocks are expensive and limited in animals' ratios. Currently, soybean meal, after soy oil extraction, is the main source of vegetable protein for animal feed [2]. In particular, approximately 97% of the globally produced soybean meal is used as animal feed [3]. However, feeding animals with soybean proteins and other grains increases the dependence of animal production on human-edible plants, negatively affecting human food security and sustainability [4].

Therefore, there is an urgent need to use alternative, more sustainable protein feed sources to partly replace the current supply chains and reduce the environmental footprint of animal products [2]. Moreover, such a necessity arises from the fact that traditional sources of protein for animal feed compete with human food for the use of high-quality land for agricultural uses. Thus, alternative animal feed proteins, such as protein-rich food industry by-products, macro-algae, insect meal, and single cell proteins (SCPs), have a high potential as sustainable protein sources, since they have a low footprint, and can be produced locally on low-quality agricultural land. [5].

YLP is an oleaginous, non-pathogenic yeast, which can be utilized as a SCP in animal feed [6]. Among others, this is due to its capability to produce considerable levels of proteins and lipids from low cost substrates [7]. In addition, these strains can grow in biofuel-produced by-products, such as glycerol, to produce a yeast biomass rich in proteins and/or lipids [8]. The annual production of SCPs reaches 1200 tons of dry mass, that contains approximately between 41% and 45% *w*/*w* of protein [9]. Available research evidence shows that the YLP biomass has a high nutritional value. It is a source of high quality proteins (especially essential amino acids), minerals, vitamins and polyunsaturated fatty acids [8,10]. Currently, the American Food and Drug Administration, has examined the safety issues of YLP yeast, and has characterized both the yeast and its various fermentation products as being generally recognized as safe (GRAS) [11]. Moreover, a YLP strain grown on raw glycerol has been considered safe for use as a high value foodstuff (EU 2017/1017) [12]. Finally, the European Food and Safety Authority (EFSA) declared the YLP yeast biomass as a novel food (NF) safe for use, pursuant to Regulation (EU) 2015/2283 [13] on dietary supplements intended for the general population over three years of age [14].

Previous investigations using YLP as a feed supplement for turkeys [15–20], piglets [8,21–23], calves [24], rats [25], fish [26–29], crustaceans [30,31] and mollusks [32,33] have been reported, giving promising results, in terms of their health and performance. More specifically, the incorporation of YLP in the diet of productive animals has been shown to beneficially affect: (1) weight gain, (2) feed conversion ratio, (3) intestinal and ruminal microbiomes, (4) intestinal morphology, (5) antioxidant status and immune response, (6) erythropoietic processes and hematobiochemical profiles, (7) survival rate, (8) digestibility, (9) pathogen elimination (10) the fatty acid composition of fish fillets and (11) the metabolic status of the organism [34]. However, to our knowledge, there are no indications in the available literature regarding the dietary effect of YLP yeast in broilers' growth performance or meat quality. Moreover, no data are available on whether the use of feed with YLP yeast affects the welfare and behavior of productive animals, in general, and broiler chickens, in particular.

This study aims to assess the dietary effect of YLP yeast on broilers' growth performance and meat quality, as well as certain welfare and quality behavior indicators of broilers. Moreover, different inclusion rates of YLP were tested to specify the optimal inclusion rate of YLP in chickens' feed.

2. Materials and Methods

2.1. Yeast Biomass Production

The yeast biomass, which was used in the feeding trials, was obtained through a series of semi-continuous fermentations of the wild type strain YPL MUCL 28849 (BCCM/MUCL, (Agro) Industrial Fungi & Yeasts Collection, Belgium). All submerged fermentations were conducted using sterilized (121 °C for 20 min) synthetic media supplemented with 46 g/L crude glycerol (≥90–92% *w*/*w* purity) that derived as a by-product of the biodiesel production of the Fytoenergeia/NewEnergy S.A. industrial plant, sited in Northern Greece (Paralimnio, Serres). A modified synthetic medium composition was used [35], which comprised (g/L): $(NH_4)_2SO_4$, 3.0; KH_2PO_4, 2.0; $Na_2HPO_4 \times 2\ H_2O$, 2.6; $MgSO_4 \times 7\ H_2O$, 1.0; $CaCl_2 \times 2\ H_2O$, 0.2; $FeCl_3$, 0.02; (mg/L): H_3BO_3, 0.5; $CuSO_4 \times 5\ H_2O$, 0.06; KI, 0.1; $MnSO_4 \times H_2O$, 0.45; $ZnSO_4 \times 7\ H_2O$, 0.71; and $Na_2MoO_4 \times 2\ H_2O$, 0.23. YPL was transferred from a YPG-agar culture (yeast extract 10 g/L; peptone 20 g/L; glycerol 2% *v*/*v*) to the first flask preculture for 24-h incubation (30 °C, 150 rpm, 0.1 L). The first preculture content was centrifuged and the pellet was used to inoculate 2 L of a fermentation culture in a bench-scale bioreactor (BioFlo120, Eppendorf, Hamburg, Germany). The bioreactor, which was sterilized at 121 °C for 20 min, was equipped with sensors for optical density, pH, temperature and dissolved oxygen recording. Aeration was kept constant at 0.75 vvm, the agitation rate was set at 500 rpm, the pH was controlled at 6.0, through the addition of 5 N NaOH and/or 5 N H_2SO_4 and the temperature was constant at 30 °C.

Following 20 h of cultivation (middle exponential growth phase), the dry biomass concentration reached approximately 14 g/L, and 1.5 L of the fermentation culture was collected and used as inoculum for the main fermentation in an industrial-scale (140 L) pilot bioreactor (working volume 100 L) that was designed by NRRE Lab (CPERI/CERTH, Thessaloniki, Greece) and operated at the premises of the Fytoenergeia/NewEnergy S.A. industrial plant; a schematic representation of the industrial-scale pilot plant is presented in Figure S1. The reactor was sterilized at 121 °C for at least 20 min, and cooled down at 28 °C, prior to its inoculation. The aeration rate was constant at 1 vvm, the agitation speed was 150 rpm, the pH was controlled at 6.0 through the addition of 30% *w*/*w* NaOH, and the temperature was set to 28 °C. The main operating parameters, i.e. dissolved oxygen, pH, temperature and bioreactor volume, were monitored through on-line sensors. The bioreactor operated in batch mode for the first 24 h of cultivation; at that point, the dry biomass reached a concentration of approximately 15 g/L. Thereafter, a semi-continuous operation was employed, through the withdrawal of 50 L of culture, and the addition of 50 L of fresh fermentation medium every 24 h. The yeast biomass was recovered from the withdrawn culture through gravity filtration using a fiber-filter bag with a 5 μm pore size (Fluxflo G1PE5-S, Envirogen Group, Alfreton, UK). The wet yeast cake (moisture content approximately 85% *w*/*w*) was scraped off the filter's surface and subsequently air-dried in an oven at 60 °C, till constant weight. All of the dried yeast biomass collected each day was thoroughly mixed/ homogenized and was used in the feeding trials.

A sample of the homogenized dry yeast biomass was analyzed for its physicochemical properties (nutrients analysis), heavy metals (arsenic, cadmium, lead and mercury), fatty acid and amino acid profiles, and microbiological characterization (Table 1). The physicochemical analysis for the basic nutrients was performed according to the European Commission's (EC) Regulation No 152/2009 [36] for the following parameters: dry matter, crude protein, ash, crude fats and crude fiber; the carbohydrate content and the gross energy were calculated, based on the proximate analysis. The sugar content was measured using an enzymatic method used for the analysis of D-glucose, D-fructose and sucrose in plant and food products, employing a Megazyme K-SUFRG 04/18 assay kit [37]. The amino acid profiles (excluding tryptophan) were also determined, according to the EC's Regulation No 152/2009 [36].

The concentrations of the mono-, poly-unsaturated, and saturated fatty acids were measured through a GC-FID analysis. The extracted fatty acids, via the Soxhlet extraction method [38] (Soxtherm SOX412-MACRO by Gerhardt), were analyzed by GC-FID after transesterification employing a methanol-potassium hydroxide solution. The chromatographic analyses were carried out with a Shimadzu GC-2010 Plus High-End gas chromatography system equipped with an FID detector, employing Supelco SP2560, a 100 m × 0.25 mm × 0.20 μm column, and helium (grade 99.999%) as a carrier gas, at a flow rate of 2 mL/min, according to a previously described analytical protocol [39].

Heavy metals, namely arsenic, cadmium, lead and mercury were determined with ICP-MS. The analysis of heavy metals was carried out via an inductively coupled plasma mass spectrometer (ICP-MS) (Agilent 7850), based on the U.S. Food and Drug Administration [40]. An amount of HNO_3 and H_2O_2 was added to an aliquot of the sample, followed by a digestion process at 210 °C. Once digestion was completed, the sample was diluted and measured by the ICP-MS method.

The dried yeast biomass and the feed ratios were also microbiologically characterized; the following analyses were made: Enterobacteriaceae (ISO 21528:2004) [41], *E. coli* (ISO 16649:2001) [42], yeasts and molds (ISO 7954:1987) [43], *Salmonella* spp. (ISO 6579:2002) [44] and *Listeria monocytogenes* (ISO 11290:2017) [45].

Table 1. Nutrients analysis, heavy metals, fatty acid composition and microbiological characterization of the dried YLP used in the study.

Items	Dried YLP
Moisture and Volatiles (g/100 g)	3.21
Ash (g/100 g)	10.96
Fat (g/100 g)	5.52
Proteins (g/100 g)	48.77
Crude Fibers % (g/100 g)	2.20
Carbohydrates (g/100 g)	29.34
Sugars (g/100 g)	<0.30 *
Energy (kcal/100 g)	362.1
Lysine %	11.4
Threonine %	5.6
Fatty Acids (% of total fats)	
Eicosapentaenoic acid (C20:5 n3)	ND
Behenic acid (C22:0)	ND
Linoleic acid (C18:2 n6c)	35.2%
Arachidic acid (C20:0)	ND
Eicosenic acid (C20:1)	0.9%
α -linolenic acid (C18:3 n3)	4.5%
Margaric acid (C17:0)	0.5%
Heptadecenoic acid (C17:1)	1.6%
Stearic acid (C18:0)	2.2%
Palmitic acid (C16:0)	9.6%
Elaidic acid (C18:1 n9t)	0.5%
Oleic acid (C18:1 n9c)	42.5%
Pentadecanoic acid (C15:0)	0.5%
Palmitoleic acid (C16:1)	1.8%
Heavy Metals (mg/kg)	
Arsenic	0.23
Cadmium	0.004
Lead	<0.04 *
Mercury	<0.10 *
Microbiological Characterization (cfu/g)	
Enterobacteriaceae	4.8×10^6
E. coli	2.1×10^5
Yeasts and Molds	6.8×10^8
Salmonella spp.	ND
Listeria monocytogenes	ND

* This value is the detection limit of the assay. ND: not detected.

2.2. Animals, Diets and the Experimental Design

A total of 108 13 day-old Ross 308 male broilers, with a starting body weight (BW) of 274.68 ± 1.46 g were used in the present study. The chicks were randomly allocated in 9 consecutive floor pens (12 birds/pen) in a close commercial poultry house in Greece. Each pen was equipped with nipple drinkers and a bell feeder and its floor was covered with rice husk. Throughout the experimental period of 29 days in total, the broiler feed was offered ad libitum, in mash form, and the birds were allowed free access to fresh water. The stoking density in each pen was in agreement with the instructions of EU Directive 2007/43/EC [46]. The temperature, lighting and relative humidity were controlled following the Ross 308 management guidelines (Aviagen 2018) [47].

The broilers were randomly and equally divided into 3 dietary groups CON, YLP3 and YLP5, according to the incorporation rate of YLP in their feed (0%, 3% and 5%, respectively) with 36 chicks/group, 3 replicate-pens/group, 12 chicks/replicate-pen. A three-phase feeding program was used in each dietary treatment, which included a grower diet fed from 13 to 20 days of age, a finisher 1 and a finisher 2 diet, fed from 21 to 32 days of age and from 33 to 41 days of age, respectively. In total, 9 ratios were formulated, 1 per feeding period/dietary treatment (Table 2). In the YLP-diets, the dried yeast in the form of flour mainly replaced the soybean meal and a small quantity of sunflower oil from the control diet, so as all ratios were isonitrogenous and isocaloric.

Table 2. Formulation and nutrient composition of the diets containing YLP, compared with the control diet (CON).

	Grower (13–20 Days)			Finisher 1 (21–32 Days)			Finisher 2 (33–41 Days)		
Items	**CON**	**YLP3**	**YLP5**	**CON**	**YLP3**	**YLP5**	**CON**	**YLP3**	**YLP5**
Ingredients									
Wheat	47.15	46.88	47.03	36.583	36.743	36.593	40.46	40.59	39.15
Corn	15.0	15.2	15.03	30.0	29.9	30.08	30.0	29.95	31.4
Soybean Meal 46.5%	30.43	27.7	25.85	25.93	23.1	21.23	21.93	19.1	17.31
Yarrowia Lipolytica	0	3.0	5.0	0	3.0	5.0	0	3.0	5.0
Sunflower Oil	4.4	4.2	4.07	5.0	4.77	4.61	5.0	4.75	4.53
MCP *	0.4	0.4	0.4	0.18	0.18	0.18	0.14	0.14	0.14
Limestone	1.08	1.08	1.08	0.89	0.89	0.89	0.92	0.92	0.92
NaCl	0.27	0.27	0.27	0.27	0.27	0.27	0.27	0.27	0.27
Dl-Methionine	0.34	0.34	0.34	0.29	0.29	0.29	0.25	0.25	0.25
L-Lysine	0.28	0.28	0.28	0.26	0.26	0.26	0.25	0.25	0.25
L-Threonine	0.07	0.07	0.07	0.07	0.07	0.07	0.06	0.06	0.06
RONOZYME® HiPhos	0.02	0.02	0.02	0.015	0.015	0.015	0.01	0.01	0.01
Premix [1]	0.4	0.4	0.4	0.4	0.4	0.4	0.4	0.4	0.4
Mycotoxins Binder	0.1	0.1	0.1	0.01	0.01	0.01	0.2	0.2	0.2
Enzymes	0.01	0.01	0.01	0.001	0.001	0.001	0.01	0.01	0.01
Coccidiostat	0.05	0.05	0.05	0.001	0.001	0.001	0.05	0.05	0.05
Lipidol Ultra 0.075%				0.05	0.05	0.05	0.05	0.05	0.05
Avatec 150 (150 g/kg)				0.05	0.05	0.05			
Total	100	100	100	100	100	100	100	100	100
Chemical Analysis									
Crude Protein (%)	20.97	20.99	21.01	18.80	18.80	18.79	17.32	17.32	17.31
Crude Fiber (%)	2.91	2.77	2.68	2.81	2.67	2.58	2.70	2.57	2.48
Fat (%)	6.1	5.90	5.77	7.05	6.82	6.67	7.05	6.80	6.62
Ash (%)	3.0	3.22	3.37	2.67	2.89	3.04	2.67	2.89	3.04
Calculated Analysis									
ME (kcal/kg)	2939	2941	2943	3050	3051	3052	3077	3077	3077
Lysine (%)	1.28	2.19	2.79	1.13	2.04	2.64	1.02	1.93	2.53
Methionine (%)	0.64	0.84	0.97	0.57	0.77	0.90	0.51	0.71	0.84
Ca (%)	0.84	0.96	1.04	0.72	0.84	0.92	0.70	0.82	0.90
P (%)	0.64	0.75	0.83	0.56	0.67	0.74	0.51	0.63	0.70
Na (%)	0.14	0.62	0.94	0.14	0.62	0.94	0.14	0.62	0.94

[1] Premix Finisher contains (per kg of product): vitamin A, 2,500,000 IU; vitamin D3, 1,250,000 IU; vitamin E, 20,000 mg; vitamin K3, 1500 mg; biotin, 35,000 mcg; folic acid, 300 mg; vitamin B1, 1500 mg; vitamin B2, 1500 mg; vitamin B6, 750 mg; vitamin B12, 6000 mcg; niacinamide, 7500 mg; calcium D-pantothenate, 3750 mg; choline chloride, 150,000 mg; carbonate (siderite), 12,500 mg; copper as copper sulphate pentahydrate, 2500 mg; manganic oxide, 27,500 mg; zinc oxide, 20,000 mg; calcium iodate anhydrous, 300 mg; coated granulated sodium selenite, 75 mg; citric acid, 14 mg; orthophosphoric acid, 3.50 mg; butylhydroxytoluene (BHT), 35 mg; butylated hydroxyanisole (BHA), 8.75 mg; calcium carbonate, 55.90%; calcium, 22.21%; phosphorous, 0.01%; * MCP: monocalcium phosphate.

At the end of the trial (day 41), all birds were transported and slaughtered at the local abattoir. Prior to their transportation, 18 chickens (6 birds per group) were randomly selected for meat quality analyses and sensor evaluation testing (SET) and were individually marked (leg bands) for identification. At the abattoir, the carcasses from the selected birds were scalded at 61–65 °C for 60 s, defeathered in a rotary drum picker for 25 s and the whole carcasses (head, feet, blood, without intestines) were air-chilled at 4 °C. Following the chilling process, the selected carcasses were weighed 24 h post-mortem. From each carcass, half of the breast and one thigh were used for meat quality analyses and the other half breast and thigh were used for SET. This way, 12 final breast meat samples/group (6 for meat quality and 6 for SET) and 12 final thigh meat samples/group (6 for meat quality and

6 for SET) were formulated. All meat samples were individually packed in sealable food bags and frozen at −18 °C temperature until the day of the meat quality analyses and SET.

2.3. Production Traits

The broilers' BW was determined at the onset of the experiment (at 13 days old) and at 20, 27 and 41 days old. The body weight gain (BWG) and feed consumption (FC) per bird were measured every week and calculated at weekly intervals, as well as for the whole experimental period (13–41 days of age). Based on the FC and BWG, the feed conversion ratio (FCR) per bird was calculated both at weekly intervals and during the entire experiment. The mortality rate was recorded daily.

2.4. Welfare and Behavior Indicators

At the age of 41 days, the chickens from each treatment group were individually evaluated for feather cleanliness, foot pad dermatitis (FPD), hock burn, as well as for quality behavior traits, according to the Welfare Quality (2009) protocol [48]. Initially, observations of the quality behavior characteristics (active, eating, fearful, calm, friendly and pecking behaviors) were undertaken in order to avoid confounding data due to handling stress. For behavioral quality traits, individual visual observations were made lasting 2 min for each pen. In every group, the number of birds exhibiting a particular type of behavior was noted and then divided by the total number of live birds in that group and multiplied by 100. Thus, the data % was calculated. The same methodology was used to estimate the welfare characteristics for each score category/welfare parameter. All observations were made by the same evaluator at 10 a.m.

Then, each bird was gently caught by one person from the research team (same for all assessments) and was examined for the feather cleanliness assessment, by scoring on a 3-point scale: Score 0-completely clean feathers; Score 1-slight feather soiling; Score 2-moderate feather soiling; and Score 3-severe feather soiling. The percentage of chicks presenting each score was then calculated. Both legs of the birds were examined for the presence of FPD (swelling-bubble foot) or hock burn and the scores were estimated according to the following scale: (a) FPD: Score 0-no evidence of FPD; Score 1 and 2-minimal evidence of FPD, Score 3 and 4-evidence of FPD (b) Hock burn: 0-no evidence of hock burn; Score 1 and 2-minimal evidence of hock burn, Score 3 and 4-evidence of hock burn. The percentage of birds with each scoring category was then recorded. Apart from day 41, the birds' quality behavior traits were also estimated at 20 and 27 days of age.

2.5. Meat Analysis

2.5.1. Materials and Reagents

The following reagents were used in the analyses: sulfuric acid (H_2SO_4) 98% for analysis (PanReac AppliChem, Barcelona, Spain, Belgium), sodium hydroxide NaOH (Merck, Darmstadt, Germany), boric acid (H_3BO_3) (PanReac AppliChem, Darmstadt, Germany), nitric acid (HNO_3) 65% (ChemLab, Zedelgem, Belgium), Tashiro's indicator solution (Honeywell Fluka, Munchen, Germany), hydrochloric acid (HCl) 0.1 N (VWR Chemicals BDH, Rosny-sous-Bois cedex, France), petroleum ether 40–60 ar (ChemLab), boron trifluoride methanol solution (FLuca), n-hexane pesticide grade (ChemLab, Zedelgem, Belgium), tablets Kjeldahl Cu (Gerhardt, Königswinter, Germany), potassium hydroxide (KOH) 85% (Panreac, Barcelona, Spain), methanol (VWR chemicals, Radnor, PA, USA), 2-thiobarbituric acid (Sigma-Aldrich, Darmstadt, Germany), butylated hydroxytoluene (Sigma Aldrich, Darmastadt, Germany) and trichloroacetic acid (Merck, Darmstadt, Germany) [49].

The reference standards used included: a 37 component mixture of fatty acids methyl esters, FAME mix C_6–C_{24} (SIGMA 18919-1AMP, certified reference material), a 13 polyunsaturated fatty acids mixture, PUFA No.1 (marine source analytical standard from SIGMA, 47033 with C14:0, C16:0, C16:1 n7, C18:1 n9, C181 n7, C18:2 n6, C20:1 n9, C18:4 n3, C22:1 n11, C22:1 n9, C20:5 n3, C22:5 n3, C22:6 n3), a 14 polyunsaturated fatty acids mixture PUFA No.2 (animal source analytical standard, SIGMA 47015-U with C14:0, C16:0, C16:1 n7, C18:0,

C18:1 n9, C18:1 n7, C18:2 n6, C18:3 n6, C18:3 n3 C20:1 n9, C20:2 n6, C20:3 n6, C20:4 n6, C20:5 n3, C22:4 n6, C22:5 n3 C22:6 n3) and a linoleic acid, conjugated methyl ester standard (CLA), purchased from Sigma Aldrich (SIGMA 05632) (St. Louis, MO, USA) [49].

2.5.2. Physicochemical Analysis

The chicken meat samples were analyzed for pH, protein, total nitrogen and lipid oxidation. The pH was measured with an electronic pH-meter (Consort, Belgium). The total nitrogen (TN) of the chicken meat samples was analyzed using the Kjeldahl method, according to the AOAC official method [50]; the raw protein content was estimated as 6.25 times the TN. A duplicate analysis was performed for all parameters under consideration.

2.5.3. Lipid Oxidation

Lipid oxidation was determined, based on the formation of malondialdehyde (MDA), using a selective third-order derivative spectrophotometric method [51]. The samples were blended in a small food processor. The subsamples (2 g) were homogenized with 8 ml of aqueous trichloroacetic acid (5% *w*/*v*) and 5 Ml of butylated hydroxytoluene in hexane (0.8% *w*/*v*) and the mixture was centrifuged. The top hexane layer was discarded and the bottom aqueous layer was transferred to a volumetric flask (10 mL) and analyzed, according to Ioannidou et al. [49]. Lipid oxidation is expressed as ng of MDA per g of muscle meat.

2.5.4. Fatty Acid Composition

Tissue fat was extracted, according to the Soxhlet method, using a Soxtec 2050 (Foss, Tecator, Denmark) automated system. The fatty acid methyl esters were prepared with boron trifluoride in a methanol solution. An appropriate quantity of the extracted fat was saponified by the addition of NaOH in methanol, followed by heating at 100 °C for 15 min. Fatty acid methyl esters were prepared by incubation at 100 °C for 5 min in a boron trifluoride methanol reagent. The produced fatty acid methyl esters were extracted by the addition of 1 mL hexane, followed by the addition of a saturated solution of potassium hydroxide and vigorous agitation [52]. The fatty acid methyl esters were removed and placed in GC vials. The prepared fatty acid methyl esters were analyzed using an HP 5890 (Hewlett-Packard) gas chromatograph equipped with a split/splitless injector (split mode) and a flame ionization detector (FID), according to Ioannidou et al. [49].

Based on the proportions of particular FAs and their groups, the health quality of the meat breast and thigh lipids was estimated by calculating the atherogenic index (AI), thrombogenic index (TI) and the ratio between the hypocholesterolemic (h)/hypercholesterolemic (H) fatty acids (h/H). The following equations were used to calculate these indices:

Atherogenic index [53]:

$$\text{AI} = (4 \times \text{C14:0} + \text{C16:0} + \text{C18:0})/(\Sigma\text{MUFA} + \Sigma\text{PUFA-n-6} + \Sigma\text{PUFA-n-3}) \quad (1)$$

Thrombogenic index [54]:

$$\text{TI} = (\text{C14:0} + \text{C16:0} + \text{C18:0})/(0.5 \times \Sigma\text{MUFA} + 0.5 \times \Sigma\text{PUFA-n-6} + 3 \times \Sigma\text{PUFA-n-3} + \Sigma\text{PUFA-n-3}/\Sigma\text{PUFA-n-6}) \quad (2)$$

Ratio between the hypocholesterolemic and hypercholesterolemic fatty acids [55]:

$$\text{h/H} = \text{C18:1n9c} + \text{C18:2n6c} + \text{C18:3n3c} + \text{C18:3n6c} + \text{C20.2n6} + \text{C20:3n6} + \text{C20:4n6} + \text{C22:6n3}/\text{C14:0} + \text{C16:0} \quad (3)$$

where:

Σ = Summatory,
MUFAs = monounsaturated FAs and
PUFAs = polyunsaturated FAs

2.6. Sensory Evaluation Testing

2.6.1. Participants

All participants took part voluntarily and signed their informed consent before the SET. All data collected during the SET were irreversibly anonymous, e.g., the individual persons could not be identifiable from the collected data sets. Two groups of consumers were randomly formulated from an initial group of participants to evaluate the broilers' breast and thigh meat, respectively. All participants declared that: (i) their smell and taste were not debilitated (e.g., due to an illness) at the time of the SET; (ii) they consumed broiler meat regularly; and (iii) that they were not allergic to broiler meat or any of the foods used during the SET. The first group comprised 6 persons (4 males and 2 females) aged between 28 and 65 years of age; the second group consisted of 5 persons (3 males and 2 females) aged between 28 and 65 years. All participants were informed about the aims and scope of the SET and were provided with a short explanation of the SET procedure and with instructions on the completion of the evaluation sheet and the interpretation of the 5-point hedonic scale that was used to evaluate the meat samples.

2.6.2. Meat Samples

The meat samples were defrosted overnight in a commercial refrigerator at 4 °C and subsequently cooked unseasoned with the skin removed and any visible external fat trimmed off, in a convection oven at approximately 200 °C. Cooking lasted approximately 45 min until the internal temperature of the meat samples reached 72 °C; the temperature was measured with an electronic hand-held thermometer (model: pH-meter CP-411, company: ELMETRON, country: Poland). Once cooked, the broiler's breasts and thighs were cut into rather uniform meat samples (approximately weight 15 $\pm$ 5 g), containing no bones, they were randomly placed on a white ceramic food plate, and were immediately served to the participants after reaching a temperature of approximately 60 °C.

2.6.3. Sensory Evaluation Testing

The SET took place the same day for both the breast and thigh meat at the premises of a commercial restaurant in Pieria, Northern Greece. The tasting area was used exclusively for the SET, and each participant was individually and simultaneously served a food plate that contained 3 meat samples (either breast or thigh meat) from each of the three experimental treatments, i.e., CON, YLP3 and YLP5. The placement of the meat samples on the food plate was the same for all participants, who were not aware of the meat samples' identity (i.e., to which experimental treatment they belonged). The participants were allowed to randomly select the order of evaluating the nine meat samples. Following the assessment of each sample, the participants neutralized their taste with a bite of wheat toast containing no salt or sugar, and a sip of natural mineral water. A standard evaluation form was provided to all participants, who manually completed their scores employing a 5-point hedonic scale, from 1: extreme dislike to 5: extreme like. The following characteristics were evaluated for each meat sample: color, flavor, tenderness, juiciness and overall impression.

2.7. Statistical Analysis

The statistical analysis of the data was performed using Jeffreys's Amazing Statistics Program JASP (JASP v 0.16.3) software [56]. The significance of the differences of the welfare and behavior indicators among the dietary groups was assessed by a Chi-square test. The normality of the data for the analysis of the broilers' growth performance (BW, BWG, FC and FCR), for the meat quality traits (MDA, meat protein and fat content, fatty acid composition and health lipid indices of the meat) and for the SET were tested employing a Shapiro–Wilk test; the homogeneity of variance was evaluated with Levene's test. A one-way ANOVA was used to compare the average values of the parameters evaluated among the dietary treatments. A post hoc analysis was performed using Tukey's test. When the distribution was not normal, the non-parametric tests Kruskal–Wallis and Mann–Whitney were used to make the comparisons at a significance level of $p \leq 0.05$. A multiple regression analysis (linear and quadratic) was applied for the evaluation of the relationships between all examined sensory characteristics and the broiler meat's overall acceptance (both for breast and thigh meat). These results are presented as standardized regression coefficients.

3. Results

3.1. Performance Parameters

The supplementation of YLP in the broilers' diet, did not significantly affect the BWG of the birds ($p > 0.05$), both at weekly intervals and for the whole experimental period (13–41 days of age). The overall growth performance was also not affected by the addition of YLP in their diets, since similar BW results of the chicks at 41 days of age, and FC and FCR, for the whole experimental period ($p > 0.05$), were observed among the groups (Table 3). The highest BW at 20 days of age, was recorded in the broilers fed with YLP at the rate of 3%, followed by the controls and the chickens fed diets with 5% YLP. The observed differences in the BW at 20 days of age were significant only between the YLP3 and YLP5 groups ($p < 0.05$). During the grower phase, the YLP-fed broilers presented a better FCR, compared to the CON chicks, however the observed differences were significant between the CON and YLP3 groups ($p < 0.05$). In the same period, the FC did not differ among the groups ($p > 0.05$). From 21 to 27 days of age, the YLP-fed broilers consumed significantly less feed, compared to the control groups ($p < 0.05$). However, during the grower phase, a similar FCR was recorded among the groups ($p > 0.05$). Moreover, the addition of YLP in the broilers' feed did not affect their BW at 27 days of age ($p > 0.05$). From 28 to 41 days of age, none of the evaluated growth parameters differed among the dietary groups ($p > 0.05$).

Table 3. Growth performance of the broilers fed the control and diets containing different levels of YLP. Data are presented as mean ± SE.

	CON	YLP3	YLP5
Body weight (g)			
Day 13	274.03 ± 2.74	273.89 ± 2.52	276.11 ± 2.38
Day 20	687.08 ± 12.31 ^{ab}	719.31 ± 9.40 ^{a}	679.03 ± 11.49 ^{b}
Day 27	1297.92 ± 25.50	1320.83 ± 18.84	1262.36 ± 20.10
Day 41	2692.92 ± 54.77	2717.43 ± 47.26	2627.00 ± 48.83
Body weight gained (g)			
13–20 d	413.06 ± 13.62	445.42 ± 15.22	402.92 ± 9.17
21–27 d	610.833 ± 15.34	601.528 ± 13.66	583.333 ± 12.09
28–41 d	1395.00 ± 37.37	1402.29 ± 36.19	1364.57 ± 34.78
Total period (13–41 d)	806.30 ± 150.06	817.28 ± 149.43	784.11 ± 148.18
Feed consumption (g)			
13–20 d	582.92 ± 10.49	587.36 ± 28.32	537.08 ± 5.29
21–27 d	872.50 ± 11.47 ^{a}	821.53 ± 14.87 ^{b}	810.42 ± 8.32 ^{b}
28–41 d	2005.14 ± 18.24	2008.03 ± 13.50	2001.84 ± 41.61
Total period (13–41 d)	1153.52 ± 217.08	1138.97 ± 220.11	1116.45 ± 225.18
FCR			
13–20 d	1.41 ± 0.03 ^{a}	1.32 ± 0.02 ^{b}	1.33 ± 0.02 ^{ab}
21–27 d	1.43 ± 0.03	1.37 ± 0.04	1.39 ± 0.04
28–41 d	1.44 ± 0.02	1.44 ± 0.04	1.47 ± 0.01
Total period (13–41 d)	1.43 ± 0.01	1.37 ± 0.04	1.40 ± 0.02

^{a,b} Means within a row at a particular age with different superscripts differ significantly ($p < 0.05$).

3.2. Welfare and Behavior Indicators

Data regarding the dietary impact of YLP on the broilers' feather cleanliness, FPD and hock burns at the 41st day of their life, are presented in Table 4. The percentage of broilers with completely clean feathers (Score 0) or slight feather soiling (Score 1) did not significantly differ among the dietary groups ($p > 0.05$). However, a significantly higher percentage of birds with moderate feather soiling (Score 2) was recorded in the YLP5 group, compared to that observed in the CON and YLP3 groups ($p < 0.05$). Interestingly, a significantly higher percentage of broilers with no evidence of FPD (Score 0) was observed in the

YLP-fed broilers, compared to the controls ($p < 0.05$). The evaluation of the 41 day-old chicks for evidence of hock burns revealed no significant differences among the experimental groups ($p > 0.05$).

Table 4. Percentage of broilers scoring for the welfare parameters (feather cleanliness, FPD, hock burn) at the age of 41 days, in regard to the three dietary treatments (CON, YLP3, YLP5).

Score [1]	Day 41 CON	Day 41 YLP3	Day 41 YLP5
		Feather Cleanliness	
0	52.78	51.43	31.42
1	44.44	48.57	34.29
2	2.78 ^{a}	0.00 ^{a}	34.29 ^{b}
3			
		Foot Pad Dermatitis	
0	63.89 ^{a}	91.43 ^{b}	82.86 ^{a}
1	5.56	5.71	11.43
2	19.44	2.86	5.71
3	11.11	0.00	0.00
4			
		Hock Burn	
0	61.11	62.86	45.71
1	27.78	25.71	25.71
2	5.56	8.57	14.29
3	2.78	0.00	5.72
4	2.77	2.86	8.57

^{a,b} Means within a row at each score category with different superscripts differ significantly ($p < 0.05$). [1] Feather cleanliness: score 0: indicates completely clean feathers, score 1: indicates slight feather soiling, score 2: indicates moderate feather soiling and Score s: indicates severe feather soiling; foot pad dermatitis: score 0: indicates no evidence of FPD, score 1 and 2: indicate minimal evidence of FPD, score 3 and 4: indicate evidence of FPD; hock burn: score 0: indicates no evidence of hock burn, score 1 and 2: indicate minimal evidence of hock burn, score 3 and 4: indicate evidence of hock burn.

Table 5 displays the data of the quality behavior parameters that were evaluated on the 20, 27 and 41 day-old broilers from all three dietary groups. The incorporation of YLP yeast in the broilers' diet significantly affected the feeding behavior of the 20 day-old chicks ($p < 0.05$). In particular, a higher percentage of 20 day-old chicks from the CON group were recorded feeding, in comparison to that observed in the YLP groups, with significant differences noticed between the CON and YLP3 groups ($p < 0.05$). At 27 and 41 days of age, the chicks of all groups expressed similar quality behavioral traits ($p > 0.05$).

Table 5. Percentage of broilers observed in the three dietary treatments (CON, YLP3, YLP5), scoring for quality behavior characteristics (active, feeding, fearful, calm, friendly and pecking behaviors) at the age of 20, 27 and 41 day.

Quality Behavior Traits	Day 20 CON	Day 20 YLP3	Day 20 YLP5	Day 27 CON	Day 27 YLP3	Day 27 YLP5	Day 41 CON	Day 41 YLP3	Day 41 YLP5
Active	36.11	41.67	27.78	36.11	19.44	27.78	5.55	17.14	8.57
Feeding	22.22 ^{a}	2.78 ^{b}	5.55 ^{a}	2.78	2.78	2.78	2.78	2.86	0
Fearful	0	0	0	0	0	0	0	0	0
Calm	27.78	47.22	50.00	58.33	72.22	63.89	91.67	80.00	85.71
Friendly	13.89	8.33	13.89	2.78	5.56	5.55	0	0	0
Pecking	0	0	2.78	0	0	0	0	0	5.72

^{a,b} Means within a row at a particular age for each type of behavior with different superscripts differ significantly ($p < 0.05$).

3.3. Meat Analysis

As demonstrated in Table 6, the breast and thigh meat chemical composition did not differ among the dietary groups, regarding fat and protein ($p > 0.05$). Moreover, a similar breast and thigh meat pH was recorded among all dietary groups ($p > 0.05$) (Table 6). Moreover, lipid oxidation analysis of the breast and thigh meat (Table 7) revealed significant differences among the dietary treatments ($p < 0.05$). Thigh meat MDA was found significantly higher ($p < 0.05$) in the CON group, compared to the YLP groups. Breast meat MDA was also higher in the CON group, in comparison to the YLP groups, however significant differences ($p < 0.05$) were found only between the CON and YLP3 groups.

Table 6. Protein and fat content in the breast and thigh meat of the broilers of the three dietary groups. Data are presented as mean ± SE.

Breast	CON	YLP3	YLP5
Protein %	23.18 ± 0.34	22.58 ± 0.62	23.23 ± 0.47
Fat %	0.46 ± 0.20	0.43 ± 0.11	0.33 ± 0.06
pH	5.68 ± 0.03	5.66 ± 0.09	5.78 ± 0.05
Thigh			
Protein %	20.11 ± 0.30	20.31 ± 0.22	19.85 ± 0.26
Fat %	0.58 ± 0.09	0.78 ± 0.10	0.51 ± 0.10
pH	6.00 ± 0.05	6.03 ± 0.06	6.04 ± 0.04

No significant differences were detected among groups ($p > 0.05$).

Table 7. Effect of the CON and YLP diets on the broiler chickens' meat oxidative stability. Data are presented as mean ± SE.

MDA ppb (ng/gr)	CON	YLP3	YLP5
Thigh meat lipid oxidation			
Day 1 of storage	87.63 ± 9.37 ^{a}	57.36 ± 4.72 ^{b}	52.71 ± 8.33 ^{b}
breast meat lipid oxidation			
Day 1 of storage	46.33 ± 8.27 ^{a}	24.81 ± 3.97 ^{b}	30.08 ± 5.68 ^{ab}

^{a,b} Means within a row with different superscripts differ significantly ($p < 0.05$). MDA: malondialdehyde.

The fatty acid (FA) composition analysis of the breast meat revealed that MUFAs, PUFAs, n-6 PUFAs and SFAs, as well as the PUFA/SFA and n-6 PUFA/n-3 PUFA ratios were significantly affected ($p < 0.05$) by the dietary treatments (Table 8). In particular, the YLP3 group presented significantly higher ($p < 0.05$) MUFAs and PUFAs and significantly lower ($p < 0.05$) SFAs, compared to the CON and YLP5 groups, respectively. Similarly, the breast meat of the YLP3 broilers had significantly higher n-6 PUFAs ($p < 0.05$) than that found to the breast meat of the other two groups. Moreover, a numerical increase of n-3 PUFAs was recorded in the YLP groups, compared to CON but the observed differences were not significant ($p > 0.05$). Moreover, the breast meat of the YLP3 group had a significantly higher n-6 PUFA/n-3 PUFA ratio than that recorded in the other two groups ($p < 0.05$). The significant differences of the MUFA, PUFA and SFA levels observed in the breast meat of all investigated groups, resulted in relevant differences of the PUFA/SFA ratio. More specifically, the CON and YLP5 groups had a significantly lower PUFA/SFA ratio than the YLP3 group ($p < 0.05$). The incorporation of YLP in the diet of broilers significantly affected the health lipid indices of the breast meat ($p < 0.05$). As shown in Table 8 the lowest atherogenic and thrombogenic indices were recorded in the YLP3 group and the highest ones in the CON and YLP5 groups ($p < 0.05$). In contrast, the highest h/H ratio was observed in the breast meat of YLP3 group ($p < 0.05$).

Table 8. The effect of the YLP supplementation on the broiler breast meat FA composition and health lipid indices. Data are presented as mean ± SE.

Item	CON	YLP3	YLP5
MUFA %	36.78 ± 0.65 [b]	41.97 ± 1.81 [a]	35.46 ± 0.76 [b]
PUFA %	2.84 ± 1.18 [b]	16.66 ± 4.71 [a]	4.39 ± 0.92 [b]
SFA %	60.38 ± 1.17 [a]	41.37 ± 6.03 [b]	60.15 ± 1.23 [a]
PUFA/SFA	0.05 ± 0.02 [b]	0.52 ± 0.18 [a]	0.07 ± 0.02 [b]
PUFA n6	2.72 ± 1.10 [b]	16.10 ± 4.55 [a]	4.04 ± 0.88 [b]
PUFA n3	0.12 ± 0.09	0.35 ± 0.09	0.29 ± 0.07
PUFA n6/ PUFA n3	12.92 ± 1.08 [b]	44.13 ± 8.86 [a]	12.33 ± 2.18 [b]
AI	1.62 ± 0.08 [a]	0.85 ± 0.23 [b]	1.64 ± 0.10 [a]
TI	2.90 ± 0.16 [a]	1.55 ± 0.42 [b]	2.80 ± 0.14 [a]
h/H	0.80 ± 0.06 [b]	2.07 ± 0.44 [a]	0.85 ± 0.06 [b]
Fatty acids			
Myristic acid (C14:0)	1.930 ± 0.234 [a]	1.017 ± 0.179 [b]	2.415 ± 0.497 [a]
Myristelic acid (C14:1)	0.168 ± 0.059	0.092 ± 0.011	0.112 ± 0.025
Pentadecanoic acid (C15:0)	0.410 ± 0.054 [a]	0.227 ± 0.029 [b]	0.477 ± 0.036 [a]
Pentadecenoic acid (C15:1)	ND	ND	ND
Palmitic acid (C16:0)	40.922 ± 1.420 [a]	28.717 ± 3.921 [b]	39.470 ± 1.384 [a]
Palmitoleic acid (C16:1)	3.024 ± 0.333 [a]	3.088 ± 0.221 [a]	1.878 ± 0.081 [b]
Margaric acid (C17:0)	0.512 ± 0.071	0.372 ± 0.152	0.568 ± 0.077
Heptadecenoic acid (C17:1)	ND	ND	ND
Stearic acid (C18:0)	15.090 ± 0.841 [ab]	10.358 ± 1.889 [b]	15.520 ± 0.746 [a]
Elaidic acid (C18:1n9t)	0.424 ± 0.027 [b]	0.413 ± 0.046 [b]	0.713 ± 0.084 [a]
Oleic (C18:1n9c)	31.244 ± 0.389 [b]	36.413 ± 1.650 [a]	30.752 ± 0.666 [b]
Vaccenic acid (C18:1n7)	1.586 ± 0.104	1.562 ± 0.092	1.428 ± 0.149
Linolelaidic (C18:2n6t)	ND	ND	ND
Linoleic acid (C18:2n6c)	2.578 ± 1.037 [b]	15.507 ± 4.407 [a]	3.658 ± 0.841 [b]
γ-Linolenic acid (C18:3n6)	0.140 ± 0.069	0.238 ± 0.065	0.377 ± 0.072
α-Linolenic acid (C18:3n3)	0.090 ± 0.064	0.328 ± 0.101	0.250 ± 0.070
Conjugated linoleic acid CLA	ND	0.035 ± 0.087	0.067 ± 0.163
Stearidonic acid (C18:4n3)	ND	ND	ND
Arachidic acid (C20:0)	0.494 ± 0.099 [a]	0.252 ± 0.034 [b]	0.502 ± 0.066 [a]
Gondoic acid (C20:1n9)	0.338 ± 0.150	0.358 ± 0.030	0.305 ± 0.098
Eicosadienoic acid (C20:2)	ND	0.135 ± 0.053	ND
Eicosadienoic acid (C21:0)	0.120 ± 0.062 [ab]	0.045 ± 0.021 [b]	0.232 ± 0.045 [a]
Dihomo-γ-linolenic acid (C20:3n6)	ND	0.090 ± 0.041	ND
Arachidonic acid (C20:4n6)	ND	0.223 ± 0.081	ND
Eicosatrienoic acid (C20:3n3)	ND	ND	ND
Behenic acid (C22:0)	0.704 ± 0.169 [ab]	0.302 ± 0.035 [b]	0.643 ± 0.031 [a]
Eicosapentaenoic acid (EPA)(C20:5n3)	ND	ND	ND
Docosenoic acid (C22:1n11)	ND	ND	ND
Erucic acid (C22:1n9)	ND	ND	0.290 ± 0.096
Docosadienoic acid (C22:2)	ND	ND	ND
Tricosanoic acid (C23:0)	ND	ND	ND
Docosatetraenoic acid (C22:4n6)	ND	0.043 ± 0.028	ND
Lignoceric acid (C24:0)	0.196 ± 0.090 [ab]	0.082 ± 0.031 [b]	0.323 ± 0.082 [a]
Docosapentaenoic acid (C22:5n3)	ND	ND	ND
Nervonic acid (C24:1)	ND	ND	ND
Docosahexaenoic acid (DHA) (C22:6n3)	ND	ND	ND

[a,b] Means within a row with different superscripts differ significantly (p <0.05), ND: Not Detected,. AI: Atherogenic Index, TI: Thrombogenic Index, h/H: hypocholesterolemic (h)/hypercholesterolemic (H) fatty acids.

The supplementation of YLP in the broilers' diet also affected the individual breast meat FAs. The differences observed among the dietary treatments are presented in detail in Table 8. It seems that most of the individual breast meat FAs, except myristoleic acid (C14:1), margaric acid (C17:0), vaccenic acid (C18:1n7), γ -linolenic acid (C18:3n6), α-linolenic acid (C18:3n3) and gondoic acid (C20:1n9), differed significantly ($p < 0.05$). In general, the most abundant fatty acid among SFAs recorded in the breast meat of all investigated groups, was the palmitic acid (C16:0), followed by stearic acid (C18:0). The breast meat from the YLP3 group had a significantly lower concentration of palmitic acid, compared to the CON and YLP5 groups ($p < 0.05$). Moreover, the breast meat from the CON and YLP5 groups had higher levels of stearic acid, compared to the YLP3 group however, significant differences were observed only between the YLP groups ($p < 0.05$). A data analysis revealed that oleic acid (C18:1n9c) was the most abundant of the MUFAs detected in the breast meat of all dietary treatments and its concentration was found significantly higher in the YLP3 group ($p < 0.05$), compared to the other two groups. According to the results, linoleic acid (C18:2n6c) was the most abundant of the PUFAs recorded in the breast meat of all groups. The highest concentration of linoleic acid was found in the breast meat of the YLP3 group and differed significantly ($p < 0.05$) from that recorded in the CON and YLP5 groups.

FA composition analysis of the thigh meat demonstrated that the incorporation of YLP in broilers diet did not affect the concentration of MUFA, SFA, PUFA and PUFA n6 FAs as well as the PUFA/SFA and PUFA n6/PUFA n3 ratios ($p > 0.05$) (Table 9). However, significant higher levels of PUFA n3 were detected in the thigh meat of YLP groups compared to CON group ($p < 0.05$). Consequently, few significant differences of individual FAs in the thigh meat of birds were observed among groups. In particular, significantly higher concentration of α-Linolenic acid was detected in the thigh meat of YLP groups compared to CON ($p < 0.05$). CLA and γ-Linolenic acid were detected only in the thigh meat of YLP groups and their concentration was found significantly lower in YLP3 compared to YLP5 group ($p < 0.05$). From the MUFAs detected in the thigh meat of birds, palmitoleic acid (C16:1) was significantly lower in YLP5 compared to CON group ($p < 0.05$). Finally, the health lipid indices of thigh meat did not differ significantly among dietary treatments ($p > 0.05$).

Table 9. The effect of YLP supplementation on broiler thigh meat FA composition and health lipid indices. Data are presented as mean ± SE.

Item	CON	YLP3	YLP5
MUFA %	38.39 ± 1.38	36.93 ± 4.56	35.54 ± 3.08
PUFA %	2.71 ± 0.72	12.11 ± 4.49	13.77 ± 5.37
SFA %	58.90 ± 2.03	50.97 ± 8.63	50.69 ± 7.83
PUFA/SFA	0.05 ± 0.01	0.37 ± 0.16	0.42 ± 0.21
PUFA n6	2.65 ± 0.72	11.45 ± 4.35	13.03 ± 5.10
PUFA n3	0.06 ± 0.04 [a]	0.45 ± 0.13 [b]	0.49 ± 0.11 [b]
PUFA n6/ PUFA n3	25.58 ± 18.23	28.49 ± 7.30	21.98 ± 5.39
AI	1.56 ± 0.14	1.65 ± 0.61	1.46 ± 0.49
TI	2.80 ± 0.24	2.58 ± 0.84	2.48 ± 0.81
h/H	0.83 ± 0.08	1.49 ± 0.48	1.61 ± 0.53
Fatty acids			
Myristic acid (C14:0)	1.998 ± 0.286	2.265 ± 0.647	1.727 ± 0.393
Myristelic acid (C14:1)	0.114 ± 0.013	0.210 ± 0.083	0.112 ± 0.016
Pentadecanoic acid (C15:0)	0.432 ± 0.120	0.303 ± 0.061	0.410 ± 0.065
Pentadecenoic C15:1)	ND	ND	ND
Palmitic acid (C16:0)	41.054 ± 1.738	36.408 ± 6.136	34.575 ± 5.296
Palmitoleic acid (C16:1)	3.368 ± 0.236 [a]	3.463 ± 0.606 [ab]	2.562 ± 0.118 [b]
Margaric acid (C17:0)	0.312 ± 0.082	0.323 ± 0.085	0.417 ± 0.080
Heptadecenoic acid (C17:1)	ND	ND	ND
Stearic acid (C18:0)	13.972 ± 0.750	10.928 ± 2.162	12.372 ± 1.859
Elaidic acid (C18:1n9t)	0.582 ± 0.167	0.385 ± 0.093	0.412 ± 0.046
Oleic (C18:1n9c)	32.530 ± 1.217	30.955 ± 3.870	30.623 ± 3.002
Vaccenic acid (C18:1n7)	1.450 ± 0.089	1.373 ± 0.157	1.295 ± 0.097
Linolelaidic (C18:2n6t)	ND	ND	ND

Table 9. *Cont.*

Item	CON	YLP3	YLP5
Linoleic acid (C18:2n6c)	2.574 ± 0.681	11.113 ± 4.284	12.402 ± 5.059
γ-Linolenic (C18:3n6)	ND	0.188 ± 0.057 [a]	0.492 ± 0.101 [b]
α-Linolenic (C18:3n3)	0.060 ± 0.040 [a]	0.365 ± 0.103 [b]	0.487 ± 0.105 [b]
Conjugated linoleic acid CLA	ND	0.163 ± 0.077 [a]	0.203 ± 0.142 [b]
Stearidonic acid (C18:4n3)	ND	ND	ND
Arachidic acid (C20:0)	0.468 ± 0.091	0.287 ± 0.049	0.385 ± 0.070
Gondoic acid (C20:1n9)	0.342 ± 0.172	0.490 ± 0.111	0.422 ± 0.122
Eicosadienoic acid (C20:2)	ND	0.047 ± 0.031	0.047 ± 0.031
Eicosadienoic acid (C21:0)	ND	ND	0.112 ± 0.051
Dihomo-γ-linolenic acid (C20:3n6)	ND	ND	ND
Arachidonic acid (C20:4n6)	ND	0.120 ± 0.054	0.105 ± 0.067
Eicosatrienoic acid (C20:3n3)	ND	ND	ND
Behenic acid (C22:0)	0.430 ± 0.185	0.312 ± 0.084	0.500 ± 0.143
Eicosapentaenoic acid (EPA)(C20:5n3)	ND	ND	ND
Docosenoic acid (C22:1n11)	ND	ND	ND
Erucic acid (C22:1n9)	ND	ND	ND
Docosadienoic acid (C22:2)	ND	ND	ND
Tricosanoic acid (C23:0)	ND	ND	ND
Docosatetraenoic acid (C22:4n6)	ND	ND	ND
Lignoceric acid(C24:0)	0.238 ± 0.129	0.115 ± 0.047	0.197 ± 0.083
Docosapentaenoic acid (C22:5n3)	ND	ND	ND
Nervonic acid (C24:1)	ND	ND	ND
Docosahexaenoic acid (DHA) (C22:6n3)	ND	ND	ND

[a,b] Means within a row with different superscripts differ significantly ($p < 0.05$), ND: Not Detected, AI: Atherogenic Index, TI: Thrombogenic Index, h/H: hypocholesterolemic (h)/hypercholesterolemic (H) fatty acids.

3.4. Sensory Evaluation Test

3.4.1. Breast Meat

The breast meat of broilers fed with different proportions of dried yeast biomass added in their feeding ratios showed some minor differences ($p > 0.05$), as scored by the taste panel, on a scale of 1 to 5 (Figure 1 and Table 10). The highest overall acceptance for breast meat, i.e., with a value of 3.67 ± 0.91, was recorded for broilers fed with 3% of dried yeast biomass, slightly lower than the corresponding acceptance for thigh meat, which scored a value of 4.00 ± 0.76, in the case of broilers fed with 5% of dried yeast biomass.

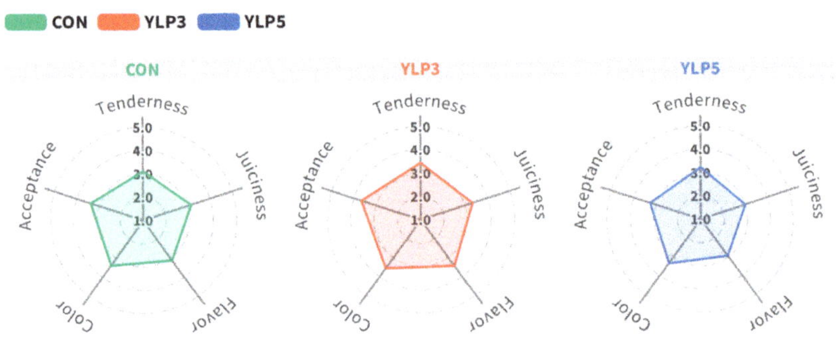

Figure 1. Sensory characteristics of broiler breast meat. Characteristics were judged by a panel on a scale from 1 to 5.

A data analysis showed no statistically significant differences ($p > 0.05$), for any sensory characteristics, among the different feeding scenarios for the breast meat. Regarding flavor, the YLP3 broilers scored the highest value (3.44 ± 0.86), meaning that their meat had a slightly better flavor, compared to the other two treatments. Similarly, the breast fillets of the YLP3 broilers were scored to be more tender (3.44 ± 1.10), juicier (3.33 ± 0.97), and with a better color (3.56 ± 0.70) than the other two treatments; however, the differences cannot be considered statistically significant under the SET conditions ($p > 0.05$). Consequently, regarding breast meat, the YLP3 treatment scored the highest in the overall acceptance and had a higher score in every individual characteristic.

Table 10. Data for the statistical means and variability for the broiler meat evaluation.

	Type of Meat				
	Tenderness	Juiciness	Flavor	Color	Overall Acceptance
			Breast meat		
CON	3.11 ± 0.18	3.17 ± 0.20	3.11 ± 0.20	3.39 ± 0.18	3.39 ± 0.18
YLP3	3.44 ± 1.26	3.33 ± 0.23	3.44 ± 0.20	3.56 ± 0.17	3.67 ± 0.21
YLP5	3.22 ± 0.27	3.00 ± 0.23	2.94 ± 0.15	3.33 ± 0.18	3.28 ± 0.23
			Thigh meat		
CON	3.40 ± 0.29	3.27 ± 0.25	3.27 ± 0.21	3.20 ± 0.28	3.33 ± 0.21 ^{b}
YLP3	3.87 ± 0.26	3.67 ± 0.23	3.40 ± 0.21	3.73 ± 0.27	3.53 ± 0.26 ^{ab}
YLP5	4.13 ± 0.22	3.93 ± 0.27	3.53 ± 0.22	3.87 ± 0.24	4.00 ± 0.20 ^{a}

^{a,b} Means within a column with different superscripts differ significantly ($p < 0.05$). All values are expressed as mean ± S.E.

3.4.2. Thigh Meat

Evaluating the broilers' thigh meat, the taste panel found statistically significant differences between the three treatments ($p < 0.05$) (feeding scenarios) (as shown in Figure 2 and Table 10). The broilers fed with 5% of dried yeast biomass had the highest overall acceptance, reaching the value of 4.00 ± 0.76, while the broilers fed with a CON diet scored the lowest overall acceptance (3.33 ± 0.82). The differences between the other sensory characteristics existed, but they were not statistically significant ($p > 0.05$). Regarding tenderness, the YLP5 broilers scored higher (4.13 ± 0.83), meaning their meat was more tender than those from the other two treatments. Similarly, the thigh fillets of the broilers fed with 5% dried yeast biomass were found to be juicier and with a better color, compared to the other two groups. Summarizing, the YLP5 treatment scored the highest values for all sensory characteristics (only overall acceptance was statistically significant), followed by the YLP3 feeding scenario, while the CON group scored the lowest values. These observations show that the addition of 5% of dried yeast biomass to the feeding ratios of broilers has a positive effect on the sensory characteristics and overall acceptance of the thigh meat. Comparing the breast to thigh meat, the latter scored higher in most cases regardless of the sensory characteristics and feeding treatment.

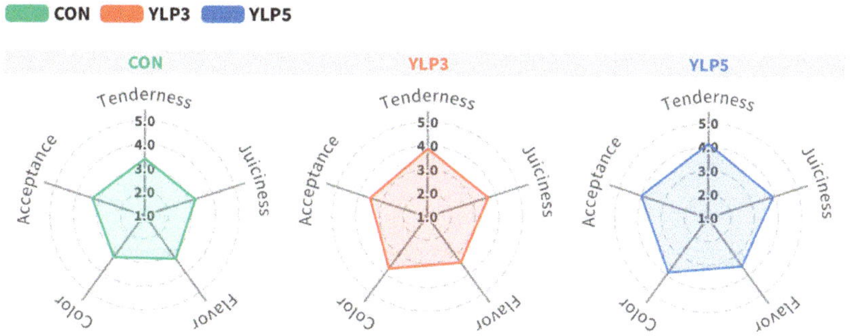

Figure 2. Sensory characteristics of broiler leg meat. Characteristics were judged by a panel on a scale from 1 to 5.

3.4.3. Regression Analysis

Multiple regression analyses (linear and quadratic) were performed to assess the correlations between the overall acceptance of the broiler breast and thigh meat assortment and its other analyzed sensory characteristics. As shown in Table 11, the tenderness and flavor had the greatest influence on the overall acceptance of broiler breast meat ($p < 0.001$ for linear regression), according to the taste panel. Furthermore, tenderness and color had a considerable effect on the overall acceptance of thigh meat, while juiciness was the least significant factor. In the case of broiler thigh meat, the highest effect upon the overall acceptance of the meat was attributed to meat tenderness ($p < 0.05$ for linear regression) followed by its color ($p = 0.05$ for linear regression). The other sensory characteristics had a lesser influence on the overall acceptance assessment of the meat; however, the flavor sensation had a greater impact than the juiciness. The results from the quadratic regression analysis are summarized in Table S1; however, it did not provide a statistically significant regression between the analyzed sensory characteristics and the overall acceptance.

Table 11. Results of the linear regression for the relationships with the broiler overall acceptance (for all types of meat).

	Breast Meat		**Thigh Meat**	
	$R^2 = 0.792$		**$R^2 = 0.629$**	
	b [1]	p [2]	b [1]	p [2]
Tenderness	0.489	<0.001	0.44	0.015
Juiciness	0.093	0.297	0.192	0.311
Flavor	0.342	<0.001	0.151	0.227
Color	0.156	0.078	0.23	0.05

[1] Standardized partial coefficient of the linear regression. [2] p-value based on the linear regression analysis.

4. Discussion

This is the first report that investigates the potential use of YLP yeast in the diet of broilers. Relative previous studies in birds have been carried out only in turkey hens, aged 1–16 weeks old [15–20]. According to our results, the incorporation of YLP did not affect the overall growth performance of chicks, as indicated by the final BW at 41 days-old, and the BWG, FC and FCR during the whole experimental period (13–41 days of age). Only some differences in the BW, FC and FCR of broilers were recorded at weekly intervals among groups. The higher BW of the YLP3 chicks at 20 days of age, compared to the other two groups, could be attributed to the numerical increase of the FC and a net numerical increase of the BWG of birds in this group during the growing period. Moreover, the reduced FC recorded in the YLP groups, compared to the CON group, between 21 and 27 days of age, could be due to an adaption of the birds to the finisher 1 diet. Finally, the observed differences in the feed efficiency between the CON and YLP3 groups during the growing period is the net result of the numerical differences in the FC and BWG seen between the groups during this time period. Similar to our findings, previous reports demonstrated that the supplementation of 3% YLP in turkeys' diet had no impact in the final BW, BWG, FC [17,19] and FCR [17]. However, Czech et al. [19] noted an amelioration of the FCR in YLP3-fed birds, compared to the controls. Moreover, Merska et al. [17] found that increasing the incorporation rate of YLP to turkeys' feed from 3% to 6%, resulted in a lower BW of birds, compared to the CON group and those fed with 3% YLP diets. The higher percentage of CON chicks that were recorded feeding at 20days of age, in comparison to that observed in the YLP groups is probably not related to the YLP supplementation. This is consistent with the non-significant difference in food consumption during the growing phase of birds of all groups. Thus, given the instant character of the quality behavior traits recording procedure, this finding could be considered random.

The incorporation of YLP in broilers' diet at both studied levels (3% and 5%), had a positive effect on FPD, as evaluated at 41 days of age. This finding is considered of critical importance for chickens' health and welfare, but also for the farmers' financial income. FPD is a multifactorial problem with litter quality, nutrition and gut health as some of the factors implicated in its incidence [57–59]. However, litter condition is considered the most important risk factor for the development of FPD [60]. The litter moisture and ammonia concentration from accumulated fecal material can burn and weaken the dermis of the footpad [61], with an increased severity of FPD resulting from the prolonged exposure of feet to wet litter. Moisture causes the outer layer of the dermis to soften, posing a risk of microbial contamination, leading to necrosis [62]. FPD is a very common and well recognized problem in the broiler industry [58] that negatively affects birds' productivity and welfare [57] and it has been associated with a reduced mobility, lameness and consequently with behavioral restrictions of birds [59,63]. FPD has also been shown to be highly associated with systemic bacterial infections, as pathogens can enter chickens' bodies through the damaged epithelium in the foot pads [64]. Furthermore, financial losses due to FPD are basically attributed to the abattoir condemnation of carcasses with contact dermatitis lesions [58]. Chicken legs is a highly profitable industrial by-product, and poor footpad conditions due to FPD degrade the product quality, leading to rejections and loss of income [59].

The exact mechanism involved in the positive nutritional effect of YLP on FPD observed in this study is presently unknown and requires further investigation as there are no similar studies in the available literature. Two possible explanations could be given for the interpretation of this result. According to the first one, it is possible that YLP ameliorated the gut health of YLP birds, resulting in a reduction of their fecal moisture and consequently reduced litter moisture, which is one of the major causative factors of FPD. The beneficial effect of YLP in intestinal function and gut microbiota has been previously demonstrated in turkey hens [19,20]. The addition of 3% YLP in turkeys' diet increased the intestinal villus length, as well as the ratio of the villus length to the crypt depth and the intestinal muscular layer thickness, thus improving the birds' intestinal health [19]. In another study it was proved that feeding turkey hens with 3% YLP favorably influenced the intestinal microbiota, since it reduced the number of fungi and coliforms, including *Escherichia coli* [20]. In similar studies performed on growing piglets, the incorporation of YLP in the piglets' diet decreased the total number of coliform bacteria and *Escherichia coli* in their enteric contents [22]. The beneficial dietary effect of YLP on gut health has been linked to the presence of β-glucans in the yeast cell wall which protects the gastrointestinal tract against the colonization by dangerous pathogens, such us *Salmonella enterica* [20]. In support of this, previous studies have demonstrated that β-glucans and mannans, the two major components of the yeast cell wall, are bioactive components with potential benefits for the development of the intestinal immune system in animals [23]. They mitigate the release of pro-inflammatory cytokines and prevent the colonization of pathogens in intestinal mucosa.

According to a second explanation, the higher percentage of birds with no evidence of FPD recorded in the YLP groups, compared to the CON group, could also be due to the beneficial properties of the bioactive compounds of YLP for healthy skin, such as polyunsaturated fatty acids (PUFAs). YLP used in the present study consisted of approximately 39.7% PUFA, mainly linoleic acid (35.2%). It has been previously demonstrated that oils containing high levels of essential fatty acids improve skin hydration, regenerate the damaged epidermal lipid barrier and regulate skin metabolism [65]. Moreover, omega-3 and omega-6 FAs are significant components of the cell membrane, necessary for the function of the epidermal barrier; they display anti-inflammatory and anti-allergic effects by enhancing the repair processes and soothing irritation [65].

In the present study, the broilers of all experimental groups were evaluated with very good scores for hock burn, indicating no evidence of such a welfare issue. Regarding the feather cleanliness results, even though a higher percentage of birds with moderate feather soiling (Score 2) was recorded in the YLP5 group, compared to that observed in the CON and YLP3 groups, this finding was not reflected in the FPD or hock burn outcomes. In a number of previous studies however, dirty feathers and FPD were highly correlated [66–68].

In current investigation, a significant antioxidant effect was detected in both thigh and breast meat of the broilers that were fed YLP diets, compared to the CON-fed birds, as indicated by the MDA results. This finding is very important because oxidation of the lipid components in muscle tissues is a major cause of quality deterioration and short shelf life after slaughter. The high concentration (42.5%) of oleic acid in YLP used in this study might have played a key role in this result. It has been previously demonstrated that oleic acid reduces oxidative stress and inflammatory response by activating the peroxisome proliferator-activated receptors in animals [69]. Complementarily, the antioxidant effect of YLP in broiler meat could also be attributed to the bioactive components of the yeast cell wall, such as β-glucans and mannans. Previous research evidence indicated that adding mannan oligosaccharides and β-glucans in poultry feed decreases the accumulation of the lipid oxidation end product (MDA) in the tissues of broiler chickens [70]. The beneficial dietary impact of YLP on the antioxidant status of turkey hens has been demonstrated in previous reports [16,18]. According to Merska et al. [16], supplementing turkeys' feed with 3% and 6% YLP yeast, increases the catalase activity (CAT) and decreases the plasma concentrations of lipid peroxidation products, such as hydroperoxide (LOOH) and MDA. Those results were confirmed in a later study by Czech et al. [18] who indicated that the role of YLP in the activation of antioxidant enzymes and the reduction of peroxidation products, when incorporated in turkeys' diet at 3% inclusion rate.

The supplementation of broilers' diet with 3% YLP yeast ameliorated the lipid profile of breast meat, as indicated by the increased concentration of MUFAs and PUFAs and the decreased levels of SFAs observed in this group. The reduced SFA concentration recorded in the breast meat of YLP3 birds could be primarily attributed to a reduction of palmitic acid, which was the most abundant among the SFAs recorded in the breast meat of all dietary treatments and, to a lesser degree, to a similar decrease of the stearic acid concentrations. The bioactive components of the yeast cell wall, such as β-glucans and mannans, might have played a key role to the reduction of the SFA levels in the breast meat of YLP3 birds, since they act as antioxidants and could suppress meat lipid oxidation. The increased level of PUFAs in the breast meat of the YLP3 group was mainly linked to an elevation of linoleic acid, the most abundant fatty acid among PUFAs found in the breast meat of all dietary treatments. Taking into consideration the high level of linoleic acid (35.2%) in the composition of the YLP used in the present study, it could be supported that the addition of YLP at a 3% incorporation rate in broilers' diet resulted in a concomitant increase of linoleic acid concentration in the breast meat of birds of this group. However, this beneficial dietary effect of YLP was not proportional to the incorporation rate of YLP in the birds' diet, as it was revealed by the breast lipid profile results of the YLP5 birds, which were similar with the CON group. It has been previously documented that the FA profile of chicken meat is affected by the birds' diet [71] and genetic factors [72]. Moreover, a FA diet composition, fat metabolism and fat deposition in edible tissues are often correlated in monogastric animals. Dietary supplementation with polyunsaturated FAs, such as linoleic, α-linolenic and arachidonic acids is often associated with increased levels of these acids in the muscle and adipose tissues, both through direct incorporation and modification of the unsaturated FA synthesis in these tissues [73,74]. However, the mechanisms involved are complicated and affect the lipogenic genes' expression [75,76].

The supplementation of broilers' diet with 3% YLP, positively affected the nutritional quality of their breast meat, as indicated by the improved PUFA/SFA ratio and the better health lipid indices recorded in the breast meat of YLP3 birds. The PUFA/SFA ratio is the most-commonly used index for estimating the effect of a certain food on cardiovascular health, considering that all PUFAs are capable to decrease low-density lipoprotein cholesterol and serum cholesterol, whereas all SFAs could contribute to the elevation of serum cholesterol [77]. Thus, higher values indicate a better (positive) effect given by a particular meat or meat product intake. The dietary ratios of PUFA to SFA, that are higher than

0.45, are regarded as safe for human consumption [78], and suitable to protect against the development of ischemic heart disease [79]. The optimal PUFA/SFA ratio from a nutritive point of view was achieved with the 3% incorporation rate of YLP, compared to the CON and YLP5 groups.

The most frequently used index for estimating the nutritional value of dietary foods is the PUFA/SFA ratio. However, it is regarded too general and inappropriate for evaluating the atherogenicity of foods [80] as specific SFAs and PUFAs have different metabolic effects [79]. Thus, AI was established [54] in order to estimate the atherogenic potential of the FAs in food. A lower value indicates better nutritional characteristics of the food [77]. Another important and commonly used index to further characterize the thrombogenic potential of FAs is TI [80]. This index points out the trend to form clots in blood vessels and indicates the contribution of different FAs, showing the relationship between the pro-thrombogenic FAs (C12:0, C14:0 and C16:0) and the anti-thrombogenic FAs (MUFAs and the n-3 and n-6 families) [54]. It has been documented that animal products with a low index of thrombogenicity reduce the threat of atrial fibrillation [81]. Summarizing, AI and TI can be employed for the estimation of the potential impact of the FA composition on cardiovascular health (CVH). An FA composition with a lower AI and TI, has a better nutritional quality, thus its consumption may decrease the risk of coronary heart disease (CHD), but the recommended values for the AI and TI have not been provided yet [80]. The results of this study indicated that the addition of 3% YLP in broilers' diet decreased significantly both the AI and TI, compared to the CON and YLP5 groups. At the same time, this incorporation rate of YLP yeast significantly reduced the h/H ratio of the produced breast meat. Santos-Silva et al. [82] supported that the higher the ratio between the hypocholesterolemic and the hypercholesterolemic fatty acids, the more the oil or fat is suitable to human nutrition. These findings indicate the health benefit potential related to the fat intake from breast meat produced from YLP3-fed broilers, since they presented a higher nutritional quality than the CON and YLP5 breast meat.

The FA profile analysis of the thigh meat revealed a significant elevation of n-3 PUFAs in the YLP-fed groups, compared with the CON group. This elevation is attributed to the higher concentration of α-linolenic acid observed in the thigh meat of the YLP groups, compared to the CON group, possibly related to the α-linolenic acid concentration of YLP used in the study. This finding is very important from the consumers' point of view. Intake of the recommended amounts of PUFAs, and particularly n-3 acids, is absolutely necessary for ensuring the correct functioning of the human body and essential for the prevention and mitigation of several diseases, such as cardiovascular diseases, skin diseases, autoimmune conditions and certain forms of cancer—breast, colon and prostate cancer [83–86]. Differences in the breast and thigh meat FA profile observed in the current study could be attributed to the different roles of FAs in these tissues or to their different phospholipid contents. The PUFAs are preferentially incorporated into phospholipids [87], and more phospholipids are in found in breast than in thigh muscles [88]. Since there is lack of research evidence regarding the effect of YLP in the lipid profile of breast and thigh meat, as well as on the relative health lipid indices when supplementing broilers' feed, the comparison of our results with those in literature is not possible. However, similar studies curried out on Atlantic salmon, have shown a beneficial effect of YLP on the FA composition in fish fillets [27].

In the present study, the response of the panelists for the breast and thigh meat of broilers fed diets supplemented with YLP yeast did not influence the sensory attributes of the broiler meat. However, according to the SET results, the thigh meat of YLP5 chicks scored the highest values for all sensory characteristics, leading to significantly higher overall acceptance scores, compared to the other dietary treatments. These findings are very important from the consumers' point of view, since they imply that YLP does not compromise the organoleptic characteristics of broiler meat; on the contrary, it increases the overall likeability of thigh meat. One of the main concerns when using foods rich in polyunsaturated lipids is the deterioration of the sensory quality of the poultry products. The results of several studies have demonstrated that the meat of animals containing more

PUFAs is more susceptible to oxidative processes [89], which has a negative impact on its organoleptic characteristics and shelf life [90]. In the present study however, despite the high concentrations of PUFAs in breast meat of the YLP3 group, as well as the higher concentration of n-3 PUFAs in the thigh meat of the YLP-fed broilers, compared to the CON group, the tenderness, juiciness, flavor, color and overall acceptance of YLP meat were not reduced. Instead, a positive effect on the overall acceptance of thigh meat from the YLP groups was detected with significant differences observed in the YLP5 group. This positive effect could be attributed to the antioxidant activity of oleic acid (present in high concentration 42.5%) in YLP used in this study, as well as to the bioactive components of the yeast cell wall, such as β-glucans and mannans, that have antioxidant properties protecting meat from lipid peroxidation. This is also supported by the lower values of MDA recorded in the breast and thigh meat of YLP birds, compared to the CON birds.

5. Conclusions

The present study, which is the first one carried out on broilers, shows that the supplementation of broilers' diet with dried YLP at two different levels (3% and 5%) positively affected FPD without any adverse effect on the overall growth performance of chicks, feather cleanliness and the qualitative behavior characteristics evaluated. These observations denote potential health and welfare benefits for YLP-fed chickens, while the producers can possibly benefit from a reduction of the production cost. These data also suggest that the broilers are able to adapt and efficiently consume diets containing YLP yeast at both concentrations tested (3% and 5%) at various growth stages. A meat quality analysis revealed a beneficial antioxidant dietary effect of YLP on both thigh and breast meat of YLP-fed broilers, as indicated by the reduced MDA values, compared to the CON group. This finding implies a potential beneficial effect of YLP to the quality and shelf life of meat after slaughter. Moreover, the supplementation of broilers' diet with 3% YLP yeast ameliorated the lipid profile of the breast meat, which also presented a superior nutrient quality, compared to the CON and YLP5 groups, as indicated by the better PUFA/SFA ratio, decreased values of AI, TI and the increased value of the h/H ratio. These observations demonstrate the health beneficial potential associated with the fat intake from breast meat produced from YLP3-fed broilers. Furthermore, an FA profile analysis of the thigh meat revealed a significant elevation in n-3 PUFAs in the YLP-fed groups, compared to the CON group. According to the SET results, tenderness, juiciness, flavor, color and the overall acceptance of YLP-fed broilers' meat were not reduced. Instead, a positive effect on the overall acceptance of thigh meat from YLP broilers was detected with significant differences observed in the YLP5 group. Finally, from the evaluated dietary levels of YLP, 3% seems to be more advantageous for the consumer, in terms of meat nutrition quality, so it is highly recommended.

Supplementary Materials: The following supporting information can be downloaded at: https://www.mdpi.com/article/10.3390/su15031924/s1, Figure S1: Schematic representation of the industrial-scale bioreactor pilot plant designed by NRRE Lab (CPERI/CERTH) used for the production of the yeast biomass. Table S1: Results of the quadratic regression for the relationships with broiler's overall acceptance (for all types of meat).

Author Contributions: Conceptualization, A.D., S.I.P. and E.N.S.; methodology, A.D., S.I.P., K.N.K., M.I. and A.Z.; software, A.D., M.-Z.K. and K.N.K.; validation, A.D., S.I.P. and E.N.S.; formal analysis, A.D. and K.N.K.; investigation, A.D., M.-Z.K. and M.I.; resources, S.I.P. and E.N.S.; data curation, A.D. and E.N.S.; writing—original draft preparation, A.D.; writing—review and editing, E.N.S., S.I.P. and K.N.K.; visualization, M.-Z.K., K.N.K., M.I. and A.Z.; supervision, E.N.S.; project administration, A.D., S.I.P., E.N.S. and A.Z.; funding acquisition, S.I.P. and E.N.S. All authors have read and agreed to the published version of the manuscript.

Funding: This research has been co-financed by the European Regional Development Fund of the European Union and Greek national funds through the Operational Program Competitiveness, Entrepreneurship and Innovation, under the call RESEARCH–CREATE–INNOVATE (project code: T1EDK-02871).

Institutional Review Board Statement: The study was conducted in accordance with the Declaration of Helsinki and approved by the Research Ethics Committee of the Hellenic Agricultural Organization DIMITRA (Reference Number 2340/57419/24-10-2022). The Research Ethics Committee of the Hellenic Agricultural Organization-DIMITRA has approved the experimental protocol and implemented animal care procedures of this study (Reference Number 2340/57419/24-10-2022).

Data Availability Statement: The data presented in this study are available upon request from the corresponding author. The data are not publicly available due to privacy.

Acknowledgments: We sincerely thank the DIMITRIADIS CHRISTOS S.A. poultry farm, who valuably contributed to this research by offering the experimental animals and facilities. Asimina Tsirigka (Chem. Eng. Dep., AUTh, Greece), Michail Sountourlis and Evangelos Tziaras (Fytoenergeia S.A., Serres, Greece) are greatly acknowledged for their technical contribution during the yeast biomass production process.

Conflicts of Interest: The authors declare no conflict of interest.

References

1. Alexandratos, N.; Bruinsma, J. *World Agriculture towards 2030/2050: The 2012 Revision*; ESA Working paper No. 12-03; FAO: Rome, Italy, 2012.
2. Kim, S.W.; Less, J.F.; Wang, L.; Yan, T.; Kiron, V.; Kaushik, S.J.; Lei, X.G. Meeting Global Feed Protein Demand: Challenge, Opportunity, and Strategy. *Annu. Rev. Anim. Biosci.* **2019**, *7*, 221–243. [CrossRef] [PubMed]
3. HighQuest Partners. *How the Global Oilseed and Grain Trade Works*; U.S. Soybean Export Council: St. Louis, MI, USA, 2008.
4. Mottet, A.; de Haan, C.; Falcucci, A.; Tempio, G.; Opio, C.; Gerber, P. Livestock: On Our Plates or Eating at Our Table? A New Analysis of the Feed/Food Debate. *Glob. Food Secur.* **2017**, *14*, 1–8. [CrossRef]
5. Tallentire, C.W.; Mackenzie, S.G.; Kyriazakis, I. Can Novel Ingredients Replace Soybeans and Reduce the Environmental Burdens of European Livestock Systems in the Future? *J. Clean. Prod.* **2018**, *187*, 338–347. [CrossRef]
6. Patsios, S.I.; Dedousi, A.; Sossidou, E.N.; Zdragas, A. Sustainable Animal Feed Protein through the Cultivation of YARROWIA Lipolytica on Agro-Industrial Wastes and by-Products. *Sustainability* **2020**, *12*, 1398. [CrossRef]
7. Bourdichon, F.; Casaregola, S.; Farrokh, C.; Frisvad, J.C.; Gerds, M.L.; Hammes, W.P.; Harnett, J.; Huys, G.; Laulund, S.; Ouwehand, A.; et al. Food Fermentations: Microorganisms with Technological Beneficial Use. *Int. J. Food Microbiol.* **2012**, *154*, 87–97. [CrossRef]
8. Czech, A.; Smolczyk, A.; Ognik, K.; Kiesz, M. Nutritional Value of Yarrowia Lipolytica Yeast and Its Effect on Growth Performance Indicators n Piglets. *Ann. Anim. Sci.* **2016**, *16*, 1091–1100. [CrossRef]
9. Rywińska, A.; Juszczyk, P.; Wojtatowicz, M.; Robak, M.; Lazar, Z.; Tomaszewska, L.; Rymowicz, W. Glycerol as a Promising Substrate for Yarrowia Lipolytica Biotechnological Applications. *Biomass Bioenergy* **2013**, *48*, 148–166. [CrossRef]
10. Michalik, B.; Biel, W.; Lubowicki, R.; Jacyno, E. Chemical Composition and Biological Value of Proteins of the Yeast *Yarrowia Lipolytica* Growing on Industrial Glycerol. *Can. J. Anim. Sci.* **2014**, *94*, 99–104. [CrossRef]
11. Groenewald, M.; Boekhout, T.; Neuvéglise, C.; Gaillardin, C.; van Dijck, P.W.M.; Wyss, M. *Yarrowia Lipolytica*: Safety Assessment of an Oleaginous Yeast with a Great Industrial Potential. *Crit. Rev. Microbiol.* **2014**, *40*, 187–206. [CrossRef]
12. European Commission. *Commission Regulation (EU) 2017/1017 of 15 June 2017 Amending Regulation (EU) No 68/2013 on the Catalogue of Feed Materials*; European Commission: Brussels, Belgium, 2017; pp. 48–119. Available online: https://eur-lex.europa.eu/eli/reg/2017/1017/oj (accessed on 2 July 2022).
13. European Commission. *Regulation (EU) 2015/2283 of the European Parliament and of the Council of 25 November 2015 on Novel Foods, Amending Regulation (EU) No 1169/2011 of the European Parliament and of the Council and Repealing Regulation (EC) No 258/97 of the European Parliament and of the Council and Commission Regulation (EC) No 1852/2001*; European Commission: Brussels, Belgium, 2015; pp. 1–22. Available online: http://data.europa.eu/eli/reg/2015/2283/oj (accessed on 27 July 2022).
14. EFSA Panel on Nutrition, Novel Foods and Food Allergens (NDA); Turck, D.; Castenmiller, J.; de Henauw, S.; Hirsch-Ernst, K.; Kearney, J.; Maciuk, A.; Mangelsdorf, I.; McArdle, H.J.; Naska, A.; et al. Safety of Yarrowia Lipolytica Yeast Biomass as a Novel Food Pursuant to Regulation (EU) 2015/2283. *EFS2* **2019**, *17*, e05594. [CrossRef]
15. Czech, A.; Merska, M.; Ognik, K. Blood Immunological and Biochemical Indicators in Turkey Hens Fed Diets With a Different Content of the Yeast Yarrowia Lipolytica. *Ann. Anim. Sci.* **2014**, *14*, 935–946. [CrossRef]
16. Merska, M.; Czech, A.; Ognik, K. The Effect of Yeast Yarrowia Lipolytica on the Antioxidant Indices and Macro-and Microelements in Blood Plasma of Turkey Hens. *Pol. J. Vet. Sci.* **2015**, *18*, 709–714. [CrossRef] [PubMed]

17. Merska, M.; Czech, A.; Ognik, K. The Effect of Different Doses of Dried Yeast Yarrowia Lipolytica on Production Effects of Turkey Hens and Hematological Indicators of Blood. *Ann. Univ. Mariae Curie-Skłodowska. Sect. EE Zootech.* **2013**, *31*, 35–41.
18. Czech, A.; Merska-Kazanowska, M.; Całyniuk, Z. Redox Status, Biochemical Parameters and Mineral Elements Content in Blood of Turkey Hens Fed a Diet Supplemented with Yarrowia Lipolytica Yeast and Two Bacillus Species. *Animals* **2020**, *10*, 459. [CrossRef]
19. Czech, A.; Merska-Kazanowska, M.; Ognik, K.; Zięba, G. Effect of the Use of Yarrowia Lipolytica or Saccharomyces Cerevisiae Yeast with a Probiotic in the Diet of Turkey Hens on Growth Performance and Gut Histology. *Ann. Anim. Sci.* **2020**, *20*, 1047–1063. [CrossRef]
20. Czech, A.; Sembratowicz, I.; Zieba, G. Effect of the Use of Yarrowia Lipolytica and Saccharomyces Cerevisiae Yeast with a Probiotic in the Diet of Turkeys on Their Gut Microbiota and Immunity. *Vet. Med.* **2020**, *65*, 174–182. [CrossRef]
21. Czech, A.; Smolczyk, A.; Grela, E.R.; Kiesz, M. Effect of Dietary Supplementation with *Yarrowia Lipolytica* or *Saccharomyces Cerevisiae* Yeast and Probiotic Additives on Growth Performance, Basic Nutrients Digestibility and Biochemical Blood Profile in Piglets. *J. Anim. Physiol. Anim. Nutr.* **2018**, *102*, 1720–1730. [CrossRef]
22. Czech, A.; Smolczyk, A.; Ognik, K.; Wlazło, Ł.; Nowakowicz-Dębek, B.; Kiesz, M. Effect of Dietary Supplementation with Yarrowia Lipolytica or Saccharomyces Cerevisiae Yeast and Probiotic Additives on Haematological Parameters and the Gut Microbiota in Piglets. *Res. Vet. Sci.* **2018**, *119*, 221–227. [CrossRef]
23. Cheng, Y.-C.; Duarte, M.E.; Kim, S.W. Effects of Yarrowia Lipolytica Supplementation on Growth Performance, Intestinal Health and Apparent Ileal Digestibility of Diets Fed to Nursery Pigs. *Anim. Biosci.* **2022**, *35*, 605–613. [CrossRef]
24. Stefańska, B.; Komisarek, J.; Stanisławski, D.; Gąsiorek, M.; Kasprowicz-Potocka, M.; Frankiewicz, A.; Nowak, W. The Effect of Yarrowia Lipolytica Culture on Growth Performance, Ruminal Fermentation and Blood Parameters of Dairy Calves. *Anim. Feed. Sci. Technol.* **2018**, *243*, 72–79. [CrossRef]
25. Michalik, B.; Jacyno, E.; Lubowicki, R.; Biel, W. Biological Evaluation of the Protein Nutritional Value in the Diets of Rats Based on Cereals and the Yeast *Yarrowia Lipolytica* Growing on Industrial Glycerol. *Acta Agric. Scand. Sect. A Anim. Sci.* **2013**, *63*, 163–168. [CrossRef]
26. Berge, G.M.; Hatlen, B.; Odom, J.M.; Ruyter, B. Physical Treatment of High EPA *Yarrowia Lipolytica* Biomass Increases the Availability of n-3 Highly Unsaturated Fatty Acids When Fed to Atlantic Salmon. *Aquacult. Nutr.* **2013**, *19*, 110–121. [CrossRef]
27. Hatlen, B.; Berge, G.M.; Odom, J.M.; Mundheim, H.; Ruyter, B. Growth Performance, Feed Utilisation and Fatty Acid Deposition in Atlantic Salmon, Salmo Salar L., Fed Graded Levels of High-Lipid/High-EPA Yarrowia Lipolytica Biomass. *Aquaculture* **2012**, *364–365*, 39–47. [CrossRef]
28. Caruffo, M.; Navarrete, N.; Salgado, O.; Díaz, A.; López, P.; García, K.; Feijóo, C.G.; Navarrete, P. Potential Probiotic Yeasts Isolated from the Fish Gut Protect Zebrafish (Danio Rerio) from a Vibrio Anguillarum Challenge. *Front. Microbiol.* **2015**, *6*, 1093. [CrossRef] [PubMed]
29. Alamillo, E.; Reyes-Becerril, M.; Cuesta, A.; Angulo, C. Marine Yeast Yarrowia Lipolytica Improves the Immune Responses in Pacific Red Snapper (Lutjanus Peru) Leukocytes. *Fish Shellfish. Immunol.* **2017**, *70*, 48–56. [CrossRef] [PubMed]
30. Hamza, F.; Vaidya, A.; Apte, M.; Kumar, A.R.; Zinjarde, S. Selenium Nanoparticle-Enriched Biomass of Yarrowia Lipolytica Enhances Growth and Survival of Artemia Salina. *Enzym. Microb. Technol.* **2017**, *106*, 48–54. [CrossRef]
31. Licona-Jain, A.; Campa-Córdova, Á.; Luna-González, A.; Racotta, I.S.; Tello, M.; Angulo, C. Dietary Supplementation of Marine Yeast Yarrowia Lipolytica Modulates Immune Response in Litopenaeus Vannamei. *Fish Shellfish. Immunol.* **2020**, *105*, 469–476. [CrossRef]
32. Aguilar-Macías, O.L.; Ojeda-Ramírez, J.J.; Campa-Córdova, A.I.; Saucedo, P.E. Evaluation of Natural and Commercial Probiotics for Improving Growth and Survival of the Pearl Oyster, Pinctada Mazatlanica, during Late Hatchery and Early Field Culturing. *J. World Aquac. Soc.* **2010**, *41*, 447–454. [CrossRef]
33. Wang, M.; Wang, B.; Jiang, K.; Liu, M.; Shi, X.; Wang, L. A Mitochondrial Manganese Superoxide Dismutase Involved in Innate Immunity Is Essential for the Survival of Chlamys Farreri. *Fish Shellfish. Immunol.* **2018**, *72*, 282–290. [CrossRef]
34. Guardiola, F.A.; Esteban, M.Á.; Angulo, C. Yarrowia Lipolytica, Health Benefits for Animals. *Appl. Microbiol. Biotechnol.* **2021**, *105*, 7577–7592. [CrossRef]
35. Egermeier, M.; Russmayer, H.; Sauer, M.; Marx, H. Metabolic Flexibility of Yarrowia Lipolytica Growing on Glycerol. *Front. Microbiol.* **2017**, *8*, 49. [CrossRef] [PubMed]
36. European Commission. *Commission Regulation (EC) No 152/2009 of 27 January 2009 Laying down the Methods of Sampling and Analysis for the Official Control of Feed*; European Commission: Brussels, Belgium, 2009; pp. 1–130. Available online: http://data.europa.eu/eli/reg/2009/152/oj (accessed on 4 September 2022).
37. Megazyme. *Validation Report: Sucrose/D-Fructose/D-Glucose Assay Kit (Cat. No. K-SUFRG)*; Validation Report; Ireland, 2018; pp. 1–7. Available online: https://www.megazyme.com/documents/Validation_Report/K-SUFRG_Validation_Report.pdf (accessed on 5 October 2022).
38. Gerhardt, C. Soxhlet Extraction Method. Available online: https://www.gerhardt.de/en/analysis-methods/chemical-extraction/ (accessed on 12 October 2022).
39. Dedousi, A.; Kritsa, M.-Z.; Đukić Stojčić, M.; Sfetsas, T.; Sentas, A.; Sossidou, E. Production Performance, Egg Quality Characteristics, Fatty Acid Profile and Health Lipid Indices of Produced Eggs, Blood Biochemical Parameters and Welfare Indicators of Laying Hens Fed Dried Olive Pulp. *Sustainability* **2022**, *14*, 3157. [CrossRef]

40. U.S. Food and Drug Administration. *Guidelines for the Validation of Chemical Methods for the FDA Foods Program*, 3rd ed.; U.S. Food and Drug Administration: Silver Spring, MD, USA, 2019. Available online: https://www.fda.gov/ (accessed on 10 September 2022).
41. *ISO 21528:2004*; Microbiology of Food and Animal Feeding Stuffs—Horizontal Methods for the Detection and Enumeration of Enterobacteriaceae. International Organization for Standardization (ISO): Ginebra, Switzerland, 2004.
42. *ISO 16649:2001*; Microbiology of Food and Animal Feeding Stuffs—Horizontal Method for the Enumeration of Beta-Glucuronidase-Positive Escherichia Coli. International Organization for Standardization (ISO): Ginebra, Switzerland, 2001.
43. *ISO 7954:1987*; Microbiology—General Guidance for Enumeration of Yeasts and Moulds—Colony Count Technique at 25 Degrees C. International Organization for Standardization (ISO): Ginebra, Switzerland, 1987.
44. *ISO 6579:2002*; Microbiology of Food and Animal Feeding Stuffs—Horizontal Method for the Detection of Salmonella Spp. International Organization for Standardization (ISO): Ginebra, Switzerland, 2002.
45. *ISO 11290:2017*; Microbiology of the Food Chain—Horizontal Method for the Detection and Enumeration of Listeria Monocytogenes and of Listeria Spp. International Organization for Standardization (ISO): Ginebra, Switzerland, 2017.
46. EUR-lex. Council Directive 2007/43/EC of 28 June 2007 Laying down Minimum Rules for the Protection of Chickens Kept for Meat Production. 2007. Available online: https://eur-lex.europa.eu/legal-content/EN/TXT/HTML/?uri=CELEX:32007L0043&from=EN (accessed on 25 October 2022).
47. Aviagen, W. Ross 308: Broiler's Management and Nutrition Specification. ROSS An Aviagen Brand 2018. Available online: https://en.aviagen.com/assets/Tech_Center/Ross_Broiler/Ross-BroilerHandbook2018-EN (accessed on 20 September 2022).
48. Welfare Quality. *Welfare Quality®Assessment Protocol for Poultry (Broilers, Laying Hens)*; Welfare Quality®Consortium: Lelystad, The Netherlands, 2009.
49. Ioannidou, M.D.; Maggira, M.; Samouris, G. Physicochemical Characteristics, Fatty Acids Profile and Lipid Oxidation during Ripening of Graviera Cheese Produced with Raw and Pasteurized Milk. *Foods* **2022**, *11*, 2138. [CrossRef] [PubMed]
50. *Nitrogen in Meat; Kjeldahl Method*; Association of Official Analytical Chemists (AOAC): Rockville, MD, USA, 1974.
51. Botsoglou, N.A.; Fletouris, D.J.; Papageorgiou, G.E.; Vassilopoulos, V.N.; Mantis, A.J.; Trakatellis, A.G. Rapid, Sensitive, and Specific Thiobarbituric Acid Method for Measuring Lipid Peroxidation in Animal Tissue, Food, and Feedstuff Samples. *J. Agric. Food Chem.* **1994**, *42*, 1931–1937. [CrossRef]
52. Eleftheriadou, A.; Samouris, G.; Ioannidou, M.; Kasapidou, E. Changes in Lipid Oxidation and Fatty Acid Composition in Pork and Poultry Meat during Refrigerated and Frozen Storage. *Arch. Lebensm.* **2010**, *61*, 173–178. [CrossRef]
53. Omri, B.; Chalghoumi, R.; Izzo, L.; Ritieni, A.; Lucarini, M.; Durazzo, A.; Abdouli, H.; Santini, A. Effect of Dietary Incorporation of Linseed Alone or Together with Tomato-Red Pepper Mix on Laying Hens' Egg Yolk Fatty Acids Profile and Health Lipid Indexes. *Nutrients* **2019**, *11*, 813. [CrossRef]
54. Ulbricht, T.L.V.; Southgate, D.A.T. Coronary Heart Disease: Seven Dietary Factors. *Lancet* **1991**, *338*, 985–992. [CrossRef]
55. Ko, E.-Y.; Saini, R.K.; Keum, Y.-S.; An, B.-K. Age of Laying Hens Significantly Influences the Content of Nutritionally Vital Lipophilic Compounds in Eggs. *Foods* **2020**, *10*, 22. [CrossRef]
56. JASP. *Computer Software*, Version 0.16.3; JASP Team: Amsterdam, The Netherlands, 2020.
57. Swiatkiewicz, S.; Arczewska-Wlosek, A.; Jozefiak, D. The Nutrition of Poultry as a Factor Affecting Litter Quality and Foot Pad Dermatitis - an Updated Review. *J. Anim. Physiol. Anim. Nutr.* **2017**, *101*, e14–e20. [CrossRef]
58. Amer, M.M. REVIEW: Footpad Dermatitis (FPD) in Chickens. *Korean J. Food Health Converg.* **2020**, *6*, 11–16. [CrossRef]
59. Freeman, N.; Tuyttens, F.A.M.; Johnson, A.; Marshall, V.; Garmyn, A.; Jacobs, L. Remedying Contact Dermatitis in Broiler Chickens with Novel Flooring Treatments. *Animals* **2020**, *10*, 1761. [CrossRef] [PubMed]
60. Grimes, J.L.; Smith, J.; Williams, C.M. Some Alternative Litter Materials Used for Growing Broilers and Turkeys. *World's Poult. Sci. J.* **2002**, *58*, 515–526. [CrossRef]
61. Kaukonen, E.; Norring, M.; Valros, A. Effect of Litter Quality on Foot Pad Dermatitis, Hock Burns and Breast Blisters in Broiler Breeders during the Production Period. *Avian Pathol.* **2016**, *45*, 667–673. [CrossRef] [PubMed]
62. Mayne, R.K. A Review of the Aetiology and Possible Causative Factors of Foot Pad Dermatitis in Growing Turkeys and Broilers. *World's Poult. Sci. J.* **2005**, *61*, 256–267. [CrossRef]
63. Choi, J.; Kim, W.K. Dietary Application of Tannins as a Potential Mitigation Strategy for Current Challenges in Poultry Production: A Review. *Animals* **2020**, *10*, 2389. [CrossRef]
64. Olsen, R.H.; Christensen, H.; Kabell, S.; Bisgaard, M. Characterization of Prevalent Bacterial Pathogens Associated with Pododermatitis in Table Egg Layers. *Avian Pathol.* **2018**, *47*, 281–285. [CrossRef]
65. Michalak, M.; Pierzak, M.; Kręcisz, B.; Suliga, E. Bioactive Compounds for Skin Health: A Review. *Nutrients* **2021**, *13*, 203. [CrossRef]
66. de Jong, I.C.; Gunnink, H.; van Harn, J. Wet Litter Not Only Induces Footpad Dermatitis but Also Reduces Overall Welfare, Technical Performance, and Carcass Yield in Broiler Chickens. *J. Appl. Poult. Res.* **2014**, *23*, 51–58. [CrossRef]
67. Saraiva, S.; Saraiva, C.; Stilwell, G. Feather Conditions and Clinical Scores as Indicators of Broilers Welfare at the Slaughterhouse. *Res. Vet. Sci.* **2016**, *107*, 75–79. [CrossRef]
68. Göransson, L.; Yngvesson, J.; Gunnarsson, S. Bird Health, Housing and Management Routines on Swedish Organic Broiler Chicken Farms. *Animals* **2020**, *10*, 2098. [CrossRef]

69. Rehman, K.; Haider, K.; Jabeen, K.; Akash, M.S.H. Current Perspectives of Oleic Acid: Regulation of Molecular Pathways in Mitochondrial and Endothelial Functioning against Insulin Resistance and Diabetes. *Rev. Endocr. Metab. Disord.* **2020**, *21*, 631–643. [CrossRef] [PubMed]
70. Li, X.H.; Chen, Y.P.; Cheng, Y.F.; Yang, W.L.; Wen, C.; Zhou, Y.M. Effect of Yeast Cell Wall Powder with Different Particle Sizes on the Growth Performance, Serum Metabolites, Immunity and Oxidative Status of Broilers. *Anim. Feed. Sci. Technol.* **2016**, *212*, 81–89. [CrossRef]
71. Gou, Z.Y.; Cui, X.Y.; Li, L.; Fan, Q.L.; Lin, X.J.; Wang, Y.B.; Jiang, Z.Y.; Jiang, S.Q. Effects of Dietary Incorporation of Linseed Oil with Soybean Isoflavone on Fatty Acid Profiles and Lipid Metabolism-Related Gene Expression in Breast Muscle of Chickens. *Animal* **2020**, *14*, 2414–2422. [CrossRef]
72. Zanetti, E.; De Marchi, M.; Dalvit, C.; Molette, C.; Remignon, H.; Cassandro, M. Carcase Characteristics and Qualitative Meat Traits of Three Italian Local Chicken Breeds. *Br. Poult. Sci.* **2010**, *51*, 629–634. [CrossRef] [PubMed]
73. Betti, M.; Perez, T.I.; Zuidhof, M.J.; Renema, R.A. Omega-3-Enriched Broiler Meat: 3. Fatty Acid Distribution between Triacylglycerol and Phospholipid Classes. *Poult. Sci.* **2009**, *88*, 1740–1754. [CrossRef]
74. Hernández-Sánchez, J.; Amills, M.; Pena, R.N.; Mercadé, A.; Manunza, A.; Quintanilla, R. Genomic Architecture of Heritability and Genetic Correlations for Intramuscular and Back Fat Contents in Duroc Pigs1. *J. Anim. Sci.* **2013**, *91*, 623–632. [CrossRef]
75. Gregory, M.K.; Gibson, R.A.; Cook-Johnson, R.J.; Cleland, L.G.; James, M.J. Elongase Reactions as Control Points in Long-Chain Polyunsaturated Fatty Acid Synthesis. *PLoS ONE* **2011**, *6*, e29662. [CrossRef]
76. Ogłuszka, M.; Szostak, A.; te Pas, M.F.W.; Poławska, E.; Urbański, P.; Blicharski, T.; Pareek, C.S.; Juszczuk-Kubiak, E.; Dunkelberger, J.R.; Horbańczuk, J.O.; et al. A Porcine Gluteus Medius Muscle Genome-Wide Transcriptome Analysis: Dietary Effects of Omega-6 and Omega-3 Fatty Acids on Biological Mechanisms. *Genes Nutr.* **2017**, *12*, 4. [CrossRef]
77. Dal Bosco, A.; Cartoni Mancinelli, A.; Vaudo, G.; Cavallo, M.; Castellini, C.; Mattioli, S. Indexing of Fatty Acids in Poultry Meat for Its Characterization in Healthy Human Nutrition: A Comprehensive Application of the Scientific Literature and New Proposals. *Nutrients* **2022**, *14*, 3110. [CrossRef]
78. Wołoszyn, J.; Haraf, G.; Okruszek, A.; Wereńska, M.; Goluch, Z.; Teleszko, M. Fatty Acid Profiles and Health Lipid Indices in the Breast Muscles of Local Polish Goose Varieties. *Poultry Sci.* **2020**, *99*, 1216–1224. [CrossRef]
79. Simopoulos, A.P. The Importance of the Ratio of Omega-6/Omega-3 Essential Fatty Acids. *Biomed. Pharmacother.* **2002**, *56*, 365–379. [CrossRef] [PubMed]
80. Chen, J.; Liu, H. Nutritional Indices for Assessing Fatty Acids: A Mini-Review. *IJMS* **2020**, *21*, 5695. [CrossRef]
81. Watson, T.; Shantsila, E.; Lip, G.Y. Mechanisms of Thrombogenesis in Atrial Fibrillation: Virchow's Triad Revisited. *Lancet* **2009**, *373*, 155–166. [CrossRef] [PubMed]
82. Santos-Silva, J.; Bessa, R.J.B.; Santos-Silva, F. Effect of Genotype, Feeding System and Slaughter Weight on the Quality of Light Lambs. *Livestock Production Sci.* **2002**, *77*, 187–194. [CrossRef]
83. Zdanowska-Sasiadek, Z.; Michalczuk, M.; Marcinkowska-Lesiak; Damaziak, K. Factors Determining the Course of Lipid Oxidation in Poultry Meat. *Zesz. Probl. Post. Nauk. Roln.* **2013**, *574*, 77–84.
84. Zheng, J.-S.; Hu, X.-J.; Zhao, Y.-M.; Yang, J.; Li, D. Intake of Fish and Marine N-3 Polyunsaturated Fatty Acids and Risk of Breast Cancer: Meta-Analysis of Data from 21 Independent Prospective Cohort Studies. *BMJ* **2013**, *346*, f3706. [CrossRef]
85. Gerber, M. Omega-3 Fatty Acids and Cancers: A Systematic Update Review of Epidemiological Studies. *Br. J. Nutr.* **2012**, *107*, S228–S239. [CrossRef]
86. Bosire, C.; Stampfer, M.J.; Subar, A.F.; Park, Y.; Kirkpatrick, S.I.; Chiuve, S.E.; Hollenbeck, A.R.; Reedy, J. Index-Based Dietary Patterns and the Risk of Prostate Cancer in the NIH-AARP Diet and Health Study. *Am. J. Epidemiol.* **2013**, *177*, 504–513. [CrossRef]
87. Hulan, H.W.; Ackman, R.G.; Ratnayake, W.M.N.; Proudfoot, F.G. Omega-3 Fatty Acid Levels and General Performance of Commercial Broilers Fed Practical Levels of Redfish Meal. *Poult. Sci.* **1989**, *68*, 153–162. [CrossRef]
88. Ratnayake, W.M.N.; Ackman, R.G.; Hulan, H.W. Effect of Redfish Meal Enriched Diets on the Taste Andn-3 Pufa of 42-Day-Old Broiler Chickens. *J. Sci. Food Agric.* **1989**, *49*, 59–74. [CrossRef]
89. Betti, M.; Schneider, B.L.; Wismer, W.V.; Carney, V.L.; Zuidhof, M.J.; Renema, R.A. Omega-3-Enriched Broiler Meat: 2. Functional Properties, Oxidative Stability, and Consumer Acceptance. *Poult. Sci.* **2009**, *88*, 1085–1095. [CrossRef] [PubMed]
90. Konieczka, P.; Czauderna, M.; Rozbicka-Wieczorek, A.; Smulikowska, S. The Effect of Dietary Fat, Vitamin E and Selenium Concentrations on the Fatty Acid Profile and Oxidative Stability of Frozen Stored Broiler Meat. *J. Anim. Feed Sci.* **2015**, *24*, 244–251. [CrossRef]

sustainability

MDPI

Article

Sustainable Broilers Production Performance under High Stocking Condition through Colocynth Seed Supplementation

Farid S. Nassar [1,2,*], Ahmed O. Abbas [1,2], Abdulaziz A. Alaqil [1] and Nancy N. Kamel [3,*]

1 Department of Animal and Fish Production, College of Agricultural and Food Sciences, King Faisal University, P.O. Box 420, Al-Ahsa 31982, Saudi Arabia
2 Department of Animal Production, Faculty of Agriculture, Cairo University, Giza P.O. Box 12613, Egypt
3 Department of Animal Production, National Research Center, El Buhouth St., Dokki, Giza P.O. Box 12622, Egypt
* Correspondence: fsmahmoud@kfu.edu.sa (F.S.N.); nn.kamel@nrc.sci.eg (N.N.K.)

Abstract: The negative impacts of high stocking density on the commercial poultry industry compromise sustainable birds' productivity and welfare. Thus, we investigated the potential of colocynth seed (CC) supplementation in alleviating the negative impacts of high stocking density on broilers' performance, immunity, inflammation, and redox status. A total of 648 one-day-old male Cobb 500 chicks were recruited and distributed into 2 × 2 factorial arrangements. The treatment groups were assigned based on stocking density as low stocking density (9 birds/m^2; LSD) or high stocking density (19 birds/m^2; HSD), and CC supplementation as without (0 g/kg feed; −CC) or with (1 g/kg feed; +CC) supplementation. Data were collected from week three to week five of age. Production performance was monitored and meat quality was assessed. Blood samples were collected to measure stress markers, humoral immune response, inflammatory cytokines, and antioxidant activity levels. The results indicated that HSD induced production performance reduction, immunosuppression, and imbalance redox status, along with elevation in inflammation and stress markers levels. Breast meat weight and yield were reduced in the HSD groups by 9 and 1%, respectively, compared to LSD groups. However, CC supplementation to HSD birds was able to slightly improve daily weight gain, body weight gain, and breast weight, showing no significant difference compared to the LSD-CC group, and significantly increased breast yield. Furthermore, CC supplementation significantly reduced inflammatory cytokines and stress markers levels. Under HSD, both cell-mediated and humeral immune responses were elevated with CC supplementation compared with the non-supplemented group. It can be concluded that HSD is a detrimental factor in the commercial poultry industry, which generates oxidative and inflammatory responses and, subsequently, immunosuppression and impaired performance. Nevertheless, dietary CC supplementation can be used as a natural antioxidant source to mitigate the negative impacts of HSD on broilers' production performance, as well as physiological competency.

Keywords: colocynth seed; broilers performance; meat quality; redox status; immune response; inflammatory cytokines; stocking density

Citation: Nassar, F.S.; Abbas, A.O.; Alaqil, A.A.; Kamel, N.N. Sustainable Broilers Production Performance under High Stocking Condition through Colocynth Seed Supplementation. *Sustainability* **2023**, *15*, 5102. https://doi.org/10.3390/su15065102

Academic Editors: George K. Symeon and Vassilios Dotas

Received: 17 February 2023
Revised: 7 March 2023
Accepted: 11 March 2023
Published: 14 March 2023

1. Introduction

The sustainable development goals consist of ending hunger, achieving food security, and promoting sustainable agriculture. The perception of the intensive poultry production system is employed to achieve greater productivity per area, attaining the highest profitability. Admittedly, high stocking density (HSD) is an inevitable practice to accomplish this goal. The official journal of the European Union defined stocking density as "the total live weight of chickens which are present in a house at the same time per square meter of useable area". Thus, increasing stocking density is applied to increase the profitability of the square meter. Nevertheless, there are thoughtful concerns about compromising broilers' welfare, as well as productivity, when applying HSD. A decade ago, Verspecht et al. [1] stated that maintaining profitability demands a stocking density of approximately 46 kg/m^2. Recently, numerous challenges have faced the commercial broiler production system regarding inputs availability and costs, as well as consumer demand and sales in response to global crises such as the COVID-19 pandemic [2–4] and climate change [5,6], as well as the ongoing conflicts. These production-challenging conditions justify the breeder urge to increase stocking density in order to meet the market demands and earn an adequate profit margin for maintaining sustainable poultry production.

There is an ongoing dispute about the influence of high stocking density (HSD) on broilers' welfare and production performance. Stocking density elevating from 28 to 42 kg/m^2 was reported to negatively impact broiler weight gain and intake, provoking liver injury and elevating body surface temperature [7]. Further, a stocking density of 20 broilers/m^2 adversely influences feed consumption and daily weight increase, generating oxidative damage and an increased mortality rate [8,9]. Additionally, HSD (18 birds/m^2) exhibited a negative impact on broiler's growth performance, wing feathering score, and footpad dermatitis score compared with lowered stocking density (12 birds/m^2) [10]. Under heat stress conditions, elevating the stocking density from 16 and 18 birds/m^2 to 23 and 26 birds/m^2 induced oxidative stress response and reduced the broiler's meat quality [11]. Furthermore, regarding high stocking density (17 birds/m^2), Guardia et al. [12] linked the impairment of broilers' production performance (i.e., the ratio of the feed conversion and daily weight increase) to the changes in the digestive microbiota fingerprint throughout the episode of 32–39 days of age. Bilal et al. [13] indicated that the adverse consequences of HSD not only included growth performance reduction with abnormal gait scores, but also showed immunosuppression and decreased profitability. Conversely, from the veterinary point of view and with the use of the annual mortality, rate of feed conversion, and antimicrobial usage in broiler farms with high (39 kg/m^2) or low (33 kg/m^2) stocking densities, Tarakdjian et al. [14] suggested that stocking density exhibits a minor role in broiler health and welfare. Interestingly, Bergeron et al. [15] stated that daily weight gain is positively linked to stocking density under commercial broiler systems. Furthermore, Nasr et al. [16] reported no difference between low (14 birds/m^2) and medium (18 birds/m^2) stocking densities on carcass features, oxidative stress markers, growth performance, and quality of meat in two varied broiler breeds. However, when stocking density increased to 20 birds/m^2, they reported significant negative impacts on the studied parameters. Generally, in order to alleviate the detrimental effect of oxidative stress generated under high stocking density on broilers' production and physiological performances, different dietary strategies were suggested, including different dietary feed additives (e.g., probiotics, prebiotics, synbiotics, vitamins, and plant products) [17].

Colocynth (*Citrullus colocynthis*) accounts for an herbal plant conventionally utilized in folk medicine. The highest antioxidant activities were found in colocynth seeds (CC), with a documented presence of 28 significant antioxidant active metabolites compared to other fruit parts (i.e., rinds and pulps) [18]. Our research group identified the CC seed phenolic compounds and fatty acids profiles [19]. The data revealed the presence of 18 polyphenol compounds with high concentrations of chlorogenic acid, p-hydroxy benzoic acid, and rosmarinic acid, among other antioxidant and anti-inflammatory bioactive phytochemical compounds. In addition, Bourhia et al. [20] reported that CC extract possesses antioxidant, antibacterial, and anticancer bioactivities. Furthermore, CC extracts exhibit potential pharmacological activities, which include analgesic and anti-inflammation [21–23]. Dietary CC supplementation was found to be potent immunomodulation and natural growth promoter agent for broilers reared under heat stress-induced-oxidative stress conditions [24]. Moreover, dietary CC supplementation was found to be an effective strategy to reduce the consequences of severe oxidative stress induced via paraquat on laying hens' production and immune responses [19]. Thus, we aimed to identify the HSD's negative consequences on broilers' immune response, inflammation, and redox status. Moreover, we investigated the antioxidant, immune-modulation, and anti-inflammation potentials of dietary CC supplementation to sustain broiler performance under HSD rearing conditions.

2. Materials and Methods

2.1. Birds Management and Experimental Design

Six hundred and forty-eight Cobb 500 male broiler chickens, 21 days old, have been recruited from a commercial broiler flock. The birds were allocated to a 2 × 2 factorial arrangement according to the stocking density (low (LSD; 9 birds/m^2) or high (HSD; 18 birds/m^2)) and colocynth (*C. colocynthis*) seed supplementation (CC; 0 or 1 g/kg feed) as the primary factors. The experimental groups were as follows: LSD without CC supplementation (LSD − CC), LSD with CC supplementation (LSD + CC), HSD without CC supplementation (HSD − CC), and HSD with CC supplementation (HSD + CC). Each experimental group comprises six replicate pens containing 18 birds for LSD groups and 36 birds for HSD groups. During the experiment period, water and feed were provided *ad libitum*. The basal diet of the experiment was prepared to meet the bird's requirement following the Cobb 500 rearing guide (https://www.cobb-vantress.com; accessed on 15 October 2022). The basal diet was corn-soybean meal based diet with 3150 kcal/kg metabolizable energy and 20% crude protein [24]. For CC supplemented groups, colocynth seed was daily grinded to fine powder and was gradually mixed with the basal diet. Birds were housed in 1.6 × 1.6 m pens that contained two feeders and one bell-shaped drinker, with a total mobility space of 2.052 m^2. The brooding temperature has been kept at 22 ± 3 °C during the experimental period (21 to 35 days of age) with a lighting pattern of only 4 hrs dark per day and 20 hrs light. Broilers were reared on floor pens with a deep litter floor brooder consisting of a wood shave (5 cm). Cooling pads and fans were employed to regulate the ambient temperature and ventilation.

2.2. Colocynth Phenolic Compounds Profile

Using high-performance liquid chromatography, colocynth seed phenolic components were identified (HPLC). The approach outlined in [25] was followed in the sample preparation process. HPLC equipment (Agilent 1260 infinity HPLC Series, Agilent, Palo Alto, CA, USA) was used to conduct the test methodology described in Agilent Application Note (No. 5991-3801EN; 2014). The separation was carried out through a Kinetex® 5 m EVO C18 100 (100 4.6 mm) column and an HPLC with quaternary pump (Phenomenex, CA, USA), functioned at 30 °C utilizing a ternary linear elution gradient (HPLC-grade water, 0.2% H_3PO_4 (v/v), acetonitrile, and methanol). The injection volume attained 20 μL. The phenolic compounds were detected using a variable wavelength detector (VWD) customized at 284 nm. Figure 1 lists the phenolic compounds in colocynth seed identified by HPLC.

2.3. Production Performance Parameters

The preliminary body weight was obtained on the 21st day of the age, whereas the final body weight was obtained on the 35th day. During these 14 days, feed intake (FI) was reported weekly and feed conversion (FCR) was accordingly estimated.

2.4. Meat Quality

At the end of the experiment period, one chicken from each group replicate was randomly selected, weighted, and slaughtered for breast muscle harvesting. The breast muscle weight was recorded using a sensitive digital balance of 0.1 g, and breast yield percentage was calculated relative to the preslaughter live body weight. The breast muscle color was assessed postmortem applying the CIELAB system using an automated color reader (Konica Minolta, Tokyo, Japan). The breast meat pH was determined at a depth of 1 cm after chilling for 24 h postmortem by a pH meter (Hanna Instruments, Venice, Italy). The cooking loss of muscle samples were assessed at 24 h postmortem as described by Lu et al. [26].

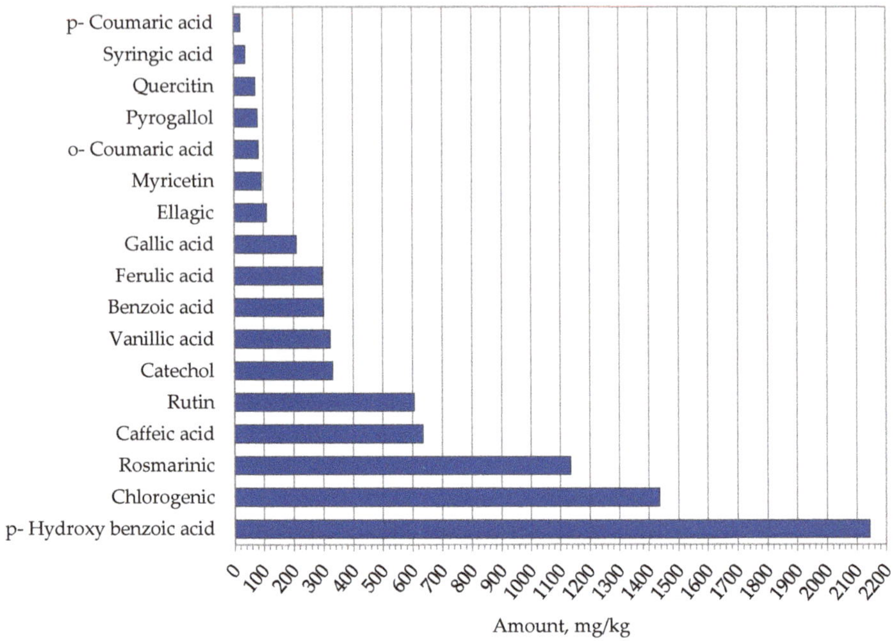

Figure 1. Phenolic acids profile of colocynth seed. Data are introduced, on a dry matter basis, as mg phenolic compound per each kg of colocynth seed.

2.5. Blood Sampling and Preparation

Samples of blood were obtained from each experimental group (n = 6, one sample per replicate) at the end of the 5th week of the age. Samples were left for 10 min in a centrifuge operated at 2000× g and 4 °C. Serum was then isolated and maintained at −20 °C for the analysis of immunoglobulins, inflammatory cytokines, and antioxidant activity.

2.6. Stress Markers Quantification

The H/L ratio was manually estimated in whole blood sample (n = 6, one sample per each group replicate) depending on the method given by Kamel et al. [27]. The H/L ratios were computed after the differential counts of a total of 200 leukocytes in six samples per experimental group utilizing a light microscope. Serum corticosterone hormone was assessed via chicken-specific quantitative competitive ELISA kits (cat#: MBS701668; MyBioSource, San Diego, CA, USA). According to the manufacturer, the assay sensitivity is <0.5 ng/mL with intra- and inter-assay precision of <8 and <10%, respectively. Serum malondialdehyde (MDA) level was estimated following the thiobarbituric acid technique by means of a commercial assay kit (Nanjing Jiancheng Bioengineering Institute, Nanjing, China).

2.7. Serum Antioxidant Activity

The quantification of superoxide dismutase (SOD), total antioxidant capacity (TAC), and catalase (CAT) activities were executed in serum samples (n = 6, one sample per each group replicate) through colorimetric kits (cat#: MBS2563691, MBS2540515, and MBS2540413, respectively; MyBioSource, San Diego, CA, USA). According to the manufacturer, the kit sensitivity, intra-assay precision, and inter-assay precision for SOD were 1.35 U/mL, 5.1% and 9.6%; for TAC, were 0.62 U/mL, 2.7% and 8.2%; and for CAT, were 0.27 U/mL, 3.1% and 5.1%, respectively.

2.8. Immune and Inflamation Response Parameters

2.8.1. Leucocytes Cell Count and Viability Measures

Total white blood cells (TWBCs) counts were executed manually in six complete blood samples per each experimental group (n = 6, one sample per group replicate) following Gehad et al. [28]. Whereas leucocyte cell viability (n = 6; one sample per replicate) was measured as Abbas et al. [29] described.

2.8.2. Immunoglobulins Concentration

The serum immunoglobulin (Ig) A, G, and M concentrations were measured (n = 6, one sample per each group replicate) using chicken ELISA kits (cat#: CSB-E11232Ch, CSB-EQ027259CH, CSB-E16200C, respectively; Cusabio Biotech Co., Ltd., Wuhan, China). According to the manufacturer of the kit, the intra- and inter-assay precision for IgA were <8 and <10%; for IgG, were <15 and <15%; and for IgM, were <8 and <10%, respectively.

2.8.3. The Humoral-Mediated Immune Response

Sheep red blood cells (SRBCs) antibody titer was estimated to assess the humoral-mediated immune response. At the age of 28 days, 12 chickens were randomly allocated from each examined group (two birds per replicate) and injected intravenously with 1 mL of 5% SRBCs saline suspension. After seven days, post-SRBCs injection, blood samples were taken, and the SRBCs antibody production was assessed through the micro-hemagglutination method [30]. The values of SRBCs antibody production were expressed when visible agglutination was observed as log2 of the reciprocal of the highest dilution.

2.8.4. The Cell-Mediated Immune Response

Wattle thickness, developed due to injection of phytohemagglutinin-P (PHA) antigen, was accomplished to examine the cell-mediated immune response, as described by Edelman et al. [31]. In brief, six birds per each examined group (one bird per replicate) were randomly assigned and injected with 50 µL of PHA solution into the center of the wattle. The wattle thickness was then calculated by subtracting the thickness 24 h after PHA injection from the thickness before injection using Schnelltester automatic caliper.

2.8.5. Inflammatory Cytokines

Blood serum levels of interleukins IL-6, IL-10, IL-1β, and tumor necrosis factor-alpha (TNF-α) were estimated (n = 6, one sample per each group replicate) by chicken-specific ELISA kits (cat#: MBS2021018, MBS701683, MBS2024496 and MBS2031870, respectively; MyBioSource, San Diego, CA, USA). According to the manufacturer, the kit sensitivity, intra-assay precision, and inter-assay precision for IL-6 were <5.6 pg/mL, <10% and <12%; for IL-10, were <0.5 pg/mL, <15% and <15%; for IL-1β, were <2.9 pg/mL, <10% and <12%; and for TNF-α, were <3.7 pg/mL, <10% and <12%, respectively.

2.9. Statistical Analysis

Data were analyzed through the two-way analysis of variance according to the general linear model (GLM) of the SAS software package. The statistical model includes the type of stocking density (SD), the CC supplementation, and their interaction as the main effects. When the interaction effect was significant, the Duncan post-hoc test was accomplished in order to determine the significance amongst the experimental groups at $p < 0.05$. Data results were presented as means ± SEM.

3. Results

3.1. Production Performance

The production performance of broilers reared under low or high stocking density with or without CC supplementation is presented in Table 1. Irrespective of CC supplementation, the results revealed a considerable decrease in average daily gain and final body weight in the HSD group relative to the LSD group. Furthermore, the feed conversion ratio was reduced in the HSD groups relative to the LSD groups, with no impact of HSD on average daily feed intake. However, the supplementation of CC substantially lessened the negative influence of HSD on broilers' body weight and body weight gain, yet it had no influence on the conversion ratio compared to the non-supplemented LSD-CC group. No interaction between SD and CC supplementation was detected.

Table 1. Production performance of broiler chickens reared under low (LSD, 9 birds/m^2) or high (HSD, 18 birds/m^2) stocking density and feed diet with or without colocynth seed supplementation (CC, 1 g/kg feed).

Parameters	LSD		HSD		SEM	p-Value		
	−CC	+CC	−CC	+CC		SD	CC	SD × CC
BW3wks, g	744	745	756	753	6.51	0.495	0.929	0.882
BW5wk, g	1894	1987	1785	1804	22.8	0.0003	0.109	0.284
ADG, g	82.1	88.7	73.5	75.1	1.69	0.0002	0.109	0.322
ADFI, g	125	133	121	121	2.17	0.068	0.422	0.368
FCR	1.53	1.49	1.65	1.61	0.01	<0.0001	0.018	0.859

Means having various superscripts significantly differ ($p < 0.05$). BW3wks: body weight after 3 weeks of age; BW5wks: body weight after 5 weeks of age; ADG: average daily gain; ADFI: average daily feed intake; FCR: feed conversion ratio.

3.2. Meat Quality

The breast meat quality parameters of broilers reared at various stocking densities with CC supplementation are displayed in Table 2. The HSD-CC group showed a significant reduction in breast weight and breast yield by 8 and 2%, respectively, relative to the LSD-CC group. Nevertheless, meat color and cooking loss percentage were not affected by the stocking density. Furthermore, the CC supplementation to broiler chickens under HSD was able to improve the breast weight and yield. Moreover, the meat pH was significantly elevated in the HSD and non-supplemented CC groups compared to the LSD- and CC-supplemented groups.

Table 2. Breast meat quality parameters of broiler chickens reared under low (LSD, 9 birds/m^2) or high (HSD, 18 birds/m^2) stocking density and feed diet with or without colocynth seed supplementation (CC, 1 g/kg feed).

Parameters	LSD		HSD		SEM	p-Value		
	−CC	+CC	−CC	+CC		SD	CC	SD × CC
Breast weight, g	169	180	156	162	2.21	<0.0001	0.009	0.374
Breast yield, %	8.94	9.05	8.77	8.97	0.03	0.037	0.011	0.402
pH, 24 h	5.77	5.82	5.77	5.78	0.01	0.026	0.002	0.074
L*	49.4	49.7	49.9	49.5	0.22	0.6800	0.8716	0.422
a*	4.39	4.54	4.40	4.50	0.04	0.822	0.131	0.740
b*	9.28	9.73	10.13	10.04	0.18	0.119	0.618	0.453
Cooking loss, %	20.7	20.9	21.0	21.0	0.22	0.671	0.848	0.876

Means having various superscripts significantly differ ($p < 0.05$). L*: lightness; a*: redness; b*: yellowness.

3.3. Stress Markers and Antioxidant Status

The stress markers and antioxidant enzyme activity of broilers reared at different stocking densities and with or without CC supplementation are presented in Table 3. Birds under HSD significantly exhibited high levels of stress markers (i.e., corticosterone, H/L ratio, and MDA), low antioxidant enzymes activity (i.e., catalase and SOD), and a reduction in total antioxidant activity. The H/L ratio, concentration of corticosterone, and MDA level were significantly elevated by 1.7-, 1.8-, and 2.2-fold in the HSD-CC group relative to the LSD-CC group, respectively. Additionally, significant 1.7-, 1.4-, and 1.4-fold reductions were noticed in SOD, catalase, and T-AOC activities, respectively, in the HSD-CC group relative to the LSD-CC group. Under HSD rearing conditions, CC supplementation significantly reduced the H/L ratio and increased the SOD activity. Furthermore, under LSD-rearing conditions, CC supplementation significantly increased the T-AOC activity.

Table 3. Stress markers and antioxidant enzymes activity of broiler chickens reared under low (LSD, 9 birds/m^2) or high (HSD, 18 birds/m^2) stocking density and feed diet with or without colocynth seed supplementation (CC, 1 g/kg feed).

Parameters	LSD		HSD		SEM	*p*-Value		
	−CC	+CC	−CC	+CC		SD	CC	SD × CC
CORT, pg/mL	4.61	4.03	8.29	7.42	0.40	<0.0001	0.029	0.653
H/L Ratio	0.37 ^{c}	0.35 ^{c}	0.63 ^{a}	0.50 ^{b}	0.03	<0.0001	0.006	0.046
MDA, μM/mL	1.98	1.95	4.35	3.58	0.20	<0.0001	0.063	0.084
SOD, U/ml	5.05	5.15	2.98	3.70	0.22	<0.0001	0.067	0.161
Catalase, U/mL	0.83	0.91	0.59	0.67	0.03	<0.0001	0.009	1.000
T-AOC, U/mL	8.34 ^{b}	10.38 ^{a}	5.88 ^{c}	6.25 ^{c}	0.40	<0.0001	0.0006	0.011

Means having various superscripts significantly differ ($p < 0.05$). CORT: corticosterone; H/L ratio: heterophil-to-lymphocyte ratio; MDA: malondialdehyde; SOD: superoxide dismutase; T-AOC: total antioxidant capacity.

3.4. Inflammation Markers

The pro-inflammatory cytokines concentrations of broilers reared under various stocking densities and with or without CC supplementation are presented in Table 4. The results indicated the induction of inflammation in response to the HSD condition. The fold increase in TNF-α, IL-6, IL-10, and IL-1β concentrations for the HSD-CC group compared with the LSD-CC group were 1.7-, 3.2-, 3.4-, and 2.1-fold, respectively. Additionally, CC supplementation to the HSD group exhibited an anti-inflammation effect and significantly decreased the pro-inflammatory cytokine concentrations compared to the non-supplemented HSD group by 16, 21, 28, and 18% for TNF-α, IL-6, IL-10, and IL-1β, respectively. A significant interaction was detected between SD and CC supplementation for cytokine levels.

Table 4. Serum pro-Inflammatory cytokines concentration of broiler chickens reared under low (LSD, 9 birds/m^2) or high (HSD, 18 birds/m^2) stocking density and feed diet with or without colocynth seed supplementation (CC, 1 g/kg feed).

Parameters	LSD		HSD		SEM	*p*-Value		
	−CC	+CC	−CC	+CC		SD	CC	SD × CC
TNF-α, pg/mL	90.8 ^{c}	84.8 ^{c}	155.7 ^{a}	130.0 ^{b}	6.40	<0.0001	0.002	0.039
IL-6, pg/mL	3.27 ^{c}	2.90 ^{c}	10.53 ^{a}	8.30 ^{b}	0.71	<0.0001	0.005	0.035
IL-10, pg/mL	2.63 ^{c}	2.38 ^{c}	9.03 ^{a}	6.53 ^{b}	0.59	<0.0001	<0.0001	0.001
IL-1β pg/mL	271 ^{c}	243 ^{c}	556 ^{a}	453 ^{b}	28.3	<0.0001	0.002	0.049

Means having various superscripts significantly differ ($p < 0.05$). TNF-α: tumor necrosis factor alpha; IL-6: interleukin 6; IL-10: interleukin 10; and IL-1β: interleukin1β.

3.5. Immune Responses

The humoral and cell-mediated immune responses of broilers reared at various stocking densities with CC supplementation are displayed in Table 5. In general, all the experimentally-considered immune-response parameters were negatively affected by HSD compared to LSD. The leucocyte cell count (TWBCs) was reduced by 31% and 22%, whereas the leucocyte cell viability percentage was reduced by 16% and 12% in the HSD-CC and HDS + CC groups relative to the LSD-CC group, respectively. Furthermore, the HSD − CC group manifested serum immunoglobulins concentration reduction by 1.7-, 1.9-, and 1.7-fold for IgA, IgG, and IgM, respectively, relative to the LSD − CC group. Additionally, the antibody production against SRBC was significantly reduced by 1.6-fold. The wattle cell-mediated immune response to PHA antigen injection was negatively affected in the HSD − CC group, showing 37% wattle thickness reduction relative to the LSD − CC group.

However, the CC supplementation to HSD birds considerably improved the serum immunoglobulins concentrations, SRBC antibody production, and cell-mediated response to PHA-wattle injection compared to the non-supplemented HSD group. It is worth noticing that, under LSD rearing conditions, CC supplementation elevated the IgA and IgM serum concentrations, and improved the birds' cell-mediated response to PHA-wattle injection. A significant interaction between SD and CC supplementation was only observed for the SRBC antibody parameter.

Table 5. Immune response parameters of broiler chickens reared at low (LSD, 9 birds/m^2) or high (HSD, 18 birds/m^2) stocking density and feed diet with or without colocynth seed supplementation (CC, 1 g/kg feed).

Parameters	LSD		HSD		SEM	*p*-Value		
	−CC	+CC	−CC	+CC		SD	CC	SD × CC
TWBCs, ×10^3/mL	45.4	48.9	31.1	35.3	1.67	<0.0001	0.025	0.825
LCV, %	101.7	107.7	85.3	89.0	2.14	<0.0001	0.034	0.590
IgA, µg/mL	62.7	71.2	36.8	46.8	3.01	<0.0001	0.001	0.758
IgG, µg/mL	137.2	151.7	72.8	87.0	7.55	<0.0001	0.043	0.980
IgM, µg/mL	299	347	172	242	15.00	<0.0001	0.0003	0.424
SRBC-Ab, log2	6.83 ^{ab}	7.17 ^{a}	4.17 ^{c}	6.00 ^{b}	0.28	<0.0001	0.0015	0.020
Wattle thickness, mm	0.35	0.39	0.22	0.28	0.01	<0.0001	<0.0001	0.399

Means having various superscripts significantly differ ($p < 0.05$). TWBC: total white blood cells; LCV: leucocyte cell viability; IgA: immunoglobulin A; IgG: immunoglobulin G; IgM: immunoglobulin M; SRBC-Ab: sheep red blood cells antibody.

4. Discussion

The stocking density elevation was proven to have massive negative impacts, not only on production performance, but also on birds' physiological systems and meat quality. The HSD employed in the current investigation was 18 birds/m^2 with an actual final weight of 32 kg/m^2. Broiler production performance was significantly reduced, in terms of body weight and ADG, and FCR was impaired under the HSD. Several researchers have reported an HSD negative consequence on broiler production performance (i.e., feed conversion, feed intake, and body weight gain) [32–36]. However, other researchers reported no effect [15]. These controversial results can be justified by the influence of other rearing conditions, such as ventilation and age, on performance. Under the current study HSD condition, CC supplementation significantly improved BW and ADG with no effect on feed intake and FCR. Such CC supplementation's positive effect is likely attributed to its antioxidant activities, which helped in mitigating the negative consequences of HSD-induced-oxidative-stress on broiler performance [19,24]. In the present study, breast meat weight and yield were reduced in the HSD-CC group with no effect on meat color and cooking loss. Nasr et al. [16] reported a reduction in breast yield and meat quality parameters associated with HSD in Arbor acres and Ross-308 broiler breeds. By contrast, other researchers indicated no effect of different stocking densities (ranged from 12.6 up to 40.91 kg/m^2) on meat quality parameters [11,37,38].

The stress markers of broilers reared under HSD showed considerably higher values than LSD. These findings agree with those presented by other researchers who documented an associated increase in corticosterone concentration [39,40] and H/L ratio [37,41,42] along with increasing stocking density. The corresponding results of antioxidant enzyme activity drop, as well as an increase in the level of lipid peroxidation metabolite observed under the current HSD rearing conditions, demonstrate a state of oxidative stress induction. Son et al. [11] found an increase in liver MDA concentration, as well as a reduction in blood SOD activity, in broilers reared under HSD. Miao et al. [8] also reported that plasma MDA level was increased, whereas the T-AOC was decreased with HSD (20 birds/m^2) compared with LSD (14 birds/m^2).

On one hand, there is a close interrelation between oxidative stress, immune response, and inflammation [43]. Cytokines, which represent the regulators and mediators of host immune responses, include pro-inflammatory cytokines (e.g., IL-6, TNF-α, and IL-1β) that are involved in the induction of fever, inflammation, and tissue destruction [44,45]. The elevated pro-inflammatory cytokines levels are one of the significant negative impacts of HSD under the current study condition and can directly be connected to the detected immunosuppression state. The anti-inflammation property of CC has been demonstrated by several researchers [21–23]. Thus, the CC supplementation to HSD birds significantly reduced pro-inflammatory cytokines levels, reflecting its anti-inflammation properties.

On the other hand, the present study supported the likelihood that HSD caused immunosuppression, with the observed decrease in both humoral (i.e., immunoglobulins and SRBC-AB levels) and cell-mediated (i.e., PHA-injection immune response) immune responses. Li et al. [46] reported that HSD induced physiological and oxidative stresses with a significant reduction in serum immunoglobulin (IgA and IgG) concentration. Moreover, Sevim et al. [47] reported a significant increase in lymphocyte DNA damage for broilers reared under HSD (18 birds/m^2). Moreover, a high corticosterone level was previously demonstrated to have an immunosuppression effect [48]. In the current experiment, the corticosterone level was 61 to 80% higher in the HSD groups relative to the LSD-CC group, which might partially justify the detected immunosuppression. Furthermore, regardless of the stressor type, oxidative stress generation was reported to induce immunosuppression in broilers [27,49,50]. Moreover, Xu et al. [45] concluded that the immune response was mediated via the level of the inflammatory cytokine due to infectious bursal disease virus (IBDV) infection in chickens. Thus, the oxidative stress generation and the high corticosterone and pro-inflammatory cytokines concentration induced by HSD in the current study could be responsible for the detected immunosuppression. However, the reported immunomodulation and antioxidant bioactive properties of CC contributed to alleviating the negative impacts of HSD, partially reducing stress markers levels, and improving broiler immune response [19,24].

5. Conclusions

Stocking density is still a controversial concern for the sustainability of the commercial poultry industry. Under intensive commercial broiler production, increasing stocking density is an inevitable stressor. The present study highlighted the HSD negative impact on broilers' production performance with a new insight into birds' immune response, inflammation induction, and antioxidant status. The evidence from this study revealed profound immunosuppression, inflammation, and imbalance redox status when rearing chickens under the presence stocking conditions. However, the supplementation of CC as a natural antioxidant and anti-inflammation agent by 1 g/kg feed revealed positive mitigation of the immune response and inflammatory cytokine levels. Further works are needed to be done to alleviate the negative impacts of HSD under commercial production conditions. Nevertheless, it can be concluded that natural antioxidant and anti-inflammatory products, such as CC seed, can effectively improve the physiological mechanisms and sustain broiler production performance under HSD rearing conditions.

Author Contributions: Conceptualization, F.S.N., N.N.K. and A.O.A.; data curation, N.N.K. and A.O.A.; formal analysis, N.N.K.; funding acquisition, F.S.N.; investigation, F.S.N., A.O.A. and A.A.A.; methodology, F.S.N., A.O.A. and A.A.A.; project administration, F.S.N.; resources, A.O.A. and A.A.A.; supervision, F.S.N.; validation, A.O.A. and A.A.A.; visualization, N.N.K.; writing—original draft, N.N.K.; writing—review and editing, A.O.A. All authors have read and agreed to the published version of the manuscript.

Funding: This research was funded by the Deputyship for Research and Innovation, Ministry of Education, Saudi Arabia, grant number INST129.

Institutional Review Board Statement: The study design and the experimental protocols were approved by the Research Ethics Committee (REC), King Faisal University, Saudi Arabia (KFU-REC-2022- AUG -ETHICS141).

Informed Consent Statement: Not applicable.

Data Availability Statement: Not applicable.

Acknowledgments: The authors are grateful to the Deputyship for Research and Innovation, Ministry of Education, Saudi Arabia, for funding this research.

Conflicts of Interest: The authors declare no conflict of interest. The funders did not contribute to the study design; the collection, analyses, and interpretation of data; manuscript writing; or in the decision about publishing the results.

References

1. Verspecht, A.; Vanhonacker, F.; Verbeke, W.; Zoons, J.; Van Huylenbroeck, G. Economic impact of decreasing stocking densities in broiler production in Belgium. *Poult. Sci.* **2011**, *90*, 1844–1851. [CrossRef]
2. Abu Hatab, A.; Liu, Z.; Nasser, A.; Esmat, A. Determinants of SARS-CoV-2 impacts on small-scale commercial broiler production systems in Egypt: Implications for mitigation strategies. *Animals* **2021**, *11*, 1354. [CrossRef] [PubMed]
3. Pawłowska, J.; Borecka, A.; Sowula-Skrzyńska, E. Effect of COVID-19 on poultry production in Poland. *World's Poult. Sci. J.* **2022**, *78*, 823–833. [CrossRef]
4. Sattar, A.A.; Mahmud, R.; Mohsin, M.A.S.; Chisty, N.N.; Uddin, M.H.; Irin, N.; Barnett, T.; Fournie, G.; Houghton, E.; Hoque, M.A. COVID-19 impact on poultry production and distribution networks in Bangladesh. *Front. Sustain. Food Syst.* **2021**, *5*, 714649. [CrossRef]
5. Kumar, M.; Ratwan, P.; Dahiya, S.P.; Nehra, A.K. Climate change and heat stress: Impact on production, reproduction and growth performance of poultry and its mitigation using genetic strategies. *J. Therm. Biol.* **2021**, *97*, 102867. [CrossRef] [PubMed]
6. Abioja, M.O.; Abiona, J.A. Impacts of climate change to poultry production in Africa: Adaptation options for broiler chickens. In *African Handbook of Climate Change Adaptation*; Oguge, N., Ayal, D., Adeleke, L., da Silva, I., Eds.; Springer International Publishing: Cham, Switzerland, 2021; pp. 275–296.
7. Abudabos, A.M.; Samara, E.M.; Hussein, E.O.S.; Al-Ghadi, M.a.Q.; Al-Atiyat, R.M. Impacts of stocking density on the performance and welfare of broiler chickens. *Ital. J. Anim. Sci.* **2013**, *12*, e11. [CrossRef]
8. Miao, Z.Q.; Dong, Y.Y.; Qin, X.; Yuan, J.M.; Han, M.M.; Zhang, K.K.; Shi, S.R.; Song, X.Y.; Zhang, J.Z.; Li, J.H. Dietary supplementation of methionine mitigates oxidative stress in broilers under high stocking density. *Poult. Sci.* **2021**, *100*, 101231. [CrossRef]
9. Jobe, M.C.; Ncobela, C.N.; Kunene, N.W.; Opoku, A.R. Effects of Cassia abbreviata extract and stocking density on growth performance, oxidative stress and liver function of indigenous chickens. *Trop. Anim. Health Prod.* **2019**, *51*, 2567–2574. [CrossRef]
10. Sevim, O.; Ahsan, U.; Tatli, O.; Kuter, E.; Khamseh, E.K.; Temiz, A.R.; Ozdemir, O.S.; Aydin, A.K.; Ozsoy, B.; Koksal, B.H.; et al. Effect of high stocking density and dietary nano-zinc on growth performance, carcass yield, meat quality, feathering score, and footpad dermatitis in broiler chickens. *Livest. Sci.* **2021**, *253*, 104727. [CrossRef]
11. Son, J.; Kim, H.J.; Hong, E.C.; Kang, H.K. Effects of stocking density on growth performance, antioxidant status, and meat quality of finisher broiler chickens under high temperature. *Antioxidants* **2022**, *11*, 871. [CrossRef]
12. Guardia, S.; Konsak, B.; Combes, S.; Levenez, F.; Cauquil, L.; Guillot, J.F.; Moreau-Vauzelle, C.; Lessire, M.; Juin, H.; Gabriel, I. Effects of stocking density on the growth performance and digestive microbiota of broiler chickens. *Poult. Sci.* **2011**, *90*, 1878–1889. [CrossRef] [PubMed]
13. Bilal, R.M.; Hassan, F.U.; Farag, M.R.; Nasir, T.A.; Ragni, M.; Mahgoub, H.A.M.; Alagawany, M. Thermal stress and high stocking densities in poultry farms: Potential effects and mitigation strategies. *J. Therm. Biol.* **2021**, *99*, 102944. [CrossRef] [PubMed]
14. Tarakdjian, J.; Capello, K.; Pasqualin, D.; Cunial, G.; Lorenzetto, M.; Gavazzi, L.; Manca, G.; Di Martino, G. Antimicrobial use in broilers reared at different stocking densities: A retrospective study. *Animals* **2020**, *10*, 1751. [CrossRef] [PubMed]
15. Bergeron, S.; Pouliot, E.; Doyon, M. Commercial poultry production stocking density influence on bird health and performance indicators. *Animals* **2020**, *10*, 1253. [CrossRef] [PubMed]
16. Nasr, M.A.F.; Alkhedaide, A.Q.; Ramadan, A.A.I.; Hafez, A.-E.S.E.; Hussein, M.A. Potential impact of stocking density on growth, carcass traits, indicators of biochemical and oxidative stress and meat quality of different broiler breeds. *Poult. Sci.* **2021**, *100*, 101442. [CrossRef] [PubMed]
17. Sugiharto, S. Dietary strategies to alleviate high-stocking-density-induced stress in broiler chickens—A comprehensive review. *Arch. Anim. Breed.* **2022**, *65*, 21–36. [CrossRef]
18. Al-Nablsi, S.; El-Keblawy, A.; Ali, M.A.; Mosa, K.A.; Hamoda, A.M.; Shanableh, A.; Almehdi, A.M.; Soliman, S.S.M. Phenolic contents and antioxidant activity of *Citrullus colocynthis* fruits, growing in the hot arid desert of the UAE, influenced by the fruit parts, accessions, and seasons of fruit collection. *Antioxidants* **2022**, *11*, 656. [CrossRef]
19. Abbas, A.O.; Alaqil, A.A.; Kamel, N.N.; Moustafa, E.S. *Citrullus colocynthis* seed ameliorates layer performance and immune response under acute oxidative stress induced by paraquat injection. *Animals* **2022**, *12*, 945. [CrossRef]

20. Bourhia, M.; Bouothmany, K.; Bakrim, H.; Hadrach, S.; Salamatullah, A.M.; Alzahrani, A.; Alyahya, H.K.; Albadr, N.A.; Gmouh, S.; Laglaoui, A.; et al. Chemical profiling, antioxidant, antiproliferative, and antibacterial potentials of chemically characterized extract of *Citrullus colocynthis* L. seeds. *Separations* **2021**, *8*, 114. [CrossRef]
21. Marzouk, B.; Mahjoub, M.A.; Bouraoui, A.; Fenina, N.; Aouni, M.; Marzouk, Z. Anti-inflammatory and analgesic activities of a new cucurbitacin isolated from *Citrullus colocynthis* seeds. *Med. Chem. Res.* **2013**, *22*, 3984–3990. [CrossRef]
22. Marzouk, B.; Marzouk, Z.; Fenina, N.; Bouraoui, A.; Aouni, M. Anti-inflammatory and analgesic activities of Tunisian *Citrullus colocynthis* Schrad. immature fruit and seed organic extracts. *Eur. Rev. Med. Pharmacol. Sci.* **2011**, *15*, 665–672. [PubMed]
23. Marzouk, B.; Marzouk, Z.; Haloui, E.; Turki, M.; Bouraoui, A.; Aouni, M.; Fenina, N. Anti-inflammatory evaluation of immature fruit and seed aqueous extracts from several populations of Tunisian *Citrullus colocynthis* Schrad. *Afr. J. Biotechnol.* **2011**, *10*, 4217–4225.
24. Alzarah, M.I.; Althobiati, F.; Abbas, A.O.; Mehaisen, G.M.K.; Kamel, N.N. *Citrullus colocynthis* Seeds: A potential natural immune modulator source for broiler reared under chronic heat stress. *Animals* **2021**, *11*, 1951. [CrossRef] [PubMed]
25. Corrales, M.; Han, J.H.; Tauscher, B. Antimicrobial properties of grape seed extracts and their effectiveness after incorporation into pea starch films. *Int. J. Food Sci. Technol.* **2009**, *44*, 425–433. [CrossRef]
26. Lu, Q.; Wen, J.; Zhang, H. Effect of chronic heat exposure on fat deposition and meat quality in two genetic types of chicken. *Poult. Sci.* **2007**, *86*, 1059–1064. [CrossRef]
27. Kamel, N.N.; Ahmed, A.M.H.; Mehaisen, G.M.K.; Mashaly, M.M.; Abass, A.O. Depression of leukocyte protein synthesis, immune function and growth performance induced by high environmental temperature in broiler chickens. *Int. J. Biometeorol.* **2017**, *61*, 1637–1645. [CrossRef]
28. Gehad, A.E.; Mehaisen, G.M.; Abbas, A.O.; Mashaly, M.M. The role of light program and melatonin on alleviation of Inflammation induced by lipopolysaccharide injection in broiler chickens. *Int. J. Poult. Sci.* **2008**, *7*, 193–201. [CrossRef]
29. Abbas, A.O.; Alaqil, A.A.; El-Beltagi, H.S.; Abd El-Atty, H.K.; Kamel, N.N. Modulating laying hens productivity and immune performance in response to oxidative stress induced by E. coli challenge using dietary propolis supplementation. *Antioxidants* **2020**, *9*, 893. [CrossRef]
30. Loa, C.C.; Lin, T.L.; Wu, C.C.; Bryan, T.; Thacker, H.L.; Hooper, T.; Schrader, D. Humoral and cellular immune responses in turkey poults infected with turkey coronavirus. *Poult. Sci.* **2001**, *80*, 1416–1424. [CrossRef]
31. Edelman, A.S.; Sanchez, P.; Robinson, M.E.; Hochwald, G.M.; Thorbecke, G.J. Primary and secondary wattle swelling response to phytohemagglutinin as a measure of immunocompetence in chickens. *Avian Dis.* **1986**, *30*, 105–111. [CrossRef]
32. Dozier, W.A.; Thaxton, J.P.; Purswell, J.L.; Olanrewaju, H.A.; Branton, S.L.; Roush, W.B. Stocking density effects on male broilers grown to 1.8 kilograms of body weight. *Poult. Sci.* **2006**, *85*, 344–351. [CrossRef] [PubMed]
33. Feddes, J.J.; Emmanuel, E.J.; Zuidhoft, M.J. Broiler performance, body weight variance, feed and water intake, and carcass quality at different stocking densities. *Poult. Sci.* **2002**, *81*, 774–779. [CrossRef] [PubMed]
34. Zuowei, S.; Yan, L.; Yuan, L.; Jiao, H.; Song, Z.; Guo, Y.; Lin, H. Stocking density affects the growth performance of broilers in a sex-dependent fashion. *Poult. Sci.* **2011**, *90*, 1406–1415. [CrossRef] [PubMed]
35. Goo, D.; Kim, J.H.; Choi, H.S.; Park, G.H.; Han, G.P.; Kil, D.Y. Effect of stocking density and sex on growth performance, meat quality, and intestinal barrier function in broiler chickens. *Poult. Sci.* **2019**, *98*, 1153–1160. [CrossRef] [PubMed]
36. Sanchez-Casanova, R.; Sarmiento-Franco, L.; Phillips, C. The effects of outdoor access and stocking density on the performance of broilers reared under tropical conditions. *Br. Poult. Sci.* **2021**, *62*, 632–637. [CrossRef]
37. Simitzis, P.E.; Kalogeraki, E.; Goliomytis, M.; Charismiadou, M.A.; Triantaphyllopoulos, K.; Ayoutanti, A.; Niforou, K.; Hager-Theodorides, A.L.; Deligeorgis, S.G. Impact of stocking density on broiler growth performance, meat characteristics, behavioural components and indicators of physiological and oxidative stress. *Br. Poult. Sci.* **2012**, *53*, 721–730. [CrossRef]
38. Thema, K.K.; Mnisi, C.M.; Mlambo, V. Stocking density-induced changes in growth performance, blood parameters, meat quality traits, and welfare of broiler chickens reared under semi-arid subtropical conditions. *PLoS ONE* **2022**, *17*, e0275811. [CrossRef]
39. Li, X.M.; Zhang, M.H.; Liu, S.M.; Feng, J.H.; Ma, D.D.; Liu, Q.X.; Zhou, Y.; Wang, X.J.; Xing, S. Effects of stocking density on growth performance, growth regulatory factors, and endocrine hormones in broilers under appropriate environments. *Poult. Sci.* **2019**, *98*, 6611–6617. [CrossRef]
40. Najafi, P.; Zulkifli, I.; Amat Jajuli, N.; Farjam, A.S.; Ramiah, S.K.; Amir, A.A.; O'Reily, E.; Eckersall, D. Environmental temperature and stocking density effects on acute phase proteins, heat shock protein 70, circulating corticosterone and performance in broiler chickens. *Int. J. Biometeorol.* **2015**, *59*, 1577–1583. [CrossRef]
41. Thaxton, J.P.; Dozier, W.A.; Branton, S.L.; Morgan, G.W.; Miles, D.W.; Roush, W.B.; Lott, B.D.; Vizzier-Thaxton, Y. Stocking density and physiological adaptive responses of broilers. *Poult. Sci.* **2006**, *85*, 819–824. [CrossRef]
42. Astaneh, I.Y.; Chamani, M.; Mousavi, S.N.; Sadeghi, A.A.; Afshar, M.A. Effects of stocking density on performance and immunity in Ross 308 broiler chickens. *Kafkas Üniversitesi Vet. Fakültesi Derg.* **2018**, *24*, 483–489. [CrossRef]
43. Lee, M.T.; Lin, W.C.; Lee, T.T. Potential crosstalk of oxidative stress and immune response in poultry through phytochemicals—A review. *Asian-Australas. J. Anim. Sci.* **2019**, *32*, 309–319. [CrossRef] [PubMed]
44. Dinarello, C.A. Proinflammatory cytokines. *Chest* **2000**, *118*, 503–508. [CrossRef] [PubMed]
45. Xu, Z.Y.; Yu, Y.; Liu, Y.; Ou, C.B.; Zhang, Y.H.; Liu, T.Y.; Wang, Q.X.; Ma, J.Y. Differential expression of pro-inflammatory and anti-inflammatory genes of layer chicken bursa after experimental infection with infectious bursal disease virus. *Poult. Sci.* **2019**, *98*, 5307–5314. [CrossRef] [PubMed]

46. Li, W.; Wei, F.; Xu, B.; Sun, Q.; Deng, W.; Ma, H.; Bai, J.; Li, S. Effect of stocking density and alpha-lipoic acid on the growth performance, physiological and oxidative stress and immune response of broilers. *Asian-Australas J. Anim. Sci.* **2019**, *32*, 1914–1922. [CrossRef] [PubMed]
47. Sevim, Ö.; Ahsan, U.; Tatlı, O.; Kuter, E.; Khamseh, E.K.; Temiz, A.R.; Özdemir, Ö.S.; Aydın, A.K.; Şahiner, H.S.; Avcıoğlu, M.; et al. Effect of dietary nano-selenium on stress indicators, immune response, and DNA damage in broiler subjected to different stocking density. *Eur. Poult. Sci.* **2021**, *85*, 345. [CrossRef]
48. Mehaisen, G.M.K.; Eshak, M.G.; Elkaiaty, A.M.; Atta, A.M.M.; Mashaly, M.M.; Abass, A.O. Comprehensive growth performance, immune function, plasma biochemistry, gene expressions and cell death morphology responses to a daily corticosterone injection course in broiler chickens. *PLoS ONE* **2017**, *12*, e0172684. [CrossRef]
49. Guo, Y.; Zhang, J.; Li, X.; Wu, J.; Han, J.; Yang, G.; Zhang, L. Oxidative stress mediated immunosuppression caused by ammonia gas via antioxidant/oxidant imbalance in broilers. *Br. Poult. Sci.* **2022**, *5*, 1–11. [CrossRef]
50. Ali, A.; Khatoon, A.; Saleemi, M.K.; Abbas, R.Z. Aflatoxins associated oxidative stress and immunological alterations are mitigated by dietary supplementation of Pichia kudriavzevii in broiler chicks. *Microb. Pathog.* **2021**, *161*, 105279. [CrossRef]

sustainability

MDPI

Article

Prevalence of Antibiotic Resistant *E. coli* Strains Isolated from Farmed Broilers and Hens in Greece, Based on Phenotypic and Molecular Analyses

Anna Xexaki [1], Dimitrios K. Papadopoulos [2], Maria V. Alvanou [2], Ioannis A. Giantsis [2,*], Konstantinos V. Papageorgiou [1], Georgios A. Delis [1], Vangelis Economou [1], Spyridon K. Kritas [1], Evangelia N. Sossidou [3] and Evanthia Petridou [1]

1 School of Veterinary Medicine, Faculty of Health Sciences, Aristotle University of Thessaloniki, 54124 Thessaloniki, Greece
2 Department of Animal Science, Faculty of Agricultural Sciences, University of Western Macedonia, 53100 Florina, Greece
3 Veterinary Research Institute, Hellenic Agricultural Organization—DEMETER, 57001 Thessaloniki, Greece
* Correspondence: igiants@agro.auth.gr

Abstract: The use of antimicrobials is beneficial for livestock health; however, their overuse and misuse may increase resistance to these compounds. Thus, the aim of the present study was the phenotypic and molecular examination of the presence of *Escherichia coli* antibiotic-resistant strains in broiler and laying hen farms. The resistance of *E. coli* strains was examined against various antibiotics, including several families of compounds such as penicillin class medications (ampicillin), cephalosporins (cefotaxime, cefoxitin, cefpodoxime and ceftazidime), sulfonamides (co-trimoxazole), quinolones (enrofloxacin and nalidixic acid), aminoglycosides (gentamicin), β-lactams (imipenem), aminoglycoside (streptomycin), and polymyxin (colistin). In total, 106 strains were investigated, sampled during the years 2016–2019 from 91 poultry farms, including 75 broiler farms and 16 laying hen farms, originating from three Regional Units in Greece. The examined isolates revealed the highest resistance rates to sulfamethoxazole (81.1%), nalidixic acid (73.6%), tetracyclin (70.8%), and streptomycin (70.8%). On the other hand, the resistance of the isolates to third generation cephalosporins was found to be at lower levels for ceftazidime (2.8%), ceftriaxone (3.7%) cefoxitin (4.7%), and cefotaxime (4.7%). Phenotypic tests showed that 13.6% and 10.2% of the isolates produced ESBL, while 2.7% and 1% produced AmpC b-lactamase, for broiler and laying hens, respectively. The prevalence of the *mcr-1* gene was found to be 22.7%, detected only in broiler isolates. Based on our results, *E. coli* antibiotic resistance represents a critical control point in poultry production that, apart from farm animals, may affect public health as well.

Keywords: *Escherichia coli*; antibiotics; resistance; poultry; laying hens; broiler

Citation: Xexaki, A.; Papadopoulos, D.K.; Alvanou, M.V.; Giantsis, I.A.; Papageorgiou, K.V.; Delis, G.A.; Economou, V.; Kritas, S.K.; Sossidou, E.N.; Petridou, E. Prevalence of Antibiotic Resistant *E. coli* Strains Isolated from Farmed Broilers and Hens in Greece, Based on Phenotypic and Molecular Analyses. *Sustainability* **2023**, *15*, 9421. https://doi.org/10.3390/su15129421

Academic Editors: George K. Symeon and Vassilios Dotas

Received: 7 March 2023
Revised: 7 June 2023
Accepted: 8 June 2023
Published: 12 June 2023

1. Introduction

As the use of antimicrobial substances in the primary sector is increasing, it leads to extensive human exposure to bacteria with antimicrobial resistance (AMR), indirect gene transfer among bacteria species, and the spread of antimicrobial-resistant bacteria into the environment [1]. In particular, the use of antibiotics in compound feeds has been a substantial part of poultry production, not only for infectious bacterial disease prevention, but also for the improvement of animal growth rates [2].

E. coli is a Gram-negative, facultative anaerobic, rod-shaped bacterium and a member of Enterobacteriaceae family, typically 2–3 μm long with a 0.5 μm diameter [3]. Its genome consists of a circular, double-stranded DNA molecule which is more than 1000 μm long and typically also has one or more plasmids, a number of which contain genes, approximately 4000–5000, with only 2000 of them being common among different strains [4]. Most *E. coli* strains are harmless; however, there are some pathogenic strains that are responsible for important public health issues [5].

E. coli strains were initially classified in four major phylogenetic groups, designated as A, B1, B2, and D [6]. However, three additional phylogenetic groups were later added—namely, C, E, and F—along with one cryptic clade, I, increasing the total number to eight [7]. More recently, one extra phylogenetic group (phylogroup G) was also characterized [8]. The phylo-group F represents a sister group to phylo-group B2, while phylo-group C is closely related to phylo-group B1 [9]. The phylo-group E is now also well recognized, with its best-known member being the O157:H7 [10].

E. coli constitutes an incredibly versatile and diverse species both genetically and morphologically, which can be further subdivided into the following categories: intestinal non-pathogenic commensal isolates, intestinal pathogenic isolates, and extraintestinal pathogenic *E. coli* (ExPEC) isolates. Extraintestinal pathogenic *E. coli* (ExPEC) belongs mainly to group B2 and, to a lesser extent, to group D, while intestinal commensal symbiotic isolates mainly belong to groups A and B1 [11].

Much like with other bacterial taxa, AMR in *E. coli* is considered a main public health threat, often observed in the form of multidrug resistance [12]. The mechanism of AMR development in *E. coli* may be intrinsic or acquired, i.e., located in the bacterial chromosome or routed by other bacteria, respectively [13]. Briefly, these include the enzymatic inactivation of the drug, the modification of the drug target, the setting of limitation mechanism in drug uptake, and the activation of the efflux pump to prohibit the insertion of the antibiotic through the cell membrane.

Resistance rates are generally relatively high worldwide, reaching 50, 60%, or higher levels, depending on the antibiotic agent [14]. In Greece, rates of resistant *E. coli* strains have been detected in human patients, varying approximately between 20 and 45% for the different antibiotics [15]. Concerning farm animals, recently, Papadopoulos et al. [16] determined very high levels of multidrug resistance in swine, reaching more than 80% of the tested samples.

Further, *E. coli* isolates can develop multidrug resistance to various antibiotics such as β-lactams, mainly through the production of β-lactamase (ESBL) and/or plasmid-mediated AmpC β-lactamases (AmpC) [17]. ESBLs confer resistance to the majority of β-lactams but especially to third-generation cephalosporins (such as cefotaxime, ceftriaxone, and ceftazimidime) and aztreonam, though not to carbapenems and cephamycins (cefotetan and cefoxitin) [18]. Bacteria exhibiting resistance towards β-lactams were first observed in humans, but since then, an increase in the detection of ESBL/AmpC-producing *E. coli* in animals, such as pigs [19], cattle, cats, dogs [20], fish [21] horses [22], and mainly broiler chickens [19,23], has been reported.

Colistin, also known as polymyxin E, was discovered in the 1940s and is a circular, polypeptide antibiotic produced by *Paenibacillus polymyxa* var. *colistinus*, with its compounds targeting the bacterial cell membrane, as it binds to the lipopolysaccharide (LPS) component of the outer membrane of the Gram-negative bacteria [24]. During the 1970s, there was a significant reduction in the clinical use of colistin due to its side effects, while during the 1980s, it was almost completely abandoned [25]. Nowadays, colistin is widely used in intensive poultry production; thus, the emergence of plasmid-mediated enzymatic resistance is a serious concern globally. Some of the genes associated with resistance are found in the plasmid, a feature that provides them mobility [26]. Mobilized resistance to colistin is increasing globally and represents a major threat to public health. In total, nine colistin resistance genes (*mcr-1* to *mcr-9*) and the variants of these genes have been described in Enterobacteriaceae [27]. These inferences have increased the level of public health concern associated with the spread of mobile colistin resistance and pointed out the necessity of extensive screenings in Enterobacteriaceae. Poultry and livestock represent a major reservoir for colistin resistance and the transmission of resistance genes [26,28]. Similarly, resistance to other antibiotics such as quinolones are also genetically associated and have been observed when point mutations occur in specific portions of *GyrA* and

ParC, known as the quinolone resistance-determining regions (QRDR) [29]. Generally, quinolone resistance has been reported in *E. coli* isolated from retail chicken products [30]. Furthermore, resistance to quinolones has emerged following their widespread use in poultry farms, and as a result, quinolone-resistant *E. coli* isolates can be spread through poultry production [31].

Greece is a country with a traditionally intensively developed poultry sector, which has played a particular role in the national economy [32]. However, data regarding the screening of resistant *E. coli* strains in Greek poultry are rather scarce. Hence, the aim of the present study was to estimate the *E. coli* resistance rates of broiler and laying hen farms by applying phenotypic and molecular identification methodologies.

2. Materials and Methods

2.1. Sample Collection and Isolation of Bacteria

Fecal samples were collected from 75 poultry farms of broilers and 16 farms of egg-producing hens during the years 2016–2019. The samples were received via a walk-through of the broiler or the hen house unit. The premises were traversed in such a way as to obtain representative samples of the entire ward. For this purpose, each ward was divided into 9 isomeric sections, from each of which a separate feces sample was obtained from the litter in the case of broiler chickens or from the manure removal belts located under the cages in the case of egg-producing hens.

For each visit, biosecurity measures were taken, including a plastic apron and disposable plastic foot pads. The broiler farms included in the study originated from the Region of Epirus (34 out of the 75), Central Macedonia (39 out of the 75), and Attiki (12 out of the 75), while 10 and 6 of the laying hen farms originated from Central Macedonia and Epirus, respectively. The flocks included in the study had an average number of 19,000 animals, while six chambers were sampled from each farm on average. From every flock, 9 fecal samples were collected from the chamber inside the poultry house. Fecal samples from the litter material were taken with a sterile cotton swab, which was stored in Stuart transport medium and transported to the laboratory within 24 h. In total, 954 samples were collected and were pooled per unit for each farm, finally forming 106 (90 broilers and 16 egg laying hen flocks) different sample pools that were included in the analysis, whereas from the 16 broiler farms, two different flocks were sampled.

On arrival at the laboratory, the fecal pool samples were directly inoculated in brain heart infusion broth (Oxoid, Basingstoke, UK) supplemented with ampicillin (10 mg/L). After overnight enrichment at 37 °C, a full loop from the enrichment culture was streaked onto the surface of selective Tryptone Bile X-glucuronide agar plates (TBX; Oxoid, Basingstoke, UK) and incubated at 44 °C for 24 to 48 h, as recommended by the manufacturer for the selective growth of *E. coli*. Plates containing blue/green colonies were counted as presumptive *E. coli* (ISO 16649-2:2001). Oxydase and indole tests were also performed to verify the identity of the bacterial strains as *E. coli*. Briefly, moistened filter papers were utilized, on which tetramethyl-ρ-phenylenediaminedihydrochloride (Merck, Darmstadt, Germany) was dropped. Afterwards, cultured cells were inoculated on these papers, and purple violet coloration indicated positive *E. coli*. For indole tests, Tryptone plates were inoculated with pure cultures and incubated at 37 °C for 24 h. Then, Kovács reagent (Merck, Darmstadt, Germany) was added, and positive *E. coli* were identified by the formation of a reddish color. *E. coli* presence was confirmed in all 106 pools.

2.2. Examination of Phenotypic Antimicrobial Susceptibility

Antimicrobial susceptibility testing was performed using the Kirby Bauer disc diffusion method on Mueller–Hinton agar plates (Merck, Darmstadt, Germany), according to the Clinical and Laboratory Standard Institute (CLSI) guidelines [33]. The isolates were tested for the following antibiotics: ampicillin (AMP; 10 µg), cefotaxime (CTX; 30 µg), cefoxitin (FOX; 30 µg), cefpodoxime (CPD; 10 µg), ceftazidime (CAZ; 30 mg), chloramphenicol (CHL; 30 mg), co-trimoxazole (STX; 1.25/23.75 µg), enrofloxacin (ENR; 5 µg), gentamicin (GMN;

10 μg), imipenem (IMP; 10 μg), nalidixic acid (NAL; 30 μg), streptomycin (SMN; 10 μg), and tetracycline (TET; 30 μg) (Oxoid Ltd., Basingstoke, UK). Colistin resistance was determined using the broth microdilution method, according to the European Committee on Antimicrobial Susceptibility Testing guidelines [34]. Results were interpreted with a resistance breakpoint of 2 μg/mL. Phenotypic characterization of the analyzed strains as susceptible, intermediate, or resistant was based on the breakpoint, i.e., the lowest concentration on which no bacterial growth was observed, according to [35,36].

The isolates were tested for ESBL production via a combination disk test (CDT), according to the CLSI guidelines. To perform a CDT, disks were used, including cefotaxime (30 μg), cefotaxime/clavulanic acid (30/10 μg), ceftazidime (30 μg), and ceftazidime/clavulanic acid (30/10 μg). The test was performed on Mueller–Hinton agar using a 0.5 McFarland inoculum, followed by incubation at 37 °C for 18 h. An increase in the diameter of the inhibition zone $\geq$ 5 mm in the presence of clavulanic acid is indicative of ESBL production.

The screening of strains for the production of AmpC β-lactamases was based on a resistance or reduced susceptibility to cefoxitin or imipenem, which acts as an inducer of antibiotic resistance. Cefotaxime (30 μg) and ceftazidime (30 μg) were placed around a strain and at a distance of 25 mm from the center of the disk. The test is positive when the repulsion of the edge of the zone of inhibition in cefotaxime or ceftazidime is induced on the side of the disc towards the inducer, according to Dunne and Hardin [36].

2.3. *E. coli Phylogeny*

Genomic DNA was extracted from 106 cultures using the NucleoSpin Microbial DNA Kit (Macherey-Nagel, Düren, Germany) according to the manufacturer's instructions. The concentration and purity of the eluted DNA were determined using a Q5000 microvolume spectrophotometer (Quawell, Thmorgan, Beijing, China). To assign the phylogeny of *E. coli* strains, all isolates were subjected to a polymerase chain reaction (PCR), targeting the *chuA* and *yjaA* genes, which can reliably identify the phylogenetic group of the *E. coli* strain, according to the methodology developed by Clermont et al. [34]. PCRs were performed in reactions with a total volume of 20 μL, containing 10 μL FastGene Taq 2X Ready Mix (NIPPON Genetics, Tokyo, Japan), 0.3 pmoL of each forward and reverse primer (Table 1), and distilled water up to the total volume. The conditions of each reaction were 95 °C for 3 min, 95 °C for 30 s, annealing temperature (Table 1) for 40 s, 72 °C for 45 s, and a final extension step at 72 °C for 5 min. The amplified products were examined by electrophoresis in agarose gel stained with ethidium bromide and photographed using a photo documentation system.

2.4. *Molecular Investigation of Antimicrobial Resistance Genes and Phylogeny*

The molecular investigation of resistance was characterized by targeting the colistin resistance genes *mcr-1* and *mcr-2*, the ESBL resistant *blaTEM* gene, the tetracycline resistant tet(X) gene, and the quinolone resistance *qnrA* gene. PCRs were performed as described in Section 2.3 using the primer pairs *mcr-1F–mcr-1R*, *mcr-2F–mcr-2R*, *blaTEM-F–blaTEM-R*, *tet(X)-F–tet(X)-R*, and *qnrA-F–qnrA-R*, as described in Yuan et al. [37], and the annealing temperatures in Table 1. In samples considered positive, the amplified gene fragment proceeded to purification. After purification of the PCR products using the commercial NucleoSpin Gel and PCR Clean up kit (Macherey-Nagel, Düren, Germany), the purified products were bidirectionally sequenced by applying the Sanger methodology in a Prism 3730XL automatic capillary sequencer from the company CeMIA (Larissa, Greece), using both forward and reverse primers.

2.5. *Statistical Analysis*

Statistical analysis of the data was carried out using the software package SPSS 23.0 (IBM Corporation, Armonk, NY, USA), using the inferential statistics method. A χ^2 test was applied to compare the phenotypic and molecular findings. All hypothesis testing was conducted at a significance level of $\alpha = 0.05$ ($p < 0.05$).

Table 1. Primers used for the amplification of the target genes.

Primer	Sequence (5′-3′)	Target Gene	Expected Band Size	Annealing Temperature	Reference
mcr-1F mcr-1F	AGTCCGTTTGTTCTTGTGGC AGATCCTTGGTCTCGGCTTG	*Colistin resistance gene 1*	320 bp	55 °C	[38]
mcr-2F mcr-2R	CAAGTGTGTTGGTCGCAGTT TCTAGCCCGACAAGCATACC	*Colistin resistance gene 2*	715 bp	55 °C	[38]
blaTEM-F blaTEM-R	CATTTCCGTGTCGCCCTTATTC CGTTCATCCATAGTTGCCTGAC	*Carbapenem resistance gene*	800 bp	60 °C	[39]
tet(X)-F tet(X)-R	GGAAACCGGCTAATGGCAT AATCCTACAAATGACAACGTCG	*Tetracycline resistance genes*	230 bp	55 °C	[40]
qnrA-F qnrA-R	AGAGGATTTCTCACGCCAGG TGCCAGGCACAGATCTTGAC	*Quinolones resistance gene*	580 bp	54 °C	[41]
ChuA.1 ChuA.2	GACGAACCAACGGTCAGGAT TGCCGCCAGTACCAAAGACA	*chuA*	279 bp	55 °C	[34]
YjaA.1 YjaA.2	TGAAGTGTCAGGAGACGCTG ATGGAGAATGCGTTCCTCAAC	*yjaA*	211 bp	55 °C	[34]
TspE4C2.1 TspE4C2.2	GAGTAATGTCGGGGCATTCA CGCGCCAACAAAGTATTACG	fragment TSPE4.C2	152 bp	55 °C	[34]

3. Results

3.1. Antimicrobial Susceptibility

From the 106 sample pools investigated, all of the obtained *E. coli* strains showed resistance to at least one antimicrobial substance. Specifically, 76.4% of the strains showed resistance to at least one of the examined β-lactam antibiotics, while 89.7% showed resistance to at least one quinolone.

The highest rates of resistance were observed in sulfamethoxazole, followed by nalidixic acid, tetracycline, piperacillin, streptomycin, and enrofloxacin. In contrast to the high rates of resistance to quinolones, resistance to third-generation cephalosporins was particularly low for aztreonam, ceftazidime, ceftriaxone, cefoxitin, and cefotaxime (Table 2). No strain showed resistance to imipenem. Eleven strains (10.3%) were resistant to colistin (MIC > 2 μg/L).

Table 2. Microbial resistance results of *E. coli* strains isolated from poultry farms.

Antimicrobial	Disk Composition	Thresholds (mm) Sensitive	Resistant	Resistant Strains–Broilers	Resistant Strains–Egg-Laying Hens	Resistant Strains–Total
Sulfamethoxazole	23.75 μg	≥15	≤11	83 (91.1%)	3 (18.8%)	86 (81.1%)
Nalidixic acid	30 μg	≥19	≤13	71 (78%)	7 (43.8%)	78 (73.6%)
Tetracycline	30 μg	≥15	≤11	68 (74.7%)	7 (43.8%)	75 (70.8%)
Piperacillin	100 μg	≥21	≤17	55 (60.4%)	12 (75%)	67 (63.2%)
Streptomycin	10 μg	≥15	≤11	69 (75.8%)	6 (3.8%)	75 (70.8%)
Enrofloxacin	5 μg	≥21	≤15	54 (59.3%)	10 (6.3%)	64 (60.3%)
Aztreonam	30 μg	≥21	≤17	1 (1.1%)	-	1 (1%)
Ceftazidime	10 μg	≥21	≤17	3 (3.3%)	-	3 (2.8%)
Ceftriaxone	30 μg	≥23	≤19	4 (4.4%)	-	4 (3.7%)
Cefoxitin	30 μg	≥18	≤14	5 (5.5%)	-	5 (4.7%)
Cefotaxime	30 μg	≥26	≤22	5 (5.5%)	-	5 (4.7%)
Imipenem	10 μg	≥23	≤19	-	-	0

3.2. Phenotypic Tests for the Detection of ESBLs and AmpC-β-Lactamases

We detected phenotypic evidence for both ESBL and AmpC β-lactamase production. In particular, phenotypic tests revealed that 13.6% and 10.2% of the isolates produced ESBL, while 2.7% and 1% produced AmpC b-lactamase, for broiler and laying hens, respectively.

3.3. Molecular Identification

All of the *E. coli* strains belonged to the B1 phylogenetic group. The *mcr-1* gene was detected in 22 out of the 106 isolates (Figure 1), all originating from broilers. The results were confirmed using Sanger sequencing, with the haplotype derived in complete homology with the reference sequence with the GenBank accession number OM839890, which was then used as a positive control. As far as the *mcr-2* gene is concerned, although a band was amplified after the PCR reaction, it failed to be confirmed after sequencing for both poultry groups. No other bands were amplified in any of the remaining examined genes. No statistical association was observed between the phenotypic and molecular data.

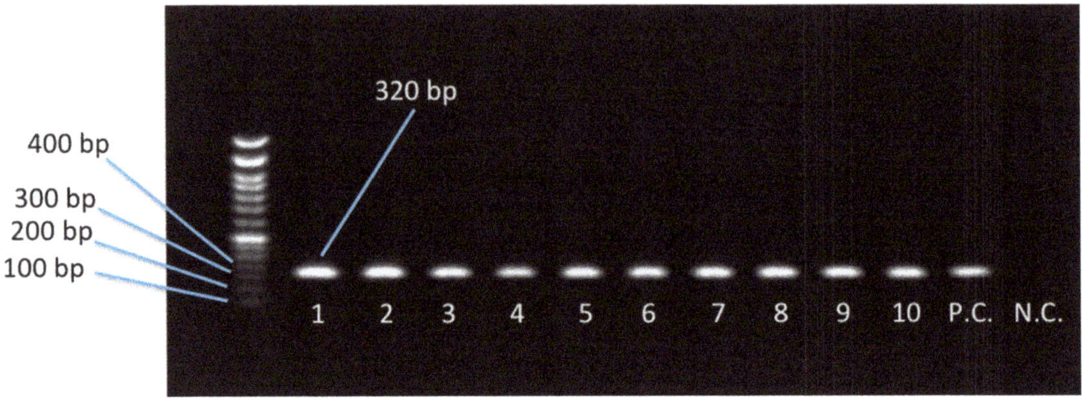

Figure 1. The PCR products of the 320 bp fragment after amplification with the *mcr-1*F–*mcr-1*R primer pair. Lanes 1–10: samples; P.C.: positive control, N.C.: negative control (no template DNA in the PCR).

4. Discussion

The animal production sector, and in particular poultry production, represents one potential source of multidrug-resistant bacteria, which possess plasmid-mediated resistance genes [42]. Under this prism, to the best of our knowledge, this is the first systematic study to assess the prevalence and the patterns of antimicrobial resistance in broilers and laying hens in Greece.

Increased rates of *E. coli*-resistant strains to quinolones isolated from farm animals have been previously reported [43], in line with the high resistance percentage to quinolones among strains both from broilers and laying hens detected in the current study. The above observations are in accordance with a report from the European Food Safety Authority (EFSA) for 2016 on antimicrobial resistance to microbial indicators in humans, food, and animals [44]. According to the EFSA, data collected from broilers among 30 countries on non-pathogenic *Escherichia coli* strains indicated higher resistance rates to quinolones, while resistance percentages to third-generation cephalosporins and colistin were lower. In laying hens, the same pattern was observed but with lower percentages, except for ceftazidime, cefotaxime, cefpodoxime, colistin, and sulfamethoxazole. Furthermore, other studies on poultry showed resistance levels to quinolone antibiotics ranging from 53% to 73% in the Czech Republic [45,46]. Nevertheless, the fact that, occasionally, resistance is only phenotypically observed, as was the case in the present study, indicates a phenotypic plasticity, in the sense that under environmental pressure such as the presence of antibiotics in microorganisms, resistant phenotypes are occasionally produced by non-resistant genotypes.

Data concerning laying hen poultry farms from Belgium, Germany, Italy, and Switzerland revealed lower resistance rates to quinolones, third-generation cephalosporins, and colistin than in the present study. More specifically, resistance to ciprofloxacin and nalidixic acid were 10.4% and 10.7, respectively; to cefpodoxime, 4.9%; and to colistin, 0.9% [47]. Similarly, resistance to quinolones was estimated at a level of 16.7%, while no resistance was found to colistin or cefpodoxime [48]. In Spain, resistance to ciprofloxacin and nalidixate were found to be 4.6% and 3%, respectively, while no resistance was revealed to colistin, ceftazidime, or cefotaxime [49]. The observed differences in antimicrobial resistance rates between broilers and laying hens may be due to the different antimicrobial substances used in each case.

Similarly, ESBL production from *E. coli* strains has been referred as a recent problem in poultry [50–52]. Although it is not very clear whether ESBL production from *E. coli* represents a major problem to the poultry sector, it might be a direct threat to public health. Keeping this in mind, the illustration of *E. coli* strains as producing ESBL/AmpC β-lactamases in the poultry sector in Greece is of high importance for β-lactam resistance screening.

In our study, none of the examined samples showed resistance to imipenem, which is most likely due to the prohibition of carbapenems in animals in Europe. A literature review covering the years 1980 to 2017 revealed that *E. coli* resistance to carbapenems remains low (<1%) among the European countries, while higher resistance rates are observed to carbapenems in Asian countries and Algeria, with the percentage resistance reaching 26% [53]. A study concerning laying hens (n = 276) in Germany showed an imipenem resistance of 1.8% [54]. These results are generally in agreement with our findings.

Furthermore, in the present study, the *mcr-1* gene was found to be present in broilers' *E. coli* isolates. Twenty-two samples (20.8%) from our survey tested positive. Colistin is an antimicrobial substance widely used in veterinary medicine to treat infections of the digestive tract, particularly in poultry and pigs. In 2015, the plasmid-mediated resistance gene *mcr-1* was first reported in poultry in China [55]. Since then, there has been an increase in research worldwide that has detected this specific gene in animals, in humans, and in foods of animal origin [56–60]. Thus far, it has mainly been detected in *E. coli* strains, as well as in other enterobacteroids belonging in the genera *Salmonella*, *Shigella*, *Klebsiella*, and *Enterobacter* [61]. Our results reveal that *E. coli* strains carrying the *mcr-1* gene are circulating in poultry farms in Greece, threatening the use of colistin as a last-resort antibiotic and highlighting the need for the surveillance of antibiotic resistance. Despite the limitation of our study regarding the detection of only one resistance gene, the fact that colistin is one of the main antibiotics for the treatment of serious human infections makes these results worth mentioning. The monitoring, collection, and presentation of all relevant information on the *mcr* gene at a global scale are of major importance in order to prevent public health threats and apply proper measures.

5. Conclusions

In conclusion, the occurrence of resistance to antimicrobials in *E. coli* in Greek poultry constitutes an issue of high importance. Antibiotic resistance to several antimicrobial substances was observed in high prevalence in *E. coli* strains. Concerning one of these substances, resistance was also reflected with the identification of a resistance gene. Although phenotypic resistance rates were high for various antibiotic compounds, the most noteworthy finding of the present study, from a public health point of view, is probably the identification of the *mcr-1* gene, which provides resistance to colistin. The circulation of this gene in Greek poultry poses risks for the transmission of resistance to other animals or to humans, as well. These results raise significant concerns regarding the use of antibiotics in Greek poultry, which traditionally plays an important role in the primary production sector. Based on these results, the proper use of antibiotics should be applied by the farm owners and operators.

Author Contributions: Conceptualization, E.P., E.N.S. and S.K.K.; methodology, A.X. and D.K.P.; software, M.V.A.; validation, G.A.D. and V.E.; formal analysis, K.V.P.; investigation, A.X. and I.A.G.; resources, E.P., E.N.S. and S.K.K.; data curation, M.V.A.; writing—original draft preparation, A.X. and M.V.A.; writing—review and editing, I.A.G., E.P. and E.N.S.; visualization, K.V.P.; supervision, E.P.; project administration, S.K.K.; funding acquisition, S.K.K. All authors have read and agreed to the published version of the manuscript.

Funding: This research received no external funding.

Institutional Review Board Statement: Not applicable.

Informed Consent Statement: Not applicable.

Data Availability Statement: Data is contained within the article.

Acknowledgments: This work is dedicated to the memory of Georgios Filiousis.

Conflicts of Interest: The authors declare no conflict of interest.

References

1. Silbergeld, E.K.; Graham, J.; Price, L.B. Industrial Food Animal Production, Antimicrobial Resistance, and Human Health. *Annu. Rev. Public Health* **2008**, *29*, 151–169. [CrossRef] [PubMed]
2. Dawadi, P.; Bista, S.; Bista, S. Prevalence of Colistin-Resistant *Escherichia coli* from Poultry in South Asian Developing Countries. *Vet. Med. Int.* **2021**, *2021*, 6398838. [CrossRef] [PubMed]
3. Skerman, V.B.D.; Sneath, P.H.A.; McGowan, V. Approved Lists of Bacterial Names. *Int. J. Syst. Evol. Microbiol.* **1980**, *30*, 225–420. [CrossRef]
4. Lukjancenko, O.; Wassenaar, T.M.; Ussery, D.W. Comparison of 61 Sequenced *Escherichia coli* Genomes. *Microb. Ecol.* **2010**, *60*, 708–720. [CrossRef] [PubMed]
5. World Health Organization. *E. coli*. 2018. Available online: http://www.who.int/mediacentre/factsheets/fs125/en/ (accessed on 7 December 2022).
6. Herzer, P.J.; Inouye, S.; Inouye, M.; Whittam, T.S. Phylogenetic distribution of branched RNA-linked multicopy single-stranded DNA among natural isolates of *Escherichia coli*. *J. Bacteriol.* **1990**, *172*, 6175–6181. [CrossRef] [PubMed]
7. Clermont, O.; Christenson, J.K.; Denamur, E.; Gordon, D.M. The Clermont *Escherichia coli* phylo-typing method revisited: Improvement of specificity and detection of new phylo-groups. *Environ. Microbiol. Rep.* **2013**, *5*, 58–65. [CrossRef]
8. Clermont, O.; Dixit, O.V.; Vangchhia, B.; Condamine, B.; Dion, S.; Bridier-Nahmias, A.; Denamur, E.; Gordon, D. Characterization and rapid identification of phylogroup G in *Escherichia coli*, a lineage with high virulence and antibiotic re-sistance potential. *Environ. Microbiol.* **2019**, *21*, 3107–3117. [CrossRef]
9. Clermont, O.; Olier, M.; Hoede, C.; Diancourt, L.; Brisse, S.; Keroudean, M.; Glodt, J.; Picard, B.; Oswald, E.; Denamur, E. Animal and human pathogenic *Escherichia coli* strains share common genetic backgrounds. *Infect. Genet. Evol.* **2011**, *11*, 654–662. [CrossRef]
10. Tenaillon, O.; Skurnik, D.; Picard, B.; Denamur, E. The population genetics of commensal *Escherichia coli*. *Nat. Rev. Microbiol.* **2010**, *8*, 207–217. [CrossRef]
11. Pitout, J.D.D. Extraintestinal Pathogenic *Escherichia coli*: A Combination of Virulence with Antibiotic Resistance. *Front. Microbiol.* **2012**, *3*, 9. [CrossRef]
12. Abdelwahab, G.E.; Ishag, H.Z.A.; Al Hammadi, Z.M.; Al Yammahi, S.M.S.; Mohd Yusof, M.F.B.; Al Yassi, M.S.Y.; Al Mansoori, A.M.A.; Al Hamadi, F.H.A.; Al Hamadi, I.A.S.; Hosani, M.A.A.A.; et al. Antibiotics Re-sistance in *Escherichia coli* Isolated from Livestock in the Emirate of Abu Dhabi, UAE, 2014–2019. *Int. J. Microbiol.* **2022**, *2022*, 3411560. [CrossRef] [PubMed]
13. Arbab, S.; Ullah, H.; Wang, W.; Zhang, J. Antimicrobial drug resistance against *Escherichia coli* and its harmful effect on animal health. *Vet. Med. Sci.* **2022**, *8*, 1780–1786. [CrossRef] [PubMed]
14. Daneman, N.; Fridman, D.; Johnstone, J.; Langford, B.J.; Lee, S.M.; MacFadden, D.M.; Mponponsuo, K.; Patel, S.N.; Schwartz, K.L.; Brown, K.A. Antimicrobial resistance and mortality following *E. coli* bacteremia. *Eclinicalmedicine* **2023**, *56*, 101781. [CrossRef]
15. Falagas, M.E.; Polemis, M.; Alexiou, V.G.; Marini-Mastrogiannaki, A.; Kremastinou, J.; Vatopoulos, A.C. Antimicro-bial resistance of *Esherichia coli* urinary isolates from primary care patients in Greece. *Med. Sci. Monit.* **2008**, *14*, CR75. [PubMed]
16. Papadopoulos, D.; Papadopoulos, T.; Papageorgiou, K.; Sergelidis, D.; Adamopoulou, M.; Kritas, S.K.; Petridou, E. Antimicrobial resistance rates in commensal *Escherichia coli* isolates from healthy pigs in Greek swine farms. *J. Hell. Vet. Med. Soc.* **2021**, *72*, 2909–2916.
17. Laube, H.; Friese, A.; von Salviati, C.; Guerra, B.; Käsbohrer, A.; Kreienbrock, L.; Roesler, U. Longitudinal Monitoring of Extended-Spectrum-Beta-Lactamase/AmpC-Producing *Escherichia coli* at German Broiler Chicken Fattening Farms. *Appl. Environ. Microbiol.* **2013**, *79*, 4815–4820. [CrossRef]
18. Cantón, R.; González-Alba, J.M.; Galán, J.C. CTX-M Enzymes: Origin and Diffusion. *Front. Microbiol.* **2012**, *3*, 110. [CrossRef]
19. Blanc, V.; Mesa, R.; Saco, M.; Lavilla, S.; Prats, G.; Miró, E.; Navarro, F.; Cortés, P.; Llagostera, M. ESBL- and plasmidic class C β-lactamase-producing *E. coli* strains isolated from poultry, pig and rabbit farms. *Vet. Microbiol.* **2006**, *118*, 299–304. [CrossRef]
20. Ewers, C.; Bethe, A.; Semmler, T.; Guenther, S.; Wieler, L. Extended-spectrum β-lactamase-producing and AmpC-producing *Escherichia coli* from livestock and companion animals, and their putative impact on public health: A global perspective. *Clin. Microbiol. Infect.* **2012**, *18*, 646–655. [CrossRef]
21. Jiang, H.-X.; Tang, D.; Liu, Y.-H.; Zhang, X.-H.; Zeng, Z.-L.; Xu, L.; Hawkey, P.M. Prevalence and characteristics of -lactamase and plasmid-mediated quinolone resistance genes in *Escherichia coli* isolated from farmed fish in China. *J. Antimicrob. Chemother.* **2012**, *67*, 2350–2353. [CrossRef]
22. Dierikx, C.M.; Van Duijkeren, E.; Schoormans, A.H.W.; Van Essen-Zandbergen, A.; Veldman, K.; Kant, A.; Huijsdens, X.W.; Van Der Zwaluw, K.; Wagenaar, J.A.; Mevius, D.J. Occurrence and characteristics of extended-spectrum-β-lactamase- and AmpC-producing clinical isolates derived from companion animals and horses. *J. Antimicrob. Chemother.* **2012**, *67*, 1368–1374. [CrossRef] [PubMed]

23. Dierikx, C.; van Essen-Zandbergen, A.; Veldman, K.; Smith, H.; Mevius, D. Increased detection of extended spectrum beta-lactamase producing *Salmonella enterica* and *Escherichia coli* isolates from poultry. *Vet. Microbiol.* **2010**, *145*, 273–278. [CrossRef]
24. Hancock, R.E.W.; Chapple, D.S. Peptide Antibiotics. *Antimicrob. Agents Chemother.* **1999**, *43*, 1317–1323. [CrossRef]
25. Biswas, S.; Brunel, J.-M.; Dubus, J.-C.; Reynaud-Gaubert, M.; Rolain, J.-M. Colistin: An update on the antibiotic of the 21st century. Expert Rev. *Anti-Infect. Ther.* **2012**, *10*, 917–934. [CrossRef]
26. Zhang, J.; Chen, L.; Wang, J.; Yassin, A.K.; Butaye, P.; Kelly, P.; Gong, J.; Guo, W.; Li, J.; Li, M.; et al. Molecular detection of colistin resistance genes (mcr-1, mcr-2 and mcr-3) in nasal/oropharyngeal and anal/cloacal swabs from pigs and poultry. *Sci. Rep.* **2018**, *8*, 3705. [CrossRef] [PubMed]
27. Borowiak, M.; Baumann, B.; Fischer, J.; Thomas, K.; Deneke, C.; Hammerl, J.A.; Szabo, I.; Malorny, B. Development of a Novel mcr-6 to mcr-9 Multiplex PCR and Assessment of mcr-1 to mcr-9 Occurrence in Colistin-Resistant *Salmonella enterica* Isolates From Environment, Feed, Animals and Food (2011–2018) in Germany. *Front. Microbiol.* **2020**, *11*, 80. [CrossRef] [PubMed]
28. Hoelzer, K.; Wong, N.; Thomas, J.; Talkington, K.; Jungman, E.; Coukell, A. Antimicrobial drug use in food-producing animals and associated human health risks: What, and how strong, is the evidence? *BMC Vet. Res.* **2017**, *13*, 211. [CrossRef]
29. Ruiz, J. Mechanisms of resistance to quinolones: Target alterations, decreased accumulation and DNA gyrase protection. *J. Antimicrob. Chemother.* **2003**, *51*, 1109–1117. [CrossRef]
30. Johnson, J.R.; Kuskowski, M.A.; Smith, K.; O'bryan, T.T.; Tatini, S. Antimicrobial-Resistant and Extraintestinal Pathogenic *Escherichia coliin* Retail Foods. *J. Infect. Dis.* **2005**, *191*, 1040–1049. [CrossRef]
31. Seo, K.; Lee, Y. Prevalence and characterization of plasmid-mediated quinolone resistance determinants qnr and aac (6')-Ib-cr in ciprofloxacin-resistant *Escherichia coli* isolates from commercial layer in Korea. *J. Microbiol. Biotechnol.* **2020**, *30*, 1180–1183. [CrossRef]
32. Dotas, V.; Gourdouvelis, D.; Hatzizisis, L.; Kaimakamis, I.; Mitsopoulos, I.; Symeon, G. Typology, Structural Char-acterization and Sustainability of Integrated Broiler Farming System in Epirus, Greece. *Sustainability* **2021**, *13*, 13084. [CrossRef]
33. CLSI. *Performance Standards for Antimicrobial Disk Susceptibility Tests*, 11th ed.; Approved Standard; Clinical and Laboratory Standards Institute (CLSI): Wayne, PA, USA, 2012.
34. Clermont, O.; Bonacorsi, S.; Bingen, E. Rapid and Simple Determination of the *Escherichia coli* Phylogenetic Group. *Appl. Environ. Microbiol.* **2000**, *66*, 4555–4558. [CrossRef] [PubMed]
35. The European Committee on Antimicrobial Susceptibility Testing. Breakpoint Tables for Interpretation of MICs and Zone Di-ameters. Version 6.0. 2016. Available online: http://www.eucast.org (accessed on 7 December 2022).
36. Dunne, W.M.; Hardin, D.J. Use of several inducer and substrate antibiotic combinations in a disk approximation assay format to screen for AmpC induction in patient isolates of *Pseudomonas aeruginosa*, *Enterobacter* spp., *Citrobacter* spp., and *Serratia* spp. *J. Clin. Microbiol.* **2005**, *43*, 5945–5949. [CrossRef]
37. Yuan, J.; Wang, X.; Shi, D.; Ge, Q.; Song, X.; Hu, W.; Wei, D.; Ge, C.; Li, X.; Hu, C. Extensive antimicrobial resistance and plasmid-carrying resistance genes in mcr-1-positive *E. coli* sampled in swine, in Guangxi, South China. *BMC Vet. Res.* **2021**, *17*, 86. [CrossRef] [PubMed]
38. Rebelo, A.R.; Bortolaia, V.; Kjeldgaard, J.S.; Pedersen, S.K.; Leekitcharoenphon, P.; Hansen, I.M.; Guerra, B.; Malorny, B.; Borowiak, M.; Hammerl, J.A.; et al. Multiplex PCR for detection of plasmid-mediated colistin resistance determinants, mcr-1, mcr-2, mcr-3, mcr-4 and mcr-5 for surveillance purposes. *Eurosurveillance* **2018**, *23*, 17-00672. [CrossRef]
39. Dallenne, C.; Da Costa, A.; Decre, D.; Favier, C.; Arlet, G. Development of a set of multiplex PCR assays for the detection of genes encoding important beta-lactamases in Enterobacteriaceae. *J. Antimicrob. Chemother.* **2010**, *65*, 490–495. [CrossRef]
40. He, T.; Wang, R.; Liu, D.; Walsh, T.R.; Zhang, R.; Lv, Y.; Ke, Y.; Ji, Q.; Wei, R.; Liu, Z.; et al. Emergence of plasmid-mediated high-level tigecycline resistance genes in animals and humans. *Nat. Microbiol.* **2019**, *4*, 1450–1456. [CrossRef]
41. Cattoir, V.; Poirel, L.; Rotimi, V.; Soussy, C.-J.; Nordmann, P. Multiplex PCR for detection of plasmid-mediated quinolone resistance qnr genes in ESBL-producing enterobacterial isolates. *J. Antimicrob. Chemother.* **2007**, *60*, 394–397. [CrossRef]
42. Leverstein-van Hall, M.A.; Dierikx, C.M.; Cohen Stuart, J.; Voets, G.M.; van den Munckhof, M.P.; van Essen-Zandbergen, A.; Platteel, T.; Fluit, A.C.; van de Sande-Bruinsma, N.; Scharinga, J.; et al. Dutch patients, retail chicken meat and poultry share the same ESBL genes, plasmids and strains. *Clin. Microbiol. Infect.* **2011**, *17*, 873–880. [CrossRef]
43. Webber, M.; Piddock, L.J. Quinolone resistance in *Escherichia coli*. *Vet. Res.* **2001**, *32*, 275–284. [CrossRef]
44. EFSA (European Food Safety Authority); ECDC (European Centre for Disease Prevention and Control). The European Union summary report on antimicrobia resistance in zoonotic and indicator bacteria from humans, animals and food in 2016. *EFSA J.* **2018**, *16*, 5182–5270.
45. Skočková, A.; Koláčková, I.; Bogdanovičová, K.; Karpíšková, R. Characteristic and antimicrobial resistance in *Escherichia coli* from retail meats purchased in the Czech Republic. *Food Control.* **2015**, *47*, 401–406. [CrossRef]
46. Hricová, K.; Röderová, M.; Pudová, V.; Hanulík, V.; Halová, D.; Julínková, P.; Dolejská, M.; Papoušek, I.; Bardoň, J. Quinolone-resistant *Escherichia coli* in Poultry Farming. *Central Eur. J. Public Health* **2017**, *25*, 163–167. [CrossRef]
47. VAN Hoorebeke, S.; VAN Immerseel, F.; Berge, A.C.; Persoons, D.; Schulz, J.; Hartung, J.; Harisberger, M.; Regula, G.; Barco, L.; Ricci, A.; et al. Antimicrobial resistance of *Escherichia coli* and *Enterococcus faecalisin* housed laying-hen flocks in Europe. *Epidemiol. Infect.* **2010**, *139*, 1610–1620. [CrossRef] [PubMed]
48. Harisberger, M.; Gobeli, S.; Hoop, R.; Dewulf, J.; Perreten, V.; Regula, G. Antimicrobial Resistance in Swiss Laying Hens, Prevalence and Risk Factors. *Zoonoses Public Health* **2011**, *58*, 377–387. [CrossRef] [PubMed]

49. Moreno, M.A.; García-Soto, S.; Hernández, M.; Bárcena, C.; Rodríguez-Lázaro, D.; Ugarte-Ruíz, M.; Domínguez, L. Day-old chicks are a source of antimicrobial resistant bacteria for laying hen farms. *Vet. Microbiol.* **2019**, *230*, 221–227. [CrossRef] [PubMed]
50. Tan, H.S.; Yan, P.; Agustie, H.A.; Loh, H.S.; Rayamajhi, N.; Fang, C.M. Characterisation of ESBL/AmpC-Producing Enterobacteriaceae isolated from poultry farms in Peninsular Malaysia. *Lett. Appl. Microbiol.* **2023**, *76*, ovac044.
51. Falgenhauer, L.; Imirzalioglu, C.; Oppong, K.; Akenten, C.W.; Hogan, B.; Krumkamp, R.; Poppert, S.; Levermann, V.; Schwengers, O.; Sarpong, N.; et al. Detection and Characterization of ESBL-Producing *Escherichia coli* From Humans and Poultry in Ghana. *Front. Microbiol.* **2019**, *9*, 3358. [CrossRef]
52. Chai, M.H.; Sukiman, M.Z.; Jasmy, N.; Zulkifly, N.A.; Yusof, N.A.S.M.; Mohamad, N.M.; Ariffin, S.M.Z.; Ghazali, M.F. Molecular Detection and Antibiogram of ESBL-Producing and Carbapenem-Resistant *Escherichia coli* from Rabbit, Swine, and Poultry in Malaysia. *Trop. Anim. Sci. J.* **2022**, *45*, 16–23. [CrossRef]
53. Köck, R.; Daniels-Haardt, I.; Becker, K.; Mellmann, A.; Friedrich, A.W.; Mevius, D.; Schwarz, S.; Jurke, A. Carbapenem-resistant Enterobacteriaceae in wildlife, food-producing, and companion animals: A systematic review. *Clin. Microbiol. Infect.* **2018**, *24*, 1241–1250. [CrossRef]
54. Schwaiger, K.; Schmied, E.-M.V.; Bauer, J. Comparative Analysis of Antibiotic Resistance Characteristics of Gram-negative Bacteria Isolated from Laying Hens and Eggs in Conventional and Organic Keeping Systems in Bavaria, Germany. *Zoonoses Public Health* **2008**, *55*, 331–341. [CrossRef]
55. Liu, Y.-Y.; Wang, Y.; Walsh, T.R.; Yi, L.-X.; Zhang, R.; Spencer, J.; Doi, Y.; Tian, G.; Dong, B.; Huang, X.; et al. Emergence of plasmid-mediated colistin resistance mechanism MCR-1 in animals and human beings in China: A microbiological and molecular biological study. *Lancet Infect. Dis.* **2016**, *16*, 161–168. [CrossRef] [PubMed]
56. Quesada, A.; Ugarte-Ruiz, M.; Iglesias, M.R.; Porrero, M.C.; Martínez, R.; Florez-Cuadrado, D.; Campos, M.J.; García, M.; Píriz, S.; Sáez, J.L.; et al. Detection of plasmid mediated colistin resistance (MCR-1) in *Escherichia coli* and *Salmonella enterica* isolated from poultry and swine in Spain. *Res. Vet. Sci.* **2016**, *105*, 134–135. [CrossRef] [PubMed]
57. Anjum, M.F.; Duggett, N.A.; AbuOun, M.; Randall, L.; Nunez-Garcia, J.; Ellis, R.J.; Rogers, J.; Horton, R.; Brena, C.; Williamson, S.; et al. Colistin resistance in *Salmonella* and *Escherichia coli* isolates from a pig farm in Great Britain. *J. Antimicrob. Chemother.* **2016**, *71*, 2306–2313. [CrossRef] [PubMed]
58. Cannatelli, A.; Giani, T.; Antonelli, A.; Principe, L.; Luzzaro, F.; Rossolini, G.M. First Detection of the mcr-1 Colistin Resistance Gene in *Escherichia coli* in Italy. *Antimicrob. Agents Chemother.* **2016**, *60*, 3257–3258. [CrossRef]
59. Zurfluh, K.; Klumpp, J.; Nüesch-Inderbinen, M.; Stephan, R. Full-Length Nucleotide Sequences of mcr-1 -Harboring Plasmids Isolated from Extended-Spectrum-β-Lactamase-Producing *Escherichia coli* Isolates of Different Origins. *Antimicrob. Agents Chemother.* **2016**, *60*, 5589–5591. [CrossRef]
60. Yang, Y.-Q.; Li, Y.-X.; Song, T.; Yang, Y.-X.; Jiang, W.; Zhang, A.-Y.; Guo, X.-Y.; Liu, B.-H.; Wang, Y.-X.; Lei, C.-W.; et al. Colistin Resistance Gene mcr-1 and Its Variant in *Escherichia coli* Isolates from Chickens in China. *Antimicrob. Agents Chemother.* **2017**, *61*, e01204-16. [CrossRef]
61. Schwarz, S.; Johnson, A.P. Transferable resistance to colistin: A new but old threat. *J. Antimicrob. Chemother.* **2016**, *71*, 2066–2070. [CrossRef]

sustainability

MDPI

Article

Comparative Study between Dietary Nanoelemental, Inorganic, and Organic Selenium in Broiler Chickens: Effects on Meat Fatty Acid Composition and Oxidative Stability

Elisavet Giamouri, Efstathios Fortatos, Athanasios C. Pappas and George Papadomichelakis *

Laboratory of Nutritional Physiology and Feeding, Department of Animal Science, School of Animal Biosciences, Agricultural University of Athens, 75 Iera Odos Str., 118 55 Athens, Greece; egiamouri@aua.gr (E.G.); s.forta@yahoo.com (E.F.); apappas@aua.gr (A.C.P.)

* Correspondence: gpapad@aua.gr; Tel.: +30-210-529-4434

Abstract: The present study investigated the impact of dietary supplementation with nano-elemental, inorganic, and organic selenium (Se) on the Se content, fatty acid (FA) composition, and oxidative stability of meat in 150 one-day-old broiler chickens. The broiler chickens were allotted into three groups: control (C), SS+SY, and SeNP. The C group received a control diet without any added Se, while the SS+SY and SeNP groups were fed diets containing 0.4 mg Se/kg from a combination of sodium selenite and selenium yeast (SS+SY at a 1:1 ratio) or elemental Se nanoparticles (SeNP), respectively. Breast meat samples were collected from 10 broiler chickens per diet group (2 per replicate) at 42 days of age for the analysis of Se content, FA composition, and oxidative stability. The findings of the study revealed that the Se levels in the breast tissue significantly increased ($p < 0.05$) and the concentrations of malondialdehyde (MDA), a marker of oxidative stress, decreased ($p < 0.05$) with the inclusion of SS+SY and SeNP in the diet. Furthermore, the levels of 22:6n − 3 (docosahexaenoic acid) and total n − 3 FA significantly increased ($p < 0.05$) in the breast meat of broiler chickens supplemented with SeNP compared to the C and SS+SY groups. In conclusion, both dietary supplementation with SeNP and SS+SY had a positive impact on the Se content and oxidative stability of the breast meat. However, SeNP supplementation resulted in a more desirable modification of the FA composition. These findings suggest that SeNP may offer a sustainable alternative to traditional forms of Se supplementation.

Keywords: breast; broiler chickens; fatty acids; oxidative stability; selenium nanoparticles; sodium selenite; selenium yeast

Citation: Giamouri, E.; Fortatos, E.; Pappas, A.C.; Papadomichelakis, G. Comparative Study between Dietary Nanoelemental, Inorganic, and Organic Selenium in Broiler Chickens: Effects on Meat Fatty Acid Composition and Oxidative Stability. *Sustainability* **2023**, *15*, 9762. https://doi.org/10.3390/su15129762

Academic Editors: George K. Symeon and Vassilios Dotas

Received: 15 May 2023
Revised: 12 June 2023
Accepted: 15 June 2023
Published: 19 June 2023

1. Introduction

The broiler chicken industry continuously strives to enhance the quality of meat by increasing its polyunsaturated fatty acid (PUFA) content, specifically focusing on long-chain n − 3 polyunsaturated fatty acids, as they play a crucial role in preventing cardiovascular diseases in humans [1–3]. However, this high PUFA content makes broiler meat susceptible to oxidation, which can have detrimental effects on taste, aroma (rancidity), storage period, and the nutritional value of meat and meat products [4]. Consequently, the preservation of oxidative stability and the extension of shelf life for broiler meat have become prominent areas of focus in poultry research, particularly as meat is predominantly sold in packaged forms in today's market. In order to prevent oxidation, antioxidants are added to broiler feeds to maintain the balance of lipids and ensure the oxidative stability of meat [5]. The inclusion of dietary selenium (Se) has long been recognized as a strategy to reduce peroxidative damage to polyunsaturated fatty acids (PUFA) and modulate fatty acid synthesis in animal tissues [6,7]. Se is typically added to animal feed in either inorganic or organic forms, each with their own advantages and disadvantages. Inorganic forms include inorganic Se salts, with sodium selenite being the most commonly used. Although these forms are cost-effective, the accumulation of Se in tissues and antioxidant activity can vary significantly [5]. On the other hand, organic Se, such as selenomethionine or Se-enriched yeast, is more efficiently incorporated into tissues and exhibits higher antioxidant activity, but it comes at a higher cost [8]. Both forms have a narrow margin between beneficial and toxic effects [9–11], leading to limitations in dietary Se supplementation in Europe to ensure feed safety [12]. However, there is a widespread concern in the animal industry that diets following current recommendations may be deficient in Se, not providing animals with adequate levels to meet the demands of intensive rearing conditions [13], consequently affecting broiler growth performance and meat quality [14]. This situation can have adverse effects on the sustainability of broiler meat production, particularly considering the increasing consumption of chicken meat compared to pork and beef. Hence, current research focuses on more sustainable alternatives with potentially higher bioavailability, bioactivity, and lower toxicity compared to commonly used inorganic and organic Se forms. The utilization of selenium nanoparticles (SeNP) has emerged as a promising approach in this regard, drawing significant attention from researchers [15–17]. SeNP refers to inorganic selenium nanoparticles that possess unique physicochemical properties, including low toxicity, making them an attractive choice for scientific investigation. These nanoparticles exhibit enhanced antioxidative activity, improved selenium absorption and retention [18], and are considered less harmful compared to other forms of selenium [19]. Studies have shown that SeNP can upregulate selenoenzymes similar to inorganic and organic forms but with reduced toxicity [20,21]. In broiler chickens, the favorable effects of SeNP on growth, oxidative stress, and selenium accumulation in tissues have already been demonstrated [22–24]. Another advantage of SeNP is their facile synthesis using sustainable and eco-friendly methods, such as biological reduction of selenite to oxyanions [25] or chemical approaches [26–28]. While the impact of dietary selenium (Se) supplementation on the fatty acid (FA) composition of meat is well-established [29–35], there is limited evidence regarding the effect of Se nanoparticles (SeNP) on the FA profile of broiler chicken meat [14]. Consequently, further investigation in this area would provide valuable insights into the potential of SeNP as sustainable alternatives to conventional Se forms. Therefore, the aim of the present study was to compare the effects of different dietary Se sources, namely sodium selenite, Se yeast, and SeNP, with a specific focus on the FA profile and oxidative stability of broiler chicken meat.

2. Materials and Methods

2.1. Animals, Diets, and Experimental Procedures

One hundred and fifty 1-day-old Ross 300 broiler chickens were purchased from a commercial hatchery. Upon arrival at the experimental facilities of the Agricultural University of Athens, the broiler chickens were randomly allotted into 3 dietary treatments, namely control (C), SS+SY, and SeNP (5 replicate pens/treatment, 10 chickens/pen), and were fed three different diets: (a) a basal diet, without any added Se (treatment C), (b) a basal diet with 0.4 mg (0.2 mg from sodium selenite, and 0.2 mg from selenium yeast) added Se/kg (treatment SS+SY), and (c) a basal diet 0.4 mg added Se/kg from elemental Se nanoparticles stabilized in chitosan (treatment SeNP). The SeNP were synthesized in our laboratory according to earlier methods [26] and their physicochemical properties have been previously assessed [36]. Sodium selenite was commercial product (anhydrous powder 99% minimum purity, Alfa Aesar, Kandel, Germany). The selenium yeast was also a commercial product in the form of Sel-Plex® (Alltech Inc., Nicholasville, KY, USA). The combinations of dietary Se forms used in this study were chosen in accordance with European recommendations [12]. These recommendations limit the addition of organic Se to 0.2 mg/kg of the diet and the total dietary Se to 0.5 mg/kg. In a typical non-supplemented diet for broiler chickens, the endogenous Se content is approximately 0.1 mg/kg. Therefore, in this study, we added 0.4 mg of Se from each tested Se form in the SS+SY and SeNP treatments to ensure that the total dietary Se content did not exceed the allowed limit of 0.5 mg/kg. The detailed ingredient, chemical, and fatty acid composition, and the Se content of the diets are given in Table 1.

Table 1. Ingredient and chemical composition of the basal diets (g/kg as fed basis), fatty acid composition (% of total fatty acids), and selenium (Se) content of the experimental diets.

	Basal Diets		
	Starter (0–10 d)	**Grower (11–24 d)**	**Finisher (25–42 d)**
Ingredient			
Maize	485	521	576
Soybean meal, 450 g CP/kg	428	390	334
Soybean oil	44.7	51.7	56.0
Monocalcium phosphate	14.3	12.3	10.6
Limestone	14.1	12.8	11.6
Sodium chloride	4.0	4.0	4.0
DL-methionine, 99%	3.6	3.1	2.8
L-lysine HCl, 80%	2.5	1.7	1.8
L-threonine	1.0	0.7	0.4
Premix [1]	2.0	2.0	2.0
Choline	0.8	0.7	0.8
Calculated chemical composition			
Dry matter	880	880	880
Crude protein	230	215	195
Ether extract	69	77	82
Lysine	14.4	12.9	11.6
Methionine + cystine	10.8	9.9	9.1
Threonine	9.7	8.8	7.8
Calcium	9.6	8.7	7.8
Available phosphorus	4.8	4.4	3.9
Metabolizable energy, MJ/kg	12.6	13.0	13.4
Fatty acid composition			
12:0	Not determined	Not determined	0.16
14:0			0.62
15:0			0.13
C16:0			27.33
C16:1			0.34
17:0			0.17
C18:0			19.98
C18:1n − 7			11.90
C18:2n − 6			30.10
C18:3n − 3			4.15
C20:0			0.49
C20:1n − 9			0.43
C20:2n − 6			0.06
C22:0			0.50
C22:1			0.36
C23:0			0.15
C24:0			0.35
C24:1			0.08

		Se content (mg/kg)	
	Experimental diets	Added [2]	Determined [3]
C	Se not determined	-	0.117 ± 0.020
SS+SY		0.40	0.492 ± 0.049
SeNP		0.40	0.488 ± 0.045

[1] Premix supplied per kg of diet: vitamin A, 10,000 IU; vitamin D3, 5000 IU; vitamin E, 75 mg; vitamin K3, 6.25 mg; thiamine, 3.25 mg; riboflavin, 8 mg; pyridoxamine, 5.25 mg; vitamin B12, 0.0275 mg; niaciamide, 55 mg; D-panthenol, 14 mg; folic acid, 2 mg; biotin, 0.2 mg; I, 1.25 mg; Fe, 20 mg; Mn, 120 mg; Cu, 16 mg; Zn, 110 mg. The premix did not contain selenium. [2] Se was added as: (a) sodium selenite and Se yeast (Sel-Plex®, Alltech Inc., Nicholasville, KY, USA) at 1:1 ratio in diet SS+SY, (b) nanoelemental Se (selenium nanoparticle-loaded chitosan microspheres) in diet SeNP; control (C) diet did not contain any supplemental Se apart from that naturally occurring in the raw materials. [3] Values represent the average of 4 samples per diet ± standard deviation.

The trial lasted for 42 days. The experimental protocol (housing, handling, care., and slaughter procedures) was approved (no. 13/16-03-2021) by the Bioethics Committee of the Agricultural University of Athens (AUA). Up to the 10th day of age, the broilers were fed a starter diet and thereafter a grower diet to the 24th day and a finisher diet to the 42nd of age. Broilers had free access to feed and water throughout the experiment. Each of the starter, grower, and finisher diets contained the same level of Se added according to the experimental treatment (Table 1). The lighting program was controlled and stocking density was in accordance with the EU legislation. On day 42 of the experiment, breast samples from the *Pectoralis major* (PM) muscle were collected from 10 broilers per treatment (2/replicate), vacuum packed, and stored at −20 °C until analyses. The right half of PM was used for Se concentration and FA composition and the left half for lipid oxidation determination.

2.2. Determination of Se Content

Selenium in feed and meat samples was analyzed by atomic absorption spectrometry (Agilent 240FS AA; Santa Clara, CA, USA) according to Pappas et al. [37]. Briefly, 0.50 g of feed or meat were digested in 10 mL of nitric acid (65% *w*/*v*, Suprapur; Merck, Germany) in a microwave-accelerated digestion system (CEM, Mars X-Press, Matthews, NC, USA). The power was ramped from 100 to 1200 W within 20 min and maintained to 1200 W for 15 min to obtain a maximum temperature of 200 °C. After cooling, the digested samples were filtered using disposable syringe filters (Chromafil, Macherey-Nagel, Germany) and were treated with hydrochloric acid solution (6 M) to reduce selenate to selenite prior to atomic absorption analysis. High purity standards were used to prepare the calibration standard solutions. For vapor generation, a reductant agent (sodium borohydride 0.6% *w*/*v*) was combined with sodium hydroxide (0.5% *w*/*v*) and hydrochloric acid (10 M) solutions. Two standard reference materials (RM8414 and RM1577c, LGC Standards Promochem, Wesel, Germany) were used to evaluate the analytical accuracy of the procedure.

2.3. Determination of Iron-Induced Lipid Oxidation in Meat

Iron-induced (via Fenton reaction; Fe^{2+}/H_2O_2) lipid oxidation was determined according to Tereninto et al. [38]. Briefly, breast tissues sample (2 g) were homogenized (X 1000D homogenizer; CAT, M. Zipperer GmbH, Ballrechten-Dottingen, Germany) in an ice bath with 20 mL of potassium chloride (KCl) buffer solution (0.15 M, pH 7.2) for 1 min at 12,000 rpm. The homogenate was centrifuged (2000× *g* for 10 min) while kept cool (at 4 °C) (Heraeus Biofuge Stratos, Langenselbold, Germany). Then, 0.5 mL of the supernatant was mixed with 0.5 mL of KCl buffer solution and 30 µL of butylated hydroxytoluene (BHT, 3 mM). Another 5 mL were incubated (37 °C) in a shaking water bath in the presence of 5 mL of iron sulphate (0.5 mM) and 50 µL of hydrogen peroxide (1 mM) for 30, 120, and 300 min. At the end of each incubation time, 1 mL was taken, in which 30 µL of 3 mM BHT were added to stop the oxidation reaction. Afterwards, the homogenate was incubated with 1 mL of a mixture containing 2-thiobarbituric acid (TBA) and trichloracetic acid (TCA) (35 mM TBA and 10% TCA in 125 mM HCl) in a boiling water bath for 30 min. After cooling the samples down to room temperature, the pink chromogen was extracted with 4 mL of n-butanol and obtained by centrifugation at 3000× *g* for 10 min (Heraeus Biofuge Stratos, Langenselbold, Germany). The absorbance of the supernatant was measured at 535 nm. The concentration of malondialdehyde (MDA) was calculated using the molar extinction coefficient of the MDA (156,000 M^{-1} cm^{-1}). Results were expressed as mg MDA per kg of wet meat.

2.4. Determination of Fatty Acid Composition

The FA of diet (samples milled through 1 mm screen; CT 293 CyclotecTM, Foss, Denmark) and meat samples were extracted and methylated directly [39]. Briefly, 1 (±0.05) g were hydrolyzed (1.5 h, 55 °C) in methanolic potassium hydroxide solution (1 N) with 0.5 mg of tridecanoic acid (C13:0) as internal standard. The free FA were methylated by sulphuric acid catalysis (24 N H_2SO_4) for 1.5 h at 55 °C. Subsequently, 3 mL of n-hexane were added and the reaction tube was vortex-mixed and centrifuged at 1100× g. The supernatant n-hexane layer containing the FA methyl esters was obtained in gas chromatography vials and kept at −20 °C, until analyzed on an Agilent 6890N gas chromatograph with a 20 m × 0.18 mm × 0.20 µm capillary column (DB-FastFame, Agilent Technologies, J&W GC columns, Santa Clara, CA, USA) and a flame ionization detector (FID). The initial oven temperature was set at 80 °C. After 0.5 min it was increased to 175 °C (rate 65 °C/min), then to 185 °C (rate 10 °C/min) and held for 0.5 min, and finally to 230 °C (rate 7 °C/min) and held for 2 min. Hydrogen was used as carrier gas. The front inlet split ratio and temperature were set at 50:1 and 250 °C, respectively. The FID temperature was constantly at 26 °C and the flow of hydrogen, air, and make-up gas (helium) were set at 40, 400, and 25 mL/min, respectively. The FA were identified by comparison with standards (FAME 37 Component and PUFA no.2; Sigma-Aldrich Co., Supelco, IL, USA) and were quantified using the known amount of internal standard (C13:0) added prior to hydrolysis. Total weights of FA (mg/100 g) in diets were calculated as the sum of areas for all FA peaks compared to area for 0.5 mg internal standard. Individual FA were expressed as % by weight of total FA.

2.5. Statistical Analysis

The IBM SPSS Statistics 23.0 [40] software was used for statistical analysis. Data are presented as means ± standard error (SEM). Prior to analysis, data were tested for normality using Kolmogorov–Smirnov's test. A two-step approach for transforming non-normally distributed variables to become normally distributed [41] was followed. Normally distributed and transformed data were analyzed by a one-way (diet) ANOVA, and differences between treatments were evaluated by carrying out Tukey's *post-hoc* tests.

To assess whether samples can be distinguished according to the diet (Se form) using the muscle fatty acids as predictors, a discriminant analysis was performed, which was followed by a stepwise discriminant analysis to identify the fatty acids which were responsible for the discrimination observed. Wilk's lambda (λ) criterion was used for selecting discriminant variables. Statistical significance was set at $p < 0.05$ for all tests.

3. Results

3.1. Growth Performance

Average daily feed intake (ADFI), average daily weight gain (ADWG), and feed conversion ratio (FCR) of broiler chickens fed the diets supplemented with 0.4 mg Se/kg from SS+SY and SeNP did not differ from those of the control ones. ADFI was 126, 124, and 120 g, ADWG was 78, 78 and 74 g in C, SS+SY, and SeNP fed broilers, respectively. As a result, the FCR was 1.62, 1.61, and 16.2 for C, SS+SY, and SeNP broiler chickens.

3.2. Breast Tissue Se Content

Muscle Se content was elevated ($p < 0.001$) by 164% and 169% in the broiler chickens fed the diets supplemented with 0.4 mg Se/kg from SS+SY and SeNP, respectively, as compared with the control ones (Figure 1). No differences between SS+SY and SeNP fed broiler chickens were observed.

3.3. Breast Tissue Malondialdehyde Content

The breast tissue MDA contents did not differ between C, SS+SY, and SeNP fed broiler chickens at the onset of iron-induced oxidation (0 min). Thereafter, large amounts of MDA were produced in breast tissue (Figure 2). No differences in the breast MDA content between treatments were found at 30 and 120 min after the induction of oxidation. However, 300 min after the onset of oxidation the MDA concentrations were greater ($p < 0.05$) in the breast of broiler chickens fed the control diet in comparison with those fed the SS+SY and SeNP diets, whereas no difference between SS+SY and SeNP treatments was observed at any time point.

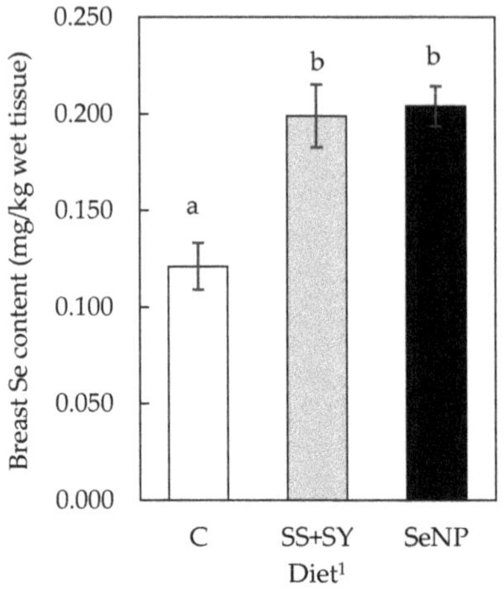

Figure 1. Effects of diet on selenium (Se) content of breast tissue in 42-day-old broiler chickens (n = 10 broiler chickens/diet). Bars on the graph represent standard error of means. Different letters denote significant difference ($p < 0.05$) [1] C, no Se added; SS+SY, 0.4 mg added Se/kg (from sodium selenite and selenium yeast at 1:1 ratio); SeNP, 0.4 mg added Se/kg (from elemental Se nanoparticles stabilized in chitosan).

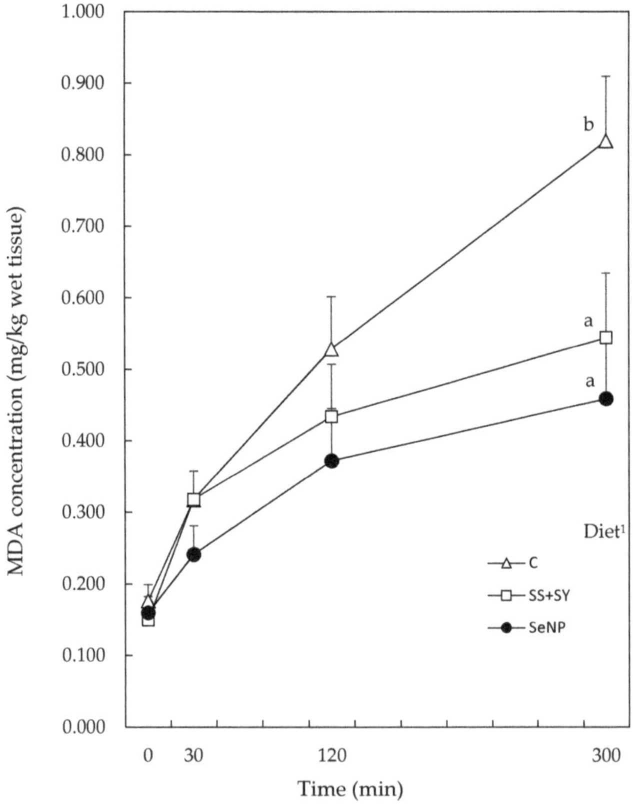

Figure 2. Effects of diet on iron-induced lipid oxidation of breast tissue in 42-day-old broiler chickens (n = 10 broiler chickens/diet). Bars on the graph represent standard error of means. Different letters

denote significant differences ($p < 0.05$). [1] C, no Se added; SS+SY, 0.4 mg added Se/kg (from sodium selenite and selenium yeast at 1:1 ratio); SeNP, 0.4 mg added Se/kg (from elemental Se nanoparticles stabilized in chitosan).

3.4. Breast Tissue Fatty Acid Composition

Total saturated FA (ΣSFA), monounsaturated FA (ΣMUFA), and polyunsaturated FA (ΣPUFA) were not affected by the diet (Table 2). On the other hand, total 22:6n − 3 (DHA) was significantly higher ($p < 0.05$) in the breast tissue of the SeNP-fed broiler chickens in comparison with C and SS+SY-fed ones thereby resulting in significantly increased ($p < 0.05$) total n − 3 FA. The total n − 6 FA were not affected by the diet and as a result the n − 6/n − 3 ratio was significantly lower ($p < 0.05$) in the breast of the SeNP compared to the C fed broilers; no difference in the n − 6/n − 3 ratio between SS+SY and SeNP broilers was observed. In addition, the total long chain (>20 carbons) n − 3 FA and elongase activity index tended to be higher ($p = 0.077$ and $p = 0.093$, respectively) in the breast of the SeNP fed as compared to C and SS+SY-fed broiler chickens (Table 2).

Table 2. Effects of diet on total fatty acid (FA) weights (mg FA/100 g wet tissue) and FA profile (% of total FA) of breast tissue in 42-day-old broiler chickens ($n = 10$ broiler chickens/diet).

	Diet [1]			SEM [2]	p-Value [3]
	C	SS+SY	SeNP		
Total FA weights	1258	1353	1285	88.5	0.552
14:0	0.28	0.27	0.26	0.010	0.159
15:0	0.05 [a]	0.06 [ab]	0.07 [b]	0.008	0.032
16:0	16.51	16.43	16.21	0.240	0.451
16:1n − 9	0.25	0.24	0.25	0.019	0.824
16:1n − 7	1.15	1.15	0.99	0.120	0.310
17:0	0.17	0.16	0.17	0.007	0.397
17:1	0.60	0.67	0.74	0.056	0.068
18:0	9.25	9.21	9.84	0.391	0.212
18:1n − 9	22.47	22.22	21.54	0.739	0.424
18:1n − 7	1.66	1.67	1.65	0.060	0.965
18:2n − 6	31.10	31.33	30.66	0.899	0.757
18:3n − 6	0.20	0.20	0.20	0.012	0.790
18:3n − 3	2.72	2.80	2.70	0.157	0.799
20:1n − 9	0.21	0.21	0.21	0.008	0.685
20:2n − 6	0.71	0.74	0.83	0.061	0.165
20:3n − 6	0.74	0.67	0.94	0.184	0.342
20:4n − 6	4.59	5.02	5.64	0.628	0.266
20:3n − 3	0.05	0.07	0.01	0.031	0.171
20:5n − 3	0.23	0.23	0.28	0.032	0.216
22:4n − 6	1.37	1.32	1.45	0.125	0.552
22:5n − 3	1.03	1.09	1.22	0.107	0.233
22:6n − 3	0.63 [a]	0.65 [a]	0.86 [b]	0.082	0.017
ΣSFA [4]	26.24	26.04	26.55	0.477	0.565
ΣMUFA [4]	26.31	26.27	25.40	0.799	0.434
ΣPUFA [4]	43.38	44.12	44.79	0.795	0.223
ΣPUFA/ΣSFA	1.66	1.70	1.69	0.049	0.650
Σn − 6 [5]	38.00	38.53	38.90	0.752	0.500
Σn − 3 [5]	4.66 [a]	4.84 [a]	5.07 [b]	0.093	<0.001
ΣLCn − 3 [6]	1.95	2.04	2.37	0.185	0.077
Σn − 6/Σn − 3	8.17 [b]	7.97 [ab]	7.69 [a]	0.180	0.038
Δ^9-desaturase index [7]	0.48	0.47	0.46	0.013	0.520

Table 2. *Cont.*

	Diet [1]			SEM [2]	*p*-Value [3]
	C	SS+SY	SeNP		
$\Delta^{5,6}$-desaturase index [8]	0.18	0.19	0.21	0.020	0.230
Elongase index [9]	0.56	0.56	0.61	0.024	0.093

Different letters denote significant differences ($p < 0.05$). [1] C, no Se added; SS+SY, 0.4 mg added Se/kg (from sodium selenite and selenium yeast at 1:1 ratio); SeNP, 0.4 mg added Se/kg (from elemental Se nanoparticles stabilized in chitosan). [2] SEM= standard error of means. [3] *p*-value of analysis of variance (ANOVA). [4] ΣSFA= total saturates (14:0 + 15:0 + 16:0 + 17:0 + 18:0), ΣMUFA= total monounsaturates(16:1n − 9 + 16:1n − 7 + 17:1 + 18:1n − 9 + 18:1n − 7 + 20:1n − 9), ΣPUFA= total polyunsaturates (18:2n − 6 + 18:3n − 3 + 18:3n − 6 + 20:2n − 6 + 20:3n − 6 + 20:3n − 3 + 20:4n − 6 + 20:5n − 3 + 22:4n − 6 + 22:5n − 3 + 22:6n − 3). [5] Σn − 6= total n − 6 fatty acids (18:2n − 6 + 18:3n − 6 + 20:2n − 6 + 20:3n − 6 + 20:4n − 6 + 22:4n − 6), Σn − 3= total n − 3 fatty acids (18:3n − 3 + 20:5n − 3 + 22:5n − 3 + 22:6n − 3). [6] ΣLCn − 3= total (≥20C) n − 3 fatty acids with carbon chain longer than 20 carbon atoms (20:3n − 3 + 20:5n − 3 + 22:5n − 3 + 22:6n − 3). [7] Total Δ^9-desaturase index calculated as 100 × [(16:1 + 18:1)/(16:1 + 16:0 + 18:1 + 18:0)]. [8] Total Δ^5-desaturase and Δ^6-desaturase index calculated as 100 × [(20:2n − 6 + 20:4n − 6 + 20:5n − 3 + 22:5n − 3 + 22:6n − 3)/(18:2n − 6 + 18:3n − 3 + 20:2n − 6 + 20:4n − 6 + 20:5n − 3 + 22:5n − 3 + 22:6n − 3)]. [9] Elongase index calculated as 18:0/16:0.

In order to investigate if the samples can be distinguished according to the diet, a discriminant analysis was carried out. All the individual FA values presented in Table 1 (22 in total) were used as predictor variables to deploy a model to distinguish the 30 meat samples. As shown in Figure 3, one canonical discriminant function (function 1) was found to be significant (p = 0.021) and distinguished the samples among the three experimental diets. This function explained the 83.90% of the observed variance. Amongst the 30 observations used to fit the model, all (100%) were classified correctly according to diet. As shown in the x-axis of Figure 3, samples from broiler chickens fed the control (C) and the SS+SY-supplemented diet were successfully separated from those fed the SeNP diet. Samples among SS+SY and C diets appeared to separate in the y-axis of Figure 3; however, this separation was only numerical and not significant (p = 0.401 for discriminant function 2). Subsequently, the stepwise discriminant analysis showed that 22:6n − 3 and 18:3n − 3, followed by 17:1 and 18:2n − 6 were the main FA responsible for the observed discrimination among the diets.

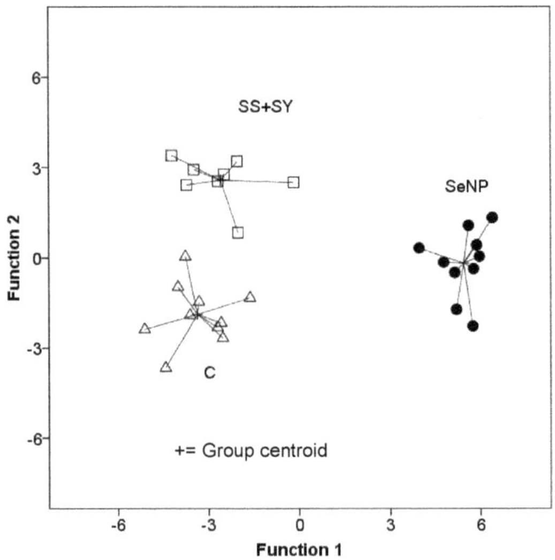

Figure 3. Discriminant plot distinguishing the samples according to diet using breast fatty acid profile in 42-day-old broiler chickens (n = 10 broiler chickens/diet). C, diet with no Se added; SS+SY, 0.4 mg added Se/kg (from sodium selenite and selenium yeast at 1:1 ratio); SeNP, 0.4 mg added Se/kg (from elemental Se nanoparticles stabilized in chitosan).

4. Discussion

The present study compared the effects of different dietary Se sources (combined sodium selenite and selenium-enriched yeast versus elemental Se nanoparticles) on meat Se content, FA composition, and oxidative strength in broiler chickens. The Se sources were added to the diet at appropriate levels in order to obtain 0.4 mg Se/kg and maintain the total dietary Se to a maximum of 0.5 mg/kg [12]. For broiler chickens, the National Research Council (NRC) recommends a dietary Se level of approximately 0.15 mg per kg [42]. This translates to a daily intake of around 18 μg of Se, assuming an average daily feed intake of 120 g. In our study, the C broilers chicken had a daily intake of 15 μg of Se, whereas the SS+SY and SeNP fed ones ingested by average 60 μg of Se on daily basis. Although the C diet which contained only the endogenous Se appears to be marginally Se-deficient, no significant differences in growth performance were found when compared to SS+SY and SeNP fed broilers. It is important to consider that Se efficacy is affected by various factors, including differences in Se sources, dosage, duration of supplementation, basal diet composition, and environmental conditions. Moreover, the Se requirements and response of broiler chickens may vary based on genetics and specific nutritional conditions [43].

Our results showed that Se deposition in the breast tissue of broiler chickens significantly increased in response to dietary Se addition regardless of the Se source. Both the SS+SY and the SeNP supplemented diets increased breast Se content to a comparable extent in comparison with the non-supplemented diet containing only the endogenous Se. Sufficient dietary selenium is a crucial factor for maintaining human health, and the recommended daily allowance is 55 μg/day, which can be increased to 75 μg/day for pregnant women [8]. Consuming 100 g of breast meat from broiler chickens fed the SS+SY and SeNP diets can provide 49.2 μg and 48.8 μg of selenium, respectively. This indicates that the meat from SeNP-fed broiler chickens can make a significant contribution to the overall dietary selenium intake in humans.

Dietary inorganic and organic Se, such as SS and SY, is known to increase the muscle Se content in broiler chickens in a dose-dependent manner when supplemented either alone [14,31,44] or in combination [44]. Regarding the ability of SeNP to increase the muscle Se content however, reports are conflicting. Some studies found that breast Se content was markedly increased by the dietary supplementation of broiler chicken diets with SeNP at levels ranging from 0.1 to 0.5 [45], 0.15 to 1.20 [44], or 0.3 to 2.0 mg Se/kg [13], with SeNP being more effective than inorganic Se [44]. Others observed that SeNP are not as effective as SS and SY in elevating muscle Se [14,46]. The current study findings are in agreement with these reporting the positive impact of SeNP on tissue Se content. However, the present results showed that supplementing diets with 0.4 mg SeNP/kg increased muscle Se content by 169% compared to the non-supplemented diet. This is in contrast to the findings of other studies [13,44,45] which reported an increase ranging from 243% to 290%, when supplementing diets with 0.3 mg SeNP/kg. The difference between the current and the earlier studies likely indicates that the characteristics of the added SeNP, may affect Se deposition in tissues of broilers, in addition to other factors (environmental, dietary or genetic). In the aforementioned studies, different preparations of SeNP were administered. Zhou and Wang [45] and Hu et al. [44] used SeNP coated (stabilized) with bovine serum albumin (BSA) whereas Cai et al. [13] and Bień et al. [14,46] tested different commercial SeNP preparations without any details about coating. In the present study, the SeNP were stabilized in chitosan (CS). The BSA is a water-soluble protein that dissolves easily in the gastrointestinal tract of animals, whereas CS is a structural polysaccharide resembling cellulose which cannot be degraded in some species of animals and humans; therefore, these two coatings control the release of Se to a different extent [26], which may explain the lower Se accumulation observed herein. The Se from SeNP is supposed to be absorbed by the broiler body more effectively than other forms of Se because nano-Se is directly incorporated into selenoproteins [47,48]; however, no such conclusion can be drawn by the present study results. This clearly indicates that there might be several factors affecting the bioavailability of the Se from different SeNP preparations and the data in the literature should be handled with care.

The literature has documented the ability of inorganic and organic forms of selenium (Se), either alone or in combination, to reduce malondialdehyde (MDA) content and enhance the oxidative stability of meat in broiler chickens [31,45]. However, when it comes to the antioxidative activity of Se from selenium nanoparticles (SeNP), conflicting reports can be found in the literature. Some studies have shown that diets containing SeNP significantly decreased meat MDA content in broiler chickens compared to control diets without added Se or diets supplemented with 0.3 mg Se/kg from sodium selenite (SS) [14,46]. Only one study reported that SeNP were more effective than SS and selenium-enriched yeast (SY) in reducing lipid peroxidation [49]. On the other hand, Cai et al. [13] observed that increasing dietary Se using SeNP did not affect meat MDA concentration compared to a diet without added Se. In the present study, the oxidative stability of meat was similar in broilers fed diets supplemented with Se (either SS+SY or SeNP), indicating that the antioxidant potential of SeNP was significant and equivalent to the combined SS+SY forms. It should be noted that the aforementioned studies measured MDA concentration at a single time point, usually 24 h post-mortem, and may not provide sufficient information about meat oxidative stability. In the present study, the induced oxidation assay was used, which is a more robust method for assessing the relative oxidative stability and shelf life potential of meat ex vivo [50].

Although it is known that Se addition to feed can modify the lipid profile of meat towards a desirable direction in several livestock species [31,32,34,35], scarce data on the effects of dietary SeNP supplementation on the FA composition of broiler chicken meat are available. Bień et al. [14] reported that total PUFA (particularly n − 6) markedly increased in the breast of broilers fed SeNP and SY compared to SS-fed chickens. However, this increase was limited mainly to the enhanced 18:2n − 6 content, whereas the long chain PUFA (with C atoms > 20) and the n − 3 FA (mainly 18:3n − 3 and 22:6n − 3) were negatively affected in the SY- and SeNP-fed broiler chickens. In contrast to the aforementioned [14], we observed increased total n − 3 (owing mainly to 22:6n − 3) FA in the breast of the SeNP fed broiler chickens compared to the SS+SY and the C ones. It was also observed that the total long chain (>20 C) n − 3 FA (ΣLCn − 3) tended to be greater in the SeNP when compared to the SS+SY- and C-fed broiler chickens thereby resulting in more desirable n − 6/n − 3 ratio. These results likely depict that SeNP, compared to SS+SY, may have had a greater in vivo protective action against the degradation of FA that are prone to peroxidation. Additionally, the increased 22:6n − 3 in meat likely depicts an inhibition of oxidation by SeNP and may reflect direct effects of SeNP on FA metabolism. The synthesis of long chain n − 3 FA includes several elongation, Δ6 desaturation, and partial β-oxidation stages [51,52]. There are data suggesting that dietary Se is involved in the peroxisomal β-oxidation [53]. Taking into account that a) n − 3 and n − 6 fatty acids compete the same enzymatic system of elongases for the addition of carbons and desaturases for the formation of double bonds in their chains [54] and b) there was a tendency to increased elongase activity index and total long chain n − 3 FA in the SeNP-fed broilers herein, it is not unlikely that dietary SeNP may have favoured the n − 3 FA synthesis. To this aspect, discriminant analysis helped to understand that SeNP diets indeed affected breast FA profile in a dissimilar manner compared to C or SS+SY diets. This discrimination was mainly owed to 22:6n − 3, followed by 18:3n − 3, and then to a lesser extent by 17:1 and 18:2n − 6. Hence, dietary supplementation with SeNP affected the breast FA profile, especially long-chain n − 3 FA, more extensively in comparison with C and SS+SY diets. These changes may be credited to a stronger impact of SeNP on both FA oxidation and metabolism.

5. Conclusions

In conclusion, both SS+SY (selenium-enriched yeast and sodium selenite) and SeNP (selenium nanoparticles) demonstrated comparable efficacy as dietary Se sources. When diets were supplemented with 0.4 mg Se/kg using either SS+SY or SeNP, the breast meat

Se content increased significantly. Moreover, both SS+SY and SeNP improved the oxidative stability of the breast meat compared to a non-supplemented diet with endogenous selenium. However, the addition of SeNP had an additional advantage by significantly favoring the composition of $n-3$ FA in the meat compared to SS+SY. These findings highlight the potential of SeNP as a sustainable dietary Se source in the broiler industry, where high-quality meat with increased polyunsaturated fatty acid content and extended shelf life is desired. Further research is essential to investigate the incorporation of Se from various SeNP preparations, thereby advancing our understanding of the factors that influence the nano-Se bioavailability and bioactivity.

Author Contributions: Conceptualization, G.P. and A.C.P.; methodology, G.P. and A.C.P.; software, G.P.; validation, E.G. and E.F.; formal analysis, E.G. and E.F.; investigation, E.G., E.F., G.P. and A.C.P.; resources, G.P.; data curation, G.P.; writing—original draft preparation, E.G. and E.F.; writing—review and editing, G.P. and A.C.P.; visualization, G.P.; supervision, G.P. and A.C.P.; project administration, G.P.; funding acquisition, G.P. All authors have read and agreed to the published version of the manuscript.

Funding: Greece and the European Union (European Social Fund) co-financed this Research Project through the Operational Program «Human Resources Development, Education and Lifelong Learning 2014–2020» (Funding number: MIS 5048474).

Institutional Review Board Statement: Bioethics Committee of the Agricultural University of Athens approved housing, handling and care conditions of the animals (Approval code: 13, Approval date: 16-03-2021).

Informed Consent Statement: Not applicable.

Data Availability Statement: The data presented in this study are available upon request from the corresponding author.

Acknowledgments: The authors are also grateful to Papafili N., Sfetsios O., and Roditi A. for their assistance in the experimental and analytical procedures, and to NUEVO S.A. (N Artaki, Euboia, Greece) for providing Sel-Plex®.

Conflicts of Interest: No conflict of interest is declared by the authors.

References

1. Kouba, M.; Mourot, J. A review of nutritional effects on fat composition of animal products with special emphasis on *n*-3 polyunsaturated fatty acids. *Biochimie* **2011**, *93*, 13–17. [CrossRef]
2. Sadeghi, A.A.; Iravani, H.; Torshizi, M.K.; Chamani, M. Fatty acids profiles in meat of broiler chicks fed diet containing corn oil switched to fish oil at different weeks of age. *World Appl. Sci. J.* **2012**, *18*, 159–165.
3. Jankowski, J.; Zdunczyk, Z.; Mikulski, D.; Juskiewicz, J.; Naczmanski, J.; Pomianowski, J.F.; Zdunczyk, P. Fatty acid profile, oxidative stability, and sensory properties of breast meat from turkeys fed diets with a different *n*-6/*n*-3 PUFA ratio. *Eur. J. Lipid Sci. Technol.* **2012**, *114*, 1025–1035. [CrossRef]
4. Sohaib, M.; Anjum, F.M.; Arshad, M.S.; Imran, M.; Imran, A.; Hussain, S. Oxidative stability and lipid oxidation flavoring volatiles in antioxidants treated chicken meat patties during storage. *Lipids Health Dis.* **2017**, *16*, 16–27. [CrossRef]
5. Surai, P.F. Natural antioxidants in poultry nutrition: New developments. In Proceedings of the 16th European Symposium on Poultry Nutrition, Strasbourg, France, 26–30 August 2007; pp. 669–676.
6. Schweizer, U.; Streckfuß, F.; Pelt, P.; Carlson, B.A.; Hatfield, D.L.; Kohrle, J.; Schomburg, L. Hepatically derived selenoprotein P is a key factor for kidney but not for brain selenium supply. *Biochem. J.* **2005**, *386*, 221–226. [CrossRef]
7. Yu, L.L.; Wang, R.; Zhang, Y.Z.; Kleemann, D.; Zhu, X.; Jia, Z. Effects of selenium supplementation on polyunsaturated fatty acid concentrations and antioxidant status in plasma and liver of lambs fed linseed oil or sunflower oil diets. *Anim. Feed Sci. Technol.* **2008**, *140*, 39–51. [CrossRef]
8. Ringuet, M.T.; Hunne, B.; Lenz, M.; Bravo, D.M.; Furness, J.B. Analysis of bioavailability and induction of glutathione peroxidase by dietary nanoelemental, organic and inorganic selenium. *Nutrients* **2021**, *13*, 1073. [CrossRef]
9. Rayman, M.P. Selenium intake, status, and health: A complex relationship. *Hormones* **2020**, *19*, 9–14. [CrossRef] [PubMed]
10. Khurana, A.; Tekula, S.; Saifi, M.A.; Venkatesh, P.; Godugu, C. Therapeutic applications of selenium nanoparticles. *Biomed. Pharmacother.* **2019**, *111*, 802–812. [CrossRef] [PubMed]
11. Araujo, J.M.; Fortes-Silva, R.; Pola, C.C.; Yamamoto, F.Y.; Gatlin, D.M.; Gomes, C.L. Delivery of selenium using chitosan nanoparticles: Synthesis, characterization, and antioxidant and growth effects in Nile tilapia (*Oreochromis niloticus*). *PLoS ONE* **2021**, *16*, 0251786. [CrossRef] [PubMed]

12. EC 2013. Commission Regulation (EC) No 427/2013 of 8 May 2013 concerning the authorisation of selenomethionine produced by Saccharomyces cerevisiae NCYC R646 as a feed additive for all Animal species and amending Regulations (EC) No 1750/2006, (EC) No 634/2007 and (EC) No 900/2009 as regards the maximum supplementation with selenised Yeast. *Off. J. Eur. Union L* **2013**, *127*, 20–22.
13. Cai, S.J.; Wu, C.X.; Gong, L.M.; Song, T.; Wu, H.; Zhang, L.Y. Effects of nano—Selenium on performance, meat quality, immune function, oxidation resistance, and tissue selenium content in broilers. *Poult. Sci.* **2012**, *91*, 2532–2539. [CrossRef] [PubMed]
14. Bień, D.; Michalczuk, M.; Szkopek, D.; Kinsner, M.; Konieczka, P. Changes in lipids metabolism indices as a result of different form of selenium supplementation in chickens. *Sci. Rep.* **2022**, *12*, 13817. [CrossRef] [PubMed]
15. Samak, D.H.; El-Sayed, Y.S.; Shaheen, H.M.; El-Far, A.H.; Abd El-Hack, M.E.; Noreldin, A.E.; El-Naggar, K.; Abdelnour, S.A.; Saied, E.M.; El-Seedi, H.R.; et al. Developmental toxicity of carbon nanoparticles during embryogenesis in chicken. *Environ. Sci. Pollut. Res. Int.* **2020**, *27*, 19058–19072. [CrossRef]
16. Kumbhar, S.; Khan, A.Z.; Parveen, F.; Nizamani, Z.A.; Siyal, F.A.; Abd El-Hack, M.E.; Gan, F.; Liu, Y.; Hamid, M.; Nido, S.A.; et al. Impacts of selenium and vitamin E supplementation on mRNA of heat shock proteins, selenoproteins and antioxidants in broilers exposed to high temperature. *AMB Express* **2018**, *8*, 112. [CrossRef] [PubMed]
17. Abdul, H.; Sivaraj, R.; Venckatesh, R. Green synthesis and characterization of zinc oxide nanoparticles from *Ocimum basilicum* L. var purpurascens Benth.-lamiaceae leaf extract. *Mater. Lett.* **2014**, *131*, 16–18. [CrossRef]
18. Gangadoo, S.; Stanley, D.; Hughes, R.J.; Moore, R.J.; Chapman, J. Nanoparticles in feed: Progress and prospects in poultry research. *Trends Food Sci. Technol.* **2016**, *58*, 115–126. [CrossRef]
19. Li, H.; Liu, D.; Li, S.; Xue, C. Synthesis and cytotoxicity of selenium nanoparticles stabilized by α-D-glucan from *Castanea mollissima* Blume. *Int. J. Biol. Macromol.* **2019**, *129*, 818–826. [CrossRef]
20. Zhang, J.S.; Wang, X.; Xu, T. Elemental selenium at nano size (Nano-Se) as a potential chemopreventive agent with reduced risk of selenium toxicity: Comparison with Se-Methylselenocysteine in mice. *Toxicol. Sci.* **2008**, *101*, 22–31. [CrossRef]
21. Wang, H.; Zhang, J.; Yu, H. Elemental selenium at nano size possesses lower toxicity without compromising the fundamental effect on selenoenzymes: Comparison with selenomethionine in mice. *Free Radic. Biol. Med.* **2007**, *42*, 1524–1533. [CrossRef]
22. Zhou, X.; Wang, Y. Influence of dietary nano elemental selenium on growth performance, tissue selenium distribution, meat quality and glutathione peroxidase activity in Guangxi Yellow Chicken. *Poult. Sci.* **2011**, *90*, 680–686. [CrossRef]
23. Saleh, A.A.; Ebeid, T.A. Feeding sodium selenite and nano-selenium stimulates growth and oxidation resistance in broilers. *S. Afr. J. Anim. Sci.* **2019**, *49*, 176–184. [CrossRef]
24. Aparna, N.; Karunakaran, R. Effect of selenium nanoparticles supplementation on oxidation resistance of broiler chicken. *Indian J. Sci. Technol.* **2016**, *9*, 1–5. [CrossRef]
25. Li, B.; Li, D.; Jing, W.; Fan, J.; Dahms, H.U.; Lee, S.C.; Wang, L. Biogenic selenium and its hepatoprotective activity. *Sci. Rep.* **2017**, *7*, 15627. Available online: https://www.nature.com/articles/s41598-017-13636-1 (accessed on 7 April 2023). [CrossRef] [PubMed]
26. Bai, K.; Hong, B.; He, J.; Hong, Z.; Tan, R. Preparation and antioxidant properties of selenium nanoparticles-loaded chitosan microspheres. *Int. J. Nanomed.* **2017**, *12*, 4527–4539. [CrossRef]
27. Ren, L.; Wu, Z.; Ma, Y.; Jian, W.; Xiong, H.; Zhou, L. Preparation and growth-promoting effect of selenium nanoparticles capped by polysaccharide-protein complexes on tilapia. *J. Sci. Food Agric.* **2021**, *101*, 476–485. [CrossRef] [PubMed]
28. Filipović, N.; Ušjak, D.; Milenković, M.T.; Zheng, K.; Liverani, L.; Boccaccini, A.R.; Stevanović, M.M. Comparative study of the antimicrobial activity of selenium nanoparticles with different surface chemistry and structure. *Front. Bioeng. Biotechnol.* **2021**, *8*, 624621. [CrossRef]
29. Zanini, S.F.; Torres, C.A.A.; Bragagnolo, N.; Turatti, J.M.; Silva, M.G.; Zanini, M.S. Effect of oil sources and vitamin E levels in the diet on the composition of fatty acids in rooster thigh and chest meat. *J. Sci. Food Agric.* **2004**, *84*, 672–682. [CrossRef]
30. González, E.; Tejeda, J.F. Effects of dietary incorporation of different antioxidant extracts and free-range rearing on fatty acid composition and lipid oxidation of Iberian pig meat. *Animal* **2007**, *1*, 1060–1067. [CrossRef]
31. Pappas, A.C.; Zoidis, E.; Papadomichelakis, G.; Fegeros, K. Supranutritional selenium level affects fatty acid composition and oxidative stability of chicken breast muscle tissue. *J. Anim. Physiol. Anim. Nutr.* **2011**, *96*, 385–394. [CrossRef]
32. Pereira, A.S.C.; Santos, M.V.; Aferri, G.; Silva Corte, R.R.P.; Silva, S.L.; de Freitas, J.E., Jr.; Leme, P.R.; Renno, F.P. Lipid and selenium sources on fatty acid composition of intramuscular fat and muscle selenium concentration of Nellore steers. *Rev. Bras. Zootec.* **2012**, *41*, 2357–2363. [CrossRef]
33. Schäfer, K.; Kyriakopoulos, A.; Gessner, H.; Grune, T.; Behne, D. Effects of selenium deficiency on fatty acid metabolism in rats fed fish oil-enriched diets. *J. Trace Elem. Med. Biol.* **2004**, *18*, 89–97. [CrossRef] [PubMed]
34. Netto, A.S.; Zanetti, M.A.; Ribeiro del Claro, G.; Pires de Melo, M.; Garcia Vilela, F.; Bertonha Correa, L. Effects of copper and selenium supplementation on performance and lipid metabolism in confined brangus bulls. *Asian-Austral. J. Anim. Sci.* **2014**, *27*, 488–494. [CrossRef] [PubMed]
35. Papadomichelakis, G.; Zoidis, E.; Pappas, A.C.; Mountzouris, K.C.; Fegeros, K. Effects of increasing dietary organic selenium levels on meat fatty acid composition and oxidative stability in growing rabbits. *Meat Sci.* **2017**, *131*, 132–138. [CrossRef] [PubMed]

36. Fortatos, E.; Giamouri, E.; Pappas, A.C.; Yannopoulos, S.N.; Papadomichelakis, G. Selenium nanoparticles-loaded chitosan microspheres as a dietary selenium source in rabbits: Impact on meat selenium content and oxidative stability. *Acta Sci. Vet. Sci.* **2021**, *3*, 27–38. Available online: https://actascientific.com/ASVS/ASVS-03-0237.php (accessed on 2 February 2023). [CrossRef]
37. Pappas, A.C.; Acamovic, T.; Sparks, N.H.C.; Surai, P.F.; McDevitt, R.M. Effects of supplementing broiler breeder diets with organoselenium compounds and polyunsaturated fatty acids on hatchability. *Poult. Sci.* **2006**, *85*, 1584–1593. [CrossRef]
38. Terevinto, A.; Ramos, A.; Castroman, G.; Cabrera, M.C.; Saadoun, A. Oxidative status, in vitro iron-induced lipid oxidation and superoxide dismutase, catalase and glutathione peroxidase activities in rhea meat. *Meat Sci.* **2010**, *84*, 706–710. [CrossRef]
39. O'Fallon, J.V.; Busboom, J.R.; Nelson, M.L.; Gaskins, C.T. A direct method for fatty acid methyl-ester synthesis: Application to wet meat tissues, oils, and feedstuffs. *J. Anim. Sci.* **2007**, *85*, 1511–1521. [CrossRef]
40. IBM Corp. *Released, IBM SPSS Statistics for Windows, Version 23.0*; IBM Corp.: Armonk, NY, USA, 2015.
41. Templeton, G.F. A two-step approach for transforming continuous variables to normal: Implications and recommendations for IS research. *Commun. Assoc. Inf. Syst.* **2011**, *28*, 41–58. [CrossRef]
42. National Research Council (NRC). *Nutrient Requirements of Poultry*, 9th ed.; National Academies Press: Washington, DC, USA, 1994.
43. Surai, P.F.; Kochish, I.I.; Fisinin, V.I.; Velichko, O.A. Selenium in poultry nutrition: From sodium selenite to organic selenium sources. *J. Poult. Sci.* **2018**, *55*, 79–93. [CrossRef]
44. Hu, C.H.; Li, Y.L.; Xiong, L.; Zhang, H.M.; Song, J.; Xia, M.S. Comparative effects of nano elemental selenium and sodium selenite on selenium retention in broiler chickens. *Anim. Feed Sci. Technol.* **2012**, *177*, 204–210. [CrossRef]
45. Ahmad, H.; Tian, J.; Wang, J.; Khan, M.A.; Wang, Y.; Zhang, L.; Wang, T. Effects of Dietary Sodium Selenite and Selenium Yeast on Antioxidant Enzyme Activities and Oxidative Stability of Chicken Breast Meat. *J. Agric. Food Chem.* **2012**, *60*, 7111–7120. [CrossRef]
46. Bien, D.; Michalczuk, M.; Łysek-Gładysinsk, M.; Józwik, A.; Wieczorek, A.; Matuszewski, A.; Kinsner, M.; Konieczka, P. Nano-sized selenium maintains performance and improves health status and antioxidant potential while not compromising ultrastructure of breast muscle and liver in chickens. *Antioxidants* **2023**, *12*, 905. [CrossRef]
47. Suzuki, K.T.; Ogra, Y. Metabolic pathway for selenium in the body: Speciation by HPLC-ICP MS with enriched Se. *Food Addit. Contam.* **2022**, *19*, 974–983. [CrossRef] [PubMed]
48. Gangadoo, S.; Dinev, I.; Chapman, J.; Hughes, R.J.; Hao Van, T.T.; Moore, R.; Stanley, D. Selenium nanoparticles in poultry feed modify gut microbiota and increase abundance of *Faecalibacterium prausnitzii*. *Appl. Microbiol. Biotechnol.* **2018**, *102*, 1455–1466. Available online: https://link.springer.com/article/10.1007%2Fs00253-017-8688-4 (accessed on 4 March 2023). [CrossRef]
49. Visha, P.; Nanjappan, K.; Selvaraj, P.; Jayachandran, S.; Thavasiappan, V. Influence of dietary nanoselenium supplementation on the meat characteristics of broiler chickens. *Int. J. Curr. Microbiol. App. Sci.* **2017**, *6*, 340–347. [CrossRef]
50. Botsoglou, N.A.; Florou-Paneri, P.; Christaki, E.; Fletouris, D.J.; Spais, A.B. Effect of dietary oregano essential oil on performance of chickens and on iron-induced lipid oxidation of breast, thigh and abdominal fat tissues. *Br. Poult. Sci.* **2002**, *43*, 223–230. [CrossRef] [PubMed]
51. Voss, A.; Reinhart, M.; Sankarappa, S.; Sprecher, H. The metabolism of 7,10,13,16,19-docosapentaenoic acid to 4,7,10,13,16,19-docosahexaenoic acid in rat liver is independent of a delta-4-desaturase. *J. Biol. Chem.* **1991**, *266*, 19995–20000. [CrossRef]
52. Sprecher, H.; Luthria, D.L.; Mohammed, B.S.; Baykousheva, S.P. Re-evaluation of the pathways for the biosynthesis of polyunsaturated fatty acids. *J. Lipid Res.* **1995**, *36*, 2471–2477. [CrossRef]
53. Glauert, H.P.; Beaty, M.M.; Clark, T.D.; Greenwell, W.S.; Chow, C.K. Effect of dietary selenium on the induction of altered hepatic foci and hepatic tumors by the peroxisome proliferator ciprofibrate. *Nutr. Cancer* **1990**, *14*, 261–271. [CrossRef]
54. Lee, J.M.; Lee, H.; Kang, S.B.; Park, W.J. Fatty acid desaturases, polyunsaturated fatty acid regulation, and biotechnological advances. *Nutrients* **2016**, *8*, 23. [CrossRef] [PubMed]

sustainability

MDPI

Article

Fatty Acid Profile and Oxidative Stability of Layers' Egg Yolk as Affected by Dietary Supplementation with Fresh Purslane and Addition of Aromatic Plant Essential Oils to Drinking Water

Vassilios Dotas [1,*], Dimitrios Gourdouvelis [1], George Symeon [2], Lampros Hatzizisis [3], Ioannis Mitsopoulos [4], Dimitrios Galamatis [5], Maria Ioannidou [6] and Evangelia Sossidou [6]

1 Department of Animal Production, Faculty of Agriculture, Aristotle University of Thessaloniki, 54124 Thessaloniki, Greece; dgourdou@agro.auth.gr
2 Research Institute of Animal Science, Hellenic Agricultural Organization Demeter, 58100 Giannitsa, Greece; gsymeon@elgo.gr
3 Faculty of Agriculture, University of Ioannina, 47100 Arta, Greece; lamprosxatz@uoi.gr
4 Department of Agriculture, School of Geosciences, International Hellenic University, 57400 Sindos, Greece; gmitsop@ihu.gr
5 Department of Animal Science, School of Agricultural Sciences, University of Thessaly, 41500 Larissa, Greece; dgalamatis@uth.gr
6 Veterinary Research Institute, Hellenic Agricultural Organization Demeter, 57001 Thessaloniki, Greece; ioannidou@vri.gr (M.I.); sossidou@vri.gr (E.S.)
* Correspondence: vdotas@agro.auth.gr; Tel.: +30-2310991735

Abstract: The objective of this study was to investigate the possibility of producing eggs enriched in omega-3 polyunsaturated fatty acids (PUFA) while also increasing the oxidative stability of egg yolk. Here, 432 68-week-old Isa Brown layers were split into two groups of 216, consisting of three subgroups of 72 each. Group C was fed a standard corn–soybean meal diet, while Group P received the same basic diet with an additional 24 g of fresh purslane. In the drinking water of hens of the three subgroups of group C and the three subgroups of group P, either no essential oil (C-0, P-0) or 100 ppm of oregano essential oil (C-ORE, P-ORE) or 100 ppm of a blend of oregano, sage, and fennel essential oils (C-BLEND, P-BLEND) was administered. The purslane supplementation resulted in increased egg weight, improved yolk color, higher levels of α-linolenic and linoleic acids, and an improved omega-6/omega-3 nutritional index. The addition of essential oils resulted in a significant increase in the oxidative stability of the egg yolk, with the BLEND being the most effective. In conclusion, the combined administration of fresh purslane and essential oils of aromatic plants could be suggested for the production of eggs enriched in omega-3 PUFA, protected with natural antioxidants of plant origin.

Keywords: layers; purslane; essential oils; egg quality traits; fatty acid profile; oxidative stability

Citation: Dotas, V.; Gourdouvelis, D.; Symeon, G.; Hatzizisis, L.; Mitsopoulos, I.; Galamatis, D.; Ioannidou, M.; Sossidou, E. Fatty Acid Profile and Oxidative Stability of Layers' Egg Yolk as Affected by Dietary Supplementation with Fresh Purslane and Addition of Aromatic Plant Essential Oils to Drinking Water. *Sustainability* **2023**, *15*, 11539. https://doi.org/10.3390/su151511539

Academic Editor: Marc A. Rosen

Received: 13 May 2023
Revised: 18 July 2023
Accepted: 24 July 2023
Published: 26 July 2023

1. Introduction

It is well documented that enriching the human diet with high levels of polyunsaturated fatty acids (PUFA) is associated with a reduced risk of early cardiovascular disease [1]. Investigating the effect of PUFA on cholesterol fractions, it has been found that these fatty acids cause a decrease in low-density lipoprotein-cholesterol (LDL-C) and an increase in high-density lipoprotein-cholesterol (HDL-C) [2]. Dietary supplementation with PUFA leads to an increased proportion of these fatty acids in plasma lipoproteins, cell and tissue lipids [3]. It has been suggested that an adequate intake level for total omega-3 fatty acids of 1.1–1.6 g/day for humans can potentially reduce the risk of diseases, such as cardiovascular disease [4].

Atherosclerosis might be reduced or slowed by long-chain omega-3 PUFA by modifying the risk factor profile [5]. Daily supplementation of healthy people with omega-3

PUFA may have protective and antioxidant effects on LDL and is supportive of the global recommendations for long-chain omega-3 PUFA for the primary prevention of coronary disease [6]. Yang et al. [7] demonstrated that a low omega-6/omega-3 PUFA ratio (1:1 up to 5:1) had a beneficial effect on cardiovascular risk factors by enhancing favorable lipid profiles, having anti-inflammatory and anti-oxidative stress effects, and improving endothelial function.

Omega-3 α-linolenic acid (ALA) is an essential fatty acid that cannot be produced by humans and, therefore, must be ingested. This fatty acid plays a crucial role in human growth, development, and disease prevention [8]. Additionally, it is the precursor to longer-chain omega-3 fatty acids such as eicosapentaenoic acid (EPA), docosapentaenoic acid (DPA), and docosahexaenoic acid (DHA), which are generally found in marine organisms [9].

Nowadays, there is an increased interest and demand for functional foods for human consumption that provide various benefits and help prevent nutrition-related diseases. This is achieved by changing the composition of foods to include certain ingredients that are beneficial to health [10,11].

In this sense, eggs, as a conventional food, can be enriched with certain nutrients through dietary manipulation to be promoted as functional foods. Egg production is also an economically sustainable activity because it offers a moderate calorie source and a high-quality protein product at a low economic cost [12]. Functional eggs are primarily enriched with omega-3 PUFA or low-cholesterol content through modification of the layer hen's diet.

Consumers have become increasingly concerned about the amount and type of animal-based foods in recent years due to their high saturated fatty acid (SFA) content and low polyunsaturated fatty acid content, which have been linked to health problems. Increasing the functional and nutritional value of egg lipid fractions may have positive effects on human health [13]. However, modifying animal diets to increase the degree of unsaturation of tissue fatty acids to meet dietary guidelines increases the risk of oxidative deterioration.

Enriching animal products such as eggs with omega-3 PUFA can be achieved with fish by-products such as fish oil and fishmeal. However, this is not a sustainable solution for conserving the Earth's precious and finite resources [14]. Additionally, the addition of fish meal or fish oil to the feed may have a negative effect on the sensory quality of eggs, especially with regard to consumer acceptance [15].

In recent years, efforts have been made in egg-laying poultry to increase the omega-3 PUFA content of eggs through alternative and sustainable plant-based raw materials. These production practices for functional foods enriched with fatty acids should also take care of their oxidative stability during storage by utilizing additional antioxidants from sustainable sources.

Purslane (*Portulaca oleracea* L.) is one of the most widespread herbaceous plants in the world and has a long history of use for human food, animal feed, and medical purposes. A nutritional characterization of purslane accessions conducted by Ezekwe et al. [16] revealed that, despite its genetic diversity, purslane remains one of the most abundant terrestrial sources of omega-3 fatty acids, which are potentially beneficial for humans and animals. These benefits include the prevention and treatment of cardiovascular disease, some autoimmune diseases, diabetes, certain types of cancer, and their significant role in neuronal development. According to Simopoulos et al. [17] and Simopoulos [18], 100 g of fresh purslane leaves contain approximately 0.3–0.4 g ALA, 1.0 mg EPA, 0.0122 g α-tocopherol (vitamin E), 0.0266 g ascorbic acid (vitamin C), 0.0019 g β-carotene (a precursor of vitamin A), and 0.0148 g glutathione. This is a remarkable amount of EPA for land-based vegetable sources. Purslane also contains DHA and DPA [19], vitamins (primarily vitamins A and C, and lesser amounts of vitamin B and carotenoids), as well as dietary minerals such as magnesium, calcium, potassium, and iron [20]. As ALA can be converted into DHA and EPA, the inclusion of ALA in feed and the metabolic efficiency of hens in bioconversion of ALA plays an important role in long-chain (EPA, DPA, DHA) fatty acid enrichment in poultry tissues and eggs [21]. Additionally, purslane contains less cholesterol than fish oils.

Essential oils are important secondary metabolites of aromatic plants [22]. Oregano (*Origanum vulgare*) and sage (*Salvia officinalis*) are aromatic plants with ornamental, culinary, and phytotherapeutic uses all over the world. In Europe, they are traditionally used in the southern countries, particularly in the Mediterranean area [23].

The essential oils of aromatic plants have been widely used in animal nutrition as additives to replace antibiotic growth promoters. Within the *Labiatae* plant family, thyme (*Thymus vulgaris*), oregano, and sage are the most popular members, as they contain phenolic compounds such as carvacrol, eugenol, thymol, p-cymene, and γ-terpinene, which possess strong antibacterial and antioxidant properties [24]. The essential oil derived from oregano is known to possess antimicrobial, antifungal, insecticidal, and antioxidant properties [25]. Regarding antibacterial properties, cultivated sage had a 2.37% extract concentration, with basic components being cis-thujone (33.80%) and trans-thujone (6.97%) [26].

Fennel (*Foeniculum vulgare*) is also a plant whose essential oils are exploited as strong antioxidants and antimicrobials due to the presence of compounds, including anethole and polyphenols [27].

Dietary herbal additives can manipulate serum cholesterol due to their total phenolic ingredients [28]. Functional foods can be made by transferring the components of active herbal ingredients to egg yolks and enhancing them with phytonutrients [29]. Lipid oxidation accelerates quality deterioration in muscle-based foods (fish, red meat, and poultry), resulting in off-odors/flavors, color problems, texture defects, and safety concerns [30]. Malondialdehyde (MDA) is one of the most widely studied degradation products of lipid hydroperoxides and serves as a marker of lipid peroxidation [31]. The antioxidant components of these active plant compounds are thought to protect lipids from oxidation, thereby preventing lipid peroxidation and improving the oxidative stability of fat and egg yolk, as well as their storage quality [32].

Many studies have previously examined the antioxidant properties of aromatic plants and their effect on the oxidative stability of egg yolks in feed [26,33–35]. However, limited studies have been conducted on essential oils of aromatic plants supplemented in water for their antioxidant activity.

The objective of this study was to investigate the possibility of producing eggs enriched in omega-3 PUFA by adding fresh leaves and stems of the plant species purslane (*Portulaca oleracea* L.) while also increasing the oxidative stability of egg yolk by adding essential oils of aromatic plants derived from *Origanum vulgare* (oregano), *Salvia officinalis* (sage), and *Foeniculum vulgare* (fennel seeds) to the drinking water of laying hens.

2. Materials and Methods

2.1. Housing

The experiment was conducted at the poultry shed of the Aristotle University of Thessaloniki Farm located in Thermi, Thessaloniki (Latitude: 40.540238861182964° N, Longitude: 22.99927815920721° E, WGS 84). The poultry shed's ventilation system, consisting of two side vents and two roof vents, is automated and facilitates effective airflow regulation. This system ensures a constant supply of fresh air, removes stale air and odors, and promotes a healthy and comfortable environment for the laying hens, enhancing their well-being, productivity, and overall conditions. The vent openings are automatically adjusted by motorized actuators controlled by a centralized system, which takes into account temperature, humidity, and air quality, thereby maintaining stable environmental conditions and optimizing airflow. The poultry shed is equipped with enriched cages housing six hens each, which meet European Union welfare standards. The dimensions of each cage are 90 cm (width) × 60 cm (depth) × 50 cm (height), which adequately ensures the minimum requirement of 750 cm^2 surface area per hen. Each cage is equipped with a linear feeder, a nipple-type drinker, a private nest area for laying eggs, a litter pad for pecking and scratching, and perches for birds to rest on.

The experiment lasted 30 days. The lighting of the building during the experiment amounted to 16 h per day. The indoor temperature averaged 24.5 ± 2.4 °C, while the relative humidity ranged between 55 and 58%.

2.2. Birds

Four hundred and thirty-two (432) 68-week-old Isa Brown layers were used in the experiment and were divided into 6 treatments of 72 hens each. Each treatment had 12 replicates (cages) of 6 hens each.

The hens were weighed at the beginning and at the end of the experiment, with the cage of 6 hens constituting the experimental unit. The egg production rate and daily mass of eggs produced per hen (egg production rate × egg weight) were determined at the end of the experiment.

During the last three days of the experiment, 144 eggs were collected each day (2 eggs from each cage). These eggs were used to determine egg quality characteristics, the fatty acid profile of the egg yolk, and the oxidative stability of the egg yolk through the MDA method.

2.3. Diets

During the 30 days of the experiment, the hens were fed a standard corn–soybean meal diet containing 2750 kcal ME, 160 g crude protein, and 44 g ether extract per kg (Table 1). The 432 hens were split into 2 groups of 216, consisting of 3 subgroups of 72 each. The hens of the first group (group C) were fed daily with an average amount of ration of 120 g/day, an amount which, based on the laying rate and weight of the hens, is considered sufficient to support the normal nutritional requirements of the hens for maintenance and laying. The hens of the second group (group P) were given the same basic control diet and an additional 24 g of finely chopped fresh leaves and stems of the plant species purslane (*Portulaca oleraceae*) collected daily from the surrounding area of the poultry house. In the drinking water of hens of the 3 subgroups of group C and the 3 subgroups of group P, either no essential oil (C-0, P-0, respectively) or 100 ppm of oregano essential oil (C-ORE, P-ORE, respectively) or 100 ppm of a blend of oregano, sage, and fennel essential oils (C-BLEND, P-BLEND, respectively) was administered. Thus, 6 nutritional treatments emerged. The oregano essential oil used is marketed as a commercial formulation by the company "Ecopharm Hellas" under the trade name "Ecodiar Liquid". It is natural oregano essential oil in liquid form (5% v/v concentration) intended for livestock use and administered through drinking water. The blend of essential oils is released as a non-patented commercial formulation by the company "Dioscurides". It is a mixture of natural essential oils of oregano, sage, and fennel in a ratio of 4:1:1, respectively (6% v/v concentration). Like the previous formulation, it is administered to the animals through drinking water. The formulations mentioned above, produced by the two companies operating in Northern Greece, contain soy lecithin. Soy lecithin is used as an emulsifier to enhance the solubility of essential oils and facilitate their absorption through the intestinal tract.

Table 1. Ingredients and nutrient content of the experimental control diet.

	Control Diet
Ingredients (%)	
Corn	63.25
Soybean meal (45% crude protein)	23.90
Soybean oil	1.70
Limestone (calcium carbonate)	9.50
Monocalcium phosphate	0.80
Salt	0.30
Sodium carbonate	0.30
DL-methionine	0.15

Table 1. *Cont.*

	Control Diet
Vitamin and mineral premix [1]	0.10
Total	100.00
Nutrient content	
Metabolizable energy (kcal/kg)	2750.00
Dry matter (%)	89.02
Crude protein (%)	16.00
Digestible protein (%)	13.32
Crude fat (%)	4.40
Crude fiber (%)	2.83
Total lysine (%)	0.82
Total methionine (%)	0.41
Ash (%)	12.73
Calcium (%)	3.92
Total phosphorus (%)	0.51
Available phosphorus (%)	0.38

[1] Supplying per kg of feed: 10,000 IU vitamin A; 3000 IU vitamin D3; 30 mg vitamin E; 2.5 mg vitamin K3; 3 mg thiamine; 5 mg riboflavin; 5.4 mg nicotinamide; 4 mg pyridoxine; 20 mcg vitamin B12; 50 mg pantothenic acid; 0.5 mg folic acid; 150 mcg biotin; 1045 mg choline; 30 mg Fe; 10 mg Cu; 30 mg Zn; 50 mg Mn; 2 mg I; 0.2 mg Se; 1.5 mg Mo, and phytase 500 FTU.

Fresh and clean, healthy water was supplied from three independent tanks with a float, where essential oil was mixed daily with a proportionate amount of water to reach a concentration of 100 ppm. Each hen consumed approximately 0.5 L of water daily, which is approximately four times the amount of dry matter consumed through their diet.

2.4. *Egg Collection and Determinations*

2.4.1. Egg Quality Traits

On the 28th day of the experiment, a total of 144 eggs were collected, i.e., 6 treatments $\times$ 12 replicates (cages)/treatment $\times$ 2 eggs/cage, which were used to determine their quality characteristics.

Quality traits measured/estimated were egg weight; longitudinal and transverse axes; specific gravity; Haugh units; yolk weight, diameter, height, pH, and color; albumen weight, height, and pH; and shell weight, thickness, and hardness.

2.4.2. Materials and Reagents

The reagents used in the analyses for fatty acid profile and oxidative stability of egg yolk included: n-hexane pesticide grade (ChemLab, Zedelgem, Belgium), sodium hydroxide p.a. (Honeywell Fluka, München, Germany), methanol (VWR chemicals, Radnor, PA, USA), sodium bisulfate pure (Fluca München, Germany), 2-thiobarbituric acid (Sigma-Aldrich, Darmstadt, Germany), butylated hydroxytoluene (Sigma Aldrich, Darmstadt, Germany), trichloroacetic acid (Merck, Darmstadt, Germany), ferrous sulfate p.a. (Merck Darmstadt, Germany), sodium sulfate p.a. (Honeywell Fluka, München, Germany), chloroform (Merck Darmstadt, Germany), ascorbic acid p.a. (Panreac, Barcelona, Spain) and 1,1,3,3 tetraethoxypropane (Sigma Aldrich, Darmstadt, Germany). A 37-component mixture of fatty acids, methyl esters, and FAME mix C_6–C_{24} (SIGMA 18919-1AMP, certified reference material) was used as the reference standard.

2.4.3. Fatty Acid Profile of Egg Yolk

On the 29th day, an additional 144 eggs were collected using the same method as described above. These eggs were then placed in a freezer at −20 °C to be subsequently used for the determination of their yolk fatty acid profile. The eggs were then thawed at room temperature, broken, the yolks were separated, and the adhering albumen was removed by rolling on a paper towel. Yolk pools were prepared from 2 eggs from each cage and mixed with a wire whisk, resulting in 72 samples.

Lipid extraction of egg yolk was performed using Folch procedure [36]. Fatty acid methyl esters (FAMEs) were prepared and determined by gas chromatography (GC). Identification of FAMEs involved comparing their retention times with the retention times of the reference standards, and the results were expressed as a percentage (%) of the total fatty acids present in the sample.

2.4.4. Oxidative Stability of Egg Yolk

On the 30th day, another batch of 144 eggs was collected using the same method as previously described. These eggs were stored in a freezer at −20 °C for the subsequent determination of the oxidative stability of the egg yolk. The same procedure as described for the fatty acid profile determination was followed to generate 72 samples for analysis.

Lipid oxidation and iron-induced lipid oxidation of egg yolks were determined on the basis of the formation of MDA using a selective third-order derivative spectrophotometric method [37].

Susceptibility of eggs to iron-induced lipid oxidation was evaluated according to a slightly modified method [38]. In brief, yolk samples were homogenized, and four 1 g sub-samples from each yolk sample were weighed into 50 mL centrifuge tubes. Then, 1.5 mL of a solution containing 1.138 mM ferrous sulfate and 0.368 mM ascorbic acid was added to three of the sub-samples, and they were incubated at 37 °C for either 50, 100, or 150 min. Following incubation, all three iron-induced sub-samples, along with the 4th non-induced sub-sample, were immediately submitted to MDA assay for assessing the extent of lipid oxidation.

2.4.5. Determination of MDA

Determination of MDA, the compound used as an index of lipid peroxidation, was carried out by a third-order derivative method slightly modified to suit egg yolk analysis [37,39]. In brief, yolk samples were homogenized (Polytron homogenizer, PCU, Malters, Switzerland) in the presence of 8 mL of aqueous trichloroacetic acid (5% w/v) and 5 mL of butylated hydroxytoluene in hexane (0.8% w/v) and the mixture was centrifuged. The top layer was discarded, and the bottom aqueous layer was transferred to a volumetric flask (10 mL). Diluted to 10 mL with an aqueous solution of trichloroacetic acid (5% w/v), a 2.5 mL aliquot from this solution was mixed with 1.5 mL of aqueous 2-thiobarbituric acid (0.8% w/v). The mixture was further incubated at 70 °C for 30 min. After incubation, the mixture was cooled in a cold-water bath and submitted to conventional spectrophotometry. Lipid oxidation was expressed as nanograms of MDA per gram of yolk sample.

2.5. Statistical Analysis

Data were subjected to analysis of variance using the SPSS 20 statistical package [40]. A significance level of $p < 0.05$ was used. Differences between means were tested with Duncan's test.

The quality characteristics of the eggs and the fatty acid profiles were analyzed using a factorial experimental design (2 × 3) that included two factors: the type of diet (C or P) and the presence or absence of essential oils in the water (0, ORE, or BLEND). The main effects of both the diet and essential oils were determined, as well as their interaction.

The data related to the oxidative stability, as determined by the MDA method, were processed using a 2 × 3 × 4 factorial experimental design that included three factors: the type of diet (C or P), the presence or absence of essential oils in the water (0, ORE, or BLEND), and the duration of the incubation period (0, 50, 100, or 150 min). The main effects of the diet, essential oils, and time were determined, as well as their two-way and three-way interactions.

3. Results

3.1. Egg Production Rate and Egg Quality Traits

No mortality was observed throughout the experiment. Furthermore, no significant difference was observed between the treatments in terms of the change in the mean body weight of the hens between initial and final weighing (1948 g vs. 1934 g, respectively).

Table 2 shows the effect of purslane supplementation and essential oil addition on egg production rate and egg quality traits. According to the table, the addition of purslane to the ration or essential oils to the drinking water of hens did not significantly affect the rate of egg production.

Table 2. Effect of purslane supplementation and essential oil addition on egg production rate and egg quality traits.

Variable	Egg Production Rate	Egg Weight	Egg Mass Production	Longitudinal Axis	Transverse Axis	Specific Gravity
	(%)	(g)	(g/hen/day)	(mm)	(mm)	(g/cm^3)
Diet (D)						
C	68.5	68.5 ^{b}	46.8 ^{b}	59.1 ^{b}	48.7 ^{b}	1.077
P	68.4	70.7 ^{a}	48.4 ^{a}	62.5 ^{a}	50.0 ^{a}	1.078
SEM	0.253	0.494	0.374	0.827	0.297	0.004
Essential Oils (EO)						
0	68.4	69.7	47.8	60.6	49.4	1.080
ORE	68.2	69.8	47.5	60.9	49.2	1.075
BLEND	68.9	69.4	47.7	60.9	49.6	1.076
SEM	0.208	0.120	0.088	0.100	0.115	0.002
D × EO						
C-0	68.5	68.2 ^{b}	46.7 ^{b}	58.9 ^{b}	49.0 ^{b}	1.080
C-ORE	68.2	68.8 ^{b}	46.8 ^{b}	59.2 ^{b}	48.4 ^{b}	1.074
C-BLEND	68.8	68.6 ^{b}	47.0 ^{b}	59.1 ^{b}	48.8 ^{b}	1.077
P-0	68.3	71.1 ^{a}	48.8 ^{a}	62.3 ^{a}	49.8 ^{a}	1.079
P-ORE	68.1	70.7 ^{a}	48.1 ^{a}	62.6 ^{a}	49.9 ^{a}	1.076
P-BLEND	68.9	70.2 ^{a}	48.4 ^{a}	62.7 ^{a}	50.3 ^{a}	1.075
SEM	0.203	0.327	0.261	0.759	0.192	0.002
***p*-value**						
D	0.372	**0.023**	**0.028**	**0.039**	**0.018**	0.817
EO	0.277	0.090	0.240	0.428	0.317	0.587
D × EO	0.130	**0.038**	**0.042**	**0.032**	**0.036**	0.256
Variable	**Haugh Units**	**Yolk Weight**	**Yolk Diameter**	**Yolk Height**	**Yolk pH**	**Yolk Color**
		(g)	(mm)	(mm)		(DSM Scale)
Diet (D)						
C	62.70	19.8	42.6	18.7	6.26	9.1 ^{b}
P	64.04	20.3	43.3	18.3	6.28	10.1 ^{a}
SEM	0.451	0.162	0.256	0.134	0.024	0.231
Essential Oils (EO)						
0	63.33	20.0	43.6	18.6	6.27	9.5
ORE	62.18	20.2	42.7	18.9	6.31	9.6
BLEND	64.60	19.9	42.7	18.1	6.24	9.7
SEM	0.699	0.088	0.300	0.233	0.020	0.058

Table 2. *Cont.*

D × EO						
C-0	62.45	19.4	42.7	19.1	6.25	9.2 [b]
C-ORE	60.73	20.0	41.4	18.8	6.32	9.0 [b]
C-BLEND	64.91	19.9	43.7	18.1	6.21	9.1 [b]
P-0	64.21	20.5	44.4	18.0	6.29	9.8 [a]
P-ORE	63.62	20.4	43.9	18.9	6.29	10.1 [a]
P-BLEND	64.29	19.9	41.6	18.0	6.26	10.3 [a]
SEM	1.187	0.362	0.852	0.306	0.031	0.137
***p*-value**						
D	0.412	0.068	0.317	0.343	0.849	**0.035**
EO	0.435	0.121	0.695	0.258	0.712	0.329
D × EO	0.347	0.062	0.512	0.480	0.743	**0.027**
Variable	**Albumen Weight**	**Albumen Height**	**Albumen pH**	**Shell Weight**	**Shell Thickness**	**Shell Hardness**
	(g)	**(mm)**		**(g)**	**(mm)**	**(N)**
Diet (D)						
C	41.9	7.5	8.75	6.7	0.45	37.6
P	43.3	7.4	8.74	6.9	0.43	36.3
SEM	0.433	0.040	0.032	0.063	0.016	0.434
Essential Oils (EO)						
0	42.7	7.5	8.69	6.9	0.44	37.3
ORE	42.4	7.6	8.83	7.0	0.45	38.3
BLEND	42.8	7.3	8.75	6.5	0.44	35.4
SEM	0.120	0.088	0.041	0.153	0.006	0.850
D × EO						
C-0	42.0	7.5	8.66	6.6	0.45	38.3
C-ORE	41.7	7.7	8.84	6.9	0.45	39.2
C-BLEND	42.0	7.3	8.76	6.5	0.44	35.4
P-0	43.3	7.5	8.71	7.1	0.42	36.3
P-ORE	43.1	7.5	8.81	7.0	0.44	37.3
P-BLEND	43.6	7.2	8.74	6.5	0.43	35.3
SEM	0.516	0.142	0.052	0.199	0.009	1.247
***p*-value**						
D	0.072	0.856	0.912	0.574	0.833	0.517
EO	0.133	0.647	0.737	0.483	0.819	0.344
D × EO	0.057	0.494	0.549	0.639	0.754	0.612

C: control diet; P: diet containing leaves and stems of purslane; C-0, P-0: no addition of essential oils in drinking water; C-ORE, P-ORE: addition of 100 ppm of oregano essential oil in drinking water; C-BLEND, P-BLEND: addition of 100 ppm of blend of oregano, sage, and fennel essential oils in drinking water; SEM: standard error of the means; [a,b] means within a column with different superscripts inside each factor differ significantly at $p < 0.05$.

Treatments supplemented with fresh, finely chopped purslane leaves and stems showed a significant ($p < 0.05$) increase in egg weight compared to those without supplementation. The addition of essential oils did not have a significant effect on egg weight. Egg mass production per hen per day significantly ($p < 0.05$) increased in the purslane treatments, similar to the observed increase in egg weight. Additionally, the length of the longitudinal and transverse axes of the eggs of hens treated with purslane significantly ($p < 0.05$) increased compared to those of hens that did not receive purslane. In contrast, the addition of essential oils did not have a statistically significant effect on these parameters. Dietary supplementation with purslane also resulted in a significant ($p < 0.05$) improvement in yolk color, whereas providing essential oils through water did not produce a similar outcome.

The supplementation of purslane and the addition of essential oils to the hens' drinking water did not have a significant impact on the other egg quality characteristics, including specific gravity; Haugh units; yolk weight, diameter, height, and pH; albumen weight, height, and pH; and shell weight, thickness, and hardness.

3.2. Fatty Acid Profile of Egg Yolk

From Table 3, it can be concluded that dietary supplementation with purslane significantly modified the content of saturated and polyunsaturated fatty acids in egg yolk. Specifically, treatments receiving purslane had significantly ($p < 0.05$) higher levels of polyunsaturated omega-3 fatty acid α-linolenic (ALA), as well as omega-6 fatty acid linoleic (LA), compared to those not receiving purslane. Correspondingly, the content of stearic acid in the egg yolk decreased significantly ($p < 0.05$) in the purslane groups. The administration of oregano essential oil (ORE) or the blend of oregano, sage, and fennel essential oils (BLEND) in the drinking water of the layers did not result in significant differences. Furthermore, statistical analysis showed no significant differences among the treatments regarding the remaining fatty acids.

Table 3. Fatty acid (FA) profile of egg yolk.

FA (% of Total FA)	C14:0 (Myristic)	C16:0 (Palmitic)	C16:1n-7 (Palmitoleic)	C18:0 (Stearic)	C18:1n-9 (Oleic)	C18:1n-7 (Vaccenic)
Diet (D)						
C	0.40	28.29	3.55	9.32 ^{a}	39.47	3.66
P	0.35	27.76	3.10	7.26 ^{b}	39.14	3.79
SEM	0.014	0.184	0.130	0.395	0.173	0.040
Essential Oils (EO)						
0	0.39	27.87	3.38	8.18	39.58	3.58
ORE	0.34	27.61	3.19	8.44	39.67	3.70
BLEND	0.39	28.60	3.41	8.26	38.67	3.89
SEM	0.017	0.296	0.069	0.077	0.319	0.090
D × EO						
C-0	0.43	28.17	3.63	9.18 ^{a}	39.74	3.56
C-ORE	0.34	27.94	3.34	9.46 ^{a}	39.83	3.54
C-BLEND	0.42	28.75	3.67	9.33 ^{a}	38.84	3.88
P-0	0.35	27.57	3.12	7.18 ^{b}	39.41	3.60
P-ORE	0.33	27.28	3.04	7.41 ^{b}	39.51	3.86
P-BLEND	0.36	28.44	3.15	7.18 ^{b}	38.49	3.90
SEM	0.029	0.383	0.190	0.405	0.356	0.121
***p*-value**						
D	0.612	0.654	0.411	**0.007**	0.797	0.487
EO	0.593	0.599	0.519	0.378	0.676	0.313
D × EO	0.585	0.613	0.427	**0.008**	0.387	0.421

FA (% of Total FA)	C18:2n-6 (Linoleic, LA)	C18:3n-3 (α-Linolenic, ALA)	C20:1n-9 (Gondoic)	C20:4n-6 (Arachidonic)	C22:4n-6 (Adrenic)	C22:6n-3 (Docosahexaenoic, DHA)
Diet (D)						
C	10.69 ^{b}	0.18 ^{b}	0.18	2.64	0.27	1.64
P	12.74 ^{a}	0.85 ^{a}	0.19	2.84	0.29	1.65
SEM	0.463	0.123	0.004	0.059	0.009	0.006
Essential Oils (EO)						
0	11.79	0.48	0.18	2.75	0.29	1.63
ORE	11.97	0.54	0.19	2.67	0.27	1.60
BLEND	11.39	0.50	0.21	2.82	0.30	1.69
SEM	0.171	0.027	0.009	0.043	0.009	0.026

Table 3. *Cont.*

D × EO						
C-0	10.60 ^{b}	0.18 ^{b}	0.18	2.57	0.27	1.62
C-ORE	11.06 ^{b}	0.17 ^{b}	0.19	2.52	0.26	1.59
C-BLEND	10.41 ^{b}	0.18 ^{b}	0.20	2.84	0.29	1.70
P-0	12.97 ^{a}	0.77 ^{a}	0.17	2.92	0.31	1.63
P-ORE	12.88 ^{a}	0.96 ^{a}	0.19	2.81	0.27	1.61
P-BLEND	12.36 ^{a}	0.81 ^{a}	0.21	2.79	0.30	1.68
SEM	0.303	0.116	0.013	0.131	0.015	0.034
***p*-value**						
D	**0.008**	**<0.001**	0.873	0.085	0.713	0.878
EO	0.135	0.221	0.855	0.150	0.694	0.139
D × EO	**0.007**	**<0.001**	0.843	0.079	0.633	0.216

C: control diet; P: diet containing leaves and stems of purslane; C-0, P-0: no addition of essential oils in drinking water; C-ORE, P-ORE: addition of 100 ppm of oregano essential oil in drinking water; C-BLEND, P-BLEND: addition of 100 ppm of blend of oregano, sage, and fennel essential oils in drinking water; SEM: standard error of the means; ^{a,b} means within a column with different superscripts inside each factor differ significantly at $p < 0.05$.

Table 4 presents the cumulative concentrations of saturated (SFA), monounsaturated (MUFA), polyunsaturated (PUFA), omega-3, and omega-6 fatty acids. From the table, it can be observed that the inclusion of purslane in the layers' diet resulted in a significant ($p < 0.05$) decrease in SFA. This reduction was accompanied by a significant ($p < 0.05$) increase in PUFA, omega-3, and omega-6 fatty acids. The concentration of MUFA appeared to remain unchanged with the addition of purslane to the diet.

Table 4. Grouped fatty acids (%) in egg yolk.

Grouped FA (%)	SFA	MUFA	PUFA	Omega-3 FA	Omega-6 FA
Diet (D)					
C	37.8 ^{a}	47.0	15.3 ^{b}	1.8 ^{b}	13.5 ^{b}
P	35.2 ^{b}	46.4	18.3 ^{a}	2.5 ^{a}	15.8 ^{a}
SEM	0.622	0.244	0.568	0.152	0.516
Essential Oils (EO)					
0	36.5	46.7	16.9	2.1	14.8
ORE	36.2	46.8	17.1	2.2	14.9
BLEND	36.9	46.5	16.6	2.2	14.4
SEM	0.203	0.088	0.145	0.033	0.265
D × EO					
C-0	37.8 ^{a}	47.1	15.1 ^{b}	1.8 ^{b}	13.3 ^{b}
C-ORE	37.4 ^{a}	47.0	15.6 ^{b}	1.8 ^{b}	13.8 ^{b}
C-BLEND	38.2 ^{a}	46.8	15.3 ^{b}	1.9 ^{b}	13.4 ^{b}
P-0	35.1 ^{b}	46.3	18.6 ^{a}	2.4 ^{a}	16.2 ^{a}
P-ORE	34.9 ^{b}	46.6	18.5 ^{a}	2.6 ^{a}	15.9 ^{a}
P-BLEND	35.5 ^{b}	46.2	17.9 ^{a}	2.5 ^{a}	15.4 ^{a}
SEM	0.453	0.250	0.531	0.112	0.356
***p*-value**					
D	**0.037**	0.156	**<0.001**	**0.007**	**0.005**
EO	0.364	0.413	0.271	0.815	0.402
D × EO	**0.042**	0.274	**<0.001**	**0.008**	**0.007**

C: control diet; P: diet containing leaves and stems of purslane; C-0, P-0: no addition of essential oils in drinking water; C-ORE, P-ORE: addition of 100 ppm of oregano essential oil in drinking water; C-BLEND, P-BLEND: addition of 100 ppm of blend of oregano, sage, and fennel essential oils in drinking water; SEM: standard error of the means; ^{a,b} means within a column with different superscripts inside each factor differ significantly at $p < 0.05$.

Table 5 shows the relationships between the aforementioned concentrations. According to the table, the inclusion of purslane led to a significant ($p < 0.05$) increase in MUFA/SFA, PUFA/SFA, and UFA/SFA ratios. Moreover, the nutritional index omega-6

FA/omega-3 FA was significantly ($p < 0.05$) improved. On the other hand, the addition of ORE or BLEND to the hens' drinking water did not seem to significantly affect the concentrations of fatty acids and the relationships between them.

Table 5. Nutritional indices of fatty acids in egg yolk.

Nutritional Indices of FA	MUFA/SFA	PUFA/SFA	UFA/SFA	Omega-6 FA/Omega-3 FA
Diet (D)				
C	1.24^{b}	0.40^{b}	1.65^{b}	7.56^{a}
P	1.32^{a}	0.52^{a}	1.84^{a}	6.34^{b}
SEM	0.020	0.026	0.047	0.299
Essential Oils (EO)				
0	1.29	0.46	1.75	7.06
ORE	1.29	0.48	1.77	6.99
BLEND	1.27	0.46	1.72	6.81
SEM	0.007	0.007	0.015	0.074
D × EO				
C-0	1.25^{b}	0.39^{b}	1.65^{b}	7.58^{a}
C-ORE	1.24^{b}	0.42^{b}	1.67^{b}	7.76^{a}
C-BLEND	1.23^{b}	0.40^{b}	1.63^{b}	7.35^{a}
P-0	1.32^{a}	0.53^{a}	1.85^{a}	6.54^{b}
P-ORE	1.34^{a}	0.53^{a}	1.87^{a}	6.21^{b}
P-BLEND	1.30^{a}	0.51^{a}	1.80^{a}	6.26^{b}
SEM	0.012	0.021	0.028	0.193
***p*-value**				
D	**0.041**	**0.006**	**0.043**	**0.030**
EO	0.673	0.483	0.224	0.184
D × EO	**0.038**	**0.006**	**0.041**	**0.033**

C: control diet; P: diet containing leaves and stems of purslane; C-0, P-0: no addition of essential oils in drinking water; C-ORE, P-ORE: addition of 100 ppm of oregano essential oil in drinking water; C-BLEND, P-BLEND: addition of 100 ppm of blend of oregano, sage, and fennel essential oils in drinking water; SEM: standard error of the means; ^{a,b} means within a column with different superscripts inside each factor differ significantly at $p < 0.05$.

3.3. Oxidative Stability of Egg Yolk

The degree of oxidation was measured by determining the concentration of malondialdehyde (MDA), which is one of the final products of peroxidation of polyunsaturated fatty acids. Oxidative rancidity was measured at four incubation periods: 0, 50, 100, and 150 min. The main effects of diet (D), essential oils (EO), and time of incubation (T), as well as their interactions on MDA concentration in egg yolk, are presented in Table 6 and Figures 1 and 2.

Table 6. Main effects of diet, essential oils, and time and their interactions on MDA concentration in egg yolk.

	p-Value
Main effects	
D	0.482
EO	**<0.001**
T	**<0.001**
Interactions	
D × EO	**<0.001**
D × T	**<0.001**
EO × T	**<0.001**
D × EO × T	**<0.001**

MDA: concentration of malondialdehyde expressed in ng per g of egg yolk; D: diet; EO: essential oils; T: time of incubation; values with $p < 0.05$ are statistically significant.

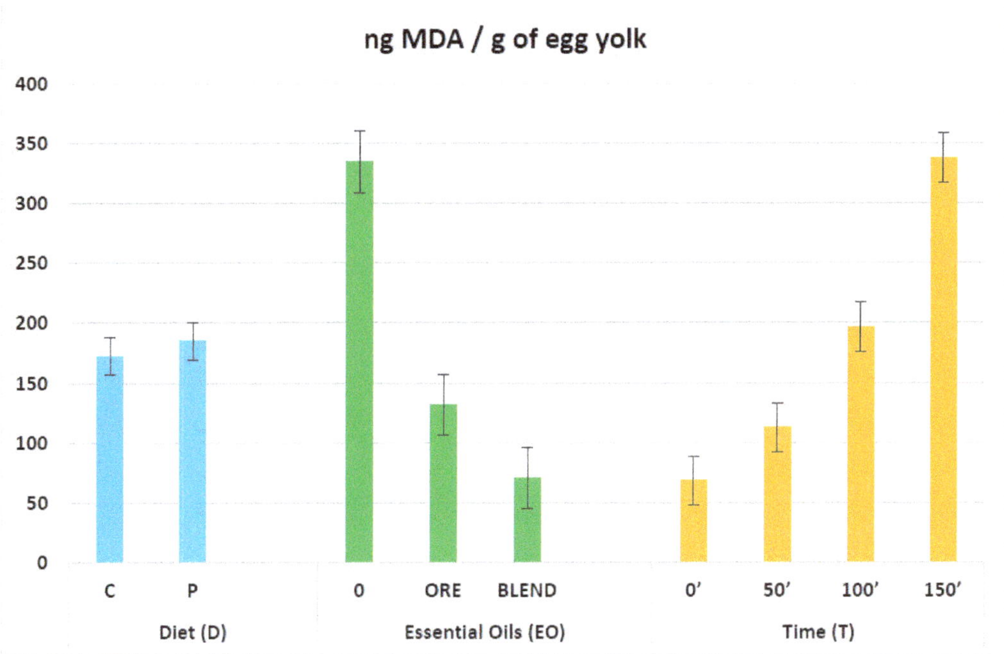

Figure 1. Main effects of diet (D), essential oils (EO), and time (T) on MDA concentration in egg yolk. MDA: concentration of malondialdehyde expressed in ng per g of egg yolk; C: control diet, P: diet containing leaves and stems of purslane; 0: no addition of essential oils in drinking water, ORE: addition of 100 ppm of oregano essential oil in drinking water; BLEND: addition of 100 ppm of blend of oregano, sage, and fennel essential oils in drinking water; 0′, 50′, 100′, 150′: time of incubation in minutes.

Based on the results, supplementation of purslane in the diet did not significantly affect MDA concentration. However, the addition of essential oils led to a significant ($p < 0.05$) reduction in MDA concentration, with BLEND being more effective than ORE. The length of incubation time had a significant ($p < 0.05$) increasing effect on MDA concentration.

All two-way interactions were significant ($p < 0.05$) on MDA concentration. Specifically, the addition of essential oils significantly ($p < 0.05$) reduced MDA concentration in both the no-purslane and purslane treatments, with BLEND being more effective. Increasing the length of incubation time significantly ($p < 0.05$) increased the concentration of MDA in treatments without purslane and in those with purslane. The length of incubation time exerted an increasing effect on MDA concentration, which was, however, significantly ($p < 0.05$) limited by the addition of essential oils, with BLEND having a greater limiting effect compared to ORE.

The D × EO × T triple interaction was significant ($p < 0.05$). While the addition of purslane and incubation time positively interacted to increase MDA concentration, the addition of essential oils dramatically limited this increase, resulting in ORE and BLEND treatments having significantly ($p < 0.05$) lower MDA concentrations than treatments without essential oils.

In summary, the results indicate that the addition of essential oils contributed to a significant ($p < 0.05$) increase in the oxidative stability of egg yolk at all time points, independent of the administration of purslane, with the blend of oregano, sage, and fennel essential oils (BLEND) being the most effective.

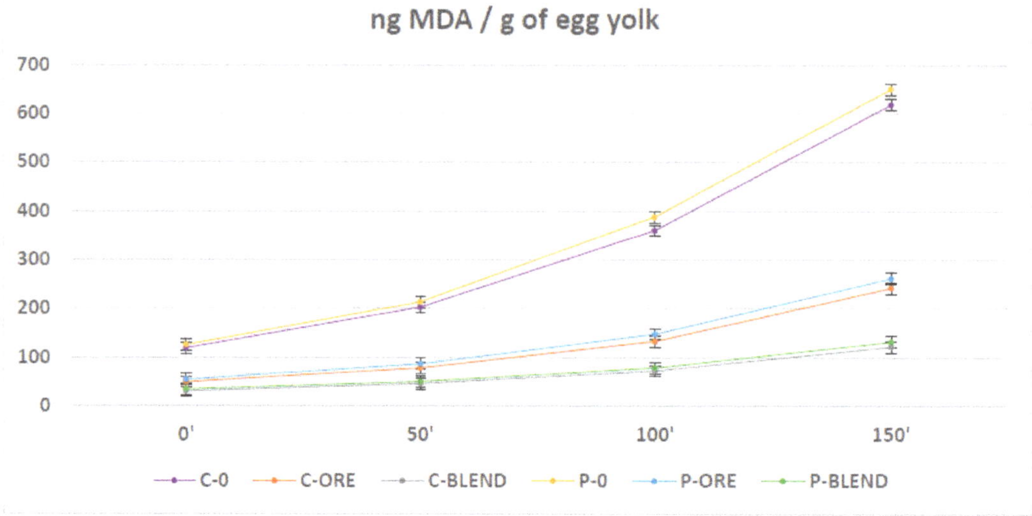

Figure 2. MDA concentration in egg yolk of the treatments following incubation periods of 0, 50, 100, and 150 min. MDA: concentration of malondialdehyde expressed in ng per g of egg yolk; C: control diet, P: diet containing leaves and stems of purslane; C-0, P-0: no addition of essential oils in drinking water; C-ORE, P-ORE: addition of 100 ppm of oregano essential oil in drinking water; C-BLEND, P-BLEND: addition of 100 ppm of blend of oregano, sage, and fennel essential oils in drinking water; 0′, 50′, 100′, 150′: time of incubation in minutes.

4. Discussion

In the present study, it was found that the addition of purslane resulted in a significant increase in egg weight and egg mass production. This increase may be attributed to the hens' increased feed and nutrient intake in the rations, which contained 24 g of fresh chopped leaves and stems of purslane per day per hen. Notably, there are no reports in the literature on the use of fresh purslane in the diets of laying hens. Previous research has primarily focused on the use of dried purslane and reported conflicting results regarding its effect on egg production rate, egg weight, and egg mass production.

The findings of the present research are consistent with the results reported by Aydin and Dogan [8] and Evaris et al. [41]. These studies found that the addition of dried purslane at rates of 10 or 20 g/kg and 100 or 200 g/kg, respectively, resulted in positive effects on egg weight and egg mass production. In the latter study, an increase in egg production was also observed. Conversely, other studies have reported positive effects only on egg weight [42] or only on egg mass production [43]. Moazedian and Saemi [9] did not find any effect on egg production or egg characteristics. On the other hand, Dalle Zotte and Pranzo [14] reported lower egg weight and total egg production in hens fed rations containing 20% dried purslane. This may be due to the reduced feed intake of the hens.

Yolk color was a trait that significantly improved in the purslane treatments, which is consistent with the findings of Kartikasari et al. [10]. An increased intensity of yolk color can be achieved by incorporating purslane, which is rich in xanthophylls and β-carotene, into the diets of laying hens [20].

The present research found that the addition of purslane significantly increased the concentration of the polyunsaturated fatty acids linoleic (LA) and α-linolenic (ALA) in the egg yolk. Similarly, Moazedian and Saemi [9] reported a significant increase in omega-3 fatty acids ALA and docosahexaenoic (DHA) in the egg yolk, which became even greater with an increase in the participation of purslane in the ration. Evaris et al. [41] also reported similar results. It is worth noting that the conversion efficiency of ALA to DHA by the hen's body is only 0–9%. However, a diet rich in ALA can still lead to an increase in yolk DHA content [10,44]. In the present study, the concentration of DHA was not significantly increased by the addition of purslane, possibly because the concentration of ALA in purslane leaves and stems was not high enough. It is also noteworthy that the concentration of LA was significantly increased in the purslane treatments, which is consistent with the findings of Moazedian and Saemi [9] and Dalle Zotte and Pranzo [14]. On the other hand, the decrease in the concentration of stearic acid observed in our study in the purslane treatments may be related to the low concentration (0.048 mg/g) of purslane leaves in stearic acid [17].

The percentage of saturated fatty acids (SFA), monounsaturated fatty acids (MUFA), and polyunsaturated fatty acids (PUFA) in the leaves of purslane are 16.4, 4.8, and 78.7%, respectively, while in the stems, it is 24.6, 3.9, and 71.5%, respectively [45]. Supplementing the diet of laying hens with feeds containing high levels of unsaturated fatty acids (UFA), such as purslane, can help reduce cholesterol in the egg yolk. This hypothesis is confirmed by Shehata and El-Krim [43], who found that *Portulaca oleracea* leaves added to the feed of layers play a very important role in reducing total lipids and total cholesterol in blood serum.

In our study, the omega-6/omega-3 ratio was significantly improved in the purslane-treated groups, a finding consistent with those of other researchers [8,9,46]. In agreement with the results of the present research, Dalle Zotte et al. [47] supplementing dried purslane meal to laying hens found a significant increase in omega-6 and omega-3 PUFA, as well as a significant improvement in the omega-6/omega-3 nutritional index. In addition, in the same study, a significant decrease in SFA content and an increase in PUFA content (LA, ALA, DHA) were observed. Similar results were found in the recent work of Dalle Zotte and Pranzo [14].

Lipid oxidation is a primary concern in the deterioration of foods like eggs. The addition of antioxidants to foods can inhibit lipid rancidity, delay the formation of toxic oxides, preserve nutritional value, and extend shelf life [48]. Purslane is known to contain various antioxidant compounds, including α-tocopherol, ascorbic acid, β-carotene, and 2,2-dyphenyl-1-picrylhydrazyl (DPPH). The total phenolic content (TPC) in different cultivars of *Portulaca oleracea*, expressed as total antioxidant capacity, ranges from 127 $\pm$ 13 to 478 $\pm$ 45 mg GAE/100 g of fresh plant weight [20]. However, in our study, the addition of purslane did not have a significant positive effect on the oxidative stability of egg yolk.

Essential oils (EO) from aromatic/medicinal plants contain active aromatic compounds that are widely used to prevent the peroxidation of egg yolk [49]. EO has antioxidant components that protect lipids from oxidation, thereby slowing lipid peroxidation [50]. The polyphenols carvacrol and thymol, which are major active compounds of oregano (*Origanum vulgare*) essential oil, can act as hydrogen donors to neutralize the excessive free radicals produced at the beginning of lipid oxidation [51]. The total phenolic content of the oregano extract amounts to 19.5 $\pm$ 0.2 mg gallic acid/g of dry sample [52]. When the diet of laying hens was supplemented with ground leaves and stems of dried oregano at a concentration of 5 g/kg feed, the oxidative stability of egg yolk was significantly increased [37]. According to the reports of Ghanima et al. [53], the addition of thymol formulation to layers' diet significantly reduced the level of MDA in egg yolk.

Among the essential oils of aromatic plants is the essential oil of the plant species sage (*Salvia officinalis*). The content of the cultivated sage in ethereal extract amounts to 2.37%, with the main active antioxidant components being cis-thujone and trans-thujone, in percentages of 33.80 and 6.97%, respectively [26]. The total phenolic content of sage extract amounts to 15.6 $\pm$ 0.1 mg gallic acid/g of dry sample [52]. Galamatis et al. [26] found a significant increase in the oxidative stability of the egg yolk when laying hens were fed diets supplemented with ground dried sage leaves and stems at levels of 0.5 and 1.0%, compared to the control, at incubation times of 50, 100, and 150 min.

Another plant species of interest for its antioxidant properties is fennel (*Foeniculum vulgare*). The essential oil of fennel seeds contains 16.81% trans-anethole (phytosterol) and 47.20% estragol, active substances with antioxidant potential that contribute to the reduction of cholesterol and triglycerides [54]. The total phenolic content of the essential oil extracted from fennel seeds is extremely high and is measured to be 70.42 mg GAE/g [55]. Gharaghani et al. [56] showed that consumption of diets containing fennel significantly reduced cholesterol and triglyceride levels in eggs. Abou-Elkhair et al. [57] found that the addition of fennel to the diets of layers had no significant effect on the MDA levels in the egg yolk, despite the strong antioxidant potential of fennel.

In our research, the addition of oregano essential oil (ORE) or the blend of essential oils (BLEND) contributed to a significant increase in the oxidative stability of the egg yolk at all incubation times, which is in general agreement with the aforementioned studies. The greater antioxidant capacity of BLEND compared to ORE is likely due to the greater total phenolic content of BLEND or possibly to a synergistic effect of the active components of the three essential oils in the mixture.

In summary, the findings of the present research provide a positive response to the question posed in the introduction regarding the possibility of producing an egg with increased omega-3 PUFA content protected by natural antioxidants. Our results indicate that adding fresh purslane to the feed and essential oils to the drinking water of laying hens can help achieve the production of such a product.

5. Conclusions

Incorporating fresh purslane into the ration of layers significantly contributes to the enrichment of the egg yolk in omega-3 PUFA. The addition of oregano essential oil, or a mixture of oregano, sage, and fennel seed essential oils, to the drinking water of laying hens results in a significant increase in the oxidative stability of the egg yolk, with a prolonged antioxidant effect over time.

The combined administration of fresh purslane and essential oils of aromatic plants could be a promising option for poultry farmers, both economically and environmentally. This approach utilizes natural raw materials from native or cultivated plants that are abundant in nature and have little impact on the cost of nutrition. It also mimics the nutritional habits of laying hens, particularly those reared in organic or free-range farms, and adapts to the physiology of their digestive system. Furthermore, this method contributes to producing eggs with an improved PUFA proportion, which are protected with natural plant-based antioxidants and potentially have an extended shelf life. Additionally, this approach meets the modern nutritional requirements of consumers and may enable product certification, leading to increased added value, farm viability, and sustainable development.

While the proposed management model has several advantages, one potential drawback is the limited availability of fresh purslane year-round. However, this can be easily addressed by cultivating the plant in a greenhouse.

Author Contributions: Conceptualization, V.D. and E.S.; methodology, I.M. and M.I.; software, G.S.; validation, V.D., D.G. (Dimitrios Gourdouvelis) and D.G. (Dimitrios Galamatis); formal analysis, G.S. and E.S.; investigation, V.D., D.G. (Dimitrios Gourdouvelis), L.H. and D.G. (Dimitrios Galamatis); resources, I.M. and M.I.; data curation, D.G. (Dimitrios Gourdouvelis) and I.M.; writing—original draft preparation, D.G. (Dimitrios Gourdouvelis) and G.S.; writing—review and editing, V.D. and E.S.; visualization, V.D.; supervision, E.S.; project administration, V.D.; funding acquisition, V.D., L.H. and E.S. All authors have read and agreed to the published version of the manuscript.

Funding: This research received no external funding.

Institutional Review Board Statement: The study was conducted according to the guidelines of Directive 2010/63/EU of the European Parliament and of the Council on the protection of animals used for scientific purposes. The research project was approved by the Institutional Review Board of the Aristotle University of Thessaloniki (Project Code: "PUR-SALORE EGG").

Data Availability Statement: All the relevant data are available in the paper.

Acknowledgments: We warmly thank the companies "Dioscurides" and "Ecopharm Hellas" for their kind supply of essential oil formulations.

Conflicts of Interest: The authors declare no conflict of interest.

References

1. Willett, W.C. Dietary fats and coronary heart disease. *J. Intern. Med.* **2012**, *272*, 13–24. [CrossRef] [PubMed]
2. Kris-Etherton, P.M.; Yu, S. Individual fatty acid effects on plasma lipids and lipoproteins: Human studies. *Am. J. Clin. Nutr.* **1997**, *65*, 1628S–1644S. [CrossRef]
3. Calzada, C.; Colas, R.; Guillot, N.; Guichardant, M.; Laville, M.; Véricel, E.; Lagarde, M. Subgram daily supplementation with docosahexaenoic acid protects low-density lipoproteins from oxidation in healthy men. *Atherosclerosis* **2010**, *208*, 467–472. [CrossRef] [PubMed]
4. Endo, J.; Arita, M. Cardioprotective mechanism of omega-3 polyunsaturated fatty acids. *J. Cardiol.* **2016**, *67*, 22–27. [CrossRef] [PubMed]
5. Kris-Etherton, P.M.; Grieger, J.A.; Etherton, T.D. Dietary reference intakes for DHA and EPA. *Prostaglandins Leukot. Essent. Fat. Acids* **2009**, *81*, 99–104. [CrossRef]
6. Calder, P.C. n-3 Fatty acids and cardiovascular disease: Evidence explained and mechanisms explored. *Clin. Sci.* **2004**, *107*, 1–11. [CrossRef]
7. Yang, L.G.; Song, Z.X.; Yin, H.; Wang, Y.Y.; Shu, G.F.; Lu, H.X.; Wang, S.K.; Sun, G.J. Low n-6/n-3 PUFA ratio improves lipid metabolism, inflammation, oxidative stress and endothelial function in rats using plant oils as n-3 fatty acid source. *Lipids* **2016**, *51*, 49–59. [CrossRef] [PubMed]
8. Aydin, R.; Dogan, I. Fatty acid profile and cholesterol content of egg yolk from chickens fed diets supplemented with purslane (*Portulaca oleracea* L.). *J. Sci. Food Agric.* **2010**, *90*, 1759–1763. [CrossRef]
9. Moazedian, M.; Saemi, F. Effects of different levels of Portulaca oleracea seed in laying hens diets containing rice bran on performance, egg quality, fatty acids, and cholesterol. *Comp. Clin. Pathol.* **2018**, *27*, 1397–1403. [CrossRef]
10. Kartikasari, L.R.; Nuhriawangsa, A.M.P.; Hertanto, B.S.; Swastike, W. Effect of supplementation purslane (*Portulaca oleracea*) as a source of alpha-linolenic acid on production performance and physical quality of egg of laying hens. *Anim. Prod.* **2015**, *17*, 149–153. [CrossRef]
11. Miranda, J.M.; Anton, X.; Redondo-Valbuena, C.; Roca-Saavedra, P.; Rodriguez, J.A.; Lamas, A.; Franco, C.M.; Cepeda, A. Egg and Egg-Derived Foods: Effects on Human Health and Use as Functional Foods. *Nutrients* **2015**, *7*, 706–729. [CrossRef]
12. Herron, K.L.; Fernandez, M.L. Are the current dietary guidelines regarding egg consumption appropriate? *J. Nutr.* **2004**, *134*, 187–190. [CrossRef]
13. Sirri, F.; Meluzzi, A. Modifying egg lipids for human health. In *Improving the Safety and Quality of Eggs and Egg Products*; Woodhead Publishing: Cambridge, UK, 2011; pp. 272–288.
14. Dalle Zotte, A.; Pranzo, G. Effects of dried Portulaca oleracea supplementation to the laying hen diet on productive performance, egg physical traits, fatty acid composition, and cholesterol content. *Czech J. Anim. Sci.* **2022**, *67*, 114–123. [CrossRef]
15. Kartikasari, L.R.; Hertanto, B.S.; Nuhriawangsa, A.M.P. External and Sensory Qualities of Brown Laying Hen Eggs Fed Diets Supplemented with Purslane (*Portulaca oleracea*) Meal Rich in Omega-3 Fatty Acids. *J. Ilmu Teknol. Has. Ternak* **2019**, *14*, 78–89. [CrossRef]
16. Ezekwe, M.O.; Omara-Alwala, T.R.; Membrahtu, T. Nutritive characterization of purslane accessions as influenced by planting date. *Plant Foods Hum. Nutr.* **1999**, *54*, 183–191. [CrossRef] [PubMed]
17. Simopoulos, A.P.; Norman, H.A.; Gillaspy, J.E.; Duke, J.A. Common purslane: A source of omega-3 fatty acids and antioxidants. *J. Am. Coll. Nutr.* **1992**, *11*, 374–382. [CrossRef]
18. Simopoulos, A.P. Essential fatty acids in health and chronic disease. *Am. J. Clin. Nutr.* **1999**, *70*, 560s–569s. [CrossRef] [PubMed]
19. Omara-Alwala, T.R.; Mebrahtu, T.; Prior, D.E.; Ezekwe, M.O. Omega-three fatty acids in purslane (*Portulaca oleracea*) tissues. *J. Am. Oil Chem. Soc.* **1991**, *68*, 198–199. [CrossRef]
20. Uddin, M.K.; Juraimi, A.S.; Hossain, M.S.; Nahar, M.A.U.; Ali, M.E.; Rahman, M.M. Purslane weed (*Portulaca oleracea*): A prospective plant source of nutrition, omega-3 fatty acid, and antioxidant attributes. *Sci. World J.* **2014**, 951019. [CrossRef]
21. Moghadam, M.B.; Cherian, G. Use of flaxseed in poultry feeds to meet the human need for n-3 fatty acids. *World's Poult. Sci. J.* **2017**, *73*, 803–812. [CrossRef]
22. Rehman, R.; Hanif, M.A.; Mushtaq, Z.; Al-Sadi, A.M. Biosynthesis of essential oils in aromatic plants: A review. *Food Rev. Int.* **2016**, *32*, 117–160. [CrossRef]
23. Fournomiti, M.; Kimbaris, A.; Mantzourani, I.; Plessas, S.; Theodoridou, I.; Papaemmanouil, V.; Kapsiotis, I.; Alexopoulos, A.; Panopoulou, M.; Stavropoulou, E.; et al. Antimicrobial activity of essential oils of cultivated oregano (*Origanum vulgare*), sage (*Salvia officinalis*), and thyme (*Thymus vulgaris*) against clinical isolates of Escherichia coli, Klebsiella oxytoca, and Klebsiella pneumoniae. *Microb. Ecol. Health Dis.* **2015**, *26*, 23289. [CrossRef] [PubMed]
24. Bozkurt, M.; Hippenstiel, F.; Abdel-Wareth, A.A.A.; Kehraus, S.; Küçükyilmaz, K.; Südekum, K.H. Effects of Selected Herbs and Essential Oils on Performance, Egg Quality and Some Metabolic Activities in Laying Hens—A Review. *Eur. Poult. Sci.* **2014**, *78*, 49. [CrossRef]

25. Skoufogianni, E.; Solomou, A.D.; Danalatos, N.G. Ecology, cultivation and utilization of the aromatic Greek oregano (*Origanum vulgare* L.): A review. *Not. Bot. Horti Agrobot.* **2019**, *47*, 545–552. [CrossRef]
26. Galamatis, D.; Papadopoulos, G.A.; Lazari, D.; Fletouris, D.; Petridou, E.; Arsenos, G.I.; Fortomaris, P. Effects of dietary supplementation of salvia officinalis l. in organic laying hens on egg quality, yolk oxidative stability and eggshell microbiological counts. *Animals* **2021**, *11*, 2502. [CrossRef]
27. Ghiasvand, A.R.; Khatibjoo, A.; Mohammadi, Y.; Akbari Gharaei, M.; Shirzadi, H. Effect of fennel essential oil on performance, serum biochemistry, immunity, ileum morphology and microbial population, and meat quality of broiler chickens fed corn or wheat-based diet. *Br. Poult. Sci.* **2021**, *62*, 562–572. [CrossRef] [PubMed]
28. Buğdayci, K.E.; Oğuz, F.K.; Oğuz, M.N.; Kuter, E. Effects of fennel seed supplementation of ration on performance, egg quality, serum cholesterol, and total phenol content of egg yolk of laying quails. *Rev. Bras. Zootec.* **2018**, *47*. [CrossRef]
29. Panaite, T.D.; Nour, V.; Saracila, M.; Turcu, R.P.; Untea, A.E.; Vlaicu, P.A. Effects of linseed meal and carotenoids from different sources on egg characteristics, yolk fatty acid and carotenoid profile and lipid peroxidation. *Foods* **2021**, *10*, 1246. [CrossRef]
30. Wu, H.; Richards, M.P.; Undeland, I. Lipid oxidation and antioxidant delivery systems in muscle food. *Compr. Rev. Food Sci. Food Saf.* **2022**, *21*, 1275–1299. [CrossRef]
31. Barriuso, B.; Astiasarán, I.; Ansorena, D. A review of analytical methods measuring lipid oxidation status in foods: A challenging task. *Eur. Food Res. Technol.* **2013**, *236*, 1–15. [CrossRef]
32. Botsoglou, N.; Florou-Paneri, P.; Botsoglou, E.; Dotas, V.; Giannenas, I.; Koidis, A.; Mitrakos, P. The effect of feeding rosemary, oregano, saffron and α-tocopheryl acetate on hen performance and oxidative stability of eggs. *S. Afr. J. Anim. Sci.* **2005**, *35*, 143–151. [CrossRef]
33. Yannakopoulos, A.L.; Tserveni-Gousi, A.S.; Yannakakis, S.; Yamoustaris, A. Yolk fatty acid composition of ω-3 eggs during the laying period. In Proceedings of the XIth European Symposium on the Quality of Eggs and Egg Products, Doorwerth, The Netherlands, 23–26 May 2005.
34. Christaki, E.; Bonos, E.; Giannenas, I.; Florou-Paneri, P. Aromatic plants as a source of bioactive compounds. *Agriculture* **2012**, *2*, 228–243. [CrossRef]
35. Giannenas, I.; Grigoriadou, K.; Sidiropoulou, E.; Bonos, E.; Cheilari, A.; Vontzalidou, A.; Karaiskou, C.; Aligiannis, N.; Florou-Paneri, P.; Christaki, E. Untargeted UHPLC-MS metabolic profiling as a valuable tool for the evaluation of eggs quality parameters after dietary supplementation with oregano, thyme, sideritis tea and chamomile on brown laying hens. *Metabolomics* **2021**, *17*, 51. [CrossRef] [PubMed]
36. Folch, J.; Lees, M.; Stanley, G.H.S. A simple method for the isolation and purification of total lipids from animal tissues. *J. Biol. Chem.* **1957**, *226*, 497–509. [CrossRef] [PubMed]
37. Botsoglou, N.A.; Florou-Paneri, P.; Nikolakakis, I.; Giannenas, I.; Dotas, V.; Botsoglou, E.N.; Aggelopoulos, S. Effect of dietary saffron (*Crocus sativus* L.) on the oxidative stability of egg yolk. *Br. Poult. Sci.* **2005**, *46*, 701–707. [CrossRef] [PubMed]
38. Kornbrust, D.J.; Mavis, R.D. Relative susceptibility of microsomes from lung, heart, liver, kidney, brain and testes to lipid peroxidation: Correlation with vitamin E content. *Lipids* **1980**, *15*, 315–322. [CrossRef]
39. Botsoglou, E.; Govaris, A.; Fletouris, D.; Iliadis, S. Olive leaves (*Olea europea* L.) and α-tocopheryl acetate as feed antioxidants for improving the oxidative stability of α-linolenic acid-enriched eggs. *J. Anim. Physiol. Anim. Nutr.* **2013**, *97*, 740–753. [CrossRef]
40. SPSS. *SPSS Statistics for Windows*; Release 20.0; IBM: Armonk, NY, USA, 2018.
41. Evaris, E.; Sarmiento-Franco, L.A.; Segura-Correa, J.; Capetillo-Leal, C. Effect of dietary inclusion of purslane (*Portulaca oleracea* L.) on yolk omega-3 fatty acids content, egg quality and productive performance of rhode island red hens. *Trop. Subtrop. Agroecosyst.* **2015**, *18*, 33–38.
42. Kartikasari, L.R.; Hertanto, B.S.; Nuhriawangsa, A.M. Egg Quality of Laying Hens Fed Different Diets Supplemented with Purslane (*Portulaca oleracea* L.) Meal Rich in Alpha-linolenic Acid (ALA): Egg Quality of Hens Fed diets with Purslane Meal. *Proc. Pak. Acad. Sci. B Life Environ. Sci.* **2020**, *57*, 27–33.
43. Shehata, M.; El-Krim, A. Role of dried purslane leaves meal and essential phospholipids in laying hen diets in reducing cholesterol biosynthesis. *J. Prod. Dev.* **2011**, *16*, 177–199.
44. Abedi, E.; Sahari, M.A. Long-chain polyunsaturated fatty acid sources and evaluation of their nutritional and functional properties. *Food Sci. Nutr.* **2014**, *2*, 443–463. [CrossRef] [PubMed]
45. Siriamornpun, S.; Suttajit, M. Microchemical components and antioxidant activity of different morphological parts of Thai wild purslane (*Portulaca oleracea*). *Weed Sci.* **2010**, *58*, 182–188. [CrossRef]
46. Najib, H.; Al-Yousef, Y.M. Essential fatty acid content of eggs and performance of layer hens fed with different levels of full-fat flaxseed. *J. Cell Anim. Biol.* **2010**, *4*, 58–63.
47. Dalle Zotte, A.; Tomasello, F.; Andrighetto, I. The dietary inclusion of Portulaca oleracea to the diet of laying hens increases the n-3 fatty acids content and reduces the cholesterol content in the egg yolk. *Ital. J. Anim. Sci.* **2012**, *4*, 157–159. [CrossRef]
48. Harlina, P.W.; Ma, M.; Shahzad, R.; Gouda, M.M.; Qiu, N. Effect of clove extract on lipid oxidation, antioxidant activity, volatile compounds and fatty acid composition of salted duck eggs. *J. Food Sci. Technol.* **2018**, *55*, 4719–4734. [CrossRef]
49. Migliorini, M.J.; Boiago, M.M.; Stefani, L.M.; Zampar, A.; Roza, L.F.; Barreta, M.; Arno, A.; Robazza, W.S.; Giuriatti, J.; Galvão, A.C.; et al. Oregano essential oil in the diet of laying hens in winter reduces lipid peroxidation in yolks and increases shelf life in eggs. *J. Therm. Biol.* **2019**, *85*, 102409. [CrossRef]

50. Florou-Paneri, P.; Nikolakakis, I.; Giannenas, I.; Koidis, A.; Botsoglou, E.; Dotas, V.; Mitsopoulos, I. Hen performance and egg quality as affected by dietary oregano essential oil and tocopheryl acetate supplementation. *Int. J. Poult. Sci.* **2005**, *4*, 449–454.
51. Botsoglou, N.A.; Govaris, A.; Botsoglou, E.N.; Grigoropoulou, S.H.; Papageorgiou, G. Antioxidant activity of dietary oregano essential oil and α-tocopheryl acetate supplementation in long-term frozen stored turkey meat. *J. Agric. Food Chem.* **2003**, *51*, 2930–2936. [CrossRef]
52. Proestos, C.; Lytoudi, K.; Mavromelanidou, O.K.; Zoumpoulakis, P.; Sinanoglou, V.J. Antioxidant capacity of selected plant extracts and their essential oils. *Antioxidants* **2013**, *2*, 11–22. [CrossRef] [PubMed]
53. Ghanima, M.M.A.; Alagawany, M.; Abd El-Hack, M.E.; Taha, A.; Elnesr, S.S.; Ajarem, J.; Allam, A.A.; Mahmoud, A.M. Consequences of various housing systems and dietary supplementation of thymol, carvacrol, and euganol on performance, egg quality, blood chemistry, and antioxidant parameters. *Poult. Sci.* **2020**, *99*, 4384–4397. [CrossRef]
54. Khan, R.U.; Fatima, A.; Naz, S.; Ragni, M.; Tarricone, S.; Tufarelli, V. Perspective, Opportunities and Challenges in Using Fennel (*Foeniculum vulgare*) in Poultry Health and Production as an Eco-Friendly Alternative to Antibiotics: A Review. *Antibiotics* **2022**, *11*, 278. [CrossRef] [PubMed]
55. Barakat, H.; Alkabeer, I.A.; Aljutaily, T.; Almujaydil, M.S.; Algheshairy, R.M.; Alhomaid, R.M.; Almutairi, A.S.; Mohamed, A. Phenolics and Volatile Compounds of Fennel (*Foeniculum vulgare*) Seeds and Their Sprouts Prevent Oxidative DNA Damage and Ameliorates CCl4-Induced Hepatotoxicity and Oxidative Stress in Rats. *Antioxidants* **2022**, *11*, 2318. [CrossRef] [PubMed]
56. Gharaghani, H.; Shariatmadari, F.; Torshizi, M.A. Effect of fennel (*Foeniculum vulgare* Mill.) used as a feed additive on the egg quality of laying hens under heat stress. *Braz. J. Poult. Sci.* **2015**, *17*, 199–207. [CrossRef]
57. Abou-Elkhair, R.; Selim, S.; Hussein, E. Effect of supplementing layer hen diet with phytogenic feed additives on laying performance, egg quality, egg lipid peroxidation and blood biochemical constituents. *Anim. Nutr.* **2018**, *4*, 394–400. [CrossRef]

sustainability

Article

The Effects of Varying Combinations of Dietary Selenium, Vitamin E, and Zinc Supplements on Antioxidant Enzyme Activity, and Developmental and Histological Traits in Testicular Tissues of 1-Year-Old Native Turkish Ganders

Hatice Baş [1], Hulüsi Ozan Taşkesen [2], Mehmet Akif Boz [2,*], Musa Sarıca [3], Kadir Erensoy [3], Vassilios Dotas [4] and George Symeon [5]

1. Department of Biology, Faculty of Arts and Science, Yozgat Bozok University, Yozgat 66900, Turkey; hatice.bas@yobu.edu.tr
2. Department of Animal Science, Faculty of Agriculture, Yozgat Bozok University, Yozgat 66900, Turkey; ozan.taskesen@yobu.edu.tr
3. Department of Animal Science, Faculty of Agriculture, Ondokuz Mayıs University, Samsun 55139, Turkey; msarica@omu.edu.tr (M.S.); kadir.erensoy@omu.edu.tr (K.E.)
4. Department of Animal Production, Faculty of Agriculture, Aristotle University of Thessaloniki, 54124 Thessaloniki, Greece; vdotas@agro.auth.gr
5. Research Institute of Animal Science, Hellenic Agricultural Organization Demeter, 58100 Giannitsa, Greece; gsymewn@yahoo.gr

* Correspondence: bozmakif@gmail.com

Abstract: The aim of this study was to determine the effects of varying combinations of dietary selenium (Se), vitamin E (Vit E), and zinc (Zn) supplements on antioxidant enzyme activity, and developmental and histological traits in testicular tissues of 1-year-old native Turkish ganders. A total of 48 animals were used and randomly assigned to 8 treatment groups (control, Se, Vit E, Zn, Se + Vit E, Se + Zn, Vit E + Zn, and Se + Vit E + Zn), with 6 birds in each group. In addition to the control (basic) diet, specific levels of supplements (0.3 mg/kg Se, 100 mg/kg Vit E, and 100 mg/kg Zn) were added to the diet of each treatment group. Antioxidative enzymes (superoxide dismutase, catalase, glutathione peroxidase, glutathione-S-transferase activities, and malondialdehyde level) were more advantageous in the testicular tissue of ganders fed with Se + Vit E + Zn. Malondialdehyde (MDA), which is an important indicator of lipid peroxidation, was not significantly affected by the dietary treatments. However, it was negatively correlated with the seminiferous tubule area (−0.34) and diameter (−0.35). Compared to the control, the highest seminiferous tubule area and germinative epithelial thickness were determined as being fed with Se + Vit E + Zn. The lowest seminiferous tubule diameter was determined in the control and Zn groups, while the highest was in the group fed with Se + Vit E + Zn and Se + Vit E. This study showed that the simultaneous supplementation of Se + Vit and E + Zn into the diet of native Turkish ganders had positive effects on the testicular tissue, by reducing oxidative damage and improving histological parameters without affecting their physiological status.

Keywords: ganders; Se; vitamin E; Zn; histology; oxidative stress

Citation: Baş, H.; Taşkesen, H.O.; Boz, M.A.; Sarıca, M.; Erensoy, K.; Dotas, V.; Symeon, G. The Effects of Varying Combinations of Dietary Selenium, Vitamin E, and Zinc Supplements on Antioxidant Enzyme Activity, and Developmental and Histological Traits in Testicular Tissues of 1-Year-Old Native Turkish Ganders. *Sustainability* **2023**, *15*, 12245. https://doi.org/10.3390/su151612245

Academic Editor: Marko Vinceković

Received: 11 July 2023
Revised: 1 August 2023
Accepted: 8 August 2023
Published: 10 August 2023

1. Introduction

In Türkiye, goose breeding is generally carried out in rural areas, and in addition to the traditional production structure of small-scale, recently, open grazing family farms and commercial intensive production have started [1–4]. However, low fertility and hatchability are major problems, especially when associated with low egg production in native Turkish geese. Goose owners often use 1-year-old ganders for breeding purposes in Türkiye. However, Boz et al. [5] reported that 1-year-old Turkish ganders had significantly lower fertility than 2-year-olds, alongside a lower semen quality.

Antioxidants are substances that prevent or delay oxidative stress, and this phenomenon is called antioxidant defense [6,7]. Cells have developed various defense mechanisms against oxidative stress, which are divided into enzymatic and non-enzymatic [8]. It is known that these defense mechanisms can be strengthened by the intake of non-enzymatic antioxidant substances, such as vitamins, Zn, and Se [9,10].

Free radicals are chemicals with an unpaired electron in their final orbit and are highly reactive. Although small amounts of free radicals are necessary during cell defense mechanisms, high levels can damage tissues and cause cell death [11]. Since free radicals easily exchange electrons with other molecules, such as lipids, proteins, and DNA, they cause changes in their structures [12], which disrupts the functioning of many organs [13].

Cells living in aerobic conditions are exposed to excessive amounts of oxidants. However, under normal conditions, there is a balance in living organisms between free radicals and antioxidant defense systems [14]. Oxidative stress is the disruption of the balance between the number of oxidants and antioxidant defense, in favor of the oxidants [15]. Moreover, owing to the tasks undertaken by antioxidant substances, cells are protected against damage by free radicals. When the number of free radicals exceeds the defense system, oxidative stress, and subsequent cell damage occur, which limits the functioning of organs [16].

Lipid peroxidation (LPO) is the major event that plays an important role in free radical toxicity. Free radicals attack the polyunsaturated fatty acids in the phospholipid layers in membranes to produce malondialdehyde (MDA), the end product of LPO [13]. Therefore, increased MDA is an important indicator of LPO [8]. Superoxide dismutase (SOD), catalase (CAT), glutathione-S-transferase (GST), and glutathione peroxidase (GPx) are enzymatic antioxidants that scavenge free radicals [10]. These enzymatic antioxidants in tissues neutralize the oxidative stress that occurs due to the formation of free radicals [17]. Therefore, if antioxidant enzyme activity in the cell is insufficient, an increase in the level of free radicals occurs, which promotes oxidative stress in cells [8]. SOD reduces the superoxide radical to hydrogen peroxide (H_2O_2) by donating an electron [12]. Then, the CAT enzyme converts H_2O_2 to water and molecular oxygen [15]. The main task of GPx, which uses GSH as a substrate, is to reduce H_2O_2 and alkyl peroxides [18]. Since GPx is responsible for the conversion of H_2O_2 to water, it reduces H_2O_2 levels [19]. GST is the enzyme responsible for the conjugation of various electrophilic substances to the GSH thiol group, meaning fewer toxic forms are created [12].

In male poultry, both testes are functional and located symmetrically on either side of the cervical roof midline in the abdominal cavity [20], which is the region of the body where they develop and perform their functions since the body temperature is 41–42 °C [21]. As in mammals, the testes in poultry have both endocrine and exocrine functions. While sperm production in the seminiferous tubules is an exocrine function, testosterone production by the Leydig cells in the interstitial tissue is an endocrine function [22].

Compared to other poultry species, overall reproductive performance is relatively low in native Turkish geese. The most important reason for this is low egg production on the female side and low sperm quality on the male side [5,23]. For this reason, by revealing the histological mechanisms in the testes where sperm production takes place, various strategies can be developed to improve sperm quality, especially in 1-year-old Turkish ganders. The aim of this study was to determine the effects of various combinations of dietary Se, Vit E, and Zn supplements on the antioxidant activity, and developmental and histological traits in testicular tissues of 1-year-old native Turkish ganders.

2. Materials and Methods

2.1. Animals and Experimental Design

This study was carried out at Yozgat Bozok University Research and Application Center, Yerköy Goose Production Farm. The animal material was from 48 1-year-old native Turkish ganders, with an average body weight of 3976.5 g. In this study, 8 treatment groups (control; Se; Vit E; Zn; Se + Vit E; Se + Zn; Zn + Vit E; Se + Vit E + Zn) were formed with 6 ganders in each.

2.2. Rearing and Feeding

The research was carried out in a house with 48 individual wire mesh cages (each 100 × 100 × 100 cm in size). The bottom of the cages was covered with plastic litter material, which does not harm the ganders. The house was naturally ventilated, yet fans were used when needed, and no additional heating was used. The temperature was kept between 18–24 °C during the study. Natural lighting was also provided through windows and no additional lighting was provided. An increase in day length was noted during the study, with daytime being approximately 11 h in March 2022 and 15 h in June 2022.

Since most of the goose production in Turkey is carried out under natural lighting conditions, the aim of this study was to simulate this situation. One feeder and one drinker were provided for each individual cage and the accumulated litter was cleaned every day. The ganders were fed the diets specified in the control and treatment groups for a total of 90 days (end of March to early June), in accordance with the reproduction periods of female geese in Turkey. At the end of this period, the ganders were sacrificed.

The ganders in the control group were fed the basic diet presented in Table 1, which was obtained from a private company. As shown in Table 2, the treatment groups were formed by adding specific levels of Se, Zn, and Vit E to the basic diet, with reference to Amem and Al-Daraji [24], Al-Daraji [25], and Jerysz and Lukaszewicz [26]. Individual ganders in each group were fed with 200 g/day feed, while water was provided ad libitum.

Table 1. Basic diet components and calculated contents *.

Ingredient	Unit	Amount
Corn	%	57.5
Sunflower seed meal	%	18.5
Soybean meal (CP 46%)	%	10.0
Limestone	%	8.0
Cotton seed meal (CP 26%)	%	5.0
Salt	%	0.75
Vit premix	%	0.25
	Analyzed nutrient content *	
Dry matter	%	88.76
Crude protein (CP)	%	15.50
ME	MJ/kg	10.29
Crude oil	%	3.30
Crude fiber	%	7.14
Crude ash	%	11.68
Se	mg/kg	0.15
Zn	mg/kg	60
Vit E	mg/kg	30

* Feed analyses were carried out at Yozgat Bozok University Science and Technology Application and Research Centre laboratory, Yozgat, Turkey.

2.3. Determination of Antioxidant Enzyme Activity

At the end of the study period, the ganders were dissected to investigate their malondialdehyde (MDA) levels and antioxidant enzymes (superoxide dismutase (SOD), catalase (CAT), glutathione peroxidase (GPx), and glutathione-S-transferase (GST)) activities. Testicular tissues were separated to determine the enzyme activities and MDA levels and were stored at −80 °C until they were studied. These testicular tissue samples stored at the time of the study were homogenized in homogenization buffer (pH 7.4) for 3 min and the concentration and activity were determined by measuring the absorbance of the samples using "Biotech Engineering/Spectroscan 60 DV" brand spectrophotometer [27]. For SOD, CAT, GPx, GST, and MDA values, absorbance values were measured by the spectrophotometer at 440 nm [28], 240 nm [29], 340 nm [30], 340 nm [31], and 532 nm [32], respectively.

Table 2. Se, Zn, and Vit E contents in the experimental groups.

Groups	Se (mg/kg)	Vit E (mg/kg)	Zn (mg/kg)
Control	0.15	30	60
Se	0.45	30	60
Vit E	0.15	130	60
Zn	0.15	30	160
Se + Vit E	0.45	130	60
Se + Zn	0.45	30	160
Zn + Vit E	0.15	130	160
Se + Vit E + Zn	0.45	130	160

Se: 67 mg/kg Se premix, 4.5% Sodium Selenite Na_2SeO_3; Vit E: 200 mg/kg E−50 Adsorbate Rovimix®, 50% Vit E; Zn: 131 mg/kg 76.4% Zn oxide.

2.4. Measurement of Body and Testicular Weights

Body weights of the ganders were measured on the experimental day of sacrifice with a 0.01 precision balance. Testes were quickly removed from the dissected ganders and weighed, after cleaning from neighboring tissues. Testicular weight was obtained by accumulating the weights of both testes (right and left testes). Relative testicular weights were calculated using the following formula:

$$\text{relative testicular weight} = (\text{testicular weight (g)}/\text{body weight (g)}) \times 100$$

2.5. Light Microscope Investigations

Testicular tissues removed from the dissected ganders were washed in buffer and placed in formaldehyde fixative for light microscopy examination. After the fixation stage, washing and dehydration procedures were performed. Then, the tissues were paraffin blocked. Sections of 6–7 μ in thickness were removed from the prepared blocks. The sections were stained with hematoxylin–eosin, examined, and photographed using a microscope with a camera attachment (Olympus BX 51 light microscope; Olympus Corp., Tokyo, Japan).

The archived images were analyzed morphometrically using the cell D Soft Imaging System (Olympus Corp., Tokyo, Japan). The internal diameter of the seminiferous tubules, the thickness of the germinative epithelium (without tunica propria), and the area of the seminiferous tubule and interstitial tissue were measured in every image. The relative area (%) of the interstitial tissue, defined as the ratio of the area occupied by the interstitium to the total area (interstitial and tubular area) within the archived view field under 200 magnification, was calculated for each individual [33].

2.6. Statistical Analysis

Statistical analysis was performed using SPSS 25.0 statistical software (SPSS, Inc., Chicago, IL, USA). Differences between varying combinations of the dietary Se, Zn, and Vit E supplements were analyzed by one-way ANOVA and Tukey's test. Significance was set at $p < 0.05$. Relationships between the various oxidative enzyme activity, and developmental and histology parameters were analyzed by calculating Pearson correlation coefficients (r). For the purposes of discussion, the following descriptors were used to describe the relative strength of the correlations: very weak (r < 0.20), weak (r = 0.20–0.39), moderate (r = 0.40–0.59), strong (r = 0.60–0.79), and very strong (r = 0.80–0.99) [34].

3. Results

The effects of varying combinations of dietary Se, Zn, and Vit E supplements on antioxidant enzyme activity in the testicular tissues of 1-year-old native Turkish ganders are presented in Table 3. Superoxide dismutase enzyme (SOD) activity, catalase enzyme (CAT) activity, glutathione peroxidase enzyme (GPx) activity, and glutathione-S-transferase enzyme (GST) activity were highest in the Se + Vit E + Zn combination and lowest in the control group (Table 3, $p < 0.05$). Malondialdehyde (MDA), which is an important indicator of lipid peroxidation, was not significantly affected by the dietary treatments.

Table 3. Effects of varying combinations of dietary Se, Vit E, and Zn supplements on antioxidant enzyme activity in testicular tissues of 1-year-old native Turkish ganders (n = 48).

Dietary Treatments	SOD (U/mg Protein)	CAT (mmol/mg Protein)	GPx (mmol/mg Protein)	GST (mmol/mg Protein)	MDA (mmol/mg Protein)
Control	4.84 ^{c}	1.21 ^{b}	5.10 ^{c}	0.83 ^{c}	0.47
Se	6.56 ^{abc}	1.25 ^{ab}	7.66 ^{abc}	1.05 ^{abc}	0.44
Vit E	6.54 ^{abc}	1.27 ^{ab}	7.83 ^{abc}	1.01 ^{abc}	0.42
Zn	5.30 ^{bc}	1.20 ^{b}	6.40 ^{bc}	0.89 ^{bc}	0.45
Se + Vit E	6.98 ^{ab}	1.29 ^{ab}	9.50 ^{a}	1.14 ^{ab}	0.41
Se + Zn	6.69 ^{abc}	1.25 ^{ab}	8.48 ^{ab}	1.00 ^{abc}	0.42
Vit E + Zn	6.62 ^{abc}	1.26 ^{ab}	8.55 ^{ab}	0.99 ^{abc}	0.37
Se + Vit E + Zn	7.71 ^{a}	1.36 ^{a}	10.27 ^{a}	1.20 ^{a}	0.39
SEM	0.464	0.027	0.641	0.065	0.032
df	7,40	7,40	7,40	7,40	7,40
F values	3.91	3.01	6.63	3.38	1.01
p values	0.002	0.012	0.000	0.006	0.438

Each treatment (n = 6 birds/treatment) is expressed as mean ± standard error of the mean (SEM) and the statistical analysis was conducted using a one-way ANOVA with Tukey's post hoc test. $^{a-c}$ Different letters in the same column are significantly different by Tukey's multiple comparison tests ($p < 0.05$).

The effects of varying combinations of dietary Se, Zn, and Vit E supplements on the body weight at sacrifice and testicular weights of 1-year-old native Turkish ganders are shown in Table 4. There was no significant effect by the varying combinations of Se, Zn, and Vit E supplements on either the right, left, or overall total testicular weights, and relative testis weight. The weight of the ganders at sacrifice in the experimental groups was also found to be similar (Table 4).

Table 4. Effects of varying combinations of dietary Se, Vit E, and Zn supplements on sacrificial and testes weights (in g) and percentages (%) in 1-year-old native Turkish ganders (n = 48).

Dietary Treatments	Slaughter Weight (g)	Right Testicular Weight (g)	Left Testicular Weight (g)	Total Testicular Weight (g)	Relative Total Testicular Percentage (%)
Control	4298.5	0.37	0.65	1.02	0.02
Se	4253.8	0.46	0.83	1.30	0.03
Vit E	4254.5	0.58	1.16	1.74	0.04
Zn	4255.6	0.82	1.68	2.50	0.06
Se + Vit E	4224.0	0.78	1.95	2.73	0.06
Se + Zn	3930.9	0.32	0.54	0.85	0.02
Vit E + Zn	4108.1	0.40	0.71	1.11	0.03
Se + Vit E + Zn	3910.7	0.47	0.80	1.27	0.03
SEM	63.768	0.179	0.526	0.699	0.02
df	7,40	7,40	7,40	7,40	7,40
F values	0.70	1.08	0.96	1.00	0.94
p values	0.670	0.391	0.475	0.448	0.485

Each treatment (n = 6 birds/group) is expressed as mean ± standard error of the mean (SEM) and the statistical analysis was performed by a one-way ANOVA.

Pearson correlation coefficients between oxidative enzyme activity, and developmental and histological traits in the testes of 1-year-old native Turkish ganders are shown in Figure 1. The SOD, CAT, GPx, and GST enzyme activities were moderately positively correlated with SEMTH, GERCLT, and SEMTD, while negatively correlated with INTTISSUEPERC. The MDA activity was negatively correlated with SEMTH (−0.34) and SEMTD (−0.35). While no correlation was found between the BW and TESTPERC traits and antioxidant enzyme activities, yet there was a weak correlation (−0.33) between BW and GERCLT. Strong correlations were revealed between testicular histological characteristics. Moreover, INTTISSUEPERC and SEMTH, GERCLT, and SEMTD traits were strongly negatively correlated at −0.84, −0.84, and −0.79, respectively. A very strong positive correlation was observed between SEMTD, SEMTH, and GERCLT, ranging from 0.82 to 0.85.

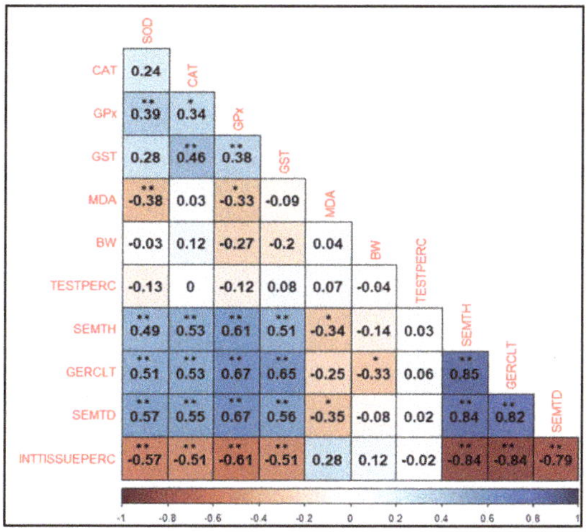

Figure 1. Pearson correlation coefficients between oxidative enzyme activity (SOD: superoxide dismutase; CAT: catalase; GPx: glutathione peroxidase; GST: glutathione-S-transferase; MDA: malondialdehyde), and histological characteristics in testicular tissue (SEMTH: seminiferous tubule area (mm^2); GERCLT: germinal cell layer thickness (mm); SEMTD: seminiferous tubule diameter (mm); INTTISSUEPERC: relative interstitial tissue (%)) and developmental traits (BW: body weight; TESTPERC: relative total testicular percentage (%)) in 1-year-old native Turkish ganders. **: $p < 0.01$; *: $p < 0.05$.

Histological images of testicular tissues in the treatment groups are shown in Figure 2 and morphometric measurements of some areas in these images are shown in Table 5. The effects of varying combinations of dietary Se, Zn, and Vit E supplements on seminiferous tubule area (mm^2), germinative epithelial thickness (mm), seminiferous tubule diameter (mm), and relative interstitial tissue area (%) were found to be significant (Table 5, $p < 0.05$). The highest seminiferous tubule area and germinative epithelial thickness were determined in ganders fed with a combination of Se + Vit E + Zn compared to the control group ($p < 0.05$). The lowest seminiferous tubule diameter was determined in ganders fed a control plus Zn diet, with the highest in the Se + Vit E + Zn and Se + Vit E combination ($p < 0.05$). The relative interstitial tissue area ratio was the highest in the control and Zn group ganders and the lowest was in ganders fed a Se + Vit E + Zn and Se + Vit E combination ($p < 0.05$).

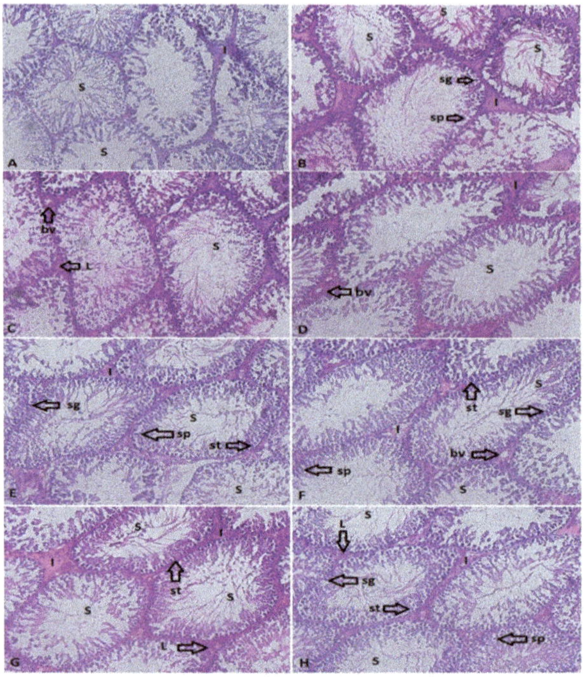

Figure 2. The histological structure of testicular tissues in 1-year-old native Turkish ganders (n = 48). (**A**) Control, (**B**) Se, (**C**) Vit E, (**D**) Zn, (**E**) Se + Vit E, (**F**) selenyum + Zn, (**G**) Vit E + Zn, and (**H**) Se + Vit E + Zn. S: seminiferous tubule; I: interstitial space; sg: spermatagonium; sp: spermatocyte; L: Leydig cell; st: Sertoli cell; bv: blood vessel. Magnification: 200×.

Table 5. Effects of varying combinations of dietary Se, Vit E, and Zn supplements on histological parameters in testicular tissues of 1-year-old native Turkish ganders (n = 48).

Dietary Treatments	Seminiferous Tubule Area (mm^2)	Germinal Cell Layer Thickness (mm)	Seminiferous Tubule Diameter (mm)	Relative Area of Interstitial Tissue (%)
Control	0.016 ^{d}	0.041 ^{d}	0.122 ^{c}	1.845 ^{a}
Se	0.019 ^{c}	0.053 ^{c}	0.131 ^{b}	1.754 ^{b}
Vit E	0.020 ^{c}	0.055 ^{c}	0.134 ^{b}	1.748 ^{b}
Zn	0.016 ^{d}	0.042 ^{d}	0.124 ^{c}	1.843 ^{a}
Se + Vit E	0.024 ^{b}	0.066 ^{b}	0.141 ^{a}	1.651 ^{c}
Se + Zn	0.019 ^{c}	0.055 ^{c}	0.131 ^{b}	1.752 ^{b}
Vit E + Zn	0.021 ^{c}	0.056 ^{c}	0.135 ^{b}	1.746 ^{b}
Se + Vit E + Zn	0.027 ^{a}	0.076 ^{a}	0.144 ^{a}	1.600 ^{c}
SEM	0.001	0.002	0.001	0.015
df	7,40	7,40	7,40	7,40
F values	36.26	40.41	28.59	30.95
p values	0.000	0.000	0.000	0.000

Each treatment (n = 6 birds/group) is expressed as mean ± standard error of the mean (SEM) and the statistical analysis was conducted by a one-way ANOVA with Tukey's post hoc test. $^{a–d}$ Different letters in the same column are significantly different by Tukey's multiple comparison tests (p < 0.05).

4. Discussion

Reactive oxygen substances (ROS) accumulate in the testes with advancing age and cause continuous oxidative stress in the cells. As a result, reproductive performance, which is of vital importance in males, decreases. Even if the testicular antioxidant capacity is low, antioxidant compounds in the testicular tissue can protect sperm against ROS [35]. Therefore, many studies have been conducted to increase and improve the antioxidant capacity in the testes [36–38]. Feeding programs provided to animals can affect reproductive activities and alter sexual behavior, morphology, and function of reproductive organs [39]. Se, Vit E, and Zn affect many biochemical and physiological systems in animal organisms, including reproduction [40–43]. Antioxidant enzymes limit the harmful effects of oxidant molecules in tissues and provide a defense against oxidative stress by scavenging free radicals [44]. Antioxidant enzymes work together to perform this task, and even small deviations in their activity can cause undesirable effects on cellular structures [45]. Therefore, it is important to determine the oxidative status in testes histology by determining the activity of SOD, CAT, GST, and GPx enzymes and measuring the MDA level.

Although MDA levels tended to be low in the control and Zn groups, SOD, CAT, GST, and GPx levels were the highest in the testes of ganders fed with a Se + Vit E + Zn supplementary diet compared to the control group. This triple combination has come to the fore as the dietary treatment that combats oxidative stress in testicular tissues of 1-year-old native Turkish ganders at the highest level. At the same time, the partial decrease in MDA values supports this situation, whereby the MDA level was already weak but significantly negatively correlated with SOD (-0.38) and GPx (-0.33) enzymes. Especially in ganders fed with the Se + Vit E combination, the MDA level tended to be slightly lower. Wan et al. [46] found lower levels of MDA in the blood serum of geese fed diets supplemented with Se. It is reported that the supplementation of Se and Vit E to the diet of roosters had a significant stimulatory effect on GPx activity in the testes [47]. Even if testicular antioxidant capacity is low, antioxidants in testicular tissue and seminal plasma can protect sperm against ROS. Improving testicular antioxidant capacity can be achieved with additional supplements to the diet, especially at older ages [7,36–38]. Since the data obtained in our study coincides with the end of the reproductive period, our findings support this situation.

In ganders, the testes are shaped like beans and located in the abdominal cavity. Their main function is to produce spermatozoa and secrete testosterone. The vasculature, size, and position of the testes vary according to whether the gander is sexually active or not [48]. Testes can reach an average weight of 1.67 g in the non-reproductive period and 12.3 g during the reproduction season. Akhtar et al. [49] determined testicular weights between 5.2 and 9.4 g and relative testicular weights between 0.40 and 0.23% in Yangzhou ganders between 181 and 227 days of age. Leska et al. [33] determined the testicular weights of ganders during the reproduction period, non-breeding period, and at the beginning of the breeding period as 12.3, 0.48, and 1.67 g, respectively. Opalka et al. [39] determined testicular weights as 5.5, 3.6, and 0.78 g in March, May, and July, respectively. In our study, total testicular weights varied between 0.8 and 2.5 g, the relative testis weight varied between 0.024 and 0.064 (%), and neither body weights nor testicular traits were affected by the dietary treatments. These low weight values are due to the fact that the ganders were at the end of the reproductive season, in line with Opalka et al. [39]. Moreover, especially in the groups where vitamins and minerals were added, they were higher than the values during the same period of other studies [33]. The differences revealed by other studies may have been influenced by many other environmental conditions, especially different dietary treatments and genotypes. Neither body weight nor testicular percentage was significantly associated with any antioxidant enzyme or histological parameter. This also indicates that various combinations of dietary Se, Vit E, and Zn supplements improved the sperm production function in ganders through antioxidant enzyme mechanisms at the cellular level, without affecting their general physiological status.

The seminiferous tubule area, germinative epithelial thickness, and seminiferous tubule diameter of the ganders fed a Se + Vit E + Zn supplementary diet seem to be advantageous in terms of continuity in the semen production compared to the control group. The strong negative correlation between these histological parameters and the relative interstitial tissue area also supports this. During the non-reproductive season, the seminiferous tubules and germinal epithelium are in a rudimentary state and develop as the reproduction season progresses. While the percentage of testicular interstitial tissue is high in the non-reproductive season, the rate of tubular tissue increases during the season [33,50]. In our study, the highest seminiferous tubule area was 0.027 mm^2, the highest germinative epithelial thickness was 0.076 mm, the highest seminiferous tubule diameter was 0.144 mm, and the lowest relative interstitial tissue area was 1.60%. Leska et al. [33] determined the seminiferous tubule area, germinative epithelial thickness, seminiferous tubule diameter, and relative interstitial tissue area characteristics during the reproductive period, non-reproductive period, and at the beginning of the reproductive period as 0.109, 0.102, 0.366, 0.17 mm^2, 0.014, 0.04, 0.127, 1.80 mm, and 0.023, 0.05, 0.148, 1.68%, respectively. Although our study was conducted at the end of the reproductive period, it found similar values to the beginning of the reproductive period in the study by Leska et al. [33], and higher values than the non-reproductive period, especially in the groups supplemented with vitamins and minerals. This highlights that the varying combinations of dietary Se, Vit E, and Zn supplements had positive results in terms of the ability to continually produce semen in native Turkish ganders, despite advancements in age.

Reproduction in geese is seasonal and cyclical, and reproductive functions are also known to decline towards the end of the laying period. This may be a result of the weakening of the physiological functions in the testicular tissue [29]. The results of our study also support this situation. In addition, Sabzian-Melei et al. [51], in their study on broiler breeder males, found higher seminiferous tubule area and seminiferous tubule diameter in birds fed with 30 mg/kg and 45 mg/kg Se added to the diet, similar to our study. The increase in Sertoli and Leydig cells probably results in higher ejaculate volume and is associated with higher spermatogenesis. This also contributes to the increase in seminiferous tubule diameter and epithelium thickness [51,52]. In our study, moderate to strong phenotypic correlations (0.49 to 0.67) observed between antioxidative enzyme activity and histological parameters revealed the stimulating effect of dietary treatments, such as Se, Vit E, and Zn, to increase and maintain an ability to produce sperm in testicular cells, despite advancements in age.

5. Conclusions

This study showed that the simultaneous supplementation of dietary Se + Vit E + Zn in native Turkish ganders had positive effects on testicular tissues, reducing oxidative damage and improving histological parameters. The fact that the seminiferous tubule area, germinative epithelial thickness, and seminiferous tubule diameter are higher than in the control group is an important finding for the continuity of sperm-producing ability. It has been shown that antioxidant vitamins and minerals (Se, Vit E, and Zn) increase sperm production ability in the testes without impairing the physiological traits of ganders. However, further studies are needed to reveal the fertilization ability and quality of this sperm produced for total reproductive efficiency in 1-year-old native Turkish ganders.

Author Contributions: Conceptualization, H.B., M.A.B. and K.E.; methodology, H.B., M.A.B. and K.E.; software, K.E.; validation, H.B., H.O.T., M.A.B., M.S. and K.E.; formal analysis, K.E. and M.A.B.; investigation, H.B., M.A.B. and H.O.T.; resources, H.B., M.A.B.; data curation, H.B., M.A.B. and K.E.; writing—original draft preparation, H.B., H.O.T., M.A.B., M.S. and K.E.; writing—review and editing, H.B., H.O.T., M.A.B., M.S., K.E., V.D. and G.S.; visualization, M.A.B. and K.E.; supervision, M.A.B. and M.S.; project administration, H.B. and M.A.B.; funding acquisition, H.B., H.O.T. and M.A.B. All authors have read and agreed to the published version of the manuscript.

Funding: This study was supported by Yozgat Bozok University (Project no: 6602b-FEN/20-412).

Institutional Review Board Statement: All procedures performed in the experiment were approved by the Erciyes University Experimental Animals Ethics Committee (Date: 1 July 2020, No: 07; Decision No: 20/095).

Informed Consent Statement: Not applicable.

Data Availability Statement: All the relevant data are available in the paper.

Acknowledgments: The authors are thankful to the Application and Research Center of Science and Technology of Yozgat Bozok University.

Conflicts of Interest: The authors declare no conflict of interest.

References

1. Boz, M.A.; Sarıca, M.; Yamak, U.S. Goose production in province Yozgat. *J. Poult. Res.* **2014**, *11*, 16–20.
2. Boz, M.A.; Sarıca, M.; Yamak, U.S. Economic evaluation of natural and artificial incubated geese in intensive and free-range production systems. *Turk. J. Agric. Food Sci. Technol.* **2016**, *4*, 981–986.
3. Önder, H.; Boz, M.A.; Sarıca, M.; Abacı, S.H.; Yamak, U.S. Comparison of growth curve models in Turkish native geese. *Eur. Poult. Sci.* **2017**, *81*, 193. [CrossRef]
4. Boz, M.A. Effect of classified rearing according to live weight on growth, carcass and some meat quality characteristics in geese. *Turk. J. Agric. Food Sci. Technol.* **2019**, *7*, 1429–1434.
5. Boz, M.A.; Baş, H.; Sarıca, M.; Erensoy, K. The effects of natural mating and artificial insemination on reproductive traits of 1-and 2-year-old native Turkish geese. *Vet. Res. Commun.* **2021**, *45*, 211–221. [CrossRef] [PubMed]
6. Çavdar, C.; Sifil, A.; Çamsarı, T. Reactive oxygen particles and antioxidant defence. *Off. J. Turk. Nephrol.* **1997**, *3*, 92–95. (In Turkish)
7. Partyka, A.; Nizanski, W. Supplementation of Avian Semen Extenders with Antioxidants to Improve Semen Quality—Is It an Effective Strategy? *Antioxidants* **2021**, *10*, 1927. [CrossRef]
8. Baş, H.; Kara, Ö.; Kara, M.; Pandır, D. Protective effect of vardenafil on ischemia-reperfusion injury in rat ovary. *Turk. J. Med. Sci.* **2013**, *43*, 684–689. [CrossRef]
9. Brenneisen, P.; Steinbrenner, H.; Sies, H. Selenium, oxidative stress, and health aspects. *Mol. Asp. Med.* **2005**, *26*, 256–267. [CrossRef]
10. Xu, Z.J.; Liu, M.; Niu, Q.J.; Huang, Y.X.; Zhao, L.; Lei, X.G.; Sun, L.H. Both selenium deficiency and excess impair male reproductive system via inducing oxidative stress-activated PI3K/AKT-mediated apoptosis and cell proliferation signaling in testis of mice. *Free Radic. Biol. Med.* **2023**, *197*, 15–22. [CrossRef]
11. Ozbek, E. Induction of oxidative stress in kidney. *Int. J. Nephrol.* **2012**, *2012*, 465897. [CrossRef] [PubMed]
12. Çaylak, E. Oxidative stress and antioxidants in the animals and the plants. *Tıp Araştırmaları Derg.* **2011**, *9*, 73–83. (In Turkish)
13. Mansour, S.A.; Mossa, A.H. Adverse effects of lactational exposure to chlorpyrifos in suckling rats. *Hum. Exp. Toxicol.* **2010**, *29*, 77–92. [CrossRef] [PubMed]
14. Codandabany, U. Erythrocyte lipid peroxidation and antioxidants in cigarette smokers. *Cell Biochem. Funct.* **2000**, *18*, 99–102. [CrossRef]
15. Derviş, E. Oral antioksidanlar. *Dermatoz* **2011**, *2*, 263–267. (In Turkish)
16. Lakshmi, B.V.S.; Sudhakar, M.; Aparna, M. Protective potential of black grapes against lead induced oxidative stress in rats. *Environ. Toxicol. Pharmacol.* **2013**, *35*, 361–368. [CrossRef] [PubMed]
17. Mates, J.M. Effects of antioxidant enzymes in the molecular control of reactive oxygen species toxicology. *Toxicology* **2000**, *153*, 83–104. [CrossRef] [PubMed]
18. Ajiboye, T.O. Redox status of the liver and kidney of 2,2-dichlorovinyl dimethyl phosphate (DDVP) treated rats. *Chem.-Biol. Interact.* **2010**, *185*, 202–207. [CrossRef] [PubMed]
19. Aly, H.A.A.; Domènech, Ò.; Banjar, Z.M. Effect of nonylphenol on male reproduction: Analysis of rat epididymal biochemical markers and antioxidant defense enzymes. *Toxicol. Appl. Pharmacol.* **2012**, *261*, 134–141. [CrossRef] [PubMed]
20. Mattsson, A.; Mura, E.; Brunström, B.; Panzica, G.; Halldin, K. Selective activation of estrogen receptor alpha in Japanese quail embryos affects reproductive organ differentiation but not the male sexual behavior or the parvocellular vasotocin system. *Gen. Comp. Endocrinol.* **2008**, *159*, 150–157. [CrossRef]
21. Kahvecioğlu, O.; Çalışlar, T. *Ürogenital ve Endokrin Sistemi. Evcil Kuşların Anatomisi*; Dursun, N., Ed.; Medisan Publication: Ankara, Türkiye, 2002; pp. 103–128. (In Turkish)
22. Deviche, P.; Hurley, L.L.; Fokidis, H.B. Avian testicular structure, function, and regulation. *Horm. Reprod. Vertebr.* **2011**, *4*, 27–69.
23. Lukaszewicz, E.; Kruszynski, W. Evaluation of fresh and frozen-thawed semen of individual ganders by assessment of spermatozoa motility and morphology. *Theriogenology* **2003**, *59*, 1627–1640.
24. Amem, M.H.; Al-Daraji, H.J. Zinc improves egg quality in Cobb500 broiler breeder females. *Int. J. Poult. Sci.* **2011**, *10*, 471–476. [CrossRef]

25. Amem, M.H.; Al-Daraji, H.J. Effect of dietary zinc on semen quality of Cobb 500 broiler breeder males. *Int. J. Poult. Sci.* **2011**, *10*, 477–482. [CrossRef]
26. Jerysz, A.; Lukaszewicz, E. Effect of Dietary Selenium and Vitamin E on Ganders' Response to Semen Collection and Ejaculate Characteristics. *Biol. Trace Elem. Res.* **2013**, *153*, 196–204. [CrossRef] [PubMed]
27. Bas, H.; Eroglu, H.E.; Dogan, H.; Uskutoglu, T.; Cosge Senkal, B.; Cesur, C. Evaluation of the chemical composition, genotoxic and cytotoxic effects of cocklebur (*Xanthium strumarium* L.) seed oil on human blood cells. *Int. J. Agric. Life Sci.* **2022**, *6*, 1–7.
28. Marklund, S.; Marklund, G. Involvement of the superoxide anion radical in the autoxidation of pyrogallol and a convenient assay for superoxide dismutase. *Eur. J. Biochem.* **1974**, *47*, 469–474. [CrossRef] [PubMed]
29. Aebi, H. Catalase in vitro. *Methods Enzym.* **1984**, *105*, 121–126.
30. Paglia, D.E.; Valentine, W.N. Studies on the quantative and qualitative characterization of glutathione peroxidase. *J. Lab. Med.* **1987**, *70*, 158–165.
31. Habig, W.H.; Pabst, M.J.; Jakoby, W.B. Glutathione-S-transferases: The first enzymatic step in mercapturic acid formation. *J. Biol. Chem.* **1974**, *249*, 7130–7139. [CrossRef] [PubMed]
32. Ohkawa, H.; Ohishi, N.; Yagi, K. Assay for lipid peroxides in animal tissues by thiobarbituric acid reaction. *Anal. Biochem.* **1979**, *95*, 351–358. [CrossRef]
33. Leska, A.; Kiezun, J.; Kaminska, B.; Dusza, L. Seasonal changes in the expression of the androgen receptor in the testes of the domestic goose *(Ancer ancer* f. *domestica). Gen. Comp. Endocrinol.* **2012**, *179*, 63–70. [CrossRef]
34. Bowker, B.; Zhuang, H. Detection of razor shear force differences in broiler breast meat due to the woody breast condition depends on measurement technique and meat state. *Poult. Sci.* **2019**, *98*, 6170–6176. [CrossRef]
35. Long, C.; Wang, Z.; Guo, Y.; Sheng, X.; Xing, K.; Ni, H.; Wang, X.; Xiao, L.; Qi, X. Research Note: Dietary supplementation with pyrroloquinoline quinone disodium (PQQ.Na2) improves oxidative status and semen quality in aging layer breeder roosters. *Poult. Sci.* **2022**, *101*, 101812. [CrossRef] [PubMed]
36. Akhlaghi, A.; Ahangari, Y.J.; Navidshad, B.; Pirsaraei, Z.A.; Zhandi, M.; Deldar, H.; Rezvani, M.R.; Dadpasand, M.; Hashemi, S.R.; Poureslami, R.; et al. Improvements in semen quality, sperm fatty acids, and reproductive performance in aged Cobb 500 breeder roosters fed diets containing dried ginger rhizomes (*Zingiber officinale*). *Poult. Sci.* **2014**, *93*, 1236–1244. [CrossRef]
37. Qi, X.; Shang, M.; Chen, C.; Chen, Y.; Hua, J.; Sheng, X.; Wang, X.; Xing, K.; Ni, H.; Guo, Y. Dietary supplementation with linseed oil improves semen quality, reproductive hormone, gene and protein expression related to testosterone synthesis in aging layer breeder roosters. *Theriogenology* **2019**, *131*, 9–15. [CrossRef] [PubMed]
38. Yan, W.; Kanno, C.; Oshima, E.; Kuzuma, Y.; Kim, S.W.; Bai, H.; Takahashi, M.; Yanagawa, Y.; Nagano, M.; Wakamatsu, J.-I.; et al. Enhancement of sperm motility and viability by turmeric by-product dietary supplementation in roosters. *Anim. Reprod. Sci.* **2017**, *185*, 195–204. [CrossRef]
39. Opalka, M.; Kaminska, B.; Piskula, M.K.; Puchajda-Skowronska, H.; Dusza, L. Effects of phytoestrogens on testosterone secretion by Leydig cells from Biłgoraj ganders (*Anser anser*). *Br. Poult. Sci.* **2006**, *47*, 237–245. [CrossRef]
40. Surai, P.F.; Blesbois, E.; Grasseau, I.; Chalah, T.; Brillard, J.P.; Wishart, G.; Cerolini, S.; Sparks, N.H.C. Fatty acid composition, glutathione peroxidase and superoxide dismutase activity and total antioxidant activity of avian semen. *Comp. Biochem. Physiol.* **1998**, *120*, 527–533. [CrossRef] [PubMed]
41. Zubair, M. Effects of dietary vitamin E on male reproductive system. *Asian Pac. J. Reprod.* **2017**, *6*, 145–150.
42. Surai, P.F. *Selenium in Poultry Nutrition and Health*; Wageningen Academic Publishers: Wageningen, The Netherlands, 2018; ISBN 978-90-8686-317-4.
43. Huang, L.; Li, X.; Wang, W.; Yang, L.; Zhu, Y. The Role of Zinc in Poultry Breeder and Hen Nutrition: An Update. *Biol. Trace Elem. Res.* **2019**, *192*, 308–318. [CrossRef] [PubMed]
44. Fouad, A.M.; Kasem El-Senousey, H.A.; Ruan, D.; Xia, W.; Chen, W.; Wang, S.; Zheng, C. Nutritional modulation of fertility in male poultry. *Poult. Sci.* **2020**, *99*, 5637–5646. [CrossRef] [PubMed]
45. Goel, A.; Dani, V.; Dhawan, D.K. Protective effects of zinc on lipid peroxidation, antioxidant enzymes and hepatic histoarchitecture in chlorpyrifos-induced toxicity. *Chem.-Biol. Interact.* **2005**, *156*, 131–140. [CrossRef]
46. Wan, X.L.; Ju, G.Y.; Xu, L.; Yang, H.M.; Wang, Z.Y. Dietary selenomethionine increases antioxidant capacity of geese by improving glutathione and thioredoxin systems. *Poult. Sci.* **2019**, *98*, 3763–3769. [CrossRef] [PubMed]
47. Surai, P.F.; Fujihara, N.; Speake, B.K.; Brillard, J.P.; Wishart, G.J.; Sparks, N.H.C. Polyunsaturated fatty acids, lipid peroxidation and antioxidant protection in avian semen. *Asian Australas. J. Anim. Sci.* **2001**, *14*, 1024–1050. [CrossRef]
48. Buckland, R.; Guy, G. Goose Production Systems; Chapter 5: Male and Female Reproduction Systems. In *Goose Production*; Buckland, R., Guy, G., Eds.; FAO: Rome, Italy, 2002; p. 17.
49. Akhtar, M.F.; Wei, Q.; Zhu, H.; Chen, Z.; Ahmad, E.; Zhendan, S.; Shi, F. The role of active immunization against inhibin a-subunit on testicular development, testosterone concentration and relevant genes expressions in testis, hypothalamus and pituitary glands in Yangzhou goose ganders. *Theriogenology* **2019**, *128*, 122–132. [CrossRef]
50. Akhtar, M.F.; Ahmad, E.; Ali, I.; Shafiq, M.; Chen, Z. The Effect of Inhibin Immunization in seminiferous epithelium of Yangzhou goose ganders: A Histological Study. *Animals* **2021**, *11*, 2801. [CrossRef]

51. Sabzian-Melei, R.; Zare-Shahneh, A.; Zhandi, M.; Yousefi, A.R.; Rafieian-Naeini, H.R. Effects of dietary supplementation of different sources and levels of selenium on the semen quality and reproductive performance in aged broiler breeder roosters. *Poult. Sci.* **2022**, *101*, 101908. [CrossRef] [PubMed]
52. Leal, M.; Becker-Silva, S.; Chiarini-Garcia, H.; Franca, L. Sertoli cell efficiency and daily sperm production in goats (*Capra hircus*). *Anim. Reprod. Sci.* **2018**, *1*, 122–128.

sustainability

MDPI

Article

The Effects of Varying Combinations of Dietary Selenium, Vitamin E, and Zinc Supplements on Semen Characteristics and Antioxidant Enzyme Activity of Spermatozoa in 1-Year-Old Native Turkish Ganders

Hulüsi Ozan Taşkesen [1], Hatice Baş [2], Mehmet Akif Boz [1,*], Musa Sarıca [3], Kadir Erensoy [3], Vassilios Dotas [4] and George Symeon [5]

1 Faculty of Agriculture, Department of Animal Science, Yozgat Bozok University, 66900 Yozgat, Türkiye; ozan.taskesen@yobu.edu.tr

2 Faculty of Arts and Science, Department of Biology, Yozgat Bozok University, 66900 Yozgat, Türkiye; hatice.bas@yobu.edu.tr

3 Faculty of Agriculture, Department of Animal Science, Ondokuz Mayıs University, 55139 Samsun, Türkiye; msarica@omu.edu.tr (M.S.); kadir.erensoy@omu.edu.tr (K.E.)

4 Department of Animal Production, Faculty of Agriculture, Aristotle University of Thessaloniki, 54124 Thessaloniki, Greece; vdotas@agro.auth.gr

5 Research Institute of Animal Science, Hellenic Agricultural Organization Demeter, 58100 Giannitsa, Greece; gsymewn@yahoo.gr

* Correspondence: bozmakif@gmail.com

Abstract: The aim of this study was to determine the effects of dietary Selenium (Se), Vitamin E (Vit E), and Zinc (Zn) and their various combinations on semen quantity, quality, and oxidative enzyme activities of spermatozoa in 1-year-old native Turkish ganders. In this study, 48 1-year-old native Turkish ganders were used. The ganders were randomly divided into 8 dietary treatment groups (Control, Se, Vit E, Zn, Se + Vit E, Se + Zn, Vit E + Zn, Se + Vit E + Zn) with 6 birds each. In addition to the control diet, specific amounts of 0.3 mg/kg Se, 100 mg/kg Vit E, and 100 mg/kg Zn were added to the diets of each treatment group. Semen volume, sperm concentration, sperm motility, sperm quality factor (SQF), and total live and normal sperm percentage were the lowest in the control group and highest in the ganders fed with the Se + Vit E + Zn combination. While the percentage of macro-cephalic and dead sperm was highest in the ganders fed with control feed, the lowest percentage of dead sperm was found in the sperm of the ganders fed with Vit E and Se + Vit E + Zn combinations. The lowest glutathione peroxidase enzyme (GPx) and glutathione-S-transferase (GST) and the highest amount of malondialdehyde (MDA) were determined in the spermatozoa of the control group ganders. This study revealed that the combined use of Se, Vit E, and Zn in the diet maintained higher semen quantity and quality in 1-year-old native Turkish gander despite the advancing reproduction season compared to the control group.

Keywords: ganders; Se; Vit E; Zn; semen quality; SQF; antioxidation

Citation: Taşkesen, H.O.; Baş, H.; Boz, M.A.; Sarıca, M.; Erensoy, K.; Dotas, V.; Symeon, G. The Effects of Varying Combinations of Dietary Selenium, Vitamin E, and Zinc Supplements on Semen Characteristics and Antioxidant Enzyme Activity of Spermatozoa in 1-Year-Old Native Turkish Ganders. *Sustainability* **2023**, *15*, 14083. https://doi.org/10.3390/su151914083

Academic Editor: Dario Donno

Received: 1 August 2023
Revised: 19 September 2023
Accepted: 20 September 2023
Published: 22 September 2023

1. Introduction

Goose farming is generally carried out in rural areas in Türkiye and has an extensive production structure on small-scale family farms [1]. In recent years, there has been an increasing interest in goose farming due to high consumer demands. Increasing demand has led to changes not only in the number of geese produced but also in the production systems. Closed barn and free-range production systems suitable for commercial production have also started to be used [2,3]. However, regardless of the system, low egg production and fertility, some hatching problems, and limited scientific research on native Turkish geese are the main factors limiting the development of goose production in Turkey [4–8].

Polyunsaturated fatty acids (PUFA) are abundant in avian spermatozoa membranes, which are highly sensitive to oxidative stress from reactive oxygen species (ROS) [9,10]. High ROS levels are known to be associated with poor sperm quality and infertility in male poultry [11]. A deficiency of antioxidative elements and too much selenium also cause ROS and reduce sperm quality [12,13]. ROS accumulates in testicles with increasing age and causes continuous oxidative stress in testicular cells. Therefore, oxidative stress causes a decrease in reproductive performance with advancing age. Even if the antioxidant capacity of testes and sperm is low, antioxidants in testicular tissue and seminal plasma can protect sperm against ROS. Improving the semen quality and antioxidant capacity of testicular and seminal plasma can be conducted with dietary supplements, especially at later ages [10,11,14]. Partyka and Nizanski [15] reported that the antioxidant system is important in protecting the sperm membranes from peroxidative damage, and there should be a balance between ROS formation and the protective effect of the antioxidant system. Se, Zn, and Vit E are also involved in many biochemical and physiological processes in human and animal organisms, including those related to reproduction [16]. They also play an active role in preventing or reducing the negative effects of lipid peroxidation on sperm cells [17,18], and are closely associated with antioxidant enzymes such as superoxide dismutase (SOD), catalase (CAT), glutathione peroxidase (GPx), and glutathione-S-transferase (GST) [19].

In addition to egg production and its quality in females, sperm quality is also important in males for good reproductive success in poultry [20,21]. One of the most important reasons for poor reproductive performance in ganders compared to other poultry species is poor sperm quality (low spermatozoa concentration, ejaculate volume, and live normal spermatozoa) [22,23]. Although genetic improvement of reproductive performance in geese can provide a persistent improvement, the low-moderate heritability of reproduction traits makes this way challenging and time-consuming [24]. Therefore, it seems more favorable to investigate alternative management tools that can contribute to reproductive performance for faster optimization of overall productivity in native Turkish geese.

We have already reported that 1-year-old native Turkish ganders had lower sperm quality than 2-year-olds [25]. This situation limits the use of 1-year-old ganders for breeding purposes. Since it is known that fertility increases as the semen quality improves [4], this study was needed to evaluate the semen quantity and quality in 1-year-old native Turkish ganders for higher reproductive efficiency. The aim of this study was to determine the effects of dietary Se (Se), Vit E (Vit E), and Zn (Zn) and their various combinations on semen quantity, quality, and antioxidant enzyme activity parameters of spermatozoa in 1-year-old native Turkish ganders.

2. Materials and Methods

2.1. Animals and Experimental Design

All procedures performed in this experiment were approved by the Erciyes University Experimental Animals Ethics Committee (Date: 1 July 2020, No. 07; Decision No. 20/096). This study started with 48 native Turkish ganders at 48 weeks of age and ended at 66 weeks. This study consisted of 8 dietary treatments (Control, Se, Vit E, Zn, Se + Vit E, Se + Zn, Vit E + Zn, Se + Vit E + Zn), each with 6 ganders. Ganders were randomly assigned to each treatment, and their average initial body weight was 3976.5 g. In this study, while the semen characteristics of ganders were examined, testicular histology was investigated in our previous study [26], and these two studies were complementary.

2.2. Rearing and Feeding

This study was carried out in the experimental house located in the Yozgat Bozok University Research and Application Center, Yerköy Goose Production Farm (34°05′–36°10′ N longitude and 38°40′–40°18′ E latitude), and the region is generally under continental climate conditions. In this study, geese were kept in wire-mesh individual cages (100 × 100 × 100 cm), and one feeder and one drinker were provided for each cage. The bottoms of the cages were covered with plastic to avoid injury to the ganders. The experimental house, where the cages are located, is naturally ventilated, fans are used when necessary, and no additional heating is applied. Lighting was provided with natural daylight through the windows, and no additional lighting was provided. This study was conducted in conditions of increased day length, and the daylight duration was approximately 11 h in March 2022 and 15 h in June 2022.

In this study, control group ganders were fed with commercial feed used for breeder poultry flocks from a private company (Table 1). As shown in Table 2, treatment groups were formed by adding specific levels of Se, Zn, and Vit E to the basic diet, with reference to Amem and Al-Daraji [27], Amem and Al-Daraji [28] and Jerysz and Lukaszewicz [29]. Ganders in each treatment were fed 200 g/day feed, and water was provided ad libitum. The ganders were fed with the diet specified in the control and treatment groups for a total of 90 days from 5 March 2022 to 3 June 2022 based on the active reproductive period.

Table 1. Basic diet components and calculated contents.

Ingredient	Unit	Amount
Corn	%	57.5
Sunflower seed meal	%	18.5
Soybean meal (CP 46%)	%	10.0
Limestone	%	8.0
Cotton seed meal (CP 26%)	%	5.0
Salt	%	0.75
Vitamin premix	%	0.25
	Analyzed nutrient content *	
Dry Matter	%	88.76
Crude Protein (CP)	%	15.50
ME	MJ/kg	10.29
Crude oil	%	3.30
Crude fiber	%	7.14
Crude Ash	%	11.68
Se	mg/kg	0.15
Zn	mg/kg	60
Vit E	mg/kg	30

* Feed analyses were carried out at Yozgat Bozok University Science and Technology Application and Research Centre laboratory, Yozgat, Turkey.

Table 2. Se, Zn, and Vit E contents in the experimental groups.

Groups	Se (mg/kg)	Vit E (mg/kg)	Zn (mg/kg)
Control	0.15	30	60
Se	0.45	30	60
Vit E	0.15	130	60
Zn	0.15	30	160
Se + Vit E	0.45	130	60
Se + Zn	0.45	30	160
Zn + Vit E	0.15	130	160
Se + Vit E + Zn	0.45	130	160

Se: 67 mg/kg Se Premix, 4.5% Sodium Selenite Na2SeO3; Vit E: 200 mg/kg E-50 Adsorbate Rovimix®, 50% Vit E; Zn: 131 mg/kg 76.4% Zn oxide.

2.3. Sperm Collection and Determination of Quality Characteristics

Sperm quality characteristics started to be determined 35 days after the feeding program started. Dorso-abdominal massage was applied once a week starting on March 5 to obtain the ganders accustomed to the semen collection process. Semen was collected from ganders in all groups once a week in the morning (09:00–11:00) by the same person. Then, sperm quality characteristics were determined in nine different periods from April to June

(10 April, 17 April, 24 April, 1 May, 8 May, 13 May, 22 May, 29 May, and 3 June 2022), and semen was evaluated for quality characteristics within 30 min [4,25,30].

Semen samples were analyzed for specific parameters such as semen ejaculate volume (mL), sperm motility (%), sperm concentration (n $\times$ 10^6/mL), sperm quality factor (SQF), and sperm morphology traits. The SQF values for each gander in the treatment groups were calculated according to the following equation [4,23,25]:

SQF = ejaculate semen volume (mL) $\times$ sperm concentration (n $\times$ 10^6 mL^{-1}) $\times$ live and normal morphology sperm number (%)/100.

The time to first semen ejaculation (s) is the time from the start of dorso-abdominal massage to the first semen ejaculation in a gander. The average semen ejaculation duration (s) is the time from the first semen ejaculation to the end of semen ejaculation.

Spermatozoa morphological traits were determined after eosin-nigrosin and giemsa staining [30] and categorized as follows: normal (spindle head and regular structured acrosome), macrocephalic (shapeless and enlarged head), bent-neck (more than 90% angle between neck and tail), deformed midpiece (swollen, bumpy, or absence of middle part), immature spermatozoa, other (spermatozoa not included in any of the previous 5 forms), and dead spermatozoa [22].

Three hundred spermatozoa per slide were evaluated using the 1000$\times$ magnification Olympus E-330 light microscope (Olympus Corp., Tokyo, Japan). Sperm concentration (n $\times$ 10^6 mL^{-1}) was measured with a hemocytometer, and motility was determined using the hanging drop method at 400$\times$ magnification [30]. The semen volume (mL) was measured with a semen collection cup at a scale of 10 μL [4]. For eosin-nigrosin staining, 1 drop of semen was placed on a slide, and 2 drops of 1% aqueous Eosin-Y were added and mixed thoroughly with a wooden mixer for 15 s. Then, 2 drops of 10% aqueous Nigrosin were added, 10–20 μL of this mixture was taken, and mixed thoroughly with a wooden mixer, and the smear was prepared by placing it on another slide, and the slides were left to dry. For Giemsa staining, first, a drop of semen was dropped on the slide, then a peripheral smear was made with another slide at an angle of 45 degrees and left to dry in a horizontal position. After the sample dries, the slide is placed on an oily surface for staining, and Giemsa dye is dripped with a pipette to cover the slide. After keeping it in Giemsa paint for 10–15 min, it was immersed in water. After the washed slide dried, immersion oil was dripped onto the slide [25].

2.4. Determination of Antioxidative Enzyme Activities

Malondialdehyde (MDA) level and antioxidant enzyme [superoxide dismutase (SOD), catalase (CAT), glutathione peroxidase (GPx), glutathione-S-transferase (GST)] activities were analyzed at nine different times (10 April, 17 April, 24 April, 1 May, 8 May, 13 May, 22 May, 29 May, and 3 June 2022) on semen samples taken for quality characteristics. The samples were stored at −80 °C until the analysis. These stored samples were homogenized in homogenization buffer (pH 7.4) for 3 min and the amount and activity of the samples were determined by measuring the absorbance of the samples with a "Biotech Engineering/Spectroscan 60 DV" brand spectrophotometer [31]. The amount of MDA was measured as the end product of lipid peroxidation reacting with thiobarbituric acid (TBA). The absorbance of the mixture to which TBA was added was read at a wavelength of 532 nm in the spectrophotometer [32]. While determining the superoxide dismutase enzyme activity (SOD), Tris-EDTA buffer and different volumes of supernatant were added to the cuvettes, and the enzyme source was added to them. Then, pyrogallol was added to these mixtures, and the absorbance value was read at 440 nm in the spectrophotometer [33]. Catalase enzyme (CAT) activity was determined by the method introduced by Aebi [34]. At first, Triton X-100 was added to the supernatant to reveal the CAT in the peroxisomes, then H_2O_2 was added, and the absorbance was read at 240 nm. Glutathione peroxidase enzyme (GPx) activity was based on the principle of measuring absorbance at 340 nm in a spectrophotometer by GR oxidizing nicotinamide-adenine-dinucleotide hydrogen phosphate (NADPH) [35]. Glutathione-S-transferase enzyme (GST) activity was measured depending on the oxidation of GSH by conjugating 1-chloro-2,4-dinitrobenzene (CDNB) with reduced glutathione (GSH) by the GST enzyme, and the absorbance value was read at 340 nm [36].

2.5. Statistical Analysis

The data obtained were analyzed with ANOVA and Tukey's HSD multiple comparison test using SPSS 25.0 statistical software (SPSS, Inc., Chicago, IL, USA). The statistical significance difference was declared at $p \leq 0.05$. All data were tested by one-way analysis of variance for the mean effect of dietary treatments and separately for each time point: April 10, April 17, April 24, May 1, May 8, May 13, May 22, May 29, and June 3. However, analysis of variance was applied at only five different time points for the time to first semen ejaculation and semen ejaculation duration. The effect of dietary treatments on whether the ganders produce semen or not was tested with Pearson's chi-square. Shapiro-Wilk and Levene tests were used for the normality and homogeneity of variance in all traits, respectively. Since SQF, percent sperm quality, and morphological characteristics did not have a normal distribution, log-transformation was applied before analysis. However, results are given as back-transformed observed averages. While the overall average results of the treatment effects are given in the tables, the results of each time point and age-related trends are illustrated in the figures.

3. Results

In this study, the effect of dietary Se, Vit E, and Zn and their various combinations on the time to first semen ejaculation, the mean semen ejaculation time, and the number of ganders producing and not producing semen were found to be insignificant (Table 3).

Table 3. Average time to first semen ejaculation (s), semen ejaculation duration (s), and total number (n) of ganders producing or not-producing semen in 1-year-old native Turkish ganders fed with dietary supplemented Se, Vit E, Zn, and their various combinations.

Dietary Treatments	Time to First Semen Ejaculation (s)	Semen Ejaculation Duration (s)	Total n of Ganders Producing or Not-Producing Semen	
			Not-Producing	Producing
Control	53.8	46.0	11 (20.4%)	43 (79.6%)
Se	52.1	40.7	8 (14.8%)	46 (85.2%)
Vit E	45.4	47.9	8 (14.8%)	46 (85.2%)
Zn	47.5	45.3	7 (13.0%)	47 (87.0%)
Se + Vit E	43.3	42.0	2 (3.7%)	52 (96.3%)
Se + Zn	48.6	53.9	9 (16.7%)	45 (83.3%)
Vit E + Zn	52.4	38.1	9 (16.7%)	45 (83.3%)
Se + Vit E + Zn	49.7	44.3	8 (14.8%)	46 (85.2%)
SEM	1.511	1.354		
F values	0.803	1.438	Pearson's chi-square, $\chi2 = 7.156$	
df	7,153	7,153	7	
p values	0.586	0.194	0.413	

The values for each treatment (n = 54 ganders/treatment) are expressed as Estimated Marginal Means. SEM: Standard error of the means. *df*: degree of freedom.

As the reproduction season progressed, it was determined that the time to first semen ejaculation was delayed regardless of dietary treatments (Figure 1A, $p < 0.001$). Moreover, at the end of the season (June 3), the Vit E + Zn group ganders were the most delayed to give the first semen ($p < 0.001$). While the age effect was not significant for semen ejaculation duration, the differences between treatments were significant on the measurement days of 8 May, 22 May and 3 June (Figure 1B, $p < 0.05$). Especially on June 3, the last measurement day, semen was collected for about 20 s in Vit E + Zn group ganders, while this time was over 90 s in Se + Zn group ganders ($p < 0.05$).

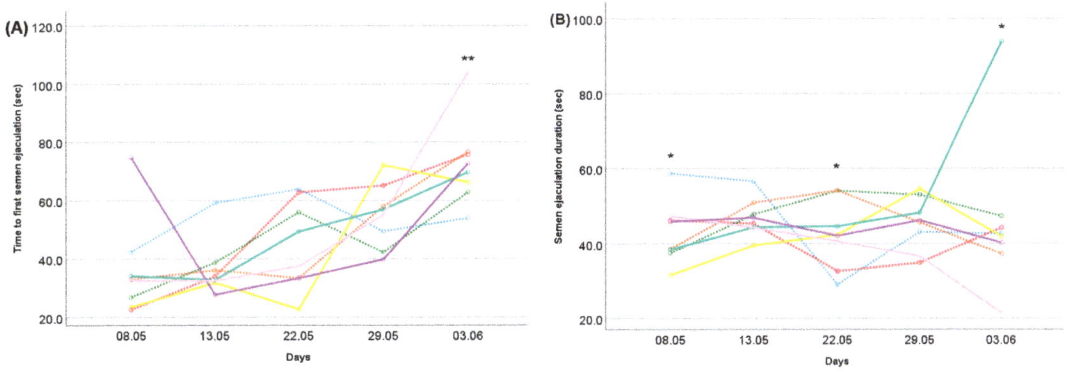

Figure 1. Change trends of (**A**) the time of first semen ejaculation (s) (Age effect $p < 0.001$) and (**B**) semen ejaculation duration (s) (Age effect *ns*) in 1-year-old native Turkish ganders (n = 54) fed with dietary supplemented Se, Vit E, and Zn and their various combinations (: Control, : Se, : Vit E, : Zn, : Se + Vit E, : Se + Zn, : Vit E + Zn, : Se + Vit E + Zn). Differences between treatments indicated by the asterisk are significant according to the ANOVA results within each measurement day. (**: $p < 0.001$; *: $p < 0.05$).

The effect of dietary Se, Vit E, and Zn and their various supplements on semen volume, sperm concentration, sperm motility, and SQF was significant (Table 4; $p < 0.001$), and these values were lowest in the control group and highest in the ganders fed with dietary Se + Vit E + Zn supplements.

Table 4. Average semen volume [1] (mL), sperm concentration [2] (n $\times$ 10^6/mL), sperm motility [3] (%), and sperm quality factor [4] (SQF) in 1-year-old native Turkish ganders fed with dietary supplemented Se, Vit E, Zn, and their various combinations.

Dietary Treatments	Semen Volume (mL)	Sperm Concentration (n $\times$ 10^6 mL^{-1})	Sperm Motility (%)	SQF
Control	0.15 ^{e}	228.12 ^{d}	42.12 ^{d}	10.69 ^{e}
Se	0.19 ^{bc}	251.87 ^{b}	46.27 ^{b}	15.92 ^{bc}
Vit E	0.18 ^{c}	251.44 ^{b}	45.45 ^{b}	15.41 ^{c}
Zn	0.17 ^{d}	241.80 ^{c}	44.38 ^{c}	13.10 ^{d}
Se + Vit E	0.20 ^{b}	258.59 ^{b}	47.31 ^{b}	17.77 ^{b}
Se + Zn	0.19 ^{c}	251.32 ^{b}	46.33 ^{b}	15.50 ^{c}
Vit E + Zn	0.18 ^{c}	253.15 ^{b}	45.84 ^{b}	15.49 ^{c}
Se + Vit E + Zn	0.21 ^{a}	263.40 ^{a}	48.93 ^{a}	19.16 ^{a}
SEM	0.001	0.415	0.099	0.099
F values	83.798	83.463	51.056	86.301
p values	<0.001	<0.001	<0.001	<0.001

The values for each treatment (n = 6 birds/treatment) are expressed as Estimated Marginal Means. $^{a-e}$ Values with different superscript letters in the same column are significantly different by Tukey's multiple comparison test ($p < 0.05$). Degree of freedom between 7 and 298. [1]. Semen volume (mL) was measured by a semen collection cup to a minimum of 10 µL. [2]. Sperm concentration (n $\times$ 10^6/mL) was measured by a hemocytometer (Micro Cell Counting Chamber). [3]. Sperm motility (%) was estimated by using the hanging drop method at 400$\times$ magnification. [4] SQF = ejaculate semen volume (mL) $\times$ sperm concentration (n $\times$ 10^6/mL) $\times$ live and normal morphology sperm number (%)/100.

The age-related trend of semen volume, sperm concentration, sperm motility, and SQF values of 1-year-old native Turkish ganders fed with diet Se, Vit E, Zn, and their various supplements is illustrated in Figure 2. The age effect was significant for all traits ($p < 0.001$). During the reproduction season, a decreasing trend was observed in all of them with advancing age. Semen volume (Figure 2A), sperm concentration (Figure 2B), and sperm motility (Figure 2C) were similar between dietary treatments during the first weeks of measurement; however, the ganders fed with dietary Se + Vit E + Zn supplementation generally had the highest values for all remaining weeks ($p < 0.05$). The ganders fed with dietary Se + Vit E + Zn supplementation were significantly superior to the control group animals at all ages in terms of SQF (Figure 2D, $p < 0.05$).

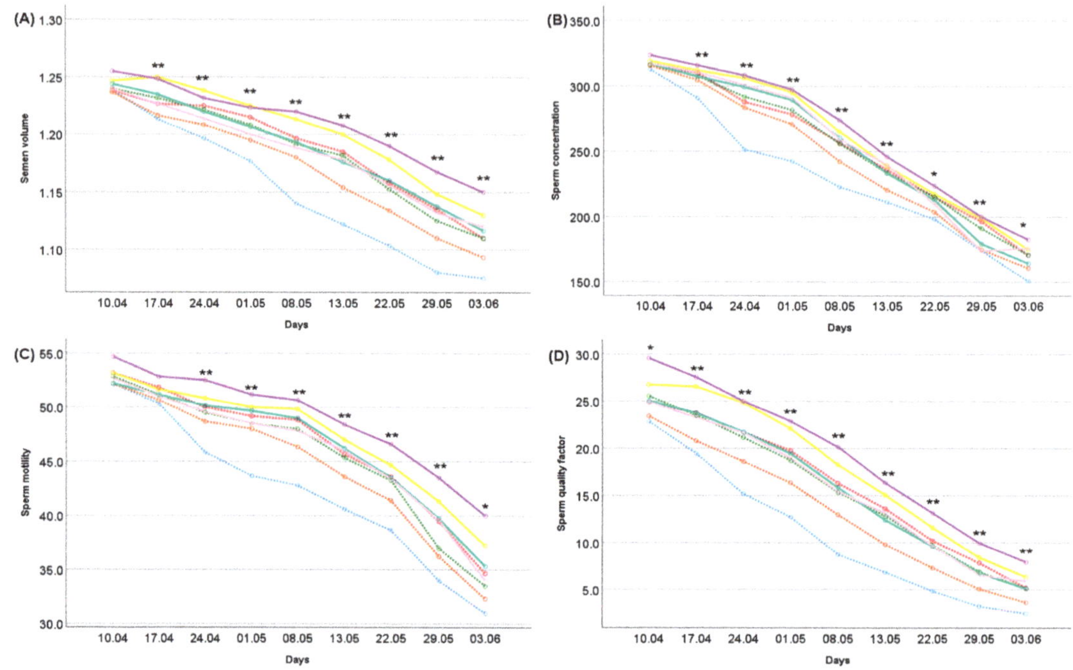

Figure 2. Change trends of (**A**) the semen volume (mL), (**B**) sperm concentration ($n \times 10^6$ mL^{-1}), (**C**) sperm motility (%), and (**D**) sperm quality factor (SQF) of 1-year-old native Turkish ganders (n = 54) fed with dietary supplemented Se, Vit E, and Zn and their various combinations (: Control, : Se, : Vit E, : Zn, : Se + Vit E, : Se + Zn, : Vit E + Zn, : Se + Vit E + Zn). (The age effect was significant at the $p < 0.001$ level for all traits). [1] Semen volume (mL) was measured by a semen collection cup to the minimum of 10 μL; [2] Sperm concentration ($n \times 10^6$/mL) was measured by a hemocytometer (Micro Cell counting chamber); [3] Sperm motility (%) was estimated by using the hanging drop method at 400× magnification; [4] SQF = ejaculate semen volume (mL) × sperm concentration ($n \times 10^6$/mL) × live and normal morphology sperm number (%)/100. Differences between treatments indicated by the asterisk are significant according to the ANOVA results within each measurement day (**: $p < 0.001$; *: $p < 0.05$).

The mean percentage of total live, normal, macrocephalic, and dead sperm was significantly different between dietary treatments (Table 5; $p < 0.05$). Total live sperm percentage was the lowest in the control group and the highest in Vit E, Se +Vit E, and Se +Vit E + Zn ($p < 0.001$). The normal sperm percentage was similarly the lowest in the control group and the highest in the ganders fed with dietary Se +Vit E + Zn supplementation ($p < 0.001$). Macro-cephalic and dead sperm percentages were determined to be the highest in the control group; the lowest dead sperm percentage was found in the ganders fed with dietary Vit E and Se +Vit E + Zn supplements ($p < 0.001$).

Table 5. Average percentage (%) of total live, normal, macro-cephalic, bent-neck, mid-piece deformed, immature, other deformities, and dead sperm deformities of 1-year-old native Turkish ganders fed with dietary Se, Vit E, Zn, and their various supplements.

Dietary Treatments	Total Live	Normal	Macro-Cephalic	Bent-Neck	Mid-Piece Deformed	Immature	Other Deformities	Dead
Control	87.2 ^{c}	27.2 ^{d}	27.3 ^{a}	16.2	6.6	3.9	6.0	12.8 ^{a}
Se	89.6 ^{ab}	31.6 ^{ab}	24.2 ^{b}	16.6	6.9	4.0	6.3	10.4 ^{bc}
Vit E	90.0 ^{a}	31.0 ^{bc}	25.2 ^{ab}	16.4	6.9	4.2	6.3	10.0 ^{bc}
Zn	88.5 ^{bc}	29.2 ^{cd}	25.2 ^{ab}	16.5	7.4	4.0	6.2	11.5 ^{ab}
Se + Vit E	90.1 ^{a}	31.8 ^{ab}	25.1 ^{ab}	16.1	7.0	3.9	6.2	9.9 ^{bc}
Se + Zn	89.3 ^{ab}	30.4 ^{bc}	25.2 ^{ab}	16.7	7.1	3.9	6.0	10.7 ^{bc}
Vit E + Zn	89.6 ^{ab}	31.3 ^{ab}	24.9 ^{ab}	16.1	7.4	3.8	6.1	10.4 ^{bc}
Se + Vit E + Zn	90.3 ^{a}	32.9 ^{a}	24.9 ^{ab}	16.2	6.9	3.8	5.6	9.7 ^{c}
SEM	0.103	0.127	0.169	0.145	0.121	0.060	0.079	0.105
F values	11.275	23.430	3.179	0.350	0.594	0.633	1.108	11.031
p values	<0.001	<0.001	0.003	0.930	0.797	0.729	0.358	<0.001

The values for each treatment (n = 6 birds/treatment) are expressed as Estimated Marginal Means. $^{a-d}$ Values with different superscript letters in the same column are significantly different by Tukey's multiple comparison test ($p < 0.05$). Degree of freedom between 7 and 298. SEM: Standard error of the means.

The age-related trend of the percentage (%) of total live, normal, macro-cephalic, bent-neck, mid-piece deformed, immature, other deformities, and dead sperm traits of 1-year-old native Turkish ganders fed with dietary Se, Vit E, Zn, and their various supplements is illustrated in Figure 3. The age effect was significant for all sperm deformity traits ($p < 0.05$), except mid-piece deformation (Figure 3E). The percentage of total live (Figure 3A) and normal sperm (Figure 3B) tended to decrease as the reproduction season progressed. Macro-cephalic (Figure 3C), bent-neck (Figure 3D), immature (Figure 3F), and dead sperm (Figure 3H) percentages generally tend to increase with advancing age; however, other sperm deformities decrease (Figure 3G). However, the effect of dietary treatments on macro-cephalic, bent-neck, mid-piece, immature, and other sperm deformities was not significant on any measurement day. Since total live or dead sperm characteristics were complementary to each other at 100, they were different between dietary treatments within the last two measurement days of the reproductive period. The percentage of normal sperm was significantly higher in the ganders fed with the dietary Se +Vit E + Zn supplement from April 8 onwards ($p < 0.05$).

In this study, dietary treatments had significant effects on the antioxidant enzyme activity parameters superoxide dismutase (SOD), catalase (CAT), glutathione peroxidase (GPx), glutathione S-transferase (GST), and Malondialdehyde (MDA) enzymes (Table 6; $p < 0.001$). While the mean SOD activity was highest in the semen of the ganders fed with dietary Vit E + Zn and Se + Vit E + Zn supplements, the lowest was in the Se + Zn group ($p < 0.001$). The highest CAT activity was in the semen of the ganders in the Se group, and the lowest in the Se + Zn group ganders. The GPx and GST enzyme activities were highest in the sperm of the ganders fed with dietary Se + Vit E + Zn, while they were lowest in the control ganders. The MDA level was also the highest in the control group.

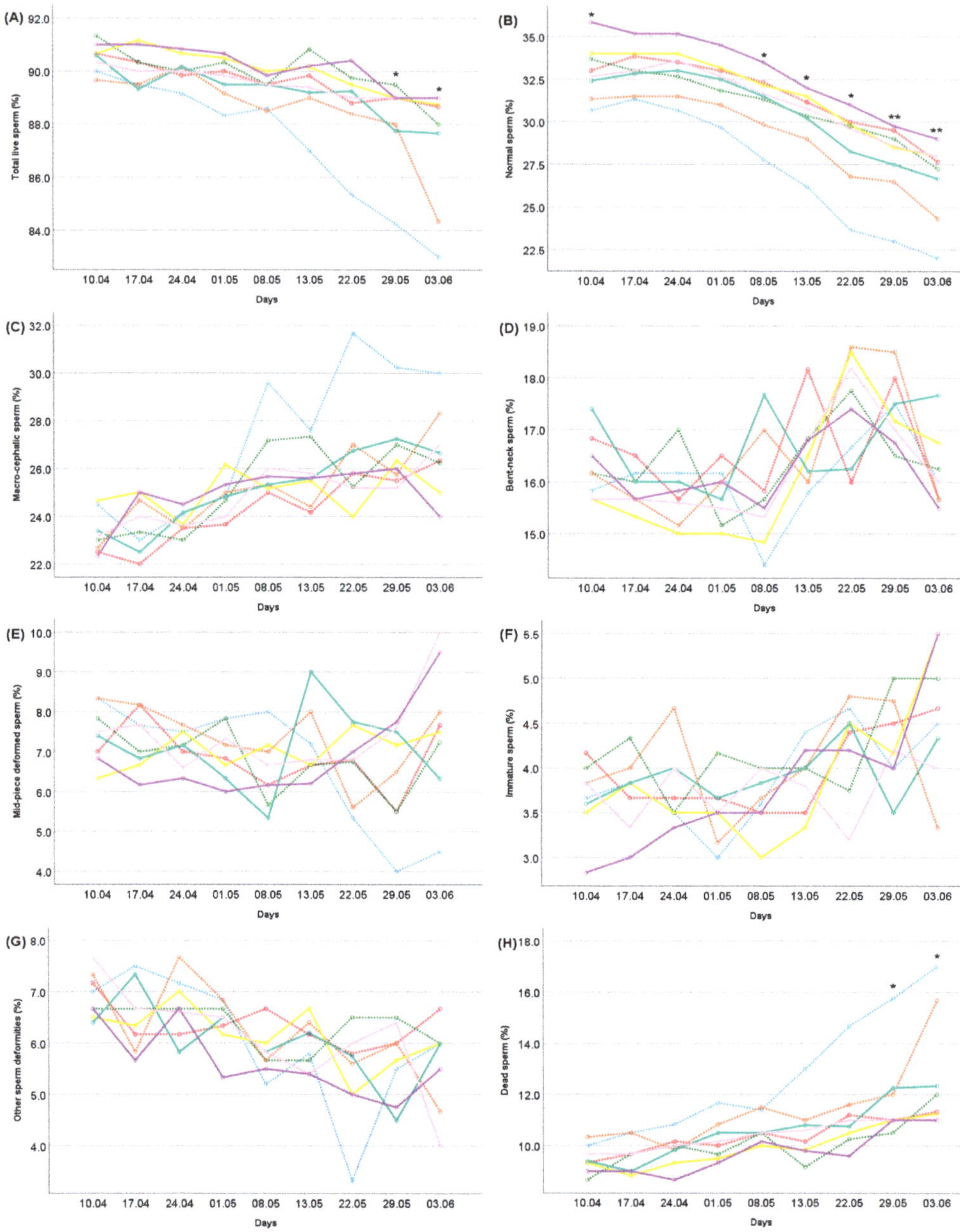

Figure 3. Change trends of the (**A**) total live (Age effect $p < 0.001$), (**B**) normal (Age effect $p < 0.001$), (**C**) macro-cephalic (Age effect $p < 0.001$), (**D**) bent-neck (Age effect $p < 0.05$), (**E**) mid-piece deformed (Age effect *ns*), (**F**) immature (Age effect $p < 0.05$), (**G**) other deformities (Age effect $p < 0.001$) and (**H**) dead sperm (Age effect $p < 0.001$) percentages (%) in 1-year-old native Turkish ganders (n = 54) fed with dietary Se, Vit E, Zn, and their various supplements (: Control, : Se, : Vit E, : Zn, : Se + Vit E, : Se + Zn, : Vit E + Zn, : Se + Vit E + Zn). Differences between treatments indicated by the asterisk are significant according to the ANOVA results within each measurement day (**: $p < 0.001$; *: $p < 0.05$).

Table 6. Average values of superoxide dismutase (SOD), catalase (CAT), glutathione peroxidase (GPx), glutathione S-transferase (GST), Malondialdehyde (MDA) enzymes in the spermatozoa of 1-year-old native Turkish ganders fed with dietary supplements Se, Vit E, Zn, and their various combinations.

Dietary Treatments	SOD (U/mg Protein)	CAT (mmol/mg Protein)	GPx (mmol/mg Protein)	GST (mmol/mg Protein)	MDA (mmol/mg Protein)
Control	2.88 ± 0.055^{bc}	4.06 ± 0.051^{bc}	2.75 ± 0.032^{c}	2.59 ± 0.033^{c}	2.14 ± 0.026^{a}
Se	3.11 ± 0.054^{ab}	4.38 ± 0.050^{a}	2.91 ± 0.031^{ab}	2.82 ± 0.032^{ab}	1.95 ± 0.026^{b}
Vit E	3.13 ± 0.053^{ab}	4.16 ± 0.049^{bc}	2.92 ± 0.031^{ab}	2.84 ± 0.032^{ab}	2.02 ± 0.025^{ab}
Zn	2.97 ± 0.051^{bc}	4.05 ± 0.047^{cd}	2.79 ± 0.029^{c}	2.73 ± 0.030^{bc}	2.01 ± 0.024^{ab}
Se + Vit E	3.10 ± 0.047^{ab}	4.32 ± 0.044^{ab}	2.93 ± 0.027^{bc}	2.86 ± 0.028^{ab}	1.95 ± 0.023^{b}
Se + Zn	2.83 ± 0.052^{c}	3.92 ± 0.048^{d}	2.89 ± 0.030^{bc}	2.82 ± 0.031^{ab}	1.95 ± 0.025^{b}
Vit E + Zn	3.20 ± 0.059^{a}	4.13 ± 0.055^{bc}	2.81 ± 0.034^{bc}	2.80 ± 0.035^{ab}	1.99 ± 0.028^{b}
Se + Vit E + Zn	3.18 ± 0.053^{a}	4.20 ± 0.049^{bc}	3.01 ± 0.031^{a}	2.91 ± 0.031^{a}	1.93 ± 0.025^{b}
F values	6.701	9.633	7.939	9.583	6.504
p values	<0.001	<0.001	<0.001	<0.001	<0.001

The values for each treatment (n = 6 birds/treatment) are expressed as Estimated Marginal Means ± SE. $^{a\text{–}d}$. Values with different superscript letters in the same column are significantly different by Tukey's multiple comparison tests ($p < 0.05$). Degree of freedom between 7 and 298.

The age-related trend of antioxidant enzyme (SOD, CAT, GPx, GST, and MDA) activities of spermatozoa in 1-year-old native Turkish ganders fed with dietary Se, Vit E, Zn, and their various supplements is shown in Figure 4. The age effect was significant for all traits ($p < 0.001$). SOD (Figure 4A), CAT (Figure 4B), GPx (Figure 4C), and GST (Figure 4D) levels decreased with advancing reproductive periods; however, MDA levels tended to increase (Figure 4E). SOD enzyme was not significant among dietary treatments in any age period. CAT and GPx enzyme activity differed between treatments on May 1 and 29, while significant differences occurred in GST enzyme activity on May 1, 8, 13, and 29. The CAT, GPx, and GST enzyme activities were lowest in the spermatozoa of the ganders fed with control feed during the aforementioned age periods. On the other hand, the MDA level was highest in the control group from 13 May onwards ($p < 0.05$).

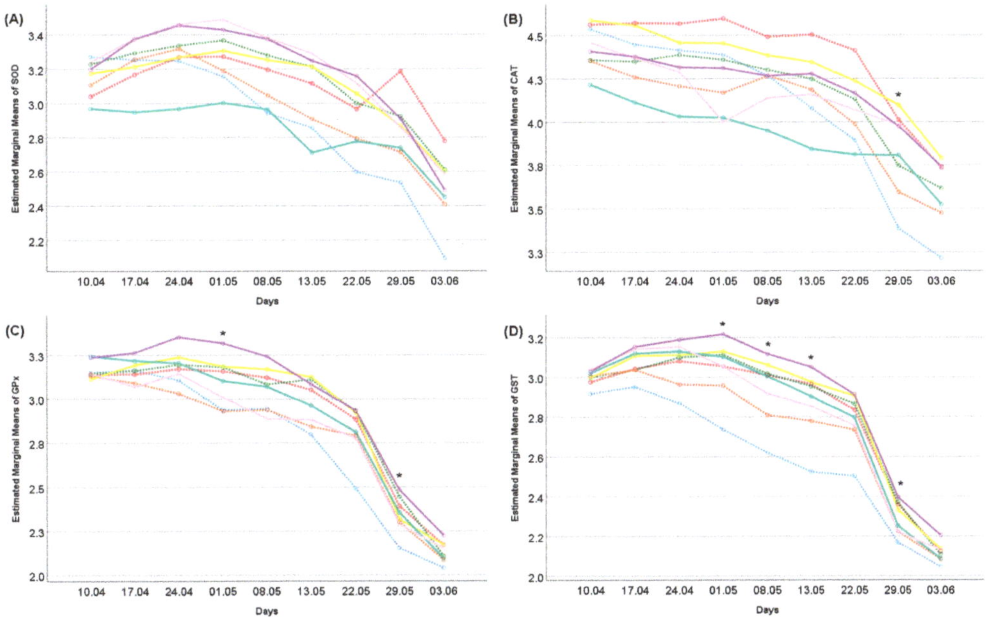

Figure 4. *Cont.*

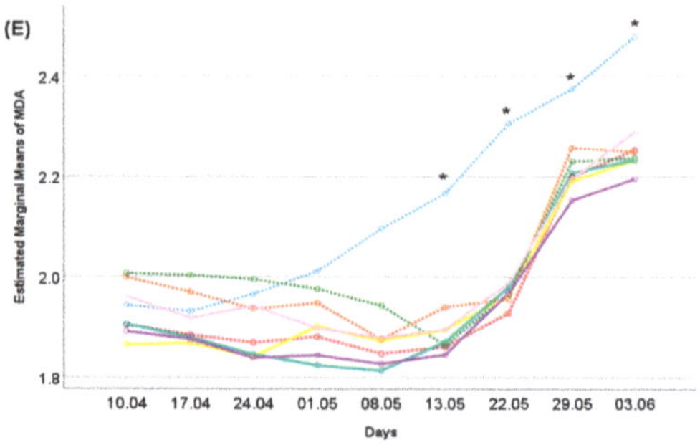

Figure 4. Change trends of (**A**) Superoxide dismutase, SOD; (**B**) Catalase, CAT; (**C**) Glutathione peroxidase, GPx; (**D**) Glutathione S-transferase, GST; and (**E**) Malondialdehyde, MDA, in the spermatozoa of 1-year-old native Turkish ganders fed with dietary supplemented Se, Vit E, and Zn and their various combinations (: Control, : Se, : Vit E, : Zn, : Se + Vit E, : Se + Zn, : Vit E + Zn, : Se + Vit E + Zn). (The age effect was significant at the $p < 0.001$ level for all traits). Differences between treatments indicated by the asterisk are significant according to the ANOVA results within each measurement day (*: $p < 0.05$).

4. Discussion

This is an initial study to reveal age-related semen characteristics and oxidative enzyme activity in 1-year-old native Turkish ganders using different antioxidative feeding strategies. In this study, dietary Se, Vit E, Zn, and their various combinations did not have a significant effect on the time to first semen ejaculation, the average semen ejaculation duration, or the number of ganders producing and not producing semen, as seen in Table 3. Jerszy and Lukaszewicz [29] reported that dietary Se and Vit E supplementation in ganders extended the reproductive period and increased positive responses to manual semen collection, and this may be related to the increase in testosterone secretion by making better use of Se and Vit E by the testes. Although the overall effect of dietary treatments was not significant in our study, the time to first semen ejaculation was prolonged in Vit E + Zn ganders at the end of the reproductive period, and the total semen collection time increased in the Se + Zn group. In addition, dietary supplementation with Vit E + Zn seems to be associated with a shortening of the total semen ejaculation time, increasing the time to first semen ejaculation, as illustrated in Figure 1. This suggests that dietary Se + Zn supplementation rather than Vit E + Zn may positively contribute to semen collection and sexual performance characteristics in 1-year-old native Turkish ganders approaching the end of the reproductive period, in accordance with Haboby et al. [36]. Although there is no significant difference between producing or not producing sperm in ganders by treatments, we observed a higher tendency to semen production in ganders dietary fed with Se + Vit E (3.7% not-producing vs. 96.7 producing), which is favorable for maintaining semen production throughout the season [29,37]. Both the semen characteristics and sexual performances of 1-year-old ganders are lower than those of 2 and 3-year-olds [25,38], making antioxidative (Vit E and Se) dietary supplements an efficient management tool for 1-year-old ganders.

Compared to other poultry species, native geese have poor reproductive efficiency due to low semen quality, egg production, fertility, and hatchability [25]. Semen quality is also an important factor affecting fertility [23,39]. Liu et al. [4] determined higher semen volume, concentration, motility, and SQF values in 2-year-old Yangzhou (*Anser cygnoides*) ganders compared to the dietary treatment groups in our study, and live and normal morphology values were similar to the Se +Vit E + Zn combination group of this study. Boz et al. [25] found the SQF values of 1- and 2-year-old native Turkish ganders to be similar to the control group of this study; however, the dietary supplementation of Se, Vit E, and Zn and their various combinations in our study significantly improved the SQF value of the gander's semen. Compared to the control group, the combined supplementation of Se + Vit E + Zn to the diet, rather than using each separately, had a synergistic effect on sperm quality characteristics and increased semen volume, sperm concentration, sperm motility, and SQF values by 40%, 15.5%, 16.2%, and 79%, respectively. This improvement in semen quality is quite consistent with our results in Baş et al. [26], the first study in which we revealed the histological parameters and oxidative enzyme activity in the testicular tissue of 1-year-old native Turkish ganders.

The antioxidant enzyme activity of the testicular tissue found in Baş et al.'s [26] study and the enzyme activity levels of the sperm cells determined in this study were quite parallel. Thus, by reducing oxidative damage in testis tissues with antioxidative feeding (Se, Vit E, Zn) in ganders, seminiferous tubule area and diameter and germinal cell layer thickness increased and the relative area of interstitial tissue decreased, which promoted the increase of sperm production and quality, which was also consistent with Malaniuk and Lukaszewicz [40], Edens and Sefton [41] and Baş et al. [26]. The variation of sperm quality and antioxidant enzyme activity parameter values obtained according to treatments in this study is also consistent with previous studies [9,29]. In our study, dietary Se + Vit E + Zn supplementation in ganders increased the percentage of live and normal sperm or vice versa for dead sperm by enhancing the morphological structure of sperm cells as well as improving semen quality characteristics. This also indicates that Se, Vit E, Zn, and their various combinations are effective in spermatogenesis and sperm maturation processes [42,43].

Lukaszewicz [39] reported that head deformations were the most common in gander's sperm cells as they were the first to be exposed to suboptimal environmental conditions, which was in agreement with our results. Ball et al. [44] reported that a low oxidative stress level has a beneficial effect on cells, while high levels lead to cell death by destroying nucleic acids, proteins, fats, and carbohydrates. Especially high ROS affects the mitochondria and sperm cell membrane, which are more vulnerable structures [45,46]. In our study, the dietary Se, Vit E, or Zn supplementation significantly reduced the percentage of macrocephalic deformity and dead sperm compared to the control group, especially in the second half of the reproductive season, as shown in Figure 3C,H.

The very high content of long-chain PUFA in poultry spermatozoa predisposes them to lipid peroxidation [37] and this increases with advancing age [29,47]. This also contributes to the loss of cell membrane integrity and may be an important indicator of the reduced fertilization ability of sperm [48]. Lipid peroxidation is expressed by the MDA level. In our study, the supplementation of Se, Vit E, and Zn to the diet of ganders, individually or in various combinations, significantly reduced the MDA level compared to the control group, especially towards the end of the reproductive period, as illustrated in Figure 4E. The GPx enzyme protects cellular membranes and organelles from oxidative damage and is involved in testicular function and spermatogenesis processes [42]. Although there was a decrease in age-related semen characteristics due to possible changes in testicular physiology with advancing reproduction age in our study [29], higher GPx levels in Se, Vit E, Zn, and their various combinations, especially in the second half of the reproductive season, compared to the control group, enhanced semen volume, concentration, and live and normal sperm percentages, enabling more efficient testicular function and spermatogenesis.

5. Conclusions

The low reproductive performance of native goose populations leads researchers to ameliorate egg-laying traits in geese and semen characteristics in ganders. The simultaneous use of dietary Se + Vit E + Zn in the diet contributed more significantly than other dietary treatments to the increase in semen quantity and quality in 1-year-old native Turkish ganders, despite the progress of the reproduction season. This study presents management perspectives to increase the reproductive performance of native Turkish ganders in the first reproduction season and to use them more effectively as breeders.

Author Contributions: Conceptualization, H.O.T., H.B., M.A.B. and K.E.; methodology, H.O.T., H.B., M.A.B. and K.E.; software, K.E.; validation, H.O.T., M.A.B., M.S. and K.E.; formal analysis, K.E. and M.A.B.; investigation, H.O.T., H.B. and M.A.B.; resources, H.O.T., H.B. and M.A.B.; data curation, H.O.T., H.B., M.A.B. and K.E.; writing—original draft preparation, H.O.T., H.B., M.A.B., M.S. and K.E.; writing—review and editing, H.O.T., H.B., M.A.B., M.S., K.E., V.D. and G.S.; visualization, M.A.B. and K.E.; supervision, H.O.T., M.A.B. and M.S.; project administration, H.O.T. and M.A.B.; funding acquisition, H.O.T., H.B., H.O.T. and M.A.B. All authors have read and agreed to the published version of the manuscript.

Funding: This study was supported by Yozgat Bozok University (Project No. 6602b-ZF/20-416).

Institutional Review Board Statement: All procedures performed in the experiment were approved by the Erciyes University Experimental Animals Ethics Committee (Date: 1 July 2020, No. 07; Decision No. 20/096).

Informed Consent Statement: Not applicable.

Data Availability Statement: All the relevant data are available in the paper.

Acknowledgments: The authors are thankful to the Application and Research Center of Science and Technology of Yozgat Bozok University.

Conflicts of Interest: The authors declare no conflict of interest.

References

1. Aral, Y.; Aydın, E. The economic importance of goose breeding and assessment possibility of goose products in Turkey. *Turk. J. Vet. Anim. Sci.* **2007**, *78*, 31–38.
2. Boz, M.A. Effect of classified rearing according to live weight on growth, carcass and some meat quality characteristics in geese. *Turk. JAF Sci. Technol.* **2019**, *7*, 1429–1434. [CrossRef]
3. Boz, M.A.; Erensoy, K.; Sarıca, M. The effects of age and daytime periods on behavioral traits of Turkish geese reared in free-range system. *J. Anim. Prod.* **2022**, *63*, 126–135.
4. Liu, S.J.; Zheng, J.X.; Yang, N. Semen quality factor as an indicator of fertilizing ability for geese. *Poult. Sci.* **2008**, *87*, 155–159. [CrossRef]
5. Tilki, M.; Yazıcı, K.; Sarı, M.; Işık, S.; Saatci, M. Effects of hatching month and sex on slaughter and carcass traits in native Turkish geese. *Kafkas Univ. Vet. Fak. Derg.* **2011**, *17*, 831–835. [CrossRef]
6. Tilki, M.; İnal, Ş. Yield traits of geese of different origins reared in Turkey I. Hatching traits. *Turk. J. Vet. Anim. Sci.* **2004**, *28*, 149–155.
7. Onk, K.; Kirmizibayrak, T. The egg production, hatchability, growing, slaughter and carcass characteristics of geese (*Anser anser*) reared under breeders conditions in Kars Province; I. Egg production and hatchability characteristics. *Turk. JAF Sci. Technol.* **2019**, *7*, 543–549.
8. Arslan, C.; Saatcı, M. Egg yield and hatchability characteristics of native geese in the Kars region. *Turk. J. Vet. Anim. Sci.* **2003**, *27*, 1361–1365.
9. Partyka, A.; Lukaszewicz, E.; Nizanski, W. Lipid peroxidation and antioxidant enzymes activity in avian semen. *Anim. Reprod. Sci.* **2012**, *134*, 184–190. [CrossRef]
10. Qi, X.; Shang, M.; Chen, C.; Chen, Y.; Hua, J.; Sheng, X.; Wang, X.; Xing, K.; Ni, H.; Guo, Y. Dietary supplementation with linseed oil improves semen quality, reproductive hormone, gene and protein expression related to testosterone synthesis in aging layer breeder roosters. *Theriogenology* **2019**, *131*, 9–15. [CrossRef]
11. Akhlaghi, A.; Ahangari, Y.J.; Navidshad, B.; Pirsaraei, Z.A.; Zhandi, M.; Deldar, H.; Rezvani, M.R.; Dadpasand, M.; Hashemi, S.R.; Poureslami, R.; et al. Improvements in semen quality, sperm fatty acids, and reproductive performance in aged Cobb 500 breeder roosters fed diets containing dried ginger rhizomes (*Zingiber officinale*). *Poult. Sci.* **2014**, *93*, 1236–1244. [CrossRef] [PubMed]
12. Xu, Z.J.; Liu, M.; Niu, Q.J.; Huang, Y.X.; Zhao, L.; Lei, X.G.; Sun, L.H. Both selenium deficiency and excess impair male reproductive system via inducing oxidative stress-activated PI3K/AKT-mediated apoptosis and cell proliferation signaling in testis of mice. *Free Radic. Biol. Med.* **2023**, *197*, 15–22. [CrossRef] [PubMed]
13. Yang, J.C.; Huang, Y.X.; Sun, H.; Liu, M.; Zhao, L.; Sun, L.H. Selenium deficiency dysregulates one-carbon metabolism in nutritional muscular dystrophy of chicks. *J. Nutr.* **2023**, *153*, 47–55. [CrossRef]
14. Yan, W.; Kanno, C.; Oshima, E.; Kuzuma, Y.; Kim, S.W.; Bai, H.; Takahashi, M.; Yanagawa, Y.; Nagano, M.; Wakamatsu, J.I.; et al. Enhancement of sperm motility and viability by turmeric by-product dietary supplementation in roosters. *Anim. Reprod. Sci.* **2017**, *185*, 195–204. [CrossRef]
15. Partyka, A.; Nizanski, W. Supplementation of Avian Semen Extenders with Antioxidants to Improve Semen Quality—Is It an Effective Strategy? *Antioxidants* **2021**, *10*, 1927. [CrossRef] [PubMed]

16. Surai, P.F.; Blesbois, E.; Grasseau, I.; Chalah, T.; Brillard, J.P.; Wishart, G.; Cerolini, S.; Sparks, N.H.C. Fatty acid composition, glutathione peroxidase and superoxide dismutase activity and total antioxidant activity of avian semen. *Comp. Biochem. Physiol. Part B Biochem. Mol. Biol.* **1998**, *120*, 527–533. [CrossRef]
17. Surai, P.F.; Brillard, J.P.; Speake, B.K.; Blesbois, E.; Seigneurin, F.; Sparks, N.H.C. Phospholipid fatty acid composition, vitamin E content and susceptibility to lipid peroxidation of duck spermatozoa. *Theriogenology* **2000**, *53*, 1025–1039. [CrossRef]
18. Khan, R.U.; Rahman, Z.; Javed, I.; Muhammad, F. Effect of vitamins, probiotics and protein on semen traits in post-molt male broiler breeders. *Anim. Reprod. Sci.* **2012**, *135*, 85–90. [CrossRef]
19. Partyka, A.; Lukaszewicz, E.; Nizanski, W. Effect of cryopreservation on sperm parameters, lipid peroxidation and antioxidant enzymes activity in fowl semen. *Theriogenology* **2012**, *77*, 1497–1504. [CrossRef]
20. Lagares, M.; Ecco, R.; Martins, N.; Lara, L.; Rocha, J.; Vilela, D.; Barbosa, V.; Mantovani, P.; Braga, J.; Preis, I. Detecting reproductive system abnormalities of broiler breeder roosters at different ages. *Reprod. Domest. Anim.* **2017**, *52*, 67–75. [CrossRef]
21. Long, C.; Wang, Z.; Guo, Y.; Sheng, X.; Xing, K.; Ni, H.; Wang, X.; Xiao, L.; Qi, X. Research Note: Dietary supplementation with pyrroloquinoline quinone disodium (PQQ.Na2) improves oxidative status and semen quality in aging layer breeder roosters. *Poult. Sci.* **2022**, *101*, 101812. [CrossRef] [PubMed]
22. Lukaszewicz, E. Effect of semen filtration and dilution rate on morphology and fertility of frozen gander spermatozoa. *Theriogenology* **2001**, *55*, 1819–1829. [CrossRef] [PubMed]
23. Lukaszewicz, E.; Kruszynski, W. Evaluation of fresh and frozen-thawed semen of individual ganders by assessment of spermatozoa motility and morphology. *Theriogenology* **2003**, *59*, 1627–1640. [PubMed]
24. Wolc, A.; Barczak, E.; Wężyk, S.; Badowski, J.; Bielińska, H.; Szwaczkowski, T. Genetic evaluation of production and reproduction traits in two selected lines of geese under multitrait animal model. *Anim. Sci. Pap. Rep.* **2008**, *26*, 71–78.
25. Boz, M.A.; Baş, H.; Sarıca, M.; Erensoy, K. The effects of natural mating and artificial insemination on reproductive traits of 1-and 2-year-old domestic Turkish geese. *Vet. Res. Commun.* **2021**, *45*, 211–221. [CrossRef]
26. Baş, H.; Taşkesen, H.O.; Boz, M.A.; Sarıca, M.; Erensoy, K.; Dotas, V.; Symeon, G. The effects of dietary Selenium, Vitamin E and Zinc supplementation and their varying combinations on antioxidant enzyme activity, developmental and histological traits in testicular tissue of 1-year-old native Turkish ganders. *Sustainability* **2023**, *15*, 12245. [CrossRef]
27. Amem, M.H.; Al-Daraji, H.J. Zinc improves egg quality in Cobb500 broiler breeder females. *Int. J. Poult. Sci.* **2011**, *10*, 471–476. [CrossRef]
28. Amem, M.H.; Al-Daraji, H.J. Effect of dietary zinc on semen quality of Cobb 500 broiler breeder males. *Int. J. Poult. Sci.* **2011**, *10*, 477–482. [CrossRef]
29. Jerysz, A.; Lukaszewicz, E. Effect of Dietary Selenium and Vitamin E on Ganders' Response to Semen Collection and Ejaculate Characteristics. *Biol. Trace Elem. Res.* **2013**, *153*, 196–204. [CrossRef]
30. Bakst, M.R.; Cecil, H.C. *Techniques for Semen Evaluation, Semen Storage, and Fertility Determination*; Poultry Science Association: Champaign, IL, USA, 1997.
31. Bas, H.; Eroglu, H.E.; Dogan, H.; Uskutoglu, T.; Cosge Senkal, B.; Cesur, C. Evaluation of the chemical composition, genotoxic and cytotoxic effects of cocklebur (*Xanthium strumarium* L.) seed oil on human blood cells. *Int. J. Agric. Life Sci.* **2022**, *6*, 1–7.
32. Ohkawa, H.; Ohishi, N.; Yagi, K. Assay for lipid peroxides in animal tissues by thiobarbituric acid reaction. *Anal. Biochem.* **1979**, *95*, 351–358. [CrossRef]
33. Marklund, S.; Marklund, G. Involvement of the superoxide anion radical in the autoxidation of pyrogallol and a convenient assay for superoxide dismutase. *Eur. J. Biochem.* **1974**, *47*, 469–474. [CrossRef]
34. Aebi, H. Catalase in vitro. *Methods Enzymol.* **1984**, *105*, 121–126.
35. Paglia, D.E.; Valentine, W.N. Studies on the quantative and qualitative characterization of glutathione peroxidase. *J. Lab. Med.* **1987**, *70*, 158–165.
36. Haboby, A.H.; Hamra, A.H.; Al-Tamemmy, M.J.; Al-Rawi, T.S. Effect of vitamin E and selenium on semen quality, sexual activity, and some blood parameters of Awassi Rams. *J. Agric. Investig.* **2004**, *2*, 58–62.
37. Surai, P.F.; Fujihara, N.; Speake, B.K.; Brillard, J.P.; Wishart, G.J.; Sparks, N.H.C. Polyunsaturated fatty acids, lipid peroxidation and antioxidant protection in avian semen. *Asian-Aust. J. Anim. Sci.* **2001**, *14*, 1024–1050. [CrossRef]
38. Varga, A.; Barna, J.; Almasi, A. Sperm analysis and sexual characteristics of frizzled Hungarian ganders (preliminary study). *Allatten Takarman* **2003**, *2*, 167–172.
39. Lukaszewicz, E. An effective method for freezing White Italian gander semen. *Theriogenology* **2002**, *58*, 19–27. [CrossRef]
40. Malaniuk, P.; Lukaszewicz, E. Effect of feed supplementation with organic selenium and vitamin E on quantitative and qualitative characteristics of Japanese quails (*Coturnix japonica*). *Zesz. Nauk. UP Wroclawiu Biol. Hod Zw.* **2006**, *548*, 99–109. (In Polish)
41. Edens, F.W.; Sefton, A.E. Sel-Plex® improves spermatozoa morphology in broiler breeder males. *Int. J. Poult. Sci.* **2009**, *8*, 853–861. [CrossRef]
42. Marin-Guzman, J.; Mahan, D.C.; Pate, J.L. Effect of dietary selenium and vitamin E on spermatogenic development in boars. *J. Anim. Sci.* **2000**, *78*, 1536–1543. [CrossRef] [PubMed]
43. Wu, S.H.; Oldfield, J.E.; Whanger, P.D.; Weswig, P.H. Effect of selenium, vitamin E, and antioxidants on testicular function in rats. *Biol. Reprod.* **1973**, *8*, 625–629. [CrossRef] [PubMed]
44. Ball, B.A.; Medina, V.; Gravance, C.G.; Baumbe, J. Effect of antioxidants on preservation of motility, viability and acrosomal integrity of equine spermatozoa during storage at 5 °C. *Theriogenology* **2001**, *56*, 577–589. [CrossRef] [PubMed]

45. Brouwers, J.F.H.M.; Gadella, B.M. In situ detection and localization of lipid peroxidation in individual bovine sperm cells. *Free Radic. Biol. Med.* **2003**, *35*, 1382–1391. [CrossRef]
46. Agarwal, A.; Prabakaran, S.A.; Said, T.M. Prevention of oxidative stress injury to sperm. *J. Androl.* **2005**, *26*, 654–660. [CrossRef]
47. Douard, W.; Hermier, D.; Magistrini, M.; Blesbois, E. Reproductive period affects lipid composition and quality of fresh and stored spermatozoa in turkeys. *Theriogenology* **2003**, *59*, 753–764. [CrossRef]
48. Blesbois, E.; Grasseau, J.; Hermier, D. Changes in lipid content of fowl spermatozoa after liquid storage at 2 to 5 °C. *Theriogenology* **1999**, *52*, 325–334. [CrossRef]

sustainability

Review

Positive Welfare Indicators and Their Association with Sustainable Management Systems in Poultry

Maria Papageorgiou [1], Michael Goliomytis [1], Ouranios Tzamaloukas [2], Despoina Miltiadou [2] and Panagiotis Simitzis [1,*]

[1] Department of Animal Science, Agricultural University of Athens, 11855 Athens, Greece; stud219174@aua.gr (M.P.); mgolio@aua.gr (M.G.)

[2] Department of Agricultural Sciences, Biotechnology and Food Science, Cyprus University of Technology, 50329 Limassol, Cyprus; ouranios.tzamaloukas@cut.ac.cy (O.T.); despoina.miltiadou@cut.ac.cy (D.M.)

* Correspondence: pansimitzis@aua.gr; Tel.: +30-2105-294-427

Abstract: Animal welfare is a key and distinct component of sustainable agriculture and food security. People, both as citizens and consumers, have become more concerned about the husbandry of livestock species. Positive welfare goes a step further than the common welfare approach, supporting that a good life for animals is not only the alleviation of negative aspects, but also the promotion of positive affectivities. So, a sustainable management system for any livestock species should promote positive aspects in the lives of animals. Poultry is one of the species whose welfare is most impaired, and numerous concerns are raised by society. For all the above, we reviewed the positive welfare indicators that have been studied in livestock poultry and that can be used to promote positive effects and assess welfare for the most common species, i.e., broilers, laying hens, turkeys, ducks, geese, quails and ostriches. We analyzed the results categorized by species, discussed the connection of the indicators with sustainable management, and made proposals for future studies. Exploration and dustbathing have been extensively studied and seem most promising, especially in broilers and laying hens, followed by nesting and perching, and swimming for waterfowl. Qualitative behavioral assessment (QBA) is already applied in protocols for broilers and laying hens, but the results are not as promising due to the homogeneity of the flock and the difficulty in observations. Play has been studied mostly in broilers but is a behavior difficult to recognize and needs further understanding. The results are limited for all species, except broilers and laying hens.

Keywords: broilers; laying hens; turkeys; geese; ducks; positive welfare evaluation; exploration; dustbathing; pre-laying; qualitative behavioral assessment

Citation: Papageorgiou, M.; Goliomytis, M.; Tzamaloukas, O.; Miltiadou, D.; Simitzis, P. Positive Welfare Indicators and Their Association with Sustainable Management Systems in Poultry. *Sustainability* **2023**, *15*, 10890. https://doi.org/10.3390/su151410890

Academic Editors: George K. Symeon and Vassilios Dotas

Received: 14 June 2023
Revised: 6 July 2023
Accepted: 10 July 2023
Published: 11 July 2023

1. Introduction

Animal welfare is a fundamental component of sustainability, agricultural development and food security. In order for a livestock system to be sustainable, it should be sustainable economically, environmentally and ecologically, and at the same time, it should be accepted ethically and socially [1–3]. Today, a product of animal origin is not considered sustainable unless the social demand for welfare of the production animals is satisfied [1]. Animal welfare of livestock species is a common debate that attracts more and more the attention globally [2–4]. It has become one of the scientific issues that is most interesting to the public [4], especially after the United Nations Committee on World Food Safety endorsed, in 2016, animal welfare as a pillar of sustainable agricultural development, food security and nutrition [5]. In addition, animal welfare has been linked to human health, as endorsed with the "One Health" approach by the World Health Organization [6] and the "One Welfare" concept [7]. It is considered a public good, beneficial for the wellbeing of the wider society [8] and an attribute of product quality [1]. People as consumers are willing to pay more for products of higher animal welfare [9,10]. As citizens, they have become more concerned about the protection of animals and the incorporation of animal welfare in national and international legislation [1–3,8,11].

As a scientific field, animal welfare has flourished during the last two decades [1,12]. Its link to sustainability is a constant debate. Global meat consumption is about to increase by 14% by 2030 compared to the average of the 2018–2020 period [13]. The increase in livestock production is linked to intensification, impacts on the environment and impacts on the welfare of the animals. It is a challenge to satisfy the growing meat demand and simultaneously not to impair the welfare of the animals and the environment. This is why scientific research on sustainable management has a crucial role, together with the consumers' willingness to pay and the policymaking that will determine the acceptable welfare level for the animals [3,4,14,15].

Poultry meat is first in terms of global consumption, and it is expected to increase even more by 2030, representing 41% of the total global meat consumption [13]. In low-income countries, it is a low-cost protein source. In high-income countries, it is perceived as a healthy meat choice. Egg production is about to increase globally by 13% over the next decade [16]. According to OECD Agriculture statistics, the projection of average egg consumption is estimated to increase from 90,513 kt in 2019–2021 to 105,809 kt in 2031 [17]. Broilers and laying hens are highly industrialized species, and their welfare is perceived as being the worst among all farm animals. Poultry is the farming species that raises the highest public concern regarding welfare management practices [9]. Consumers are willing to pay more for eggs from cage-free or free-range systems [18–20] or eggs from more "welfare-friendly" furnished cages [21]. They are also willing to pay an extra cost for organic meat [22], meat from slow-growing chickens [23] or meat from dual-purpose farming systems, where male chickens are reared for their meat and females for their eggs [24]. Nonetheless, price is always an issue [18,19,23,25]. Especially for meat from highly welfare-friendly systems, compared to eggs, since meat is a more expensive product than eggs [24,25]. Thus, a farming system that is sustainable for the animals and the environment, and at the same time economically sustainable and morally approved by society, is a challenge.

Positive welfare (PW) is the newest approach to welfare, which, as analyzed above, is a distinct and necessary pillar of sustainable agriculture and food safety. PW promotes the experience of positive affective states in an animal's life in addition to the alleviation of negative experiences [11,26]. It promotes that the life of an animal, without negative experiences, is not necessarily a good life. Good experiences are also necessary [27]. Positive experiences and stimulation are important for a high quality of life [28,29]. The attention and interest for the evaluation of PW is steadily increasing among consumers, especially for intensively farmed animals like poultry [9]. As a result, PW indicators have already been incorporated in welfare evaluation schemes and protocols for laying hens and broilers [30,31].

The aim of the present study is to review all the positive welfare indicators that can be used for the evaluation of positive welfare in poultry, including indicators that have either been tested in practice, experimentally and on the farm level, or studied on a theoretical base. We will emphasize the behavioral indicators since the indicators that are being researched as indicators of positive affective states are mainly behavioral. Furthermore, we will make proposals for future research and for the evaluation of positive welfare on the farm level, as part of sustainable management systems, in the most important poultry species, such as broilers, laying hens, turkeys, geese, ducks, quails and ostriches. To the best of the authors' knowledge, this is the first attempt to review the positive welfare indicators of all livestock poultry species and discuss their association with the sustainable management of poultry farming systems.

2. Materials and Methods

The search was conducted in databases Google Scholar and Web of Science. The search period was 2000–2023 for two reasons: firstly, because the research on PW has been thriving mainly in recent years, and secondly, because we wanted to include the most recent publications in our review. The starting point was a general search using the following key words separately or in various combinations: "positive welfare", "positive emotion*", "positive affective state*", "positive welfare indicator*", "positive welfare evaluation", "positive welfare" AND sustainability, "positive welfare" AND "sustainable management" OR "sustainable system", broiler* OR chick*, "laying hen*", turkey*, ostrich*, quail*, duck*, geese OR goose, poultry, waterfowl, "*Gallus gallus domesticus*", "*Meleagris*", "*Anser*", "*Anatidae*", "*Anas platyrhynchos domesticus*", "*Coturnix*". The duplicates were removed from the results. For every retrieved PW indicator, a separate search was then performed. The results were also supplemented with articles that were included in the references of our research findings. Moreover, we included in the results some research findings that apply to vertebrae [28,32–35], domestic species [36,37] or livestock species [38] in general and not specifically to poultry. Additionally, we included in the results the Welfare Quality Assessment protocol for poultry [30] and the Welfare Quality Assessment protocol for laying hens, version 2.0 [31]. The whole procedure was conducted by one author.

The used criteria for including a research finding in the results was that it described a positive welfare indicator, based on studies under experimental or commercial circumstances or on theoretical analysis. Physiological studies were excluded. No other review on positive welfare indicators for any poultry species was retrieved. Reviews were included in the results only if the authors marked/concluded that a reviewed trait could indicate positive affectivity.

The above search procedure led to the inclusion of a total of 95 articles in our database. We summarize the search strategy that led to our results in Figure 1.

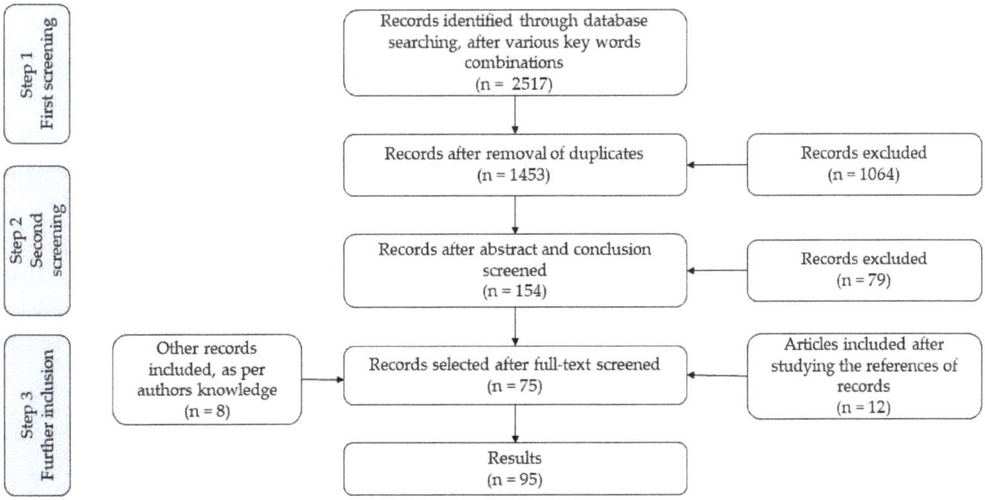

Figure 1. Flow chart of the literature selection strategy, displaying the screening and inclusion process.

3. Results

Thirteen positive welfare indicators, all animal based, were retrieved. The results are summarized in Table 1, classified by indicator, the animals that the indicators correspond to and the research findings that include all the information. The nine positive welfare indicators are summarized in the behavioral group of indicators because they are all referred to as behavioral traits.

Table 1. Positive welfare indicators categorized by poultry species.

Group	Indicator		Species	References
Behavioral parameters	Exploration and foraging		all	[28,33,34]
			broilers	[30,39–50]
			laying hens	[51–53]
			turkeys	[54,55]
			ducks	[56]
			geese	[57]
	Comfort	general	all	[27–29,34]
		dustbathing	broilers	[42,49,58,59]
			laying hens	[31,51,52,60–64]
			quails	[65]
		preening	all	[28,33,34]
			broilers	[40,41,44,58]
			laying hens	[31,52,61,66,67]
			turkeys	[55,68]
			ducks	[56,69,70]
		others	various	[39,40,44,46,48,51,57,59,71,72]
	Play		all	[32–35,73]
			broilers	[30,41,47–49,58,59,74]
			laying hens	[51]
			ducks	[75]
			ostriches	[76]
	Behavioral synchronization		all	[33–35]
			broilers	[77–79]
			laying hens	[80,81]
			ducks	[61,82]
	Swimming and access to water		ducks	[69,70,82]
			geese	[71,83]
	Pre-laying		laying hens	[60,84–87]
			ducks	[88,89]
	Perching		broilers	[40,61]
			laying hens	[51,61,90–92]
	Maternal care		all	[34]
			laying hens	[93–95]
			geese	[57]
			quails	[96,97]
	Anticipatory behavior		laying hens	[72,98–100]
	Vocalizations		all	[33,38]
			laying hens	[38,72,98,101]
	Qualitative behavioral assessment		broilers	[30,47,102]
			laying hens	[31,103–105]
	Positive human–animal interaction		all	[106–108]
			broilers	[30,109,110]
			laying hens	[31,86,111,112]
			quails	[113]
			ostriches	[114]
	Cognitive bias		all	[115,116]
			broilers	[117,118]
			laying hens	[100,119–121]
			quails	[122]

3.1. *Behavioral Paremeters*

The behavioral indicators that have been retrieved are exploration, comfort, play, synchronization, swimming, pre-laying, perching, maternal care and anticipation. Comfort behavior in poultry is mainly performed as dustbathing and preening, as will be analyzed in Section 3.1.2. All these indicators are animal based, indicative of the behavioral response of the animals to the resources of their environment and/or the management practices. Dustbathing, swimming, pre-laying and perching can be categorized as both resource-based and animal-based indicators, since they are performed under the provision of dustbath, water, nesting and perching substrates, respectively. Nonetheless, they are species-specific behaviors also referred to as behavioral traits, and they indicate the affective states of the animals, as will be described in Sections 3.1.2 and 3.1.5–3.1.7, respectively.

3.1.1. Exploration and Foraging

Exploratory behavior is considered one of the best candidates as a positive welfare indicator in various animals [28,33,34], including livestock species like dairy animals [123,124] and the domestic pig [125]. It is a strongly motivated natural behavior of most livestock species, including poultry, and is correlated to feeding behavior [34]. Animals explore their environment in order to acquire information and feel safe [34]. According to Mellor [34], the animal is performing this behavior repeatedly because it is self-rewarding. The reward is the pleasure of expecting or acquiring the feed, and in addition, the pleasure of tasting it. This is also reinforcing, and so the animal keeps performing the behavior continuously. In addition, exploration indicates the absence of fear or any type of threats to survival [28]. In vertebrae, it is a goal-directed behavior [28] that indicates positive affective states [28,33,34]. Nonetheless, it should be interpreted with caution and by taking into consideration the whole context of the situation, since sometimes it can also indicate fear or uncertainty [33]. There are two types of exploration: inquisitive and inspective exploration [33]. Inquisitive exploration is always performed due to curiosity and is an initiative of the animal. On the contrary, inspective exploration can be driven both by curiosity or fear and is a response of the animal to a change in its environment. So, inquisitive exploration always indicates positive affective states, but inspective exploration can also indicative negative affective states [33].

In poultry, exploration is correlated to foraging and feeding behavior, performed by pecking and scratching on the ground and, sometimes, pecking the environment in general [126]. Slow-growing breeds of broilers explore their environment more compared to fast-growing commercial broilers, are more active, forage more, and use more the provided environmental enrichment [39,41,44,47,48]. Rayner et al. [47] observed that slow-growing broilers ground-scratched more than fast-growing broilers under commercial conditions, and the behavior decreased as stocking density increased or the age of the animals increased. The authors conclude that this finding could be attributed to the fact that higher stocking density leads to worse litter quality and, as a consequence, a worse substrate for pecking behavior. The same observation was reported by Baxter et al. [41], but in farms of higher welfare standards. The reduced available area for scratching, when increased stocking density is applied, may also be responsible for the reduced rate of the scratching behavior by broiler chickens. Abeyesinghe et al. [39] propose exploratory behavior as a positive welfare indicator of slow-growing broilers that should be incorporated in welfare assessment protocols. Moreover, they support that it is also an indicator of better health status, since it is negatively correlated with breast plumage dirtiness, leg weakness and hockburn in commercial conditions. Exploration has been positively linked to environmental enrichment and stimulation in various studies with broilers [40,42,43,46,48–50] and negatively linked to stocking density [42,48–50]. Elevated platforms with ramps [40], pecking stones [42], straw bales [42] and barriers [43,50] are an effective way of promoting exploration and pecking on the ground and the environment. Also, enriching the litter with mealworm can induce the satisfaction of foraging pecking [46]. This enrichment can be combined with outdoor access or access to covered verandas in an attempt to promote exploration and positive affective states to a greater extent [45]. The Welfare Quality protocol for broilers [30] measures inquisitive exploration as part of the positive emotions criterion. This criterion is measured by QBA, and one on the terms that is used to describe the animals during the evaluation is "inquisitive".

Laying hens display a stronger motivation of foraging explorative behavior compared to broilers due the fact that broilers are more efficient converters of feed, are heavier, grow faster and spend more time resting [126]. Campbell et al. [51] observed that free-range hens reared in a complex environment with navigating and perching structures expressed more foraging and environmental exploration compared to hens reared in a control environment. In this study, both groups had equal quantity of feed indoors, enough to cover their nutritional needs, which indicates that the foraging and explorative behavior was a result of stimulation to explore the environment. Shimmura et al. [52] compared the total pecking behavior of hens kept in single-tiered aviaries with or without access to an outdoor area. The total pecking behavior of the two groups was the same, but the specific types were different. The layers with outdoor access performed more foraging and exploring pecking, while those reared in the aviaries were more aggressive and showed an increased incidence of feather pecking. Furthermore, inquisitive exploration [53] and foraging behavior [127] during the laying period increases when hens are reared in enriched environments as chicks. Environmental enrichment also increases the exploratory behavior of laying chickens in Y-maze tests [128]. Contrary to the Welfare Quality protocol for broilers [30], the respective one for laying hens [31] does not include the term "inquisitive" in the QBA assessment of animals' emotions.

Exploration is also linked with foraging behavior in the domestic turkeys. Animals spend most of their time foraging when raised outdoors [54]. Pecking at the environment is increased when the stocking density decreases [55], but the influence is not as strong as in broilers [41,44,47]. Regarding geese and ducks, exploration is not only correlated to foraging but also to swimming, and it is promoted by the depth and the whole swimming surface area [61]. Ducklings raised in an environment enriched with troughs with water and strings, objects or grain mixture, or just enriched with various types of feed and no water, forage more than ducklings without these provisions and perform less feather pecking [56]. At the same time and interestingly, naturally hatched geese show more foraging behavior compared to artificially hatched geese [57]. Moreover, geese in free-range systems forage and explore the environment more compared to geese kept intensively, although they show more fear to explore [57]. Geese also perform more inquisitive exploration on muddy compared to plastic floor types [71].

3.1.2. Comfort

An animal that displays behaviors associated with comfort is an animal that is presumably experiencing pleasure [27–29]. Comfort behavior is rewarding and accompanied with positive affective states [34]. After reviewing the existing literature, we concluded that two main behaviors indicate comfort in poultry and are proposed as positive welfare indicators: dustbathing and preening. Dustbathing is performed by hens, quails and ostriches [63]. Preening is a comfort behavior performed by all poultry species [61]. These behaviors will be analyzed below separately in Section 3.1.2, respectively.

Comfort behavior is described as preening and dustbathing in most findings concerning domestic hens (e.g., [39–42,48,52]). Preening while dustbathing is defined as dustbathing according to the ethograms in all studies. The Welfare Quality protocol for laying hens evaluates comfort behavior as preening and dustbathing [31], but comfort is not assessed in the respective protocol for broilers [30]. This is likely due to the short lifespan of broiler chickens, i.e., 42 days in an intensive production system. For most of the rearing period, feathers are not fully developed for adequate expression of preening and dustbathing behaviors. In addition to these two main comfort indices, some behaviors that are used as indicators of comfort are: leg stretching [39,40,44,46], wing stretching [39,40,59], wing and tail wagging [48,72] and scratching body [72]. All of these are indicators that are affected by space allowance and stocking density. Preening [57,71], feather shaking and wing flapping are applied in geese [57], whereas preening and stretching are comfort indicators for ducks [51]. Swimming is also described as comfort behavior for geese in one study [57], but this behavior will be analyzed separately in Section 3.1.5.

Dustbathing

Dustbathing is a complex behavior of domestic hens, quails and ostriches and their wild ancestors, often accompanied with sunbathing when performed in outdoor systems [61,63]. Dustbathing is suggested as a behavior that assists the bird to remove ectoparasites and excess fat from its body and feathers [63]. The full behavioral sequence in hens starts with the bird scratching and bill-raking in the substrate, then it erects the feathers and sits down. The second part consists of vertical wing-shaking, rubbing of the head, bill-raking and scratching using only one leg [63]. Then, the feathers are flattened against the body, and as described by Olsson and Keeling [63], the bird spends some time lying or rubbing on the side. During the last phase, it stands up, shakes out the dust and returns to its activities. According to Nicol et al. [61], the fact that dustbathing is a behavior that returns the motivation to the baseline after being fully performed indicates that it is an indicator of positive welfare. As reviewed by Olsson and Keeling [63] and Hemsworth and Edwards [60], although the motivation towards dustbathing in the domestic hen is well studied, the results are conflicting. The behavior is also influenced by environmental factors, which makes the understanding even more complex [60]. Hens deprived of a dustbathing substrate can perform a behavioral rebound when offered again the chance to dustbathe [60,63]. And, hens kept on wire-floored cages perform sham dustbathing [129]. Although the motivational need of hens to dustbathe is known, the value of this need for the bird compared with other needs is unknown due to conflicting results of various studies, as analyzed by Nicol et al. [61], Olsson and Keeling [63] and Hemsworth and Edwards [60]. Still, it is a motivational and rewarding behavior, often described as comfort. According to Widowski and Duncan [64], dustbathing occurs when the animal is given the opportunity to perform it and elicits positive motivational affecting states, meaning that it is not a need of the animal that leads to suffering if not fulfilled. In the study by Widowski and Duncan [64], most hens strive to obtain a dustbath when they could see one. In other words, the animal is not in need of performing the behavior but has the motivation to do so when it gets the chance and acquires pleasure by it. Another aspect that makes the behavior a good candidate as positive welfare indicator is that it is often synchronized, especially when performed outdoors [62]. According to Campbell et al. [51], chicks reared indoors with structural environmental enrichments exhibit more foraging and dustbathing behavior as adults, indicators of positive welfare according to the authors. Moreover, hens kept in aviaries with outdoor access perform more dustbathing behavior than hens kept indoors [52].

In broilers, dustbathing has been studied as part of environmental enrichment of the birds, in combination with other provisions and a more complex environment in general. Baxter et al. [58] compared the positive effects of an enrichment consisted of platform perches or platform perches and a dustbath, compared to a control environment, in commercially housed broilers. The birds showed the shortest flight distances to humans in the environment with a dustbath, but the amount of play was not altered. In a recent study by Vas et al. [49], increased environmental complexity also promoted dustbathing behavior in broilers, which was used as a positive welfare indicator. The authors defined comfort behavior in their study as vertical wing shaking, also an indicator of dustbathing, according to the analysis of the full behavioral repertoire by Olsson and Keeling [63]. Vasdal et al. [59] also observed that broilers with environmental enrichment including a dustbath were more active compared to commercially housed birds, and they showed more wing flapping, stretching, body shaking and play behavior. Nonetheless, the enrichment used in this study was a combination of peat, bales of hay and elevated platforms, and it cannot be concluded how important the contribution of the peat was in improving welfare compared to the other types of enrichments. Bergman et al. [42], on the other hand, observed that both fast- and slow-growing broilers in an enriched environment with a complex structure of perches, straw bales, pecking stones and outdoor access spend only a small amount of time dustbathing, but they attributed this to the fact that the litter material was not attractive.

The effects of dustbathing as environment enrichment have also been studied on quails, though again combined with other provisions [65,130], and the results indicate that it is used extensively. Miller and Mench [65] compared the effect of pecking and foraging (ropes, plastic blades, beads and jingle-bells), three-dimensional structural (elevated platform, tubes) enrichment or novel object (naturalistic objects) enrichment on the dustbathing behavior of Japanese quails; the authors observed that in all scenarios, the enrichment that was used most extensively and consistently by quails was the dustbath. Additionally, the birds in the study performed a variety of behaviors in the sand, like preening and pecking, but also egg laying, so the authors concluded that the provision for nest boxes was not adequate. Regarding domestic turkeys, although they perform dustbathing when given the opportunity, and the behavior is usually synchronized [131], no further data have been retrieved.

Preening

Preening behavior is performed by all domesticated birds with the intention of grooming [61]. The bird runs its beak through its feathers in order to maintain good plumage condition and distribute oil from the uropygial glad [132]. The behavior is often performed together with dustbathing, for the removal of the excess oil, but not necessarily. In addition to the hygienic function, this behavior is an indicator of pleasure and comfort, since both grooming and allogrooming are strong candidates as positive welfare indicators [28,33,34]. Still, preening can also be increased in environments with poor hygiene or ectoparasites [132]. In addition to this, due to stress, individuals can perform extensive preening, including self-directed feather-pecking and wounding. Thus, attention is needed when it is used for the assessment of positive welfare [33].

Environmental enrichment has been observed to increase preening in commercial broilers [44,58]. In the study by Baxter et al. [58], the enrichment was a platform and a dustbath. Preening while dustbathing was classified as dustbathing according to the ethogram. In the study by Dawson et al. [44], the enrichment was ropes, ramps and pecking stones. Dustbathing occurred infrequently and could not be analyzed. So, the increase in preening that the authors mention is separate from the preening performed while the birds are dustbathing. In other studies referring to the effects of enrichment on broilers, although dustbathing was used as positive welfare indicator, preening was either not studied at all or not mentioned in the definition of dustbathing in the ethogram [49,59], or it was just studied as part of the dustbathing behavior [42]. Bach et al. [40] have observed increased comfort behavior, as a total of dustbathing, preening and stretching of broilers on elevated platforms, but the percentage of each behavior was low to include in the statistical analysis separately. Both fast- and slow-growing broilers perform more sitting- than standing-preening with age [44], due to the rapid body growth, and the difference is more intense for the fast-growing breeds. But, the total amount of preening behavior is the same for both slow- and fast-growing breeds, and so preening cannot be used as a positive welfare indicator of the slow- compared to the fast-growing broilers [41,44]. As reviewed by Nicol et al. [61] and Freire and Crowling [66], preening behavior is performed at a higher level from laying hens in furnished compared to conventional cages, and so is stretching, although behaviors that require more space, such as wind flapping, can be constrained. According to Widowski et al. [67], preening is also linked to space availability, and it can be restricted below 542–750 cm^2/hen in cage systems, together with wind and leg stretching. So, preening can be both an indicator of positive affectivity and space availability. On the contrary, Shimmura et al. [52] observed that hens kept in aviaries without outdoor access perform more preening compared to hens with provision to outdoor access, although the opposite was observed regarding dustbathing.

Preening is also performed by ducks, including water preening [61]. Mi et al. [69] have observed that ducks with access to water pool perform more general preening and water preening compared to ducks with access only to drinking water. On the contrary, in a study by Riber and Mench [56], preening performance was the same between ducklings kept with access to water and the control group. Nonetheless, in this study, the birds did not have access to a water pool but to troughs with water and cup-type drinkers. Babington and Campbell [70] observed a gentle conspecific-directed behavior, described as allopreening, ignored by the recipient and provoking no response. In turkeys, preening increased on wet compared to dry litter, albeit occurring in a standing position [68]. It has also been found to be increased with increasing stocking density [55]. On the contrary, a linear decrease in this behavior was observed as lighting was increased for fourteen to twenty-three hours per day. So, although the literature results on ducks support its potentiality as positive welfare indicator, the results about turkeys are conflicting.

3.1.3. Play

Play has been studied in various species as positive welfare indicator due to the immediate pleasure that it elicits and, as a result, to its immediate impact on welfare [32–34,73]. It can be pleasurable, exciting and relaxing [34] and has a self-rewarding basis [32,34]. Biological and neurological findings support the opioid-mediated positive affective states that it generates [32]. Furthermore, another aspect that makes it promising as positive welfare indicator, is that it is usually suppressed under threats to fitness [32–34], and for this, it is considered a "luxury behavior" [32]. In addition to the immediate positive affect that is elicits, it also has long-term positive effects on welfare. As a behavior, it has significant functional benefits for the individuals engaging in playing, developing and acquiring somatic skills and competencies that will help the animals adjust to stressful situations in the future [32]. According to Špinka [35], the behavior is "emotionally contagious" and so by using only a few playful individuals, the positive emotions can be spread to the whole group. Nonetheless, it is a complex behavior with high variability between species, sexes, ages and even individuals, performed mostly by young animals and declining with age [32–34], so attention is needed for the observation, interpretation and association of the behavior to positive welfare.

All of the aforementioned benefits also apply to domesticated poultry, although the behavior has been studied mostly in mammals and is more easily recognized when performed in mammalian species [45]. As recently reviewed by Jacobs et al. [45], there are four types of spontaneously occurring behaviors that can be characterized as play in poultry: worm or food running (object play), sparring (solitary play), frolicking and jumping (locomotory play).

According to the existing literature, these behaviors have mostly been studied in broilers. Baxter et al. [58] observed that enrichment with perches and a dustbath did not have an effect on the total play behavior of broiler chicks, compared to birds housed in non-enriched environment. But, it was observed that when the observer was walking and creating space between the birds, this extra space stimulated play, especially frolicking and sparring. In accordance with the findings of Baxter et al. [58], Liu et al. [74] did not observe an increase in spontaneous play behavior in broiler chicks enriched with a ramp, platform, weighing scale, peak stone and feeder with wood shavings, compared to the control group. Additionally, in the study by Liu et al. [74], the enriched birds were less responsive during the tests that aimed at stimulating play behavior. The authors suggested that this was observed due to the fact that enriched birds were kept under an already stimulating environment and thus showed less interest. Only one study [49] mentions a positive effect of environmental enrichment on play behavior. A positive correlation was observed between the extent of provided means of environmental enrichment and the occurrence of play fighting. Furthermore, running and jumping were positively correlated with the increase in space allowance. Vasdal et al. [59] observed that play fighting increases in enriched chicks with wooden boxes for dustbathing, platforms, ramp and bales. Worm running, i.e., running while carrying a small object in the beak, also increased at young age, due to bales, but decreased with age. Baxter et al. [41], van der Eijk et al. [48] and Rayner et al. [47] compared welfare parameters between fast- and slow-growing breeds in commercial farms. In all studies, slow-growing chicks displayed more play behavior. In the studies of van der Eijk et al. [48] and Rayner et al. [47], play behavior was more frequent in the groups of chicks with less stocking density. Play is also used as indicator of positive emotions in the Welfare Quality protocol for broilers. The term "playful" is used as one of the terms that describe an animal in positive state during the QBA assessment [30].

In laying chickens, Campbell et al. [51] showed no difference in the total play behavior of birds enriched with a structural complex of perches or novel objects, compared to non-enriched birds, with the exception of running, which was more frequent during the first weeks of age of the non-enriched groups. Apart from this, no other research findings were identified. The Welfare Quality protocol for laying hens, in contrast to the one for broilers, does not use the term "playful" in QBA [31], probably because it is a behavior that is positively correlated with the younger age of chickens.

Regarding ducks, Chen et al. [75] observed an increase in play behavior in ducklings reared with enrichment (perches, colored balloons and ribbons). Amado et al. [76] studied the behavior of ostriches, kept according to commercial standards in Brasil, from ten days to five months of age. The authors proposed that the running and dancing behavior of the birds is presumably a play-like behavior, and it was also observed that it decreased with age.

3.1.4. Behavioral Synchronization

Behavioral synchronization has been proposed as a promising positive welfare indicator for all group-housed social animals [33–35]. An individual that belongs in a group experiences cohesion, companionship and safety, and so behavioral synchronicity is experienced as rewarding for the individual [33]. Multiple individuals perform the same behavior simultaneously and the behavior is spread to the whole group. This behavioral contagion is also an emotional contagion, since social animals can experience empathy and develop affiliative connections to each other [35]. It also has a buffering effect to stress in social species [35]. An advantage of behavioral synchronization is that an individual can be used to spread a positive behavior in the whole group, and so positive welfare can be promoted in a whole group by using just a few individuals [34,35]. A disadvantage is that synchronization is a group phenomenon, although welfare is an aspect of an individual [33].

Domestic poultry species are social group-housed species, and so promoting behavioral synchronization in the flock is a means of promoting positive affectivity. Several of their behaviors are synchronized, due to strong intrinsic motivation, especially behaviors that would make animals more vulnerable to predators in the wild, like feeding, drinking, dustbathing and resting [133]. In a study by Eklund and Jensen [134], White Leghorns were less synchronized during perching and comfort behavior, and they maintained a longer inter-individual distance after the comfort and perching bouts, compared to red junglefowl. Additionally, according to Lopes Carvalho et al. [135], social learning is another factor that contributes to behavioral synchronicity, especially in chickens. Broilers tend to synchronize their resting behavior and have longer resting bouts, when offered artificial brooders as resting place. When resting behavior is less synchronized, the birds have shorter resting bouts and keep changing resting groups [79]. Collins and Sumpter [78] have observed that feeding behavior is synchronized in broilers. The birds cluster at the feeder when kept in low stoking density, indicating that the feeding bursts emerge due to social facilitation. Moreover, the light regime affects broilers' synchronicity. According to Alvino et al. [77], broilers perform higher synchronicity when offered 16 h day of high intensity and 8 h of dark per day, with longer synchronized and uninterrupted resting and sleeping periods. Behavioral synchronization also occurs more in hens of the same cage than hens between different cages, indicating that this behavioral synchronicity is promoted by social factors [80]. But, to a lesser extent, synchronicity also occurs between birds of different cages [80]. Keeling et al. [81] observed that in laying hens, preening is the most synchronized behavior and daytime perching the least synchronized. Feeding is the more clustered one regarding space. In their study, synchronicity and clustering decreased as the number of birds increased. The authors concluded that it is important to provide adequate space and resources to smaller flocks, since behavioral synchronization is stronger in smaller flocks.

Synchronization also occurs in waterflow, with swimming being a social and synchronized behavior [61]. In a study by Waitt et al. [82], Pekin ducks synchronized their bathing behavior in baths and troughs and especially in showers, where they also had the opportunity to socialize on the wetted area around the bath.

3.1.5. Swimming and Access to Water

Goose and ducks are waterfowl, and many of their natural behaviors like preening, bathing, dabbing, foraging and reproductive behavior are strongly correlated to water and swimming. So, swimming is essential for the expression of a full behavioral repertoire [61]. Providing access to open water, in a depth that allows them not only to wet their feathers and perform head-dipping but also full body immersion, is important for the expression of swimming and all water-related behaviors, including thermoregulation and feather cleanness, and promotes positive affective states [70]. Liao et al. [71] observed that geese with access to a swimming pool in addition to swimming increased exploratory preening and moving and reduced feather pecking and sitting. In a study by Fattah et al. [83], Egyptian geese performed more feeding, locomotion, preening and flaying in an environment enriched with a swimming pool. The results of swimming in ducks are in accordance with those of geese. Sanshui White ducks with access to a swimming pool perform more preening behavior compared to control group [69]. Additionally, swimming is a synchronized behavior in ducks, and so it enables animals to experience social interactions, safety and cohesion. Waitt et al. [82] observed that swimming in Pekin ducks was synchronized in baths, showers and troughs, and ducks tended to swim more when other ducks were also performing the behavior simultaneously, especially in showers. On the contrary, the provision of water to Pekin ducks via misting does not seem to importantly affect their general behavior [136].

3.1.6. Pre-Laying (Nesting)

Pre-laying behavior of hens is a strong natural behavior linked to ovulation. Hours before the ovulation, the bird becomes restless and active while searching for the adequate place for nest-building [101,126]. Laying hens deprived of nesting vocalize more frequently and produce mostly gakels, i.e., vocalizations that express frustration and generally negative emotions [101]. They also perform stereotypic pacing, indicating stress [137]. The absence of nest boxes or the prevention of hens from expressing pre-laying behavior, and the negative behavioral impact that this has on behavior, has been studied extensively, as reviewed by Hemsworth and Edwards [60] and Cronin et al. [84]. Hens strive to access a nest box, and this motivation is the most studied one in different types of housing systems [60,138]. Pre-laying behavior is affected by the rearing system, the environment and the location of the nests. Hens kept in aviary systems find the nest less attractive than birds kept in cages, but they experience more active and aggressive pre-laying behavior [87]. Higher stocking density also induces competition for access to nests [139]. According to Engel et al. [140], hens kept in cage systems choose a nest compared to feed, in maze preference testing. In the same study, two groups of animals were studied: one kept in cages with provision to nests after sixteen weeks of age as adults, and one that was trained and exposed to the nests only during the training before the maze testing. Hens kept as adults with access to nest boxes choose the nest over food more frequently than hens kept without access. Still, hens kept without access to boxes also chose the nest box over feed, albeit infrequently. These findings are an indication that nesting behavior presumably induces positive affective states in hens [60,86].

Apart from these two references [60,86], no other research mentions directly nesting as a possible positive welfare indicator, although, as described above, the motivation and the preference of the hens for nests in various systems has been widely studied. In addition, studies on stress physiology, as reviewed by Hemsworth and Edwards [60] and Cronin et al. [84] do not indicate that hens housed without next boxes show physiological evidence of short-term or long-term stress, but only behavioral evidence of negative emotions, like frustration. Nonetheless, some results indicate that one function of nesting may be to provide a specific location of laying for hens, where they perform the pre-laying and sitting phase of ovulating undisturbed [84,85]. This means that hens with the pre-laying behavior are looking for a safe and quiet place [84]. Cronin et al. [85] observed that disturbed sitting before egg laying leads to higher corticosteroid levels in blood plasma and egg albumin, possibly indicating disturbance and stress to the hen that is trying to stay calm and ovulate. If a behavior promotes safety, then it promotes positive states and is a good candidate as positive welfare indicator [28,29,33,34]. Moreover, in a study by Hunniford and Widowski [141], nests areas enclosed with plastic curtains induced more settled pre-laying behavior and less aggression, even to hens that were accustomed to lay before the study with no access to the curtain enclosed nests.

Regarding other poultry species, two studies indicate pre-laying behavior as positive welfare indicator for ducks, albeit indirectly [88,89]. Ducks are highly motivated to nesting and prefer the use of manipulatable substrates that can also be used for nest building, as indicated after preference testing [142]. Barrett et al. [88] studied the difference in pre-laying behavior between Pekin ducks that were floor- and nest-layers. It was observed that floor-layers preferred specific floor locations in order to avoid agonistic interactions and lay quietly. When the behavior was performed in the nest, for both groups, the pre-laying behavior was similar. So, the only difference for the two groups is the location. This indicates that in ducks, as in hens [85,141], nesting provides safety to the bird and thus indicates positive affectivity [28,29,33,34]. Moreover, Makagon et al. [89] have observed that Pekin ducks prefer a nest box enclosure, preferably also with a curtain, as has been observed in hens by Hunniford and Widowski [141], also supporting the theory of promotion and safety of the nest to the layers.

3.1.7. Perching

Perching is a strongly intrinsic motivated behavior of the domestic hen, also linked to roosting behavior [61,90]. It is an antipredator-related behavior. Despite the domestication process and the protection from predators, the motivation of the birds to perform this natural behavior is still powerful. Laying hens, with increasing group size and the same stocking density, spend less time perching, are less vigilant while perching and spend more time on the floor preening, as would be estimated by the antipredator hypothesis [92]. Furthermore, although they spend less time preening, they still prefer to occupy the higher perches [92]. Additionally, laying hens prefer the higher perches for night roosting in aviary cages and multi-tier systems [143]. Hens strive to gain access to perches [62,90], especially at night [61,90,144], and experience frustration if the access is prevented [145]. Perching is also common during the day, due to the feeling of security that it promotes [61]. It is a complex behavior to study, because although the motivational need is established, further research is needed to understand which need is of highest priority for the bird: seeking daytime or nighttime elevation, grasping, or using a perch specifically instead of other structures like ramps and platforms [61]. Still, we propose it as an indicator of positive emotional state due to the feeling of safety that it promotes [61,90]. Moreover, access to perches during the rearing period of chicks enhances the spatial cognitive and physical abilities of birds later in life [91] and is thus a positive cognitive enrichment. Structural enrichment with perches and navigation structures on pullets reared indoors increases the positive behaviors of foraging and dustbathing when kept outdoors as free-range laying hens, from 16 weeks of age and on, indicating that perches in chicks can have a long-term effect on positive welfare [51]. In addition, it contributes to the comfort of the hens [146].

Perching behavior is a stronger motivating need in laying hens compared to broilers, due to the fast increase in broilers' body mass and the locomotory and leg issues that they suffer from [61]. This leads to perching decreasing with age as the body weight increases [61,147]. The decreased perch use in broilers can be a result only of limited physical ability of the birds and not motivation [147]. Studies indicate that broilers are highly motivated to use aerial perches and perches of aviary tiers [148]. Slow-growing broilers perch more than fast-growing breeds [39,41,44,47]. The use of perches also increases as the stocking density increases [147,149]. Broilers, in contrast to laying hens, due to impaired physical ability prefer to use other types of perching substrate, rather than common perches, like platforms [147] or straw balls [150]. According to the findings of Bach et al. [40], the comfort behavior of broilers is high on elevated platforms. The design, the material and the structure of the perching substrate is very important. Although no direct data have been shown for the use of perches as a positive welfare indicator for broilers versus laying hens, since they are motivated to use a perching substrate of structures adjusted to their physical ability (as reviewed by Nicol et al. [61]), perching can promote the same positive affective states in broilers as in laying hens. Thus, we also suggest perching as a positive welfare indicator for broilers, although further research is needed.

Ducks, geese, quails and ostriches do not perch, with the exception of Muscovy ducks [61], but no data have been retrieved.

3.1.8. Maternal Care

According to Mellor [34], maternal care is a positive welfare indicator due to the strong bonding that it elicits between the mother and the young. It is based on physiological alterations that induce positive affectivity both in mother and offspring. The promotion of mother–young bonding also means the simultaneous promotion of peer-bonding. Maternal care is not generally promoted in the poultry industry, especially in intensive systems, due to the high number of animals, the short lifespans of the animals and the impact on productivity and profitability that it would cause [95]. Chicks are hatched in incubators and reared artificially. Nonetheless, this positive welfare parameter has important positive effects not only on the mother, but especially on the chicks, as reviewed by Edgar et al. [95], like teaching them food preferences and reducing fearfulness. This is important especially for the chicks that will have outdoor access as adults [151]. The mother also has a social buffering effect on stress for the young, since chickens in the presence of their mother respond with reduced stress to an aversive stimulus [152], indicating that they feel safe and protected. Moreover, mother hens show empathic responsiveness towards their chicks, when the later are subjected to an aversive stimulus in front of the hens [93]. Additionally, brooded chicks show higher behavioral synchronicity, which is also a positive welfare indicator, although in the first days of their lives, this also has thermoregulating properties [95].

Fearfulness is also reduced in geese when naturally hatched and brooded [57]. Geese chicks naturally hatched and reared with their mothers until three days of age displayed less fearfulness compared to artificially hatched geese chicks [57]. In quails, the development of exploratory [97] and social behavior [96] of the chicks is also influenced by the mother, being higher in brooded chicks.

3.1.9. Anticipatory Behavior

The anticipatory behavior of an animal in the expectancy of reward can be used as a mean to evaluate its emotional state. The animal is initially trained to associate a signaling cue to the reward, and then its emotional state is evaluated between the conditioned stimuli of the signaling and the unconditioned stimuli of receiving the expected reward [153]. The less the anticipation, the higher the welfare level, since the lower the difference between the real and the expected level, and there is a balance between the negatives and positives in the animal's life [153]. Anticipatory behavior has been studied in various farm animals to evaluate emotional states, mostly negative, but also positive [154]. It should be interpreted with attention since it is often a result of frustration of the animal while anticipating a reward and not a result of positive emotions [154]. Anticipatory behaviors can be variable, with behavioral transitions of the animal being the most consistent parameter for the observer [154]. Nonetheless, although anticipation is a bias towards evaluating positive emotional states, it is necessary to examine them also as positive welfare indicator [72].

McGrath et al. [98] trained laying hens to signal sound cues to two different feed rewards, a dustbathing substrate as reward, or no reward at all. The conditioned animals had the ability to associate each cue with the separate reward/no reward. The anticipation for the dustbathing was higher, and the frequency of vocalizations of positive arousal were more frequent compared to the feed rewards. Zimmerman et al. [72] conditioned laying hens to a negative, a positive and a neutral cue. Again, the birds had the ability to discriminate the cues. They showed more locomotion and head movements and vocalizations indicating frustration in the anticipation of the negative event. On the contrary, the prevalence of comfort behavior was specifically associated with the anticipation of the positive event. In a study by Wichman et al. [100], laying hens were conditioned to associate light signaling cues with the arrival of feed or not. There was no difference regarding the anticipatory behavior from enriched and basic pens. According to the authors, this was observed due to the fact that the number of social interactions in the pen influenced the emotional state of the animals and their anticipation. Furthermore, the anticipation of feed reward in hens is combined with a decrease in the comb surface temperature, and as the peripheral temperature decreases, it indicates a change in positive emotional arousal [99].

3.2. *Vocalizations*

Vocalizations have been proposed as a feasible and non-invasive positive welfare indicator for all farm animals [33,38]. The digitalization of animal farming and automated microphones make their application effective. Still, either by using microphones or evaluating the indicator live on the farm level, it is easier to detect the vocalizations of a group of animals, and not of a specific individual. Vocalizations have long been studied as indicators of emotions in animals, mostly negative emotions, but today, they are also considered strong candidates as positive welfare indicators [33,38]. A vocalization indicates the emotion of the bird at the moment that it is expressed [33], so it is a short-term indicator of emotion. Usually, they are not considered reliable indicators of welfare alone, at least not yet, but they should be used in combination with other positive welfare indicators [33].

As reviewed by Laurijs et al. [38], chickens produce eight types of vocalizations: food calls (produced only by roosters), single, double, fast and gakel clucks, whines, singing and mixed vocalizations that cannot be categorized in the previous types. Fast clucks and food calls are associated with positive emotions [38]. In a study by McGrath et al. [98], laying hens, in the anticipation of food or dustbathing as a reward, after Pavlovian conditioning, produced mostly food calls and fast clucks with the signaling cue. It was concluded that these types could be indicators of emotions of positive valence that could be used for on-farm welfare evaluation. Zimmerman et al. [72], on the other hand, have observed that laying hens, in the anticipation of a negative event, produce mainly gakel clucks. Gakel vocalizations also increase when hens are deprived of food or nesting and expressing frustration [101]. No literature data have been retrieved for vocalizations as positive welfare indicators in other poultry species.

3.3. *Qualiative Behavioral Assessment (QBA)*

The qualitative assessment of behavior was firstly proposed by Wemelsfelder et al. [155] as a tool to evaluate animal welfare. Until then, it had been widely used in the study of animals' temperament, but no research had been conducted on its use as a measurement of animal welfare. Wemelsfelder et al. [155] applied QBA in the study of spontaneous expressions of growing pigs, using untrained observers. The behavioral expressions were described with high inter-observed reliability. Further studies revealed that the method also had high intra-observer reliability [156], and so they proposed QBA as a novel method of integrative animal welfare evaluation. QBA has been proposed as a part of the effective implementation of an animal welfare assessment program [157]. As a welfare indicator, it is feasible, easy, and economic to apply. It is also relatively easy to train the assessors, it is widely accepted, and it is not invasive for the animals [158].

For the above reasons, QBA has been incorporated as a both a positive and negative welfare indicator in the Welfare Quality Assessment protocol for poultry [30], for evaluating the welfare of broilers, and in the Welfare Quality Assessment protocol for laying hens, version 2.0 [31]. Both protocols are built on calculating the total welfare score of a farm by answering four main welfare principles, one of which is the appropriate behavior principle, in answer to the question, "Does the behavior of the animals reflect optimized emotional states?" [30,31]. In turn, each welfare principle compromises some welfare criteria, all independent of each other. The principle appropriate behavior compromises the criteria expression of social behaviors, of other behaviors, of good human–animal relationships and of emotional states. The last one is assessed by QBA in free-range hens and novel object test in laying hens in cages [31] and by QBA in broilers [30]. Both protocols define the emotional state criterion as the avoidance of negative emotions, like fear or distress, and the promotion of positive emotions as security or contentment. The assessor observes the interactions of the animals with each other and their environment via scan sampling for a period of time, replicates for different group of animals, and finally scores one-word terms that describe the birds' behavioral repertoire. Ten of these terms indicate positive affective states: calm, content, comfortable, inquisitive, positively occupied, confident, energetic, playful, friendly and active. Additionally, when calculating the total score, not all description terms are of the same weight. Among the positive terms, comfortable and content have the highest weight [30].

Rayner et al. [47] applied QBA as positive welfare indicator to commercial farm systems of slow- and fast-growing breeds of broilers. The slow-growing breeds displayed higher scores of "happy/active" and lower scores of "stressed/flat". QBA has also been applied on broilers by Muri et al. [102]. It was concluded that QBA may be more adequate for larger animals kept in smaller groups, because the homogeneity of the flock makes the observation of the different qualitative behaviors difficult. The authors concluded that it cannot stand as the sole tool for the on-farm welfare assessment of broilers, but it can give important supplementary information together with other measurements. Vasdal et al. [105] used QBA in laying hens and reached the same conclusion. Van Niekerk et al. [104] applied the Welfare Quality protocol for laying hens [31] to 122 flocks and, as was expected, found low scores for caging systems, high scores for aviary organic systems, but surprisingly low scores for conventional floor systems. QBA has also been studied on parent flocks of white and brown laying hens to investigate behavioral differences in birds of different breeds [103]. Again, due to the homogeneity of the flock, only three out the twenty behavioral description terms that were used could be applied by the observers: comfort, distressed and active.

Regarding all other livestock poultry species, no research has been retrieved about the proposal of QBA as a positive welfare indicator. The AWIN welfare assessment protocol for turkeys [159], which is based on the principles and criteria of Welfare Quality, mentions the emotion states criterion, but it states there is yet no available indicator for its assessment.

3.4. Positive Human–Animal Relationship (Positive HAR)

There is a reciprocal relationship between the attitude of the stockperson that handles an animal and the behavioral reaction of the animal towards this person [106]. Farm animals have the ability to discriminate among handlers that treat them differently [106,107]. A positive human–animal interaction is beneficial not only for the animal, due to the positive experiences and effects that it elicits, but also for the caregiver, since it facilitates the response of the animal to handling [37,106,107]. There are also positive effects on productivity, as reviewed by Mota-Rojas et al. [106] and Zulkifli [107]. HAR in farm animals is usually measured by a station/ passive human-test, when an animal approaches the stationary man voluntarily or by approaching/active human-test, when the human approaches the animal, and the avoidance distance is measured [37,106]. All methods of assessing HAR have been reviewer by Waiblinger et al. [107]. According to Rault et al. [106], the passive human-test is the more adequate indicator for positive HAR, while the human avoidance tests are usually used to measure fear. An animal that voluntarily approaches a human, or in general, an animal that responds positively to human handling, is an animal that experiences comfort, pleasure and anticipation and finds this interaction rewarding [37]. Positive HAR is also an indicator of long-term positive welfare, improved health and resilience for the animals [37,108].

Visual human contact of laying pullets during rearing reduces the avoidance distance to humans during adulthood [86]. In the same study by Edwards et al. [86], adult hens with close proximity to humans show a shorter avoidance distance and lower levels of plasma corticosterone concentration during handling. Avoidance distance and the stationary person test seem valid and with high correlation to each other both for cage [160] and free-range systems [112,160]. The touch test, where the assessor attempts to touch the birds, has also been studied [111,161]. Bertin et al. [111] observed that positive human visual and acoustic contact for three minutes per day and gentle daily stroking for thirty seconds of adult laying hens led to a reduction in human avoidance, compared to hens that experienced negative human contact. Additionally, the fertility of the positive HAR group was higher, yolk hormones levels were modified, and chicks' social skills were improved, with chicks preferring a familiar conspecific to a stranger. There was no transgenerational effect on chicks' avoidance to humans; still, according to the authors, a positive HAR in the hens influenced the filial imprinting. Graml et al. [112] concluded that only fifteen minutes additional human contact twice a day, with positive visual and acoustic stimulation, even touching, decreases the avoidance distance in free-range laying hens and increases the proportion of touched birds, compared to the group of hens that is subjected only to routine daily management. In laying hens, a visual positive HAR is particularly effective compared to other species [86,111,112], and the general positive HAR stimulation can have important effects even if it lasts for a few minutes daily [111,112]. Data reported in the literature are also similar in broilers. Zukifli et al. [110] observed that broiler chicks that had been subjected to ten minutes visual contact twice per day, from the first day of their life until twenty-one days of age, showed less fear of humans, less stress during handling and improved antibody response at forty-two days of age. The same results were also observed by Al-Aqil et al. [109], but in addition, the chicks were stroked for thirty seconds daily. The Welfare Quality protocol for poultry [30] and Welfare Quality Assessment protocol for laying hens, version 2.0 [31], measure both the good human–animal relationship criteria via the avoidance distance test.

Regarding positive HAR as a positive welfare indicator, studies also exist for quails [113] and ostriches [114]. Positive habituation of quails to humans, involving stroking at hand feeding for ten seconds, twice per day, affected the egg hormone level, immunoreactivity and led to heavier, stronger egg shells and heavier offspring [113]. The positive HAR was also transgenerational, with chicks from the different groups performing differently in behavioral tests. Ostriches that receive extensive human care from young age are more docile towards humans and willing to associate with them at later stages of life, compared to ostriches raised with standard commercial practices of foster parents [114]. Furthermore, human-imprinted ostriches present higher survival levels to four weeks of age compared to conventionally reared birds without human or foster parent imprinting [162].

3.5. Cognitive Bias

Cognitive bias tests have been used as an indicator of both positive and negative emotions in animals, based on the fact that changes in cognitive functions can be indicators of emotional states, as in humans [116]. In particular, indicators of change in emotional valence (positivity or negativity) rather than arousal [115,116] can be generalized across animal species, livestock included [116], and are promising as a tool for the assessment of positive emotions [115,116]. There are two types of cognitive bias test: judgement bias and attention bias [115]. When a judgement bias test is performed, the animal is firstly trained to associate one cue with a positive event and another cue with a negative/less positive event [115,116]. Then, an ambiguous cue is presented to the animal. If the animal is in a positive affective state, it will interpret the ambiguous cue as positive (optimistic response), and if not, as negative (pessimistic response). When an attention bias task is performed, the animal allocates its attention between a positive and a negative stimulus [115]. Judgement biases are more popular, and this approach has mostly been used in livestock animals [115,116].

Witchman et al. [100] trained laying hens on spatial cues, i.e., feed bowls that contained feed or not, depending on their spatial position. Then, the trained birds were subjected to a judgement bias test with bowls in ambiguous positions, and the latency to approach was recorded. The hypothesis of the authors was that birds kept in higher environmental enrichment would perform more optimistic responses, approaching the ambiguous cues faster. Nonetheless, no differences were found, indicating that the emotional states for the two groups did not differ enough to bias the birds' judgement. Still, the same conclusion was also reached via the anticipation test, and so the authors concluded that individual factors of the birds, like motivation to feed and positive social interactions present in both groups, influenced the judgement and the anticipation. Hernandez et al. [120] also observed that stress occurring immediately before a judgment test does not affect the cognitive judgment in laying hens. It was also observed that the birds approached ambiguous cues faster when they were tested immediately after rewarding events, and so the order in which the cues are presented during a judgment bias in hens matters [120]. However, in the studies of Witchman et al. [100] and Hernandez et al. [120], the manipulation of birds' affective state by the authors did not influence the birds' judgement, as was expected, while a study by Deakin et al. [119] concluded the opposite. A novel cognitive bias task was used, namely, a "screen-peck" task. The hens were trained so that when they pecked a high/low saturation orange circle cue, they obtained a feed reward, and when they pecked an oppositely saturated orange circle, they received a negative air puff. Following the training, the authors changed the temperature for some birds to near twenty-nine degrees Celsius, a temperature that, according to the authors, is pleasant for hens and induces a positive affective state. As was expected, the animals tested under higher temperature judged ambiguous cues and orange circles of intermediate saturation more positively. In a study by Zidar et al. [121], the exposure of female young chicks to cold as a stressor did not bias the judgement of the birds, but it was observed that exposure to environmental enrichment maintained the optimistic judgement in a second judgement bias test that followed. Judgement bias tests have also been performed in broilers. In a study by Anderson et al. [119], broilers housed in high-complexity pens judged optimistically and showed shorter latencies to approach ambiguous cues compared to animals housed in a low-complexity environment, indicating that the higher environmental stimulation induces positive affective states. On the contrary, in a study by Lourenco-Silva et al. [118], slow-growing broilers from groups of high and low environmental complexity performed the same in a judgement bias task. The different results in the two studies may be due to the different environmental enrichment used, enough to initiate or not a positive effect to bias the animals' judgement. Additionally, the birds in the study by Lourenco-Silva et al. [118] were tested in pairs during the task. From all the above research findings, both for laying hens and broilers, we conclude that the results in judgement bias tasks can be contradictory and can be affected by various factors. Furthermore, the stimulation that is used to manipulate the animals' emotional state should be significant in order to promote an affect-induced positive judgement. Still, there are studies where cognitive bias can assess positive affective states successfully [118,119].

Regarding the use of judgement bias tests in other poultry species, the results were retrieved only for quails. Japanese quails housed in different housing conditions, wired or deep-litter pens, were tested in spatial judgement tasks, after discriminating learning of cues associated with feed rewards or noise punishment [122]. The cues differed in shades of grey. When the birds were then presented with ambiguous cues of intermediate shades, there was no difference between the judgement responses of the two housing conditions, although it was expected that the birds kept in deep litter would have responded more optimistically, since they would had been in a more positive state.

4. Discussion

Following our literature review, we referred to thirteen animal-based indicators that have been proposed as indicators of positive affective states in domestic poultry species. The assessment of positive affectivity aims towards the evaluation of positive welfare by finding indicators that can be used as measures and applied on protocols and welfare evaluation schemes. Although a debate on the definition of welfare exists, welfare needs to be measurable so that it can be studied by scientists and used in practice [1]. In order to apply indicators in practice, they should be feasible, valid and reliable [163] and, at the same time, economically applicable. These parameters have not been evaluated in our review. The research on positive welfare is still mostly performed at the experimental level, although it is flourishing [1,12]. Our results are based mainly on experimental studies.

According to our results, explorative and foraging behavior, dustbathing and pre-laying behavior are the PW indicators that have been studied most extensively. Our results include both PW indicators that have been proposed for other species and indicators relating to poultry-specific behaviors. Exploration, foraging/feeding, comfort, play, behavioral synchronization, anticipation, QBA, vocalizations and cognitive bias have also been proposed for the domestic pig [125] and dairy ruminants [123,124,164]. Dustbathing, preening, nesting and perching have been studied specifically for poultry, and so has swimming for waterfowl.

The majority of the literature results refer to broilers (24) and laying hens (38). The results on laying hens outweigh those on broilers. This is because, as can be concluded by Table 1, some indicators have been studied only on laying hens. Comfort has been studied more on layers, while exploration has been studied more on broilers. Additionally, there is limited research on ducks (with the exception of swimming behavior) and quails, and there is even more limited research on geese and ostriches. The number of results, categorized by species, are indicated in Table 2. According to the FAO, in 2020, more than 450 million turkeys and 1.15 billion ducks were kept as production animals around the world [165]. There is a need for more studies about their welfare, their welfare assessment, the development of protocols and the incorporation of indicators in policies and legislation. Moreover, there is a need for research on breeder flocks, either broiler or laying hens. Only one result was retrieved specifically about the PW of breeders [103].

Table 2. Literature results on positive welfare indicators, categorized by poultry species.

Species	Total Number of References	References
all	14	[27,28,32–35,38,73,106–108,115,116]
broilers	25	[30,39–50,58,59,61,74,77–79,102,109,110,117,118]
laying hens	38	[31,38,51–53,60–64,66,67,72,80,81,84–87,90–95,98–101,103–105,111,112,119–121]
turkeys	3	[54,55,68]
ducks	8	[56,61,69,70,75,82,88,89]
geese	3	[57,71,83]
quails	5	[65,96,97,113,122]
ostriches	2	[76,114]

In Table 3, we summarize the positives and the negatives of each retrieved positive welfare indicator, according to our results. We include not only advantages and disadvantages described in our results, but also personal conclusions after analyzing them. Some key conclusions that can be reached by Table 3 are that the general data on exploration seem to agree on the positive affectivity that it elicits (e.g., [41,44,47,48,51,56,57]). Comfort behavior has mainly been assessed by dustbathing, which is a stronger positive welfare indicator compared to preening. Preening has been found both to be increased with environmental enrichment and space provision in some studies [44,58,61,66,67,69] and to be decreased in others [52,55], or not affected at all [56]. Nesting [84,85,88,89,141] and perching [88] promote security. A better understanding is necessary for play, since it would be a complementary aid for the on-farm welfare assessment of positive welfare compared with other indicators. Moreover, it is important to mention that a difficulty that we have faced while analyzing our results is that there is an inconsistency concerning the definition/description of behaviors in the ethograms of the studies, especially regarding comfort (as analyzed in Section 3.1.2 with details) and play. In some studies, types of play behavior were analyzed separately as running or chasing (e.g., [74]) or running and worm running (e.g., [59]). Furthermore, the fact that QBA has been found difficult to apply in commercial farms, due to the homogeneity of the poultry flock, also indicates the difficulty of observing and recognizing some behaviors in poultry [102,104,105]. Further studies are needed so that QBA can become more reliable and valid. Additionally, regarding swimming, the system of water provision is important, since ducks prefer to be able to deep their head or whole body in the water and perform all the water-related behaviors in addition to swimming, like water preening [70], and they preferably swim synchronized if there is enough space [82].

Table 3. Advantages and disadvantages of each PW indicator.

Indicator	Advantages	Disadvantages
exploration	indicates curiosity and safety [28,33,34] can be observed relatively easily on the field is positively correlated to environmental enrichment in broilers [49,52,59,65,130], laying hens [51,127] and turkeys [55] is negatively correlated to stocking density in broilers [42,48–50] can be assessed by QBA in broilers [30] is positively correlated to swimming [61] is increased in free-range systems in laying hens [51,53] and geese [57]	inspective exploration can also indicate fear [33] different enrichment in each study studied mainly in broilers slow-growing broilers perform less locomotion QBA is difficult to apply in poultry due to flock homogeneity more difficult to be promoted in intensive systems, cages, aviaries
dustbathing	motivational need of laying hens when given the opportunity [64] is positively associated with environmental enrichment in laying hens [51,52], broilers [49,58,59] and quails [65] is positively linked to outdoor access [52] can be synchronized [62]	conflicting results on the value of this need compared to other needs [60,61,63] different environmental enrichment in the results different definitions of dustbathing in the results
preening	grooming is a strong candidate as a positive welfare indicator [28,33,34] positive association to space availability in hens [67] increases seen in furnished compared to conventional cages [61,66] allopreening performed by geese [70]	can increase due to ectoparasites [132] overgrooming indicates stress [33] conflicting results when studied in connection to environmental enrichment in broilers [44,52,55,56,61], hens [66,67] and ducks [69] decreases with outdoor access in hens [52] included in dustbathing definition in some studies [42] preening while dustbathing defined as dustbathing

Table 3. *Cont.*

Indicator	Advantages	Disadvantages
play	immediate and long-term positive effect [32–34,73] emotionally contagious [35] high variability and ontogeny [32] is positively linked to space in broilers [49] can be assessed by QBA in broilers [30]	difficult to observed in poultry [45] wide range of definitions in the ethograms studied mostly in broilers studied mostly in chicks conflicting results in relation to environmental enrichment in broilers [49,51,58,59] QBA is difficult to apply in poultry due to flock homogeneity
behavioral synchronization	indicates group cohesion [33–35] buffering effect on stress [35] various behaviors synchronized in poultry	group phenomenon, difficult to observe an individual [33] not all behaviors equally synchronized [81] enough space is required occurs more frequently in smaller flocks [81]
swimming	increases preening [69,71] and exploration [69] synchronized [82]	the type of water provision system affects the results hygiene is a challenge requires space and labor
pre-laying	motivation well studied in laying hens [60,84,86,89,140] induces safety and undisturbed ovulation [84,85,88,89,141]	more studied needed for other species no stress physiological data when hens thrived access to nests [60,84] the type and place of nest is important and differs between studies
perching	motivation well studied in laying hens [61,90,144,145] promotes safety [61,90], comfort [146] promotes spatial cognitive abilities of chicks [91] slow-growing breeds perch more [39,41,44,47]	rapid decrease with body weight increase in broilers, with specific perching substrate needed [61,147] the perching type, structure, and height differs in the study more studied needed on broilers
maternal care	promotes mother–young and peer-bonding [35] positive social development of the young [57,95–97,151,152]	in commercial circumstances mostly artificial hatching
anticipation	laying hens [72,98–100] can be trained successfully	can be biased [72,154] can be affected by ontogeny and indicate frustration [154] difficult to apply under commercial circumstances since it requires training and time
vocalizations	feasible, no invasive observation [33,38] automatization, digitalization	more studies needed use only in combination with other indicators
QBA	feasible, economic, and easy to train the assessor [158] already incorporated in protocols [30,31]	due to poultry flock homogeneity, difficult to observe the various behaviors [102,104,105] more studies needed under commercial circumstances
positive human–animal interactions	facilitates animal handling [37,106,107] increases productivity [106,107] short daily visual contact effective in chicks [86,111,112]	requires training of the stockpersons and farmers
cognitive bias	laying hens [117,118], broilers [100,119–121], and quails [122] can be trained successfully to discriminate cues	conflicting results as PW indicator difficult to apply under commercial circumstances since it requires training and time

According to the literature, slow-growing broiler breeds are experiencing more positive welfare aspects compared to fast-growing breeds, including exploration [39,41,44,47,48], comfort [41,44], play [41,47,48] and the use of the environmental enrichment [40,41,44]. All studies mention the higher locomotion of these birds. In other words, they are healthier and experience more positive emotions and experiences.

Various findings also support the promotion of positive behaviors through environmental enrichment such as exploration [42,43,46,48–51,54,56], dustbathing [49,51,58,59,65], preening [44,58,61] and play [49]. Perching is also promoted in more complex environments for laying hens and with provisions like elevated platforms and bales in broilers, as reviewed by Nicol et al. [61]. Although different environmental enrichment is provided to the birds in each study and so a generalization is difficult, and further research is needed, it can be concluded that environmental enrichment can promote PW on the farm level. Moreover, further findings indicate the negative link between increasing stocking density and the promotion of the positive effects of exploration [55], preening [52,66,67], swimming [82], perching [92] and behavioral synchronicity [78]. Applying high standards of environmental enrichment and space availability for the birds would require drastic changes in the production systems, especially in intensive ones. This would also mean management changes and a decrease in productivity and profitability. Thus, it would be simultaneously necessary that the consumer be willing to pay for products of higher cost. Still, providing relatively simple environmental enrichment like perching substrate or enrichment that promotes exploration, play and comfort can be an economical starting point, requiring limited labor and management change. And, step by step, positive welfare can be improved further.

Today, animal welfare is a prerequisite of sustainable development [5]. Positive welfare supports the minimizing of negative aspects in an animal's life and the promotion of positive affective states and experiences so that an animal, being sentient, can have a good life [28,34]. Consumers today are willing to pay more for products from animals kept under higher welfare standards; however, this willingness is affected by various aspects [3,4,9,14,15], such as the type of product, type of species, the price, the demographic and cultural characteristics [9,10]. Consumers' willingness to pay for products of animal origin, especially poultry, would require another literature review, analyzing how strong this willingness is, all the influencing factors and the price that they are willing to pay. Another complex issue is also to interpret this price in higher animal and environmentally friendly management changes that can additionally maintain profitability for farmers. Nonetheless, animal welfare is a crucial component of sustainability, and sustainable livestock management promotes positive welfare, not only for poultry but for all livestock species. Especially regarding broilers and laying hens, the consumer's concern is high, and various studies have shown their willingness to pay more [18,22,23]. Although it is simplistic to say that consumers are willing to pay more, a gradual change towards more sustainable agriculture, including higher welfare standards and positive affective states for the animals, is the ongoing direction, directed by policy making, legislation and consumers, at both national and international levels.

5. Conclusions

Positive welfare indicators are associated with sustainable poultry production systems because animal welfare is a key and distinct component of sustainable agricultural development, which in turn is a part of sustainable development for the environment, society, the economy and humanity as a whole. The positive welfare indicators for poultry that have been studied mostly and seem most promising for both laying hens and broilers are exploration and dustbathing, a poultry-specific behavior indicating comfort. Changes in management that promote pre-laying behaviors are also important for promoting positive affectivity in laying hens. Swimming is indicated as an index of positive welfare in ducks. Most of the positive indicators in poultry are positively affected by environmental enrichment and negatively by high stocking density. Whilst there is research on broilers and laying hens, there are limited studies on the other domestic poultry species. Although research in animal welfare has been thriving in recent years, further research on positive welfare indicators would be beneficial in order to determine reliable and countable indices for a positive welfare assessment scheme.

Author Contributions: Conceptualization, M.P. and P.S.; writing—original draft preparation, M.P.; writing—review and editing, M.G., O.T., D.M. and P.S.; supervision, P.S. All authors have read and agreed to the published version of the manuscript.

Funding: This research received no external funding.

Institutional Review Board Statement: Not applicable.

Informed Consent Statement: Not applicable.

Data Availability Statement: Not applicable.

Conflicts of Interest: The authors declare no conflict of interest.

References

1. Broom, D.M. Farm animal welfare: A key component of the sustainability of farming systems. *Vet. Glas.* **2021**, *75*, 145–151. [CrossRef]
2. Buller, H.; Blokhuis, H.; Jensen, P.; Keeling, L. Towards farm animal welfare and sustainability. *Animals* **2018**, *8*, 81. [CrossRef] [PubMed]
3. Dwyer, C.M. Can improving animal welfare contribute to sustainability and productivity? *BSJ Agric.* **2020**, *3*, 61–65.
4. Marchant-Forde, J.N. The science of animal behavior and welfare: Challenges, opportunities, and global perspective. *Front. Vet. Sci.* **2015**, *2*, 16. [CrossRef]
5. United Nations Committee. *Proposed Draft Recommendations on Sustainable Agricultural Development for Food Security and Nutrition Including the Role of Livestock*; Animal Health and Welfare: Easton, PA, USA, 2016; Article VIII.
6. Pinillos, R.C.; Appleby, M.C.; Manteca, X.; Scott-Park, F.; Smith, C.; Velarde, A. One Welfare—A platform for improving human and animal welfare. *Vet. Rec.* **2016**, *179*, 412–413. [CrossRef]
7. Destoumieux-Garzón, D.; Mavingui, P.; Boetsch, G.; Boissier, J.; Darriet, F.; Duboz, P.; Fritsch, C.; Giraudoux, P.; Le Roux, F.; Morand, S.; et al. The one health concept: 10 years old and a long road ahead. Front. *Vet. Sci.* **2018**, *5*, 14. [CrossRef]
8. Nurse, A. Beyond the property debate: Animal welfare as a public good. *Contemp. Justice Rev.* **2016**, *19*, 174–187. [CrossRef]
9. Alonso, M.E.; González-Montaña, J.R.; Lomillos, J.M. Consumers' Concerns and Perceptions of Farm Animal Welfare. *Animals* **2020**, *10*, 385. [CrossRef] [PubMed]
10. Lagerkvist, C.J.; Hess, S. A meta-analysis of consumer willingness to pay for farm animal welfare. *Eur. Rev. Agric. Econ.* **2011**, *38*, 55–78. [CrossRef]
11. Mellor, D.J. Updating animal welfare thinking: Moving beyond the "Five Freedoms" towards "a Life Worth Living". *Animals* **2016**, *6*, 21. [CrossRef]
12. Walker, M.; Díez-León, M.; Mason, G. Animal Welfare Science: Recent Publication Trends and Future Research Priorities. *Int. J. Comp. Psychol.* **2014**, *27*, 80–100. [CrossRef]
13. OECD/FAO. *Agricultural Outlook 2021–2030: Meat*; OECD Publishing: Paris, France, 2021. [CrossRef]
14. Clark, B.; Stewart, G.B.; Panzone, L.A.; Kyriazakis, I.; Frewer, L.J. Citizens, consumers and farm animal welfare: A meta-analysis of willingness-to-pay studies. *Food Policy* **2017**, *68*, 112–127. [CrossRef]
15. Vigors, B. Citizen' and farmers' framing of positive animal welfare and the implications of farming positive welfare in communication. *Animals* **2019**, *9*, 147. [CrossRef] [PubMed]
16. OECD; FAO. *OECD/FAO Agricultural Outlook 2020–2029*; FAO: Rome, Italy; OECD Publishing: Paris, France, 2020. [CrossRef]
17. OECD; FAO. *OECD/FAO Agricultural Outlook, OECD Agriculture Statistics (Database)*; FAO: Rome, Italy; OECD Publishing: Paris, France, 2022. [CrossRef]
18. Heng, Y.; Peterson, H.H.; Li, X. Consumer Attitudes toward Farm-Animal Welfare: The Case of Laying Hens. *J. Agric. Resour. Econ.* **2013**, *38*, 418–434. Available online: http://www.jstor.org/stable/44131305 (accessed on 10 May 2023).
19. Ochs, D.S.; Wolf, C.A.; Widmar, N.J.O.; Bir, C. Consumer perceptions of egg-laying hen housing systems. *Poult. Sci.* **2018**, *97*, 3390–3396. [CrossRef] [PubMed]
20. Pettersson, I.C.; Weeks, C.A.; Wilson, L.R.M.; Nicol, C.J. Consumer perceptions of free-range laying hen welfare. *Br. Food J.* **2016**, *118*, 1999–2013. [CrossRef]
21. Rohlf, V.I.; Howell, T.J.; Coleman, G.; Rault, J.-L. Engagement through online discussion: Perceptions of laying hen welfare in furnished cages. *Animal* **2019**, *13*, 1999–2006. [CrossRef]
22. De Jonge, J.; van Trijp, H.C.M. The impact of broiler production system practices on consumer perceptions of animal welfare. *Poult. Sci.* **2013**, *92*, 3080–3095. [CrossRef]
23. Lusk, J.L. Consumer preferences for and beliefs about slow growth chicken. *Poult. Sci.* **2018**, *97*, 4159–4166. [CrossRef]
24. Gangnat, I.D.M.; Mueller, S.; Kreuzer, M.; Messikommer, R.E.; Siegrist, M.; Visschers, V.H.M. Swiss consumers' willingness to pay and attitudes regarding dual-purpose poultry and eggs. *Poult. Sci.* **2018**, *97*, 1089–1098. [CrossRef] [PubMed]
25. Vanhonacker, M.; Verbeke, W. Buying higher welfare poultry products? Profiling Flemish consumers who do and do not. *Poult. Sci.* **2009**, *88*, 2702–2711. [CrossRef] [PubMed]
26. Webster, J. Animal welfare: Freedoms, Dominions and "A Life Worth Living". *Animals* **2016**, *6*, 35. [CrossRef]

27. McMillan, F.D. Quality of life in animals. *J. Am. Vet. Med. Assoc.* **2000**, *216*, 1904–1912. [CrossRef] [PubMed]
28. Boissy, A.; Manteuffel, G.; Jensen, M.B.; Moe, R.O.; Spruijt, B.; Keeling, L.; Winckler, C.; Forkman, B.; Dimitrov, I.; Langbein, J. Assessment of positive emotions in animals to improve their welfare. *Physiol. Behav.* **2007**, *92*, 375–397. [CrossRef]
29. Yeates, J.W.; Main, D.C.J. Assessment of positive welfare: A review. *Vet. J.* **2008**, *175*, 293–300. [CrossRef] [PubMed]
30. Welfare Quality. *Welfare Quality Assessment Protocol for Poultry (Broilers, Laying Hens)*; Welfare Quality Consortium: Lelystad, The Netherlands, 2009.
31. Welfare Quality Network. *Welfare Quality Assessment Protocol for Laying Hens 2019, Version 2.0*; Welfare Quality® Consortium: Lelystad, The Netherlands, 2019.
32. Held, S.; Špinka, M. Animal Play and Animal Welfare. *Anim. Behav.* **2011**, *81*, 891–899. [CrossRef]
33. Keeling, L. Indicators of good welfare. In *Encyclopaedia of Animal Behavior*, 2nd ed.; Chun, C.J., Ed.; Elsevier: London, UK, 2019; pp. 134–140.
34. Mellor, D.J. Positive animal welfare states and encouraging environmental-focused and animal-to-animal interactive behaviors. *N. Z. Vet. J.* **2015**, *63*, 9–16. [CrossRef] [PubMed]
35. Špinka, M. Social dimension of emotions and its implication for animal welfare. *Appl. Anim. Behav. Sci.* **2012**, *138*, 170–181. [CrossRef]
36. Rault, J.-L. Be kind to others: Pro-social behaviors and their implication for animal welfare. *Appl. Anim. Behav. Sci.* **2019**, *210*, 113–123. [CrossRef]
37. Rault, J.-L.; Waiblinger, S.; Boivin, X.; Hemsworth, P. The Power of a Positive Human—Animal Relationship for Animal Welfare. *Front. Vet. Sci.* **2020**, *7*, 590867. [CrossRef]
38. Laurijs, K.A.; Briefer, F.E.; Inonge, R.; Webb, L.E. Vocalizations in farm animals: A step towards positive welfare assessment. *Appl. Anim. Behav. Sci.* **2021**, *236*, 105264. [CrossRef]
39. Abeyesinghe, S.M.; Chancellor, N.M.; Hernandez Moore, D.; Chang, Y.-M.; Pearce, J.; Demmers, T.; Nicol, C.J. Associations between behaviour and health outcomes in conventional and slow-growing breeds of broiler chicken. *Animal* **2021**, *15*, 100261. [CrossRef]
40. Bach, M.H.; Tahamtani, F.M.; Pedersen, I.J.; Rider, A.B. Effects of environmental complexity on behaviour in fast-growing broiler chickens. *Appl. Anim.Behav. Sci.* **2019**, *209*, 104840. [CrossRef]
41. Baxter, M.; Richmond, A.; Lavery, U.; O'Connell, N.E. A comparison of fast-growing broiler chickens with a slower-growing breed type reared on Higher Welfare commercial farms. *PLoS ONE* **2021**, *16*, e0259333. [CrossRef]
42. Bergmann, S.; Schwarzer, A.; Wilutzky, A.; Louton, H.; Bachmeier, J.; Schmidt, P.; Erhard, M.; Rauch, E. Behavior as welfare indicator for the rearing of broilers in an enriched husbandry environment—A field study. *J. Vet. Behav.* **2017**, *19*, 90–101. [CrossRef]
43. Bizeray, D.; Estevez, I.; Leterrier, C.; Faure, J. Effects of increasing environmental complexity on the physical activity of broiler chickens. *Appl. Anim. Behav. Sci.* **2002**, *79*, 27–41. [CrossRef]
44. Dawson, L.C.; Widowski, T.M.; Liu, Z.; Edwards, A.M.; Torrey, S. In pursuit of a better broiler: A comparison of the inactivity, behavior, and enrichment use of fast- and slower growing broiler chickens. *Poult. Sci.* **2021**, *100*, 101451. [CrossRef] [PubMed]
45. Jacobs, L.; Blatchford, R.A.; de Jong, I.C.; Erasmus, M.A.; Levengood, M.; Newberry, R.C.; Regmi, P.; Riber, A.B.; Weimer, S.L. Enhancing their quality of life: Environmental enrichment for poultry. *Poult. Sci.* **2023**, *102*, 102233. [CrossRef] [PubMed]
46. Pichova, K.; Nordgreen, J.; Leterrier, C.; Kostala, L.; Moeb, R.O. The effects of food-related environmental complexity on litter directed behaviour, fear, and exploration of novel stimuli in young broiler chickens. *Appl. Anim. Behav. Sci.* **2016**, *174*, 83–89. [CrossRef]
47. Rayner, A.C.; Newberry, R.C.; Vas, J.; Mullan, S. Slow-growing broilers are healthier and express more behavioural indicators of positive welfare. *Sci. Rep.* **2020**, *10*, 15151. [CrossRef]
48. Van der Eijk, J.; Gunnink, H.; Melis, S.; van Riel, J.W.; de Jong, I.C. Reducing stocking density benefits behaviour of fast- and slower-growing broilers. *Appl. Anim. Behav. Sci.* **2022**, *257*, 105754. [CrossRef]
49. Vas, J.; Ben Sassi, N.; Vasdal, G.; Newberry, R.C. Better welfare for broiler chickens given more types of environmental enrichments and more space to enjoy them. *Appl. Anim. Behav. Sci.* **2023**, *261*, 105901. [CrossRef]
50. Ventura, B.A.; Siewerdt, F.; Estevez, I. Access to Barrier Perches Improves Behavior Repertoire in Broilers. *PLoS ONE* **2012**, *7*, e29826. [CrossRef] [PubMed]
51. Campbell, D.L.M.; Belson, S.; Dyall, T.R.; Lea, J.M.; Lee, C. Impacts of Rearing Enrichments on Pullets and Free-Range Hens' Positive Behaviors across the Flock Cycle. *Animals* **2022**, *12*, 280. [CrossRef] [PubMed]
52. Shimmura, T.; Suzuki, T.; Hirahara, T.; Eguchi, Y.; Uetake, K.; Tanaka, T. Pecking behaviour of laying hens in single-tiered aviaries with and without outdoor area. *Br. Poult. Sci.* **2008**, *49*, 396–401. [CrossRef] [PubMed]
53. Taylor, P.S.; Campbell, D.L.M.; Jurecky, E.; Devine, N. Novelty during rearing increased inquisitive exploration but was not related to early ranging behavior of laying hens. *Front. Anim. Sci.* **2023**, *4*, 1128792. [CrossRef]
54. Karabayir, A.; Tolu, C.; Ersoy, E. Some behavioral traits of American Bronze and White (California) turkeys grazing on pasture. *J. Anim. Vet. Adv.* **2008**, *7*, 1113–1116.
55. Martrenchar, N.; Huonnic, D.; Cotte, J.D.; Boilettot, E.; Morisse, J.P. Influence of stocking density on behavioural, health and productivity traits of turkeys in large flocks. *Br. Poult. Sci.* **1999**, *40*, 323–333. [CrossRef]

56. Riber, A.B.; Mench, J.A. Effects of feed- and water-based enrichment on activity and cannibalism in Muscovy ducklings. *Appl. Anim. Behav. Sci.* **2008**, *114*, 429–440. [CrossRef]
57. Boz, A.M.; Sanca, M.; Yamak, U.S.; Erensoy, K. Behavioral traits of artificially and naturally hatched geese in intensive and free-range production systems. *Appl. Anim. Behav. Sci.* **2021**, *236*, 105273. [CrossRef]
58. Baxter, M.L.; Bailie, C.L.; O'Connell, N.E. Play behaviour, fear responses and activity levels in commercial broiler chickens provided with preferred environmental enrichments. *Animal* **2019**, *13*, 171–179. [CrossRef] [PubMed]
59. Vasdal, G.; Vas, J.; Newberry, R.C.; Moe, R.O. Effects of environmental enrichment on activity and lameness in commercial broiler production. *J. Appl. Anim. Welf. Sci.* **2019**, *22*, 197–205. [CrossRef] [PubMed]
60. Hemsworth, P.H.; Edwards, L.E. Natural behaviours, their drivers and their implications for laying hen welfare. *Anim. Prod. Sci.* **2020**, *61*, 915–930. [CrossRef]
61. Nicol, C.J.; Bouwsema, J.; Caplen, G.; Davies, A.C.; Hockenhull, J.; Lambton, S.L.; Lines, J.A.; Mullan, S.; Weeks, C.A. Farmed Bird Welfare Science Review. 2017. Available online: agriculture.vic.gov.au (accessed on 24 May 2023).
62. Olsson, L.A.S.; Duncan, L.J.H.; Keeling, L.J.; Widowski, T.M. How important is social facilitation for dustbathing in laying hens? *Appl. Anim. Behav. Sci.* **2002**, *79*, 285–297. [CrossRef]
63. Olsson, I.A.S.; Keeling, L.J. Why in earth? Dustbathing behaviour in jungle and domestic fowl reviewed from a Tinbergian and animal welfare perspective. *Appl. Anim. Behav. Sci.* **2005**, *93*, 259–282. [CrossRef]
64. Widowski, T.M.; Duncan, I.J.H. Working for a dustbath: Are hens increasing pleasure rather than reducing suffering? *Appl. Anim. Behav. Sci.* **2000**, *68*, 39–53. [CrossRef]
65. Miller, K.A.; Mench, J.A. The differential effects of four types of environmental enrichment on the activity budgets, fearfulness, and social proximity preference of Japanese quail. *Appl. Anim. Behav. Sci.* **2005**, *95*, 169–187. [CrossRef]
66. Freire, R.; Cowling, A. The welfare of laying hens in conventional cages and alternative systems: First steps towards a quantitative comparison. *Anim. Welf.* **2013**, *22*, 57–65. [CrossRef]
67. Widowski, T.M.; Hemsworth, P.H.; Barnett, J.L.; Rault, J.-L. Laying hen welfare I. Social environment and space. *Worlds Poult. Sci. J.* **2016**, *72*, 333–342. [CrossRef]
68. Hockling, P.M.; Wu, K. Traditional and commercial turkeys show similar susceptibility to foot pad dermatitis and behavioural evidence of pain. *Br. Poult. Sci.* **2013**, *54*, 281–288. [CrossRef]
69. Mi, J.; Wang, H.; Chen, X.; Hartcher, K.; Wang, Y.; Wu, Y. Lack of access to an open water source for bathing inhibited the development of the preen gland and preening behavior in Sanshui White ducks. *Poult. Sci.* **2020**, *99*, 5214–5221. [CrossRef]
70. Babington, S.; Campbell, D.L. Water for domestic ducks: The benefits and challenges in commercial production. *Front. Anim. Sci.* **2022**, *3*, 782507. [CrossRef]
71. Liao, S.C.; Lyu, P.X.; Shen, S.-Y.; Hsiao, C.-C.; Lien, C.-Y.; Wang, S.-D.; Lin, T.-Y.; Tu, P.-A. Effects of Swimming Pool Conditions and Floor Types on White Roman Geese's Physical Condition Scores and Behaviors in an Indoor Rearing System. *Animals* **2022**, *12*, 3273. [CrossRef] [PubMed]
72. Zimmerman, P.H.; Buijs, S.A.F.; Bolhuis, J.E.; Keeling, L.J. Behaviour of domestic fowl in anticipation of positive and negative stimuli. *Anim. Behav.* **2011**, *81*, 569–577. [CrossRef]
73. Ahloy-Dallaire, J.; Espinosa, J.; Mason, G. Play and optimal welfare: Does play behavior indicate the presence of positive affective states? *Behav. Process.* **2018**, *156*, 3–15. [CrossRef]
74. Liu, Z.; Torrey, S.; Newberry, R.C.; Widowski, T. Play behaviour reduced by environmental enrichment in fast-growing broiler chickens. *Appl. Anim. Behav. Sci.* **2020**, *232*, 105098. [CrossRef]
75. Chen, Y.; Aorigele, C.; Yan, F.; Li, Y.; Cheng, P.; Qi, Z. Effect of production system on welfare traits, growth performance and meat quality of ducks. *S. Afr. J. Anim. Sci.* **2015**, *45*, 173–179. [CrossRef]
76. Amado, M.F.; Xavier, D.B.; Boere, V.; Torres-Pereira, C.; McManus, C.; Bernal, F.E.M. Behaviour of captive Ostrich chicks from 10 days to 5 months of age. *R. Bras. Zootec.* **2011**, *40*, 1613–1618. [CrossRef]
77. Alvino, G.M.; Archer, G.M.; Mench, J.A. Behavioural time budgets of broiler chickens reared in varying light intensities. *Appl. Anim. Behav. Sci.* **2009**, *118*, 54–61. [CrossRef]
78. Collins, L.M.; Sumpter, D.J.T. The feeding dynamics of broiler chickens. *J. R. Soc. Interface* **2007**, *4*, 65–72. [CrossRef]
79. Forslind, S.; Hernandez, C.E.; Riber, A.B.; Wall, H.; Blokhuis, H.J. Resting behavior of broilers reared with or without artificial brooders. *Front. Vet. Sci.* **2022**, *9*, 908196. [CrossRef]
80. Webster, A.B.; Hurnik, J.F. Synchronization of behavior among laying hens in battery cages. *Appl. Anim. Behav. Sci.* **1994**, *40*, 153–165. [CrossRef]
81. Keeling, L.J.; Newberry, R.C.; Estevez, I. Flock size during rearing affects pullet behavioural synchrony and spatial clustering. *Appl. Anim. Behav. Sci.* **2017**, *194*, 36–41. [CrossRef]
82. Waitt, C.; Jones, T.; Dawkins, M.S. Behaviour, synchrony and welfare of Pekin ducks in relation to water use. *Appl. Anim. Behav. Sci.* **2009**, *121*, 184–189. [CrossRef]
83. Fattah, A.F.A.; Abd Elhameed, N.E.; Roushdy, E.M.; El Kholy, M.S.; Alagawany, M. Reproductive traits, behavioral and hormonal changes during breeding season in Egyptian Geese under natural photoperiod. *Rend. Fis. Acc. Lincei* **2021**, *32*, 539–547. [CrossRef]
84. Cronin, G.M.; Barnett, J.L.; Hemsworth, P.H. The importance of pre-laying behaviour and nest boxes for laying hen welfare: A review. *Anim. Prod. Sci.* **2012**, *52*, 398–405. [CrossRef]

85. Cronin, G.M.; Barnett, J.L.; Storey, T.H.; Thomson, P.C.; Hemsworth, P.H. The relationship between pre-laying activity and corticosterone concentrations, and the interpretation for laying hen welfare. In Proceedings of the 23rd Australian Poultry Science Symposium, Sydney, New South Wales, Australia, 19–22 February 2012; Roberts, J., Ed.; The Poultry Research Foundation, University of Sydney: Sydney, Australia, 2012; pp. 168–171.
86. Edwards, L.E.; Botheras, N.A.; Coleman, G.J.; Hemsworth, P.H. Behavioural and physiological responses of laying hens to humans. *Anim. Prod. Sci.* **2010**, *50*, 557–559. [CrossRef]
87. Hunniford, M.E.; Widowski, T.M. Rearing environment and laying location affect pre-laying behavior in enriched cages. *Appl. Anim. Behav. Sci.* **2016**, *181*, 205–213. [CrossRef]
88. Barrett, L.; Maloney, S.K.; Blache, D. Differences in Pre-Laying Behavior between Floor-Laying and Nest-Laying Pekin Ducks. *Animals* **2019**, *9*, 40. [CrossRef]
89. Makagon, M.M.; Tucker, C.M.; Mench, J.A. Factors affecting nest choice by Pekin ducks. *Appl. Anim. Behav. Sci.* **2011**, *129*, 121–128. [CrossRef]
90. Olsson, I.A.S.; Keeling, L.J. The Push-Door for Measuring Motivation in Hens: Laying Hens Are Motivated to Perch at Night. *Anim. Welf.* **2002**, *11*, 11–19. [CrossRef]
91. Gunnarsson, S.; Yngvesson, J.; Keeling, L.J.; Forkman, B. Rearing without early access to perches impairs the spatial skills of laying hens. *Appl. Anim. Behav. Sci.* **2000**, *67*, 217–228. [CrossRef] [PubMed]
92. Newberry, R.C.; Estevez, I.; Keeling, L.J. Group size and perching behaviour in young domestic fowl. *Appl. Anim. Behav. Sci.* **2001**, *73*, 117–129. [CrossRef] [PubMed]
93. Edgar, J.L.; Lowe, J.C.; Paul, E.S.; Nicol, C.J. Avian maternal response to chick distress. *Proc. R. Soc. B* **2011**, *278*, 3129–3134. [CrossRef] [PubMed]
94. Edgar, J.; Kelland, I.; Held, S.; Paul, E.; Nicol, C. Effects of maternal vocalisations on the domestic chick stress response. *Appl. Anim. Behav. Sci.* **2015**, *171*, 121–127. [CrossRef]
95. Edgar, J.; Held, S.; Jones, C.; Troisi, C. Influences of Maternal Care on Chicken Welfare. *Animals* **2016**, *6*, 2. [CrossRef] [PubMed]
96. Bertin, A.; Richard-Yris, M.-A. Mothering during early development influences subsequent emotional and social behaviour in Japanese quail. *J. Exp. Zool. A Comp. Exp. Biol.* **2005**, *303A*, 792–801. [CrossRef]
97. De Margerie, E.; Peris, A.; Pittet, F.; Houdelier, C.; Lumineau, S.; Richard-Yris, M.-A. Effect of mothering on the spatial exploratory behavior of quail chicks. *Dev. Psychobiol.* **2012**, *55*, 256–264. [CrossRef]
98. McGrath, N.; Dunlop, R.; Dwyer, C.; Burman, O.; Phillips, C.J.C. Hens vary their vocal repertoire and structure when anticipating different types of reward. *Anim. Behav.* **2017**, *130*, 79–96. [CrossRef]
99. Moe, R.O.; Stubsjøen, S.M.; Bohlin, J.; Flø, A.; Bakken, M. Peripheral temperature drop in response to anticipation and consumption of a signaled palatable reward in laying hens (*Gallus domesticus*). *Physiol. Behav.* **2012**, *106*, 527–533. [CrossRef]
100. Wichman, A.; Keeling, L.J.; Forkman, B. Cognitive bias and anticipatory behaviour of laying hens housed in basic and enriched pens. *Appl. Anim. Behav. Sci.* **2012**, *140*, 62–69. [CrossRef]
101. Zimmerman, P.H.; Koene, P.; van Hooff, J.A.R.A.M. The vocal expression of feeding motivation and frustration in the domestic laying hen, Gallus gallus domesticus. *Appl. Anim. Behav. Sci.* **2000**, *69*, 265–273. [CrossRef]
102. Muri, K.; Stubsjøen, S.M.; Vasdal, G.; Moe, R.O.; Granquist, E.G. Associations between qualitative behaviour assessments and measures of leg health, fear and mortality in Norwegian broiler chicken flocks. *Appl. Anim. Behav. Sci.* **2019**, *211*, 47–53. [CrossRef]
103. De Haas, E.N.; Kemp, B.; Bolhuis, J.E.; Groothuis, T.; Rodenburg, T.B. Fear, stress, and feather pecking in commercial white and brown laying hen parent-stock flocks and their relationships with production parameters. *Poult. Sci.* **2013**, *92*, 2259–2269. [CrossRef]
104. Van Niekerk, T.G.C.M.; Gunnink, H.; van Reenen, K. Welfare Quality Assessment Protocol for Laying Hens. *Results Assessment on 122 Flocks, Livestock Research Report 589*. 2012. Available online: https://edepot.wur.nl/235525 (accessed on 27 May 2023).
105. Vasdal, G.; Muri, K.; Stubsjøen, S.M.; Oppermann Moe, R.; Kittelsen, K. Qualitative behaviour assessment as part of a welfare assessment in flocks of laying hens. *Appl. Anim. Behav. Sci.* **2022**, *246*, 105535. [CrossRef]
106. Mota-Rojas, D.; Broom, D.M.; Orihuela, A.; Velarde, A.; Napolitano, F.; Alonso-Spilsbury, M. Effects of human-animal relationship on animal productivity and welfare. *J. Anim. Behav. Biometeorol.* **2020**, *8*, 196–205. [CrossRef]
107. Waiblinger, S.; Boivin, X.; Pedersen, V.; Tosi, M.-V.; Janczak, A.M.; Visser, E.K.; Jones, R.B. Assessing the human–animal relationship in farmed species: A critical review. *Appl. Anim. Behav. Sci.* **2006**, *101*, 185–242. [CrossRef]
108. Zulkifli, I. Review of human-animal interactions and their impact on animal productivity and welfare. *J. Anim. Sci. Biotechnol.* **2013**, *4*, 25. [CrossRef]
109. Al-Aqil, A.; Zulkifli, I.; Hair Bejo, M.; Sazili, A.Q.; Rajion, M.A.; Somchit, M.N. Changes in heat shock protein 70, blood parameters, and fear-related behavior in broiler chickens as affected by pleasant and unpleasant human contact. *Poult. Sci.* **2013**, *92*, 33–40. [CrossRef]
110. Zulkifli, I.; Gilbert, J.; Liew, P.; Ginsos, J. The effects of regular visual contact with human beings on fear, stress, antibody and growth responses in broiler chickens. *Appl. Anim. Behav. Sci.* **2002**, *79*, 103–112. [CrossRef]
111. Bertin, A.; Mocz, F.; Calandreau, L.; Palme, R.; Lumineau, S.; Darmaillacq, A.-S.; Dickel, L.; Arnoulf, C.; Houdelier, C. Human behaviour at the origin of maternal effects on offspring behaviour in laying hens (*Gallus gallus* domesticus). *Physiol. Behav.* **2019**, *210*, 175–183. [CrossRef]

112. Graml, C.; Wailinger, S.; Niebuhr, K. Validation of tests for on-farm assessment of the hen–human relationship in non-cage systems. *Appl. Anim. Behav. Sci.* **2008**, *111*, 301–310. [CrossRef]
113. Bertin, A.; Richard-Yris, M.A.; Houdelier, C.; Lumineau, S.; Möstl, E.; Kuchar, A.; Hirschenhauser, K.; Kotrschal, K. Habituation to humans affects yolk steroid levels and offspring phenotype in quail. *Horm. Behav.* **2008**, *54*, 396–402. [CrossRef]
114. Bonato, M.; Malecki, I.A.; Wanga, M.D.; Cloete, S.W.P. Extensive human presence at an early age of ostriches improves the docility of birds at a later stage of life. *Appl. Anim. Behav. Sci.* **2013**, *148*, 232–239. [CrossRef]
115. Baciadonna, L.; McElligott, A.G. The use of judgement bias to assess welfare in farm livestock. *Anim. Welf.* **2015**, *24*, 81–91. [CrossRef]
116. Mendl, M.; Bura, O.H.P.; Parker, R.M.A.; Paul, S.E. Cognitive bias as an indicator of animal emotion and welfare: Emerging evidence and underlying mechanisms. *Appl. Anim. Behav. Sci.* **2009**, *118*, 161–181. [CrossRef]
117. Anderson, M.G.; Campbell, A.M.; Cramp, A.; Arnott, G.; Jacobs, L. Environmental complexity positively impacts affective states of broiler chickens. *Sci. Rep.* **2021**, *11*, 16966. [CrossRef]
118. Lourenco-Silva, I.; Ulans, A.; Campbell, A.M.; Almelda Paz, I.C.L.; Jacobs, L. Cognitive bias in slow-growing broiler chickens raised in low- or high-complexity environments: Using a social-pair testing approach. *Res. Sq.* **2023**, Preprint. [CrossRef]
119. Deakin, A.; Browne, W.J.; Hodge, J.J.L.; Paul, E.S.; Mendl, M. A Screen-Peck Task for Investigating Cognitive Bias in Laying Hens. *PLoS ONE* **2016**, *11*, e0158222. [CrossRef]
120. Hernandez, C.E.; Hinch, G.; Lea, J.; Ferguson, D.; Lee, C. Acute stress enhances sensitivity to a highly attractive food reward without affecting judgement bias in laying hens. *Appl. Anim. Behav. Sci.* **2015**, *163*, 135–143. [CrossRef]
121. Zidar, J.; Campderrich, I.; Jansson, E.; Wichman, A.; Winberg, S.; Keeling, L.; Løvlie, H. Environmental complexity buffers against stress-induced negative judgement bias in female chickens. *Sci. Rep.* **2018**, *8*, 5404. [CrossRef] [PubMed]
122. Horváth, M.; Pichová, K.; Košťál, Ľ. The effects of housing conditions on judgement bias in Japanese quail. *Appl. Anim. Behav. Sci.* **2016**, *185*, 121–130. [CrossRef]
123. Mattiello, S.; Battini, M.; De Rosa, G.; Napolitano, F.; Dwyer, C. How can we assess positive welfare in ruminants? *Animals* **2019**, *9*, 758. [CrossRef]
124. Papageorgiou, M.; Simitzis, P.E. Positive Welfare Indicators in Dairy Animals. *Dairy* **2022**, *3*, 814–841. [CrossRef]
125. Lawrence, A.; Newberry, R.; Špinka, M. Positive welfare. What does it add to the debate over pig welfare? In *Advances in Pig Welfare*; Woodhead Publishing: Kidlington, UK, 2018; pp. 415–444.
126. Savory, J. Nutrition, Feeding and Drinking Behaviour, and Welfare. In *The Welfare of Domestic Fowl and Other Captive Birds*, 1st ed.; Springer: London, UK, 2010; pp. 165–187.
127. Campbell, D.L.M.; Hinch, G.N.; Downing, J.A.; Lee, C. Early enrichment in free-range laying hens: Effects on ranging behaviour, welfare, and response to stressors. *Animal* **2018**, *12*, 575–584. [CrossRef]
128. Krause, E.T.; Naguib, M.; Trillmich, F.; Schrader, L. The effects of short term enrichment on learning in chickens from a laying strain (*Gallus gallus* domesticus). *Appl. Anim. Behav. Sci.* **2006**, *101*, 318–327. [CrossRef]
129. Appleby, M.C.; Walker, A.W.; Nicol, C.J.; Lindberg, A.C.; Freire, R.; Hughes, B.O.; Elsoni, H.A. Development of furnished cages for laying hens. *Br. Poult. Sci.* **2002**, *43*, 489–500. [CrossRef] [PubMed]
130. Nordi, W.M.; Yamashiro, K.C.E.; Klank, M.; Locatelli-Dittrich, R.; Morais, R.N.; Reghelin, A.I.; Molento, C.F.M. Quail (*Coturnixcoturnix japonica*) welfare in two confinement systems. *Arq. Bras. Med. Vet. Zootec.* **2012**, *64*, 1001–1008. [CrossRef]
131. Sainsbury, D.; Sherwin, C.M. The domestic turkey, Meleagris gallopavo. *Lab. Anim.* **2001**, *35*, 84–91. [CrossRef]
132. Mench, J.A. Behavior of domesticated birds: Chickens, turkeys and ducks. In *The Ethology of Domestic Animals: An Introductory Text*, 3rd ed.; Per Jensen, J., Ed.; CABI: Wallingford, Oxfordshire, UK; Boston, MA, USA, 2017; pp. 153–168.
133. Estevez, I. Understanding social behavior for better flock management. In *Advances in Poultry Welfare*; Mench, J.A., Ed.; Woodhead Publishing Series in Food Science, Technology and Nutrition; Woodhead Publishing: Kidlington, UK, 2018; pp. 243–262. [CrossRef]
134. Eklund, B.; Jensen, P. Domestication effects on behavioural synchronization and individual distances in chickens (*Gallus gallus*). *Behav. Process.* **2011**, *86*, 250–256. [CrossRef]
135. Lopes Carvalho, C.; Romeiro de Oliveira, C.; Miotto Galli, G.; de Oliveira Telesca Camargo, N.; Martins Cavalcante Pereira, M.; Bastos Stefanello, T.; Melchior, R.; Andretta, I. Behavior of domestic chickens—Insights from a narrative review. *Rev. Cienc. Agrovet.* **2022**, *21*, 360–369. [CrossRef]
136. Campbell, D.L.M.; Belson, S.; Erasmus, M.A.; Lea, J.M. Behavior and welfare impacts of water provision via misting in commercial Pekin ducks. *J. Anim. Sci.* **2022**, *100*, skac341. [CrossRef]
137. Yue, A.; Duncan, I.J.H. Frustrated nesting behaviour: Relation to extracuticular shell calcium and bone strength in White Leghorn hens. *Br. Poult. Sci.* **2003**, *44*, 175–181. [CrossRef]
138. Cronin, G.M.; Butler, K.L.; Desnoyers, M.A.; Barnett, J.L. The use of nest boxes by laying hens in cages: What does it mean for welfare? *Anim. Sci. Pap. Rep.* **2005**, *23*, 121–128.
139. Hunniford, M.E.; Torrey, S.; Bédécarrats, G.; Duncan, I.J.H.; Widowski, T.M. Evidence of competition for nest sites by laying hens in large furnished cages. *Appl. Anim. Behav. Sci.* **2014**, *61*, 95–104. [CrossRef]
140. Engel, J.M.; Widowski, T.M.; Tilbrook, A.J.; Butler, K.L.; Hemsworth, P.H.; Engel, J.M.; Widowski, T.M.; Tilbrook, A.J.; Butler, K.L.; Hemsworth, P.H. The effects of floor space and nest box access on the physiology and behavior of caged laying hens. *Poult. Sci.* **2019**, *98*, 533–547. [CrossRef]

141. Hunniford, M.E.; Widowski, T.M. Curtained nests facilitate settled nesting behaviour of laying hens in furnished cages. *Appl. Anim. Behav. Sci.* **2018**, *202*, 39–45. [CrossRef]
142. Barrett, L.; Maloney, S.K.; Blache, D. Pekin ducks are motivated to lay in their preferred nest substrate. *Anim. Welf.* **2023**, *32*, e28. [CrossRef]
143. Campbell, D.L.M.; Makagon, M.M.J.; Swanson, J.C.; Siegford, J.M. Perch use by laying hens in a commercial aviary. *Poult. Sci.* **2016**, *95*, 1736–1742. [CrossRef] [PubMed]
144. Duncan, E.T.; Appleby, M.C.; Hughes, B.O. Effect of perches in laying cages on welfare and production of hens. *Br. Poult. Sci.* **1992**, *33*, 25–35. [CrossRef]
145. Olsson, I.A.S.; Keeling, L.J. Night-Time Roosting in Laying Hens and the Effect of Thwarting Access to Perches. *Appl. Anim. Behav. Sci.* **2000**, *68*, 243–256. [CrossRef] [PubMed]
146. Edgar, J.L.; Mullan, S.M.; Pritchard, J.C.; McFarlane, U.J.C.; Main, D.C.J. Towards a 'Good Life' for Farm Animals: Development of a Resource Tier Framework to Achieve Positive Welfare for Laying Hens. *Animals* **2013**, *3*, 584–605. [CrossRef] [PubMed]
147. Norring, M.; Kaukonen, E.; Valros, A. The use of perches and platforms by broiler chickens. *Appl. Anim. Behav. Sci.* **2016**, *184*, 91–96. [CrossRef]
148. Gebhardt-Henrich, S.G.; Toscano, M.J.; Würbel, H. Use of aerial perches and perches on aviary tiers by broiler breeders. *Appl. Anim. Behav. Sci.* **2018**, *203*, 24–33. [CrossRef]
149. Pettit-Riley, R.; Estevez, I. Effects of density on perching behavior of broiler chickens. *Appl. Anim. Behav. Sci.* **2001**, *71*, 127–140. [CrossRef] [PubMed]
150. Ohara, A.; Oyakawa, C.; Yoshihara, Y.; Ninomiya, S.; Sato, S. Effect of Environmental Enrichment on the Behavior and Welfare of Japanese Broilers at a Commercial Farm. *J. Poult. Sci.* **2015**, *52*, 323–330. [CrossRef]
151. Nordquist, R.E.; van der Staay, F.J.; van Eerdenburg, F.J.C.M.; Velkers, F.C.; Fijn, L.; Arndt, S.S. Mutilating Procedures, Management Practices, and Housing Conditions That May Affect the Welfare of Farm Animals: Implications for Welfare Research. *Animals* **2017**, *7*, 12. [CrossRef]
152. Edgar, J.; Held, S.; Paul, E.; Pettersson, I.; I'Anson Price, R.; Nicol, C. Social buffering in a bird. *Anim. Behav.* **2015**, *105*, 11–19. [CrossRef]
153. Spruijt, B.M.; Van den Bos, R.; Pijlman, F.T.A. A concept of welfare based on reward evaluating mechanisms in the brain: Anticipatory behavior as an indicator for the state of reward systems. *Appl. Anim. Behav. Sci.* **2001**, *72*, 145–171. [CrossRef]
154. Anderson, C.; von Keyserlingk, M.A.G.; Lidfors, L.M.; Weary, D.M. Anticipatory behaviour in animals: A critical review. *Anim. Welf.* **2020**, *25*, 231–238. [CrossRef]
155. Wemelsfelder, F.; Hunter, E.A.; Mendl, M.T.; Lawrence, A.B. The spontaneous qualitative assessment of behavioural expressions in pigs: First explorations of a novel methodology for integrative animal welfare measurement. *Appl. Anim. Behav. Sci.* **2000**, *67*, 193–215. [CrossRef]
156. Wemelsfelder, F.; Hunter, E.A.; Mendl, M.T.; Lawrence, A.B. Assessing the 'Whole Animal': A Free Choice Profiling Approach. *Anim. Behav.* **2001**, *62*, 209–220. [CrossRef]
157. Grandin, T. Implementing Effective Animal-Based Measurements for Assessing Animal Welfare on Farms and Slaughter Plants. In *Improving Animal Welfare a Practical Approach*, 3rd ed.; Grandin, T., Ed.; CABI: Londin, UK, 2021; pp. 105–142.
158. Fleming, P.A.; Clarke, T.A.; Wickham, S.L.; Stockman, C.A.; Barnes, A.L.; Collins, T.; Miller, D.W. The contribution of qualitative behavioral assessment to appraisal of livestock welfare. *Anim. Prod. Sci.* **2016**, *56*, 1569–1578. [CrossRef]
159. AWIN Welfare Assessment Protocol for Turkeys. 2015. Available online: AWINProtocolTurkeys.pdf (accessed on 27 May 2023).
160. Graml, I.D.M.; Niebuhr, K.; Wailinger, S. Reaction of laying hens to humans in the home or a novel environment. *Appl. Anim. Behav. Sci.* **2007**, *113*, 98–109. [CrossRef]
161. Raubek, J.; Niebuhr, K.; Waiblinger, S. Development of on-farm methods to assess the animal-human relationship in laying hens kept in non-cage systems. *Anim. Welf.* **2007**, *16*, 173–175. [CrossRef]
162. Wang, M.D.; Cloete, S.W.P.; Dzama, K.; Bonato, M.; Malecki, I.A. Foster parenting, human imprinting and conventional handling affects survival and early weight of ostrich chicks. *S. Afr. J. Anim. Sci.* **2012**, *42*, 123–130. [CrossRef]
163. Sørensen, J.T.; Fraser, D. On-farm welfare assessment for regulatory purposes: Issues and possible solutions. *Livest. Sci.* **2010**, *131*, 1–7. [CrossRef]
164. Keeling, L.J.; Whinckler, C.; Hintze, S.; Forkman, B. Towards a Positive Welfare Protocol for Cattle: A Critical Review of Indicators and Suggestion of How We Might Proceed. *Front. Anim. Sci.* **2021**, *2*, 753080. [CrossRef]
165. FAO. Gateway to Poultry Production and Products. 2023. Available online: https://www.fao.org/poultry-production-products/production/en/ (accessed on 6 March 2023).

sustainability

MDPI

Review

Utilization of Agro-Industrial By-Products for Sustainable Poultry Production

Alexandros Georganas [1], Elisavet Giamouri [1], Athanasios C. Pappas [1,*], Evangelos Zoidis [1,*], Michael Goliomytis [2] and Panagiotis Simitzis [2]

[1] Laboratory of Nutritional Physiology and Feeding, Faculty of Animal Science, Agricultural University of Athens, 11855 Athens, Greece

[2] Laboratory of Animal Breeding and Husbandry, Department of Animal Production, Agricultural University of Athens, Iera Odos 75, 11855 Athens, Greece

* Correspondence: apappas@aua.gr (A.C.P.); ezoidis@aua.gr (E.Z.); Tel.: +30-2105294437 (A.C.P.); +30-2105294415 (E.Z.)

Abstract: Agro-industrial by-products (AIBPs) that are not intended for human consumption can be used as alternatives to conventional feedstuffs in animal nutrition to produce animal products without competing for land or triggering the food-feed competition, thus leading to environmental, social, and economic sustainability. These by-products are also known to contain several bioactive compounds and have a potential to become nutraceuticals that can promote the health and well-being of poultry. The potentials of some AIBPs (e.g., fruit juice industry leftovers, oilseed industrial by-products, distillers' grain by-products, vinification by-products, olive oil industry by-products, pomegranate by-products, tomato processing by-products) and their derivative products as functional feeds for poultry, but also potential limitations of utilizing AIBPs in poultry nutrition are elaborated in the present review. The possible mechanisms through which AIBPs may improve the health status and productivity of poultry are also discussed. We suggest that nutrient variability across countries should be stabilized and potential hazards such as mycotoxins and pesticides should be eliminated, and the potential hazards present in AIBPs (e.g., mycotoxins) should be better controlled through appropriate legislation and proper application of control measures. Modern processing methods, new types/classifications, and proper developmental strategies foster the utilization of AIBPs in animal nutrition. This review focuses on the AIBPs as feeds, not only for their nutritional value but also for their contribution to sustainable practices.

Keywords: distillers' grain; fruit juice industry leftovers; olive oil industry by-products; oilseed industrial by-products; pomegranate by-products; tomato processing by-products; vinification by-products

Citation: Georganas, A.; Giamouri, E.; Pappas, A.C.; Zoidis, E.; Goliomytis, M.; Simitzis, P. Utilization of Agro-Industrial By-Products for Sustainable Poultry Production. *Sustainability* **2023**, *15*, 3679. https://doi.org/10.3390/su15043679

Academic Editor: Peng Yang

Received: 12 January 2023
Revised: 11 February 2023
Accepted: 15 February 2023
Published: 16 February 2023

1. Agro-Industrial By-Products and Sustainability

By 2050, the world population is projected to increase from 8 billion in 2022 to 9.7 billion [1]. This is translated to an additional 2 billion people with increased wealth to elevate the demand for food, including meat produced with substantial amounts of resources [2]. Furthermore, greenhouse gas emissions are forecast to rise at an average rate of 2–6% annually until 2030 [2]. At the same time, global hunger is exacerbated by the COVID-19 pandemic [3] and by conflicts in Eastern Europe in 2022 [4], which increased inflation and made healthy diets less affordable [3].

There is a challenge to address the need to be more resource-efficient due to the limitation of finite land and natural resources and the pressure to increase meat production posed by the rising population and the increase in incomes in developing countries. Although a large effort has been undertaken to address sustainability, triggering a change in food systems requires more than technical innovation [2]. All parts of a food ecosystem are crucial for this purpose, including consumers. Individual attitudes about food are formed by culture and temperament/identity [2], which often lack proper education and awareness of sustainable attitudes. For instance, a dietary change could be directed towards a low-meat diet containing meat from livestock fed on low-opportunity-cost feeds [5]. Such feeds include agro-industrial by-products (AIBPs), which result mainly from the processing of crop plants [6], food waste, and grass resources that are not intended for human consumption and can be used in animal nutrition to produce animal products without competing for land or triggering the food-feed competition [5]. Thus, a human diet containing products from low-cost livestock demands less arable land and resources than a vegan diet [5]. Van Hal et al. [7] estimated that 31 g of animal protein per EU capita per day can be produced if we incorporate low-opportunity-cost feed in animal diets. In this scenario, laying hens production should be reduced by 98%, and broiler raising should be ceased [7]. However, the contribution of poultry production to household food security and its importance to small-scale farmers, especially in the developing world [8], should not be overlooked. Although the scenario of van Hal et al. may seem utopic since it requires complete reformation of the food system and adaptation of the human diet [7], the agro-food system could be reformed towards the direction of environmental, social, and economic sustainability.

In the last years, much interest has been devoted by the agro-food sector to contemporary issues such as respect for the environment and human resources, production traceability, product quality, and food safety [9]. In academia, there is a large discussion about reducing dependence on conventional feedstuffs, i.e., corn and soybean, and using alternative feedstuffs in order to reduce the food-feed competition and the environmental impact of the animal production systems and to foster the production of healthy products. In the EU, it is estimated that approximately 1.6 million tons of AIBPs are produced annually, with Germany, the United Kingdom (UK), Italy, France, and Spain being the top producers [10]. Many agro-industrial by-products are commonly included in animal diets by-products from food processing or breweries, such as spent brewer's grains, maize gluten meal, cakes or meals, sugar beet pulp (SBP), tomato pulp, distillery products, and sunflower meal (SFM) [10,11]. On the other hand, fruit and vegetable wastes are underutilized resources [10], but they are commonly utilized in animal nutrition in developing countries in an informal way [12]. However, AIBPs are usually disposed of in landfills or are incinerated, posing a significant burden to the environment [13]. Instead, the AIBPs or the isolated bioactive compounds can be incorporated into animal diets and thus provide a market opportunity. In the developing world, food losses are much more than in the developed world due to a lack of infrastructure, and more opportunities to transform AIBPs into animal feed are provided [14]. Locally produced AIBPs is an environmentally-friendly alternative to soybean meal, the cultivation of which is one of the causes of deforestation in South America, given that bird performance is not compromised [15]. Furthermore, several studies, as reviewed in our work, examine the replacement of maize and other conventional feedstuffs with AIBPs, a practice that could abate feed-food competition. Some of the favorable properties of AIBPs that can be utilized in animal nutrition include high protein content, suitable amino acid profile, high digestibility, palatability, reduced levels of indigestible fibrous substances, starch, anti-nutrients [16], no difficulties in the handling of the materials and safety. Many AIBPs contain a plethora of bioactive compounds that are shown to have anti-inflammatory and anti-bacterial activity and to favor the antioxidant status of animals and thus improving growth performance, production quality, and endogenous antioxidant systems [17,18]. The mode of actions of bioactive compounds in poultry are reviewed by other researchers [19]. The nutritional characteristics of AIBPs have been compiled in tables by other reviewers [20] or can be found in repositories such as "Feedipedia" (http://www.feedipedia.org/ accessed on 1 December 2022) and "Feed Tables" (https://www.feedtables.com/ accessed on 1 December 2022) [21]. Information on feed terms can be retrieved from the Association of American Feed Control Officials [22]. In the current study, the inclusion of by-products in animal diets, the benefits in poultry performance and health, and the quality of the derived products are reviewed.

2. Agro-Industrial By-Products in Poultry Nutrition

2.1. Fruit Juice Industry Leftovers

The global production of apples was estimated to be about 9.3 million tons in 2021 [23]. Apple pomace (AP) is a by-product of apple processing and cider production, and it is estimated that 4 million metric tons (MMT) are produced globally every year [24]. AP is a significant by-product in many European countries, such as the UK, France, Spain, Ireland, and Germany. The most popular way of utilizing this by-product is by incorporation in animal feed [25]. AP consists of peel, core, seed, calyx, stem, and soft issue [26], and accounts for 25–35% of the weight of the processed raw material [27].

In broilers, dietary inclusion of 3–6% dried AP did not affect growth performance, gut morphometry, and histopathology but increased the weight of ileum and ceca and the ileum digesta viscosity and influenced the activity of some bacterial enzymes in the ileum [28]. However, the incorporation of higher levels of apple by-products in broiler diets provided unfavorable results. Air-dried apple peel waste at 50 g/kg of diet did not affect growth performance, while a level of 100 g/kg decreased the weight gain of broilers. Both dietary interventions had positive effects on the weight of some organs of the gastrointestinal system, blood cholesterol levels, digestibility, and the heat stress response of broilers [29]. Similarly, inclusion levels of 12–20% of dried AP adversely affected growth performance, immune response, gut development, antioxidant capacity, and blood biochemical parameters of broilers. AP was included in the feed in a dried and ground form [26].

In laying hens, dietary incorporation of dried AP up to 10% with the concomitant addition of a multi-enzyme additive at 0.05% improved laying hens' performance, egg traits, and blood parameters without influencing other traits. The AP was dried and fine-milled before incorporation in the diet [30]. Inclusion of 10–25% dried AP enhanced reproductive performance, semen quality, and fatty acid profile of spermatozoa in aging broiler breeder roosters. The AP was dried, ground, and screened prior to inclusion in the diet [31]. In a study focused on geese, 7% dried AP application resulted in enhancement of egg laying performance and vitality of goslings [32].

The global production of oranges was estimated to be around 75.57 million tons in 2021, and the global production of lemon and limes was 20.828 million tons [23]. Orange by-products (e.g., peels, seeds, and membranes) are an essential waste stream in South European countries such as Spain, Italy, Greece, and Portugal [25]. The utilization of orange processing by-products in animal nutrition is the most widely used practice [25]. The primary by-products from citrus processing are fresh citrus pulp (CP) or dried CP (DCP). Fresh CP is the residue that results from the extraction of juice, while DCP is generated by shedding, liming, pressing, and drying the peel, pulp, and seed residues [33]. Although the protein content in CP is low, enhancement was observed with ensiling to a level comparable to cereal grains. In regard to antinutritional factors, protease inhibitors, phytate, and tannins are present in citrus peel [34].

Dried sweet orange (*Citrus sinensis*) peel (DCSP) in broiler diets at levels of 0.5–2% DCSP reduced liver and abdominal fat and serum triglycerides without negative effects on feed conversion ratio (FCR) [35]. In another study, the application of 0.8% DCSP powder in the diet reduced some blood biochemical parameters (e.g., cholesterol) of broilers without adverse effects on growth performance and carcass traits [36]. Inclusion rates of 1.5–3% DCSP in broiler diets did not affect final weight and carcass characteristics [37]. Concurrently, supplementation with 3% DCSP decreased plasma cholesterol, low-density lipoprotein, and triglycerides levels [38] but reduced feed intake, body weight gain, and increased feed conversion rate during the starter and grower period [37]. Moreover, the blood biochemical parameters, such as plasma cholesterol, triglyceride, and aspartate aminotransferase of broilers, decreased linearly with increasing dried citrus waste from 2.5 to 7.5% in the diet. The citrus waste used was sun-dried and ground prior to incorporation into the diet [39]. However, the substitution of maize with DCSP at levels up to 20% in broiler diets did not influence growth performance, health, and weights of the most significant carcass cuts and internal organs, while substitution levels higher than 20% decreased body weight and some carcass cuts. In this study, the peels were sun-dried and milled before inclusion in the diet [40]. The modified blood biochemical profiles reported in the above-mentioned studies may be explained by the presence of vitamin C and other components of DCSP [35].

DCP inclusion rates of up to 10% in broiler diets did not affect intestinal morphometry [41], body weight, carcass traits, and meat quality, but it favorably decreased oxidation rate in chicken meat [41,42] and increased PUFAs, n-3-, and n-6 fatty acid contents in the breast intramuscular fat [42]. On the contrary, CP inclusion in broiler's diets at levels of 5–10% exhibited an increase in feed intake and feed conversion rates, and significantly decreased daily weight gain, while elevated small intestine relative length and decreased carcass yield were observed, with the 10% CP group recording significantly higher PUFA content [43]. The incorporation of sun-dried lemon pulp at levels of 2.5–12% in broiler diets worsened growth performance, intestinal morphology, and humoral immunity [44].

Other poultry species have also been examined in relation to CP utilization. In ostrich diets, supplementation of 20% DCP (ground prior to inclusion in the diet) elevated PUFA content and n-6/n-3 ratio of meat and decreased meat cooking loss [45]. In goose diets, the application of DCP up to 12% did not influence growth performance and carcass yield. Lime was added to the CP prior to drying [46]. In hen diets, DCP inclusion of up to 9% supported egg yolk oxidative stability but deteriorated growth performance and egg quality [47]. Conversely, the incorporation of dried DCP at 12% in hen diets did not influence performance and egg quality in the early phase of production [48]. In quail diets, dietary addition of 3–6% DCP did not affect performance but influenced egg traits [49]. Key ingredients in various agro-industrial by-products are summarized in Table 1.

Table 1. Key ingredients in agro-industrial by-products.

By-Product Origin	Agro-Industrial By-Product	Key Ingredients	References
Apple	Apple pomace	Pectin, catechins, hydroxycinnamates, phloretin glycosides, quercetin glycosides, procyanidins	[50,51]
Citrus fruits	Citrus pomace	Essential oil (mainly monoterpenes and triterpenoids), phenols (coumaric, caffeic, and ferulic acids), and flavonoids, mainly flavanones glycosides (hesperidin, naringin, and narirestin), flavones (hesperetin, naringenin), flavones aglycon (luteolin), and polymethoxylated flavones (tangeretin)	[16,52]
	Citrus seeds	Flavonoids	[53]
	Orange peels	Vitamin C (ascorbic acid), phenolic compounds, pectin, coumarin, volatile oils, flavonoids, and flavones (hesperidin, naringenin), nobiletin D-limonene and pigments (carotenoids)	[37]
	Lemon peels	Hesperidin and eriocitrin	[54]
Sunflower	Defatted cake	Peptides	[51]
Dried distillers' grain with solubles	Corn dried distillers' grain with solubles	Betaine and phenolic compounds	[55]
Grape and winery by-products	Grape pomace	Polyphenols (catechin, epicatechin, procyanidin B1, quercetin and kaempferol)	[56]
	Red grape pomace	Anthocyanins, flavonols	[57]

Table 1. *Cont.*

By-Product Origin	Agro-Industrial By-Product	Key Ingredients	References
Olive	Olive pomace	Hydroxytyrosol, tyrosol, caffeic protocatechuic, vanillic, p-coumaric and syringic acids, vanillin, oleuropein, apigenin	[51,58]
	Olive mill wastewater	Hydroxytyrosol, gallic acid, oleuropein, ligstroside isomers and derivatives, squalene, tocopherols, triterpenes, soluble sugars, polyphenols	[51]
	Olive flesh, stone, and seeds	Polyphenols, tocopherol	[51]
	Olive leaves	Polyphenols	[51]
Pomegranate	Pomegranate husks	Poly- and monomeric phenols	[51]
Tomato	Tomato skin and seeds	Lycopene, β-carotene, sterols, tocopherols, terpenes, glycoalkaloids	[51,59]
Brewing industry	Brewer's spent grains	Xylitol, cellulose, hemicelluloses, lignin, xylose, glucose, arabinose, protein, ferulic and p-coumaric acids	[51]

Based on references [51,60].

2.2. Nonconventional Oilseed Industrial By-Products

Sunflower seed global production was approximately 58.185 million tons in 2021 [23]. SFM is a by-product of oil extraction from sunflower seeds and can be used as a protein source for broilers [61]. There is variability between SFMs depending on the processing methods used to extract the oil (solvent/crush) [62]. The main antinutritional factor found in SFM is chlorogenic acid [63]. The main constraint in using high inclusion levels of SFM in broiler diets is the high fiber content, which is higher than 11–12%, including dehulled SFM [64].

In broilers, dietary levels of SFM up to 12% did not affect body weight, feed intake, feed conversion ratio, mortality, the European Production Efficiency Factor index, carcass percentage, and cut yield [65]. SFM at levels up to 140 g/kg in the diet enhanced the performance of broilers without worsening digestive enzyme activity, organ weight, and histological alterations of intestinal villi [66]. SFM inclusion at 15% in diets formulated on digestible amino acid basis improved broilers' performance and did not affect carcass and cut yields but resulted in an increase in digesta viscosity [67]. However, increment levels of SFM in the diet was found to deteriorate performance (reduced weight gain and increased feed conversion ratio) and carcass traits (linear reduction of a carcass, breast, breast fillet, and abdominal fat weights) of broilers, but the inclusion of 8% SFM and an enzyme mixture had the highest economic efficiency index among the groups [68]. SFM supplementation at 20% in the diet improved the feed/gain ratio and reduced feed intake in the starter and total period, but weight gain and carcass or cut yields were not affected [69]. High-protein SFM (45.4% crude protein) was recommended to be included in the broiler diets at different levels during the starter (up to 10%), grower (up to 20%), and finishing period (up to 23%), which did not affect growth performance. Furthermore, the substitution of 10% soybean meal with SFM can decrease diet costs by 3.74% to 4.61%, depending on the rearing period [70]. Moreover, high-protein SFM was suggested to be added in broiler diets at different levels during starter (5–15%), growing (8–20%), and finisher (12–22%) periods without affecting body weight and feed conversion ratio [71]. However, high-protein SFM incorporation at levels of between 10 and 15% in the diet deteriorated growth performance, but it was attributed to the lower than the optimum size of the particles of the feed [72]. Low-fiber SFM addition of up to 25% in diets led to growth performance and carcass traits of broilers comparable with broilers fed soybean meal-based diet [73].

On the other hand, sunflower cake inclusion reduced the performance of broilers (1–21 d), but the loss of weight was regained during 22–42 d with a corn-and-soybean-based diet; however, compromised carcass yield and intestinal morphometric development were observed [74]. Similarly, the inclusion of up to 16% SFM negatively affected performance and intestinal morphometry without influencing carcass yields, but the addition of an enzyme complex prevented the above-mentioned negative effects [75]. Likewise, weight gain was compromised by the incorporation of 10 or 16% SFM in the finisher diet of broilers, and gut viscosity was increased in the 16% SFM group, while enzyme supplementation counteracted these effects, especially during the growing period [76]. The above studies showed inconsistent results on utilizing SFM in broiler diets in terms of growth performance. In laying hens, dietary levels up to 25% SFM did not affect performance or egg fatty acid content and reduced yolk cholesterol level and production costs [77]. Similarly, the inclusion of up to 20% SFM in the diet of dwarf dam line hens did not affect egg quality characteristics apart from the Haugh unit that was increased [78].

2.3. Dried Distillers' Grain with Solubles

Dried distillers' grain with solubles (DDGS) is a co-product generated from the production of bioethanol by the extraction of starch from cereals during fermentation [79]. Approximately 40 million tons of DDGS are produced worldwide, with the USA accounting for 58% of global production [80]. DDGS is produced mainly by wheat in Europe, while in North America, it is generated mainly by maize. Maize DDGS contains high levels of crude protein (30%), but the content of lysine is low, and it is not consistent [81]. DDGS are one of the most studied co-products in poultry nutrition compared to other by-products. DDGS exhibits high nutrient variability depending on its source [82]. The main limitations of the wide use of DDGS in animal nutrition include the high variability of nutrient composition and the mycotoxin occurrence [83], which is discussed in Section 3.

In broilers' diets, the incorporation of 8, 16, and 24% DDGS increased body weight gain without other adverse effects [84]. Inclusion levels up to 160 g/kg in broiler diets from 22 to 42 d did not affect performance, carcass and cut yields, meat quality, and litter characteristics [85]. In the starter period of broilers, an 8% DDGS level did not have any effect on their growth performance. On the other hand, in the grower period, body weight gain and liver-relative weight (g of organ/kg of body weight) reduced linearly with increasing levels of DDGS, but feed conversion, mortality, ileal viscosity, and cecal *Clostridium perfringens* and *Escherichia coli* concentrations were not influenced by DDGS dietary levels (7.5, 15, 22.5, and 30%) [86]. In another study, a dietary level of 8% in the starter period of broilers increased the feed conversion ratio, while during the grower period, levels up to 14% did not affect growth, feed intake, carcass yield, and breast meat yield [87]. Corn-based DDGS supplementation at values up to 15% in broiler diets was found to affect specific meat quality traits and liver malondialdehyde production but increased PUFA/SFA ratio [88]. The meat quality of breast and thigh meat of broilers was not affected by dietary inclusion with 6 or 12% DDGS but levels greater than 12% increased total PUFAs, linoleic acid, and susceptibility of thigh meat to oxidation [89]. High-protein DDGS in the diet of broilers did not affect performance and can cover the requirements for supplemental lysine and arginine since DDGS can be a good source of digestible lysine [90]. Low-oil DDGS could be included at 10% in broiler diets providing improved performance or 20% with no adverse effects on performance traits [91]. From an environmental standpoint, although DDGS can be beneficial in poultry nutrition, considering DDGS as broiler feed was found to have the highest effect on elevating greenhouse gas emissions and fossil fuel consumption in comparison with diets containing soybean meal, corn, or synthetic amino acids [92].

In laying hens, no significant negative effects were observed on the production or egg traits of hens. More specifically, dietary levels of DDGS up to 32% resulted in darker (L* value) and redder (a* value) yolk, with the 16% DGGS group achieving the highest egg production compared to the 0, 8, and 24% DGGS groups [93]. In another study, inclusion levels up to 15% (from 24 to 46 wk.) and up to 25% (from 47 to 76 wk.) in hen diets did not have negative effects on performance characteristics, egg production, and quality traits, while nitrogen and phosphorus excretions were lower at the DDGS level of 25% [94]. Another study suggested that incorporating up to 12% dietary levels of DDGS in hens' diets may lead to better performance in terms of feed intake, feed conversion ratio, and egg production [95]. Likewise, supplementation of hens' diets with up to 10% DDGS did not have detrimental effects on laying performance, while enzyme supplementation may improve the use of DDGS at levels up to 20% [96]. Higher levels of DDGS in the diet, up to 50%, were found to elevate lutein and PUFA contents in egg yolk without affecting cholesterol and choline contents [97].

In a study assessing the sustainability of utilizing DDGS in laying hens' diets, it was indicated that substituting 25 or 50% of soybean meal in the diet reduced nitrogen and phosphorus excretion in hens and thus decreased pollution of these elements in the environment was achieved without affecting nutrient digestibility [98]. The reduction of nitrogen excretion may be due to the increased digestibility of the diet, while the reduction of phosphorus excretion is due to the increased bioavailability, which results in the heat-mediated destruction of phytate during drying. Moreover, the addition of 20% DDGS in hen diets aged 21–26 wk. was found to reduce daily NH_3 and H_2S air emissions by 24% and 58%, respectively, while egg weight, egg production, and feed intake were not affected [99]. In ducks, a study suggested that dietary inclusion of 10% corn DDGS did not have adverse effects on growth performance, carcass characteristics, serum biochemical indexes, meat physical and chemical quality, nutrient utilization, and the standardized ileal digestibility of amino acids of the diets [100].

2.4. Vinification By-Products

Global production of grapes accounts for 73.5 million tons annually [23]. Grape pomace (GP), which is the solid residue of grapes, constitutes around 20% of the total grape weight and results from the extraction of the juice for winemaking. It is estimated that more than 9 million tons of GP are generated every year [101]. GP comprises the skins, seeds, and stems of grapes [102]. The use of GP in monogastric nutrition is limited due to the high content of the lignified cell wall fraction and the high level of some antinutritional factors, such as condensed tannins and phytic acid, which are present in lower levels [102–104]. Treatment of GP with polyethylene glycol can inactivate, to some extent, condensed tannins [105,106].

In broiler diets, diet enrichment of 5–60 g/kg of GP enhanced intestinal morphology, the antioxidant capacity, and PUFA content of breast muscle and reduced serum cholesterol values without affecting feed intake, feed efficiency, growth performance traits and weights of pancreas, liver, spleen, and abdominal fat [107–110]. Similar levels of GP (20–60 g/kg) in heat-stressed broiler diets improved plasma biochemical indices and antioxidant enzyme activities without influencing growth performance, relative length of different small intestine segments, jejunal morphology, and antibody titer against sheep red blood cells [111]. The incorporation of 2.5–7.5% red GP in broiler diets did not affect weight gain, carcass traits, meat quality characteristics, blood biochemical parameters, and serum biochemistry, while an increase in meat redness was observed [112]. The final body, giblets, and breast weight of broilers were elevated after dietary enrichment with 3% of GP into the diet [113] without affecting malondialdehyde values in breast and thigh meat [114]. GP addition at levels of 2.5–7% in the diet increased PUFA levels and reduced the lipid oxidation rate of meat without affecting meat quality characteristics [115]. A higher dietary level of GP (20%) also led to raised antioxidant capacity of chicken meat without affecting the productive traits of broilers [116]. On the contrary, meat lightness and yellowness, lipid oxidation levels, and bacterial spoilage were not affected by the inclusion of lower dietary levels (2.5–10 g/kg) of GP [117]. Dietary levels of GP concentrate at 15–60 g/kg increased the antioxidant capacity of breast muscle, ileal content and excreta without affecting growth performance, crude protein ileal digestibility and the weight of pancreas, liver, spleen, and abdominal fat of broilers [118]. In another study, supplementation of 1.5% red GP improved apparent nutrient digestion, diet metabolizable energy, number of different

Lactobacillus spp. in the ileum, and plasma antioxidant activity without affecting growth characteristics [119]. The addition of GP up to 10% in the diet enhanced the antioxidant and immune responses of broilers without impairing growth performance [120].

In laying hen diets, the incorporation of 4 or 6% did not influence growth traits, egg production, egg quality indices, live weight, and liver weight, while egg weight was enhanced at a 4% dietary level [121]. The inclusion of 1–3% GP in heat-stressed laying hen diets at the end of the productive cycle improved growth performance, egg quality, serum total antioxidant capacity, and the activities of glutathione peroxidase and superoxide dismutase activities [122]. In quail diets, 2–6% GP dietary addition did not affect egg production, feed intake, and feed conversion rate, but a linear reduction of albumen weight, egg-specific gravity, and egg weight with increased levels of GP were observed [123].

2.5. Olive Oil Industry By-Products

Large quantities of olive by-products are generated during olive oil extraction, such as leaves, stones, olive mill wastewater, and the solid wastes' pomace residues and olive cake (OC). In 2020, approximately 3.373 million tons of olive oil were produced worldwide [23]. For every 1 kg of olives, around 800 g of OC is generated [124]. The improper disposal of these by-products is notorious for the impact they pose on the environment due to the phytotoxicity and high organic content of the by-products [125]. Olive pomace, due to its oil content, undergoes rancidity when exposed to oxygen and moisture. Drying may delay this chemical reaction [126].

In broiler diets, supplementation of up to 10% olive leaves or OC did not alter bone mineralization nor affected the growth performance of broilers [127]. For broilers at 1–28 d of age, the optimal production index and productive performance were the achieved with application of 5% OC combined with 0.4 g/kg of *Saccharomyces cerevisiae* yeast [128] or with 10% OC supplemented and 500 FTU/kg bacterial *E. coli* phytase [129,130]. During the final growth stage (28–49 d) of broilers, OC could be added at 10% of the diet or up to 20% supplemented with 1 g/kg citric acid without deteriorating feed conversion and health status [131]. The addition of *Bacillus licheniformis* enhanced the fat and nutrient utilization, growth performance, and antioxidant response in broilers [132]. Furthermore, the addition of fermented defatted OC may favor intestinal mucosa and cecal microbiota of broilers and thus control the dissemination of pathogenic bacteria and improve digestibility and absorption capacity [133]. Likewise, OC at levels up to 10% in diets of slow-growing broilers was found to improve productive traits, meat oxidation, and intestinal morphometric features [134]. In another study, it was concluded that OC at levels above 50 g/kg diet may affect some quality characteristics and the oxidative stability of meat, while at lower levels, the oxidative stability, oleic acid, and monounsaturated fatty acids (MUFA) of meat were increased [135]. Similarly, the addition of 50 g/kg of OC and an enzyme blend significantly increased carcass and offal weights [136] but decreased jejunum weight and length, serum triglycerides, and cholesterol levels of broilers [137]. Higher levels of OC (82.5–165 g/kg diet) seemed to enhance daily weight gain, meat antioxidant status, and oxidative stability [138]. The inclusion of 5% olive pomace in broiler diets also improved breast and thigh sensory attributes and antithrombotic properties [139].

Olive pulp (OP) inclusion of up to 5% in broiler diets did not affect final body weight, carcass yield, total antioxidant activity, and expression of selected antioxidant enzymes [140]. Furthermore, OP incorporation up to 100 g/kg did not affect growth and nutritional characteristics and the nutritional cost of the diet, and no influence of the studied parameters was observed with the addition of an enzyme blend in the diet containing OP [141]. In another study, growth performance, carcass traits, blood biochemistry parameters, humoral immunity response, and cecum microbiota were not affected by the addition of 4% olive meal in broiler diets [142]. Another by-product of olive oil extraction, olive mill wastewater, when added to broiler diets, improved total antioxidant capacity and redox status and reduced protein and lipid peroxidation rates in plasma and tissues [143]. In a recent study, the incorporation of silage from olive mill wastewater solids, grape pomace solids, and feta cheese whey solids at levels up to 10% showed promising results for growth performance and meat quality [144], and increased n-3 fatty acids and antioxidant capacity in meat [145]. The application of an olive pomace extract at 750 ppm in broiler diets positively affected animal growth and anti-inflammatory properties [146]. Likewise, 1500 ppm of an olive pomace extract mitigated some of the adverse effects of the fasting challenge [147].

In laying hens, OC dietary levels up to 16% elevated MUFA and PUFA and reduced saturated fatty acids (SFA) and cholesterol contents in egg yolk without any effects on productive performance [148]. In other studies, 9% OC addition in hens' diets increased egg and eggshell weights and decreased blood triglycerides level [149,150]. Higher levels of OC up to 20% combined with 0.1% citric acid could also be used in hens' diets without adverse effects on blood metabolites, laying performance, and egg quality [151]. In quails, the inclusion of OC at levels of 5–7.5% reduced SFA and PUFA and increased MUFA levels in breast muscle, and decreased total serum cholesterol and low-density lipoprotein cholesterol [152]. In another study, it was observed that up to 10% OC in quail diets improved the antioxidant status, immune response, and growth performance [153]. Similar inclusion levels (5–10%) of OC in laying quails diets led to an improvement in serum lipid profile and antioxidant status, egg cholesterol content, and performance during early laying periods [154].

2.6. Pomegranate By-Products

Pomegranate (*Punica granatum* L.) pulp (PP) is generated during pomegranate juice extraction and comprises outer peel, seeds, and residual pulp [155]. Global production of pomegranate is estimated at about 3 million tons, and the global generation of the by-products (peels and seeds), which account for approximately 54% of the fruit, is calculated to be around 1.62 million tons annually [156]. Fat, crude protein, and fiber contents in PP and their antioxidant properties can be useful in poultry nutrition [17]. Different processing methods have been reported for pomegranate by-products in the literature. In one study, a pomegranate by-product consisting of 80% peel and rind and 20% of seed was dried in a forced air oven (80 °C for 3 d), ground into powder employing a milling machine, and a 0.15-mm sieve, packed in polyethylene bags and stored at room temperature [157]. In another study, fermentation of pomegranate by-products, including peels, rinds, and seeds, was applied by drying in a forced air oven (80 °C, 2 d), grinding and sieving, pasteurization (85 °C, 30 min) and solid substrate fermentation [158].

In broiler diets, the incorporation of 1–2% PP resulted in favorable effects on meat fatty acid levels, increased protein content in the breast, and reduced meat cholesterol and lipid oxidation values [157]. Supplementation with higher levels of PP at 2–4 g/kg supported growth rate, blood serum metabolites, immunological parameters, and meat quality characteristics [159]. Likewise, the inclusion of 2–4 g/kg PP in broiler diets resulted in desirable effects with respect to performance, digestibility, carcass, and organ indices compared to broilers fed a diet supplemented with α-tocopherol [160]. The incorporation of 0.5% PP in broiler rations decreased ascites mortality and favored meat shelf-life without adverse effects on growth performance [161]. Fermented PP inclusion at levels of 1–2% in broiler diet increased daily weight gain and feed efficiency and decreased fecal ammonia emission [158]. Supplementation with similar levels of fermented PP (0.5–2%) also enhanced weight gain, reduced SFA, cholesterol, thiobarbituric acid reactive substances values in meat, and increased levels of n-3 meat fatty acids [162]. On the other hand, raw or fermented PP at levels of 5–10 g/kg affected adversely ileum morphology, but malondialdehyde values in breast meat and *C. perfringens* count were lowered without affecting animal performance and serum antioxidant enzyme values [163]. The inclusion of 7–10% in the diet of heat-stressed broilers enhanced growth performance, blood cholesterol, and antioxidant status [164]. Furthermore, urea-treated PP at values of 30–50 g/kg in the diet of heat-stressed broilers supported growth performance, plasma blood biochemical indices, liver function, immune response, intestinal morphology, and meat quality [165].

Pomegranate seed oil incorporation at a level of 1.5% led to a reduction of total cholesterol levels in blood without affecting liver enzyme activities and lipid contents [166], while abdominal fat level, PUFA content, and conjugated linoleic acids (CLA) deposition in breast increased as indicated in another study [167]. Moreover, the partial substitution of soybean oil with pomegranate seed oil led to elevated levels of punicic and rumenic fatty acids without affecting carcass traits, dressing percentage, and breast and thigh muscle physicochemical composition [168].

In laying hen diets, supplementation with 2–4% pomegranate peel powder improved blood antioxidant activity and reduced plasma cholesterol and triglyceride content [169]. Pomegranate seed oil in hen diets enhanced laying rate, color, and concentrations of punicic acid and CLA in egg yolk [170]. In quail diets, the incorporation of 2.7–7.5% pomegranate peel powder enhanced feed conversion rate, egg performance, and villus height-to-crypt depth ratio, while serum triglyceride, cholesterol, and glucose levels were reduced [171].

2.7. Tomato Processing By-Products

Tomato pomace (TP) is a residue by-product of the paste production industry, which comprises the seeds, skin (or peel), and a small amount of pulp. Global production of TP is estimated to be about 5.4–9.0 million tons [172]. TP accounts for 3–5% of the raw material and can be used as a protein and energy source in poultry nutrition [173,174]. More specifically, TP consists of fibers (59%), sugars (26%), proteins (19%), pectins (8%), fat (6%), minerals (4%), and antioxidants (e.g., lycopene) on a dry basis [175]. Regarding the processing of tomato pulp, in one study, drying (up to 65 °C) until a dry 900 g/kg was reached and grinding using a hammer mill was applied [176].

In broilers, TP at inclusion levels up to 15–20% in the diet improved the economic efficiency without negative effects on performance and carcass characteristics [177–179]. In another study, the incorporation level of 5% TP mitigated the adverse effects induced by heat stress in broilers [180]. Furthermore, the use of tomato waste juice up to 120 mL/day in broiler rations was found to contribute to the development of the internal organs of broilers [181].

In laying hens, dried TP up to 10% in the diet was demonstrated to improve egg quality traits without posing adverse effects on growth performance or other egg characteristics [176,182,183]. Dried TP at increasing concentrations from 5 to 10 g/kg induced a linear increase in feed intake and favored egg performance and egg quality characteristics [184]. The addition of 5% dried tomato waste in hen diets also reduced lipid peroxidation of eggs and enriched them with n-3 polyunsaturated fatty acids (PUFA), but absorption and deposition of n-3 PUFA in egg yolk decreased with increasing dietary levels from 2.5% to 7.5% of tomato waste [185]. In another study, the yolk color score increased with increasing dried TP from 0 to 19% in the hen diet, while animal performance and egg characteristics were not significantly affected [186].

In quails, dried TP can be incorporated at levels up to 4–6% in the diet without adverse effects on growth performance [187]. In the study of Botsoglou et al. [188], PUFA and PUFA/SFA ratio significantly increased in the meat of quails, which were provided with a diet containing 10% dried TP. However, Nikolakakis et al. [189] did not detect any change in the fatty acid profile and composition of quail meat at the inclusion dietary levels of 5–10% dried TP, while growth performance and carcass characteristics were not significantly influenced. Moreover, dried TP at incorporation levels of 2.5–5% alleviated some effects of heat stress in quails and favored feed intake and live weight gain [190].

2.8. Other Agro-Industrial By-Products

Sugar beet pulp (SBP), a by-product of the sugar cane industry, can be a valuable source of highly digestible fibers, pectins, and sugars [191]. Global production of sugar beet was estimated at approximately 270 million tons in 2021 [23]. Antinutritional factors contained in SBP include saponins [192]. In broilers, the incorporation of 2.5% sugar beet meal enhanced growth performance and giblet relative weight [193]. Furthermore,

30 g/kg SBP in diets of broilers from 1 to 10 days of age improved growth performance, the relative weight of gastrointestinal tract and gizzard, gizzard digesta content, and total tract digestibility [194]. Higher levels of SBP up to 50 g/kg benefited the development of the gastrointestinal tract [195], decreased feed conversion ratio, and enhanced nutrient digestibility [196]. However, 75 g/kg SBP may affect growth performance and intestinal mucosa structure [195]. This observation was supported by a study in which 75 g/kg of SBP reduced villus height and villus height-to-crypt depth ratio [197], body weight, weight gain, low-density lipoprotein, and total cholesterol serum levels [198]. However, the ileal digestibility of organic matter, crude fat, and crude protein, and total serum cholesterol concentration were reduced with increasing levels of 23–92 g/kg SBP in broiler diets [199]. Furthermore, aqueous methanolic extract of SBP at concentrations of 100–300 mg/kg showed adequate anticoccidial activity based on the criteria of feed conversion ratio, lesion score, oocyst score, and oocysts per gram of feces [200]. In laying hens, the inclusion of 3–7% SBP enhanced egg performance, egg quality traits, and egg yolk cholesterol, triglyceride, and malondialdehyde levels, and reduced serum biochemical indices, such as cholesterol [201]. Conversely, in quail diets, productive performance, egg quality criteria, and nitrogen balance were not influenced by 20–40 g/kg SBP inclusion; however, reproductive parameters and nutrients' digestibility were negatively affected [191].

Brewery by-products are unpopular in poultry diets due to the high level of the fiber fraction [202]. The main by-products generated during the brewing process include brewers' grain (85% of the generated by-products), malt sprouts, and brewers' dried yeast. Global production of brewer's spent grains is estimated to be about 39 million tons, of which 3.4 million tons are generated in the European Union (EU) [203]. Among the brewery by-products, brewers' dried yeast is commonly used in poultry diets due to its content of riboflavin, niacin, pantothenic acid, choline, and phosphorus [16,60]. Brewers condensed solubles can be used in poultry or turkey diets due to their high energy content ([204], cited by [202]). Substitution of maize with brewers dried grains up to 75% significantly increased final live weight, weight gain, and feed conversion ratio, while feed intake decreased with increased levels of replacement [205]. The addition of sand in pullets diets containing brewers dried grains enhanced digestibility, gain, and feed conversion ratio, while the inclusion of higher levels than 15–20% decreased feed conversion efficiency [206]. Fermented brewer's spent grains incorporation in laying quail diets increased gross egg production, the intensity of egg production, and reduced feed conversion ratio [207]. Furthermore, feeding coarse brewer's spent grain instead of ground to broilers feed increased feed utilization and gizzard weight with apparent metabolizable energy and ileal digestibility not being affected [208]. Significant findings on the inclusion levels of various agro-industrial by-products in poultry diets are summarized in Table 2.

Table 2. Key findings on inclusion levels of agro-industrial by-products in poultry diets.

Agro-Industrial By-Product	Poultry Species	Examined Inclusion Levels	Key Findings as Reported by Authors	References
Apple pomace (AP)	Broiler	0, 4, 8, 12, 16, or 20% and multi-enzyme	Aghili et al. indicated compromised growth performance and reduced antibody titer production, intestinal morphology, total antioxidant capacity, and blood parameters with increasing dietary levels of apple pomace.	[26]
		0, 3, 6%	Colombino et al. observed an effect on growth performance, gut morphometry, and histopathology, elevated ileum and ceca weight and ileum digesta viscosity and activities of α-glucosidase, α-galactosidase, β-galactosidase, β-glucuronidase, and xylase were influenced.	[28]
	Aging breeder roosters	0, 10, 20, 25%	Akhlaghi et al. demonstrated an improvement in sperm fertility and motility, hatchability rate, seminal total antioxidant capacity and sperm characteristics, and increased values of polyunsaturated fatty acids (PUFA) and monounsaturated fatty acids (MUFA) and integrity of the sperm plasma membrane.	[31]
	Goose	0, 7%, 10%, and 0.05% multi-enzyme supplement	Fiialovych and Kyryliv observed an enhancement in egg laying, hatchability, and vitality of goslings with 7% AP in the diet. Blood biochemical parameters, egg laying rate, and traits were improved with AP up to 10% supplemented with 0.05% multi-enzyme supplement in the diet, without any further adverse effect in laying hens.	[32]
	Laying hen	0, 5, 10, 15% AP and 0 or 0.05% multi-enzyme	Ghaemi et al. indicated that AP levels up to 10% combined with 0.05% multi-enzyme supplement led to enhanced blood biochemical parameters, egg laying rate, and traits without affecting other parameters.	[30]
Apple peel waste	Broiler	0, 50, or 100 g/kg and 0 or 500 mg/kg multi-enzyme	Heidarisafar et al. found increased gizzard and small intestine weights and high-density lipoprotein (HDL) cholesterol levels, decreased low-density lipoprotein (LDL) cholesterol and malondialdehyde content and apparent ileal protein digestibility, while 100 g/kg apple peel waste decreased weight gain of heat-stressed broilers.	[29]
Dried *Citrus sinensis* peel (DCSP)	Broiler	0, 1.5, 3.0%	Ebrahimi et al. demonstrated that final weight, hot carcass weight, and carcass yield were not affected by the incorporation of DCSP at levels 1.5–3%. DCSP addition at 3% during the starter period (1–21 d) achieved the highest values for breast and pancreas weight and ileum length, but during the whole period (1–42 d), the lowest values for breast and thigh weight were indicated at 1.5 and 3% DCSP, respectively. However, reduced feed intake, body weight (BW) gain, and increased feed conversion rate during both the starter and growing periods were indicated.	[37]
		0, 1.5, 3.0%	Ebrahimi et al. DCSP inclusion at the dose of 3% led to a reduction of plasma cholesterol, LDL, triglycerides values, and glucose.	[38]
		0, 0.5%, 1.0%, 1.5%, 2.0%	Abbasi et al. observed an elevation in feed intake and BW gain and a decrease in liver and abdominal fat content and serum triglyceride levels, with feed conversion rate not being affected.	[35]
		0, 0.8%	Alzawqari et al. observed reduced serum glucose, cholesterol, LDL and HDL, triglyceride concentration and enhanced total antioxidant status without affecting feed intake, BW gain, feed conversion rate, and carcass traits with 0.8% DCSP level.	[36]
Citrus sinensis peel	Broiler	0, 10, 20, 30, 40, 50% substitution of maize	Agu et al. indicated a reduction in BW with the substitution of maize with higher than 20% sweet orange peel with feed intake, BW gain, feed conversion rate, weights of the most important carcass cuts (thigh, breast, back, neck, and shoulder) and internal organs (kidney, liver, heart, spleen, and lung) not being influenced. Dressing percent, drumstick, and wing were significantly affected at substitution levels of maize with sweet orange peel above 30%.	[40]

Table 2. *Cont.*

Agro-Industrial By-Product	Poultry Species	Examined Inclusion Levels	Key Findings as Reported by Authors	References
Citrus pulp (CP)	Broiler	0, 5, 10%	Mourão et al. observed elevated small intestine relative length and decreased carcass yield. Furthermore, PUFA meat content, feed intake and feed conversion rate increased, but daily weight gain decreased with a 10% CP dietary level.	[43]
	Ostrich	0, 20%	Lanza et al. indicated that the *iliofibularis* muscle exhibited reduced ultimate pH and lighter color compared to the *gastrocnemius* muscle. M. *gastrocnemius* recorded higher moisture and reduced crude protein contents compared to M. *iliofibularis*. Reduced content of C14:0, C16:0, and C16:1 and increased content of C18:0, C20:2n-6, C20:4n-6, and C20:5n-3 in M. *gastrocnemius* compared to M. *iliofibularis*. The CP group exhibited elevated meat ultimate pH and decreased cooking losses, crude fat, and ash percentages compared to the control. The proportions of intramuscular saturated fatty acids (SFA) and MUFA were decreased in the CP group. The percentage of PUFA (C18:2n-6 and C20:4n-6) and n-6/n-3 ratio in the meat of the CP group was increased.	[45]
Dehydrated citrus pulp	Broiler	0, 2, 4, 6, 8, 10%	Diaz-Vargas et al. did not observe any negative effect on BW, carcass traits, meat quality, and intestinal morphometry among the dietary treatments; however, decreased oxidation rate in chicken meat was observed.	[41]
Orange pulp (OPU)	Broiler	Control (without additives), 50 g/kg OPU, 0.15 ppm Se, or 50 g/kg OPU and 0.15 ppm Se	Zoidis et al. (2022) reported that the OPU and Se-supplemented group and the Se group exhibited enhanced meat oxidative stability (assessed based on malondialdehyde (MDA) content) during frozen storage (90–210 d), with a synergistic action between OPU and Se. BW, cumulative feed intake, feed conversion ratio (FCR), carcass weight, weights of liver, heart, gizzard, fat pad, percentage of chickens standing at the feeder, and percentage of chickens standing at the drinker were not significantly different among dietary treatments. Meat lightness (L*), redness (a*), yellowness (b*), hue angle (H*), chroma (C*) were not significantly different. Meat pH and dressing percentage declined in the OPU groups. The PUFA and α-linolenic acid (ALA) contents in breast meat were elevated in the OPU groups. Feeding and drinking behaviors were not affected by the addition of OPU and/or Se.	[42]
	Laying hen	0, 9%	Goliomytis et al. observed enhanced egg yolk oxidative stability but also a deterioration of performance and egg quality with 9% OPU.	[47]
Dried citrus pulp (DCP)	Laying hen	0, 4, 8, 12, 16%	Nazok et al. stated that dried DCP up to 12% did not affect performance and egg quality in early-phase hens reducing at the same time egg cholesterol levels.	[48]
	Goose	0, 4, 8, 12, 16%	Wang et al. found that dried DCP up to 12% in the diet did not influence weight gain, feed intake, feed/gain ratio, the carcass yields (%) of breast and leg meat, subcutaneous fat and skin, and abdominal fat.	[46]
	Laying quail	0, 3, 6%	Florou-Paneri et al. reported that BW increased in the DCP groups, while BW gain was not influenced among the dietary groups. Egg production declined in the 3% DCP group compared to the control, but no effect was observed in the 6% DCP group. Hatchability increased in the 6% DCP group compared to the control. Average egg weight was higher in the DCP groups. Average specific gravity was decreased in the 6% DCP group compared to the other groups. Mortality was not affected among the groups.	[49]
Citrus waste (CW)	Broiler	0, 2.5, 5.0, 7.5% CW and multi-enzyme	Behera et al. indicated a linear reduction of plasma cholesterol, triglyceride, and aspartate aminotransferase values with increment levels of CW, while total protein, albumin, globulin, and blood urea nitrogen content were not influenced.	[39]

Table 2. *Cont.*

Agro-Industrial By-Product	Poultry Species	Examined Inclusion Levels	Key Findings as Reported by Authors	References
Dried lemon (*Citrus aurantifulia*) pulp (DLP)	Broiler	0, 2.5, 5.0, 7.5, 10, 12%	Basir et al. reported that 2.5–12% DLP dietary levels reduced final BW, daily weight gain, and deteriorated feed conversion rate. The 7.5–12% DLP levels reduced jejunal crypt depth and antibody titers against influenza disease virus and sheep red blood cells.	[44]
Sunflower meal (SFM)	Broiler	0, 4, 8, or 12%	Sangsoponjit et al. reported that BW, feed intake, FCR, mortality, and the European Production Efficiency Factor index of broilers were not significantly affected by the dietary treatments. Carcass percentage and cut yield of breast, fillet, three joint wings, thigh, and drumstick (including abdominal fat) were not different among the dietary treatments. Feed intake was higher in the 12% SFM broiler (22–42 d) group compared to the 8% SFM group.	[65]
		0, 70, 140, or 210 g/kg	Moghaddam et al. observed an enhancement with respect to BW gain, feed intake, and FCR with inclusion levels of SFM up to 140 g/kg, while 210 g SFM/kg had adverse effects on performance. Relative weights of the gastrointestinal tract and gizzard were elevated in the experimental groups. The activities of digestive enzymes (protease and α-amylase) were not affected by the inclusion of SFM in the diet. HDL cholesterol was elevated, while LDL was reduced in the experimental groups. With increasing levels of SFM in the diet, villus height decreased, and crypt depth increased in the duodenum and jejunum.	[66]
		0 or 15%	Araújo et al. reported that 15% SFM incorporation in a diet formulated on a total amino acid basis deteriorated FCR and BW gain and did not affect feed intake, while when the diet was formulated on a digestible amino acid basis, FCR was not affected among treatments. Carcass and cut yields were not affected in the SFM groups, while digesta viscosity was elevated in the 15% SFM group.	[67]
		0, 8, 16, and 24%, and three levels of enzyme blend	Araújo et al. indicated that the weight gain and FCR of broilers (21–42 d) deteriorated with increasing levels of SFM. The most favorable economic efficiency index was recorded in the 8% SFM group. Carcass, breast, breast fillet, and abdominal fat weights were reduced with increasing levels of SFM in the diet.	[68]
		0 or 20%, with or without enzyme complex	Tavernari et al. did not observe any interactions between SFM and the enzyme blend with respect to performance. In the starter phase and total period, feed intake was decreased, but weight gain did not differ. The feed/gain ratio was enhanced in the SFM groups in all phases. Weight gain was higher in the groups fed the enzyme complex in the starter phase. Dietary apparent metabolizable energy corrected for N values was not affected by the supplementation of the enzyme complex, while apparent metabolizability coefficients of P and Ca were enhanced. Carcass and cut yields were not influenced by the SFM or enzyme complex dietary addition.	[69]
		0, 4, 8, 12, or 16% SFM, with or without enzyme complex	De Oliveira et al. reported that growth performance deteriorated with the dietary addition of SFM, but weight gain and feed intake increased with the enzyme supplementation. Intestinal morphometry was impaired by the SFM inclusion in the diet, but the parameter was improved with the enzyme supplementation in the diet. With an increasing level of SFM, the villus height in the jejunum and the crypt depth in the duodenum and ileum were reduced linearly. Higher villus height in the duodenum and decreased crypt depth in the jejunum were observed in the enzyme-supplemented groups. Significant differences were observed with respect to villus height in the duodenum and ileum among the groups. Wing yields linearly increased with increment dietary levels of SFM. Thigh and leg yields were higher in the groups fed SFM than in the enzyme-supplemented groups.	[75]

Table 2. *Cont.*

Agro-Industrial By-Product	Poultry Species	Examined Inclusion Levels	Key Findings as Reported by Authors	References
		Group I (0% SFM or enzyme blend), group II (6% SFM in grower and 10% in finisher diet, with or without 0.01% enzyme blend), group III (8% in grower and 16% in finisher diet, with or without 0.01% enzyme blend)	Horvatovic et al. indicated an enhancement in weight gain and FCR by the enzyme supplementation during the grower phase, while weight gain decreased by the SFM addition in the diet during the finisher phase. Feed intake was unaffected by the inclusion of SFM or enzyme blend. The 16% SFM group had elevated ileal viscosity, and the interaction between diet and enzyme on the parameter was significant. Dressing percentage and breast, thigh and drumstick, and abdominal fat yields were not affected in the experimental groups. Decreased maltase activity was observed in the SFM groups.	[76]
	Laying hen	8.26, 16.52, or 24.84%	Shi et al. did not observe any significant differences with respect to growth performance (BW gain, egg production, daily egg mass, daily feed intake, and feed conversion) and egg quality (average egg weight, egg specific gravity, shell strength, shell color, shell thickness, shell percentage, albumen percentage, yolk percentage, yolk color, and Haugh units). C17:0 fatty acid concentration in yolk was reduced in the SFM groups, while no differences were observed for yolk SFA, MUFA, and PUFA. Egg yolk cholesterol (at 6 wk) decreased in the SFM groups.	[77]
		0, 10, 15, or 20%	Das et al. reported that there were no significant differences in egg quality traits, except for the Haugh unit, which was higher in the white dwarf line compared to the colored dwarf, and an increasing effect on the parameter was observed due to the inclusion of SFM in the diet. Significant differences in egg quality traits depending on the line (white-plumaged dwarf broiler breeder dam line or colored dwarf dam line hens) were observed.	[78]
High-protein sunflower meal (HPSFM)	Broiler	Starter diet (from 5 to 15% HPSFM and from 20 to 29.9% soybean meal (SBM)), grower diet (from 10 to 25% HPSFM and from 5.5 to 20% SBM), finisher diet (from 15 to 26.5% HPSFM and from 0 to 11.3% SBM)	Gerzilov and Petrov did not observe any differences with respect to BW among treatments. The lowest costs for 1 kg weight gain were recorded in the group fed 10% HPSFM and 24.9% SBM during the starter phase (1–10 d), 20% HPSFM and 10.5% SBM during the grower phase (11–24 d), and 23% HPSFM and 3.5% SBM during the finisher phase.	[70]
		Group I (5, 8, and 12% in starter, grower, and finisher diets, respectively), group II (15, 20, and 22%, respectively), group III (32.95, 28.55, and 26.50%, respectively)	Kyrkelanov et al. reported a significant increase in BW at d 10 in groups I and II (5 and 15% HPSFM in the starter phase, respectively). In the grower and finisher periods, live BW was not significant among treatments. FCR was improved from d 0 to 10 in group I, while group III recorded the highest value among the treatments. During the grower and finisher phases, no significant differences were observed for FCR among treatments. The daily gain was not different among treatments at d 42.	[71]
		0, 10, 15%	Chobanova indicated that live weight was decreased, and the feed/gain ratio increased in the HPSFM groups.	[72]

Table 2. *Cont.*

Agro-Industrial By-Product	Poultry Species	Examined Inclusion Levels	Key Findings as Reported by Authors	References
Low-fiber sunflower meal (LFSFM)	Broiler	0, 25, 50, or 75% substitution of SBM with LFSFM, and 0 or 0.2 g/kg phytase	Ciurescu et al. reported that substitution of SBM with LFSFM beyond 25% reduced FCR and BW gain. No interactions were observed with regard to the inclusion of LFSFM and phytase on growth performance. Carcass traits were not affected in the experimental groups. Abdominal fat was reduced in the 50 and 75% LFSFM groups, while the weight of the small intestine was elevated. Plasma HDL cholesterol and total cholesterol increased by the inclusion of LFSFM in the diet.	[73]
Sunflower cake (SC)	Broiler	0, 5, 10, 15, or 20% SC, with or without enzyme complex	Berwanger et al. indicated that, with increasing levels of SC in the diet, the weight gain, final weight, and feed intake linearly decreased during d 1–21. During the 1–21 d period, carcass yield was reduced, and abdominal fat increased in the SC groups. Thigh, breast, and carcass yield increased with the supplementation of the enzyme complex in the diet. At d 21, villus height decreased, and crypt depth increased with increasing levels of SC.	[74]
Dried distillers' grains with solubles (DDGS)	Broiler	0, 10, 40, 70, 100, 130, or 160 g/kg	Damasceno et al. reported that BW gain, feed intake, and FCR of broilers (22–42 d) were not significantly different among treatments. Serum total protein concentration, uric acid (UA), and gamma-glutamyl transferase were not affected, but there was a quadratic effect on cholesterol with the highest concentration at the 160 g/kg DDGS group. Blood glucose was elevated in the experimental groups, while serum albumin concentration and aspartate aminotransferase concentrations were higher in the 160 g/kg DDGS compared to the control. Carcass, breast, legs, and wings yield, abdominal fat percentage (42 d), meat pH, water retention capacity, cooking loss, shear force, luminosity (L*), redness (a*), and yellowness (b*) were not affected by dietary treatment. Volatilized ammonia levels, litter pH, and dry matter were not influenced by the inclusion of DDGS. Relative weights of the proventriculus, gizzard, pancreas, small and large intestine, and intestine length were not different among treatments, but relative liver weight decreased in the 10 g/kg DDGS group.	[85]
		5% conventional DDGS (control group), or 10, 15, or 20% high-protein DDGS (34% crude protein on a wet basis)	Fries-Craft and Bobeck indicated that BW was lower, feed intake was not influenced, and thus FCR was higher in the 15 and 20% high-protein DDGS groups compared to the control. The standardized ileal amino acid digestibility of lysine and methionine was found to be 80.9% and 88.6%, respectively. The N-corrected metabolizable energy of high-protein DDGS was determined to be 11.4 MJ/kg.	[90]
		0, 8, 16, 24%	Shim et al. reported that the BW gain of broilers was not significantly different at 42 d, but the parameter was elevated during 0–18 d in the DDGS groups. The percentage of the fat pad of female broilers was reduced with increasing levels of DDGS. The Pellet durability index was reduced due to the incorporation of DDGS. In the second experiment, higher BW gain in the DDGS groups compared to the control was observed.	[84]

Table 2. *Cont.*

Agro-Industrial By-Product	Poultry Species	Examined Inclusion Levels	Key Findings as Reported by Authors	References
		Starter diet (0 or 8% DDGS), grower diet (0, 7.5, 15, 22.5, or 30% DDGS)	Loar II et al. reported a linear decrease in pellet quality with increasing levels of DDGS. BW gain and relative liver weight decreased linearly with increasing levels of DDGS during the starter phase. Broilers that were not fed DDGS during the started phase had reduced feed consumption with increasing levels of DDGS during the grower phase; however, the 8% DDGS group (starter phase) was not affected by DDGS inclusion during the grower phase. Feed conversion, mortality, ileal viscosity, and cecal *Clostridium perfringens* and *Escherichia coli* concentrations were not affected in the grower DDGS groups.	[86]
		0, 5, 10, 15, 20, and 25%	Min et al. reported that increased b* (yellowness) values and shear force, decreased cooking loss, and differences in the fatty acid profiles of the breast and thigh were observed. SFA, MUFA, and PUFA were not affected, but PUFA/SFA ratio was elevated total superoxide dismutase (SOD) activity in breast meat and liver tissue decreased. Total SOD activity in breast and liver tissue decreased in the DDGS groups. Glutathione peroxidase (GPx) activity in the liver was similar between 0 and 15% DDGS groups. MDA production of breast muscle was not affected, but liver MDA increased.	[88]
		Starter and grower diet (0 or 8%), finisher diet (0, 7, 14, 21, or 28%)	Loar II et al. indicated an increase in FCR and a decrease in BW gain during 0–28 d. During the finisher phase, increasing levels of DDGS in the 14, 21, and 28% DDGS groups, BW gain, and feed intake decreased linearly in comparison with the control. Dressing percentage and breast meat yield decreased linearly with increasing dietary levels of DDGS. Large intestine and relative gizzard weights increased linearly with increasing levels of DDGS during the finisher phase. *E. coli* concentrations in the ileum exhibited a linear reduction with high levels of DDGS. Interactions between *E. coli* and *Listeria monocytogenes* in the ileum and for *L. monocytogenes* in the ceca were observed during the pre-finisher and finisher phases.	[87]
		0, 6, 12, 18, or 24%	Schilling et al. did not observe any differences in terms of cooking loss, instrumental color, and consumer acceptability of breast meat among the groups, but the shear force of breast meat from the control group was slightly reduced compared to the 18 and 24% DDGS groups. The proximate composition of breast and thigh meat was no different among treatments. Linoleic acid and PUFA increased linearly with increasing levels of DDGS, and thiobarbituric acid (TBA) values were higher in the 18 and 24% DDGS groups at d 5 in comparison with the control and 6% DDGS groups.	[89]
	Laying hen	0, 8, 16, 24, or 32%	Loar II et al. reported that egg production was higher in the 16% DDGS group compared to the 0, 8, and 24% DDGS groups, while the 32% exhibited intermediate values with no significant differences with the other treatments. Incorporation of DDGS in the diet led to darker (L*) and redder (a*) yolk compared with the control group. The flavor and overall consumer acceptability of eggs were slightly better in the case of DDGS-fed hens compared to the non-DDGS-fed hens.	[93]

Table 2. *Cont.*

Agro-Industrial By-Product	Poultry Species	Examined Inclusion Levels	Key Findings as Reported by Authors	References
		0, 5, 10, 15, 20, or 25%	Masa'deh et al. reported that daily feed intake, egg production, and overall weight gain were unaffected by the inclusion of DDGS in the diet. Egg weight decreased at dietary levels of DDGS past 15% during 24–46 wk, while the parameter was not different during 47–76 wk. Yolk color increased with increasing dietary levels of DDGS. N and P retention was higher in the 25% DDGS group, and N and P excretion decreased linearly with increasing levels of DDGS in the diet.	[94]
		0, 17, 35, or 50%	Sun et al. indicated that total PUFA increased in the DDGS groups, while choline and cholesterol contents were higher in the 50% DDGS during the beginning of the 24-wk study period but did not differ at the end of the period in comparison with the other treatments. Lutein content increased linearly with increasing dietary DDGS levels.	[97]
		0, 6, 12, or 18%, and 0 or 250 mg enzyme mixture/kg	Abd El-Hack et al. reported that the lowest egg production and daily feed intake and the worst FCR were observed in the 18% DDGS group. Shell thickness and shell percentage were increased in the 6% DDGS group. The 6% and 12% DDGS groups exhibited higher egg weights compared to the control and 18% DDGS groups. The interaction effect of DDGS and the enzyme mixture was significant in most of the egg traits. Yolk color density increased with increasing dietary levels of DDGS. Yolk cholesterol, total fat, and total USFA increased in the DDGS groups.	[95]
		0, 5, 10, 15, or 20%, with or without two different enzymes	Shalash et al. indicated that no significant differences were observed with respect to digestibility coefficient values of crude protein, ether extract, crude fiber, nitrogen-free extract, BW gain, feed intake, and egg quality by the addition of DDGS in the diet. No significant differences were observed for semen quality, fertility, hatchability, and BW of chicks in the hatch in the experimental groups. Egg production, egg number, and egg mass were elevated in the 5% DDGS group. In the 15 and 20% DDGS groups, yolk color and shell thickness increased, while egg production, egg number, egg weight, and egg mass decreased, with the FCR being the worst compared to the 0, 5, and 10% DDGS groups. Enzyme supplementation exhibited favorable results with respect to the digestibility coefficient value of ether extract and egg traits.	[96]
	Laying hen	0, 10, 20%	Wu-Haan et al. reported that egg production, egg weight, and feed intake were not influenced by dietary treatments. NH_3 emissions of hens (21–26 wk) were reduced by 24% and H_2S emissions by 58% in the 20% DDGS group compared to the control.	[99]
		0, 25, 50, 75, or 100% substitution of SBM with DDGS (corresponds to 0, 5.5, 11, 16.5, and 22% DDGS in the diet, respectively), and additives (without, 250 mg enzyme/kg, or 200 mg vitamin E/kg)	Abd El-Hack et al. reported that digestion coefficient values of nutrients were improved in the 25% DDGS substitution group of hens (22–42 wk), whereas the 100% DDGS group exhibited a reduction of the parameter. The amount of daily excreted N was reduced in the 25 and 50% DDGS groups, while N excretion was increased in the 75 or 100% DDGS groups. P excretion was reduced with increasing substitution levels of DDGS. The supplementation with the enzyme nor vitamin E did not affect the studied parameters.	[98]

Table 2. *Cont.*

Agro-Industrial By-Product	Poultry Species	Examined Inclusion Levels	Key Findings as Reported by Authors	References
	Duck	0, 5, 10, 15, 20%	Ding et al. reported a linear and quadratic decrease in BW (42 d), average daily gain, and average daily feed intake from d 11 to 42, breast meat yield, the moisture and protein content in the breast meat, and dietary dry matter and ether extract utilization with increasing levels of DDGS in the diet. b* value of the breast meat and serum total cholesterol and triglyceride concentrations exhibited a linear and quadratic increase. No negative effects were observed with respect to growth performance, carcass characteristics, serum biochemical indexes, meat physical and chemical quality, nutrient utilization, and the standardized ileal digestibility of amino acids of the diets in the 10% DDGS group compared to the control.	[100]
Low-oil distillers dried grains with solubles (LO-DDGS)	Broiler	0, 10, or 20%	Guney et al. indicated that feed efficiency (0–18 d) was enhanced in the 10% LO-DDGS group compared to the 20% LO-DDGS group. Abdominal fat pad weights were elevated in the 10 and 20% LO-DDGS groups. BW and fat pad weights varied depending on the source (sample) and levels of DDGS.	[91]
Grape pomace (GP) and fermented grape pomace (FGP)	Broiler	Basal diet (no additives), 0.25 g/kg synthetic antioxidants (AO), 15 g/kg GP, or 15 g/kg FGP	Gungor et al. reported that mortality rate, dressing percentage, and relative weights of heart, liver, gizzard, gastrointestinal tract, abdominal fat, spleen, and edible internal organs were not different among the treatments. pH and L*, a*, and b* values, MDA level, pH, and color of breast meat were not different among the treatments. Elevated serum GPx and SOD concentrations, and ileum lamina muscularis thickness were observed in the GP groups, while caecal bacterial species were not affected. Dietary inclusion of FGP increased BW, the serum catalase (CAT) level, and decreased the caecal *C. perfringens* count, while ileal morphology was not affected. The AO groups exhibited similar growth performance to the FGP group but recorded better BW and FCR than the GP group. The villus height and villus height-to-crypt depth ratio were higher in the AO group compared to the control. Lamina muscularis mucosa thickness was higher in the GP group compared to the FGP group.	[109]
Grape pomace	Broiler	5, 15, 30 g/kg	Goñí et al. observed increasing content of α-tocopherol concentration in liver with increasing GP dietary levels, but it was lower than in the case of vitamin E dietary supplementation. Furthermore, lipid oxidation of meat during refrigeration storage was reduced.	[108]
		0, 5.0, 7.5, 10 g/kg	Aditya et al. found that BW, feed intake, FCR, serum levels of glucose, triglyceride, and HDL cholesterol were not influenced. Thiobarbituric acid reactive substances (TBARS) linearly increased with increment levels of GP. Meat color values such as redness decreased.	[107]
		0, 3, 6%	Turcu et al. found a higher meat color difference for breast and thigh meat, increased meat hardness, improved meat color and texture, and decreased TBARS in thigh meat. Breast meat yellowness value increased at 6% white GP dietary inclusion, while the intensity of breast meat red color (C*) was reduced at 6% red GP dietary inclusion.	[110]
		0, 2.5, 4.5, 5.5, 7.5%	Kumanda et al. reported no effect of red GP dietary incorporation on weight gain, blood biochemical parameters, serum biochemistry, carcass traits, and meat quality characteristics except for increased meat redness, while feed conversion efficiency was higher at 5.5 and 7.5% GP dietary levels.	[112]
		0, 1, 2, 3%	Haščík et al. indicated an increase in the final body, giblets, and breast weight at a 3% GP dietary level.	[113]
		0, 1, 2, 3%	Jurčaga et al. did not observe any effect on lipid oxidation in meat.	[114]

Table 2. *Cont.*

Agro-Industrial By-Product	Poultry Species	Examined Inclusion Levels	Key Findings as Reported by Authors	References
		0, 2.5, 5, 7.5%	Bennato et al. showed an elevation in PUFA values and a reduction in the lipid oxidation rate of meat without affecting meat pH, cooking loss, and lightness.	[115]
		0, 1.5%	Lichovnikova et al. reported no negative effects on feed intake, feed/gain ratio, and an improvement in apparent nutrient digestion, diet metabolizable energy, number of the considered beneficial bacteria *Lactobacillus* spp. in the ileum, and plasma antioxidant activity.	[119]
		0% GP, 200 mg/kg vitamin E, 5% GP, 7.5%, 10%	Ebrahimzadeh et al. did not observe any negative effects on growth performance, while improved antioxidant and immune responses at dietary GP levels up to 10% were reported.	[120]
		0, 2.5, 5, 10 g/kg	Kasapidou et al. indicated no effect on meat lightness and yellowness, lipid oxidation levels, and bacterial spoilage.	[117]
	Heat-stressed broiler	0 g/kg (rearing at comfort temperatures), or 0, 20, 40, and 60 g/kg (rearing at comfort temperatures initially and heat stress application from d 25 to 42)	Hosseini-Vashan et al. reported that feed intake linearly increased with increasing levels of GP (starter and grower periods), while it linearly reduced blood concentration of triglycerides, plasma cholesterol, LDL, and enzyme activity of aspartate aminotransferase. MDA concentration decreased in the GP groups, and GPx and SOD activities increased. Blood concentration of HDL cholesterol and total protein (24 d) increased in the experimental groups. Antibody titer against sheep red blood cells, growth performance, relative length of different small intestine segments, and jejunal morphology indices were not influenced by GP inclusion in a heat-stressed broiler. Thigh, drumstick, bursa, and thymus percentages were elevated abdominal fat percentage decreased in the GP groups.	[111]
	Laying hen	0, 4, 6%	Kara et al. reported that feed intake, feed efficiency, live weight and egg production, eggshell thickness, eggshell ratio, albumen index, egg-specific gravity, egg yolk index, Haugh unit, yolk color, total protein, total cholesterol, and triglyceride levels were not affected. Egg yolk and plasma malondialdehyde and serum glucose levels decreased. Enhancement of egg weight at 4% GP inclusion and of liver weight at 4 and 6% GP was observed.	[121]
	Heat-stressed laying hen	0, 1, 2, 3%	Reis et al. found that heat-stressed hens at the end of the productive cycle showed elevated serum total antioxidant capacity and GPx and SOD activities, and improved performance, and antioxidant capacity, while it reduced lipid peroxidation rate in the yolk.	[122]
	Quail	0, 2, 4, 6%	Fróes et al. reported no effect on egg production, feed intake, FCR, Haugh unit, and eggshell thickness. Albumen weight, egg-specific gravity, and egg weight were linearly reduced with increment dietary levels of GP.	[123]
Wine-grape pomace flour (WGPF)	Broiler	0% WGPF, 20% red WGPF, or 20% white WGPF	Reyes et al. reported that BW, daily weight gain, feed intake, and FCR were not affected in the white WGPF group. FCR was higher in the red WGPF group. Ether extract of breast meat was higher in the red WGPF group due to the higher inclusion of soy oil in the diet compared to the other groups. The antioxidant capacity of breast and leg meat exhibited an increase in the white WGPF group compared to the other groups.	[116]

Table 2. *Cont.*

Agro-Industrial By-Product	Poultry Species	Examined Inclusion Levels	Key Findings as Reported by Authors	References
Grape pomace concentrate (GPC)	Broiler	Control group (without GP or additives), 15, 30, or 60 g/kg GP, or 200 mg/kg α-tocopheryl acetate (vitamin E)	Brenes et al. reported that growth performance, apparent ileal digestibility of crude protein, the relative weight of abdominal fat, liver, pancreas, and spleen, and the relative intestinal length were not affected by the incorporation of GPC in the diet. Fat digestibility was higher in the vitamin E-supplemented group. The ileal and fecal digestibility of hydrolyzable polyphenols was lower in the GPC groups. Antioxidant activity in the GPC diet, ileal content, and excreta recorded higher scavenging free radical capacity compared to the other groups. The lipid oxidation in breast meat was reduced (1, 4, and 7 d of refrigeration) in the vitamin E-supplemented group. Oxidative stability in breast meat was similar (1, 4, and 7 d) between the GPC and the vitamin E-supplemented groups. The bioavailability of hydrolyzable polyphenols was higher than that of condensed tannins.	[118]
Olive cake (OC)	Broiler	0, 5, or 10% OC and 0.2, or 0.4 g/kg yeast	Al-Harthi reported that the best BW gain, FCR, and European production efficiency index were recorded in the 5% OC plus 0.4 g/kg yeast-supplemented group. The highest survivability rate (100%) was recorded in the 5 and 10% OC plus 0.2 g/kg yeast and the 10% OC plus 0.4 g/kg yeast-supplemented groups. Carcass traits and inner organs were not affected by the addition of OC to the diet.	[128]
		0, 5, 10% OC and 0 or 500 FTU/kg of phytase	Al-Harthi et al. indicated that the growth rate, European production index, and economic efficiency of broilers (7–28 d) were not affected by OC dietary inclusion, while these parameters increased with phytase supplementation. Plasma cholesterol and triglycerides were reduced, and plasma inorganic phosphorus increased with OC and phytase addition. The economic efficiency of broilers fed 10% OC was the highest among treatments.	[130]
		0, 5, 10% OC with or without galzym or phytase	Al-Harthi et al. reported that incorporation of OC up to 10% did not affect BW gain, final BW, survival rate, FCR, dressing percentage, inner and immune organs ratios to live BW 5% OC and galzyme enzyme significantly increased feed intake. 10% OC and galzyme enzyme achieved the best FCR.	[129]
		0, 10, 20% with or without 1 or 2 g/kg citric acid	Al-Harthi and Attia indicated that 10% OC inclusion did not affect the following parameters of broilers (28–49 d): BW gain, feed intake, FCR, survival rate, European production efficiency index, meat pH, meat color, water holding capacity, meat tenderness, dressing percentage, abdominal fat, the proportions of heart, pancreas, intestine, and cecum, red blood cell characteristics, hepatocellular leakage markers; however, the liver proportion was lower compared to the control group. 20% OC and 1 g/kg citric acid did not affect FCR and the health status of broilers.	[131]
		0, 2, 4% OC with or without *Bacillus licheniformis* (BL)	Saleh et al. reported that the inclusion of OC and BL did not influence feed intake, improved weight gain, and reduced FCR, abdominal fat, or blood total cholesterol. Blood total protein, albumin, Newcastle disease titer, and HDL cholesterol were elevated in the experimental diets. Muscle oleic and linoleic acids, and vitamin E were elevated in the 4% OC and BL group, while linolenic acid was elevated in all groups but not in the BL and control groups. Liver MDA was reduced in the BL group and in the 2% or 4% OC and BL groups.	[132]

Table 2. *Cont.*

Agro-Industrial By-Product	Poultry Species	Examined Inclusion Levels	Key Findings as Reported by Authors	References
	Laying hen	0, 10, or 20% and 0, 0.1, or 0.2% citric acid	Al-Harthi and Attia reported that in the OC groups without citric acid addition, there was no effect on laying performance, egg quality, or liver function indices in laying hens (40–56 wk), but feed intake was increased, and FCR deteriorated compared to the OC-free groups. The relative weight of the liver was reduced when 0.1% citric acid was added compared with the other citric acid-supplemented groups, while the relative weight of the ovary increased compared to the control group. The 0.1% citric acid-supplemented group exhibited similar FCR to the unsupplemented control.	[151]
	Quail	0, 2.5, 5, 7.5%	Ozcan et al. found that the 5 and 7.5% OC meal groups recorded reduced serum total cholesterol and LDL cholesterol levels, and elevated cholesterol levels in the breast muscle. The OC groups had decreased SFA and PUFA, and increased MUFA and total USFA levels in the breast muscle. 5% OC meal was recommended in quail diets.	[152]
Olive cake and olive leaves (OL)	Broiler	0, 5, or 10% OC, or 0, 5, or 10% OL	Pečjak et al. did not observe significant differences in growth performance (final live weight, feed intake) and in the mineral content in the femur, tibia, and humerus among dietary treatments. Higher feed intake in the 10% OL group compared to the 5% OL group during the first wk. In the 5 and 10% OL groups, Cu content in the humerus was higher without affecting bone mineralization.	[127]
Defatted olive cake	Broiler	0 or 2%	Rebollada-Merino et al. indicated that broilers (14–35 d) had increased villus height in the duodenum and villus and crypt depth in the duodenum and the cecum, which may improve mucosal renewal.	[133]
Semi-solid olive cake	Broiler	0, 82.5, 165.0 g/kg	Branciari et al. reported that growth rate increased with increasing levels of OC, and meat antioxidant status and oxidative stability were enhanced, especially at 165 g/kg OC was applied. Meat quality parameters, such as meat color traits, pH_{24}, drip loss, cooking loss, and shear force, were not affected.	[138]
Olive pulp (OP)	Broiler	0, 5, 10%	Tufarelli et al. reported that there was no effect on growth performance, dressing percentage, breast yield, or breast meat fatty acid composition. The meat was less susceptible to lipid and protein oxidation in the experimental diets. Breast muscle pH_{24}, duodenal villus height, crypt depth, and villus height-to-crypt depth ratio, villus surface area were higher in the 10% OP diet.	[134]
		T1: 0% OP, T2: 25 or 50 g OP/kg, T3: 50 g OP/kg, T4: 50 or 80 g OP/kg	Papadomichelakis et al. reported that FCR was higher in T2 and T3 in comparison with the control group during the grower phase, while it was higher in T3 compared to T1, T2, and control groups during the finisher phase. C18:1ω9 and total MUFA contents in breast muscle were elevated in the OP diets. Decreased oxidative stability, lower pH_{24}, and an increased lightness of breast meat were observed in T3 compared to the control, T1, and T2 groups. Papadomichelakis et al. suggested that 25 g OP/kg in grower diets and 50 g OP/kg in finisher diets could be used.	[135]
		0, 2.5, 5, 8%	Pappas et al. reported that no differences were observed in terms of final BW, carcass yield, total antioxidant activity, and the values of serum glutamic oxaloacetic transaminase/aspartate aminotransaminase (SGOT/AST), serum glutamic pyruvic transaminase/alanine aminotransferase (SGPT/ALT), blood urea nitrogen (BUN), γ-glutamyl transferase (γ-GT), alkaline phosphatase, cholesterol, total protein, albumins, globulins, and hematocrit among treatments. FCR was not affected by the inclusion of up to 5% OP, while in the 8% OP group, the parameter was statistically lower.	[140]

Table 2. *Cont.*

Agro-Industrial By-Product	Poultry Species	Examined Inclusion Levels	Key Findings as Reported by Authors	References
		0, 50, and 100 g/kg processed or unprocessed OP with or without enzyme (ENZ) blend	Sayehban et al. reported that there were no significant differences in feed intake, weight gain, feed efficiency, energy intake, energy efficiency, protein intake, protein efficiency, feed cost per kg live weight, and production index between dietary treatments. The inclusion of processed OP improved feed and energy efficiencies, while the enzyme blend did not affect the studied parameters.	[141]
		50 or 100 g/kg	Sayehban et al. indicated that serum triglycerides and cholesterol levels, jejunum weight, and length were decreased by OP inclusion. Processed OP decreased jejunum weight and length, jejunum relative weight, left cecum length, serum triglycerides, and very LDL cholesterol levels. Enzyme supplementation had no effect on any parameter. 100 g/kg OP levels increased jejunum relative weight and jejunum length.	[137]
	Laying hen	2 control diets, 160 g/kg OP with or without probiotics	Afsari et al. reported that OP dietary addition did not affect egg production and egg mass, BW, and excreta pH, while feed intake, FCR ratio increased, and serum levels of cholesterol and HDL decreased. At sampling week 3, the Haugh unit, yolk color, and shell weight were reduced, while at sampling week 7, probiotic treatment of feed decreased the Haugh unit. At wk 7, yolk color decreased in the OP group. Probiotic treatment decreased egg production and egg mass.	[148]
		0, 4.5, or 9.0% OP with or without 0 or 0.05% enzyme	Zangeneh and Torki reported that experimental diets did not show any significant difference in overall egg production, egg mass, FCR, and feed intake, while eggshell weight was higher in the OP groups than in the control diet. The 9% OP group showed the highest egg weight and decreased Haugh unit compared with the other experimental diets. Enzyme supplementation did not affect egg quality characteristics.	[149]
		0 or 9% OP with or without commercial cocktail enzyme	Zarei et al. indicated a reduction in egg production and blood triglycerides levels, and an increase of the yolk index in the 9% OP group, while there was no effect on feed intake and egg mass between OP and control groups, and between enzyme-fed and control groups. Enzyme supplementation enhanced FCR during wk 6.	[150]
	Quail	0, 50, or 100 g/kg OP (irradiated or not)	Abd El-Moneim et al. suggested that 5% OP or irradiated OP has the highest live BW and daily BW gain and the lowest values of daily feed intake and FCR, followed by the 10% OP group. Digestibility coefficients such as dry matter, organic matter, and crude protein were not affected in the experimental groups except for crude fiber. No effects were recorded in serum levels of total protein, albumin, liver enzymes, UA, creatinine, and lipid constituents of quails in the OP groups, but LDL and serum MDA was reduced. Serum glutathione was reduced, and glutathione reductase was not affected in the OP groups. Antibody titer against sheep erythrocytes was increased in the OP groups.	[153]
	Laying quail	0.1% *Aspergillus awamori*, 5% OP, 5% OP and *A. awamori*, 10% OP or 10% and 0.1% *A. awamori*	Abd El-Moneim et al. reported that the experimental diets had increased egg weight, and final BW, feed consumption, FCR, and egg mass were not affected by dietary treatment. Yolk (%) and yolk:albumin ratio were improved in 5% OP and 0.1% *A. awamori*, 10% OP, and 10% OP and 0.1% *A. awamori*. All experimental groups had enhanced egg shape index except for the 10% OP group during 16–20 wk. Yolk contents of cholesterol and total lipids and serum levels of triglycerides, cholesterol, and LDL cholesterol decreased in almost all groups fed *A. awamori*-treated diets. Glutathione content and glutathione reductase activity increased, and lipid peroxidation was reduced.	[154]

Table 2. *Cont.*

Agro-Industrial By-Product	Poultry Species	Examined Inclusion Levels	Key Findings as Reported by Authors	References
Olive pulp and commercial enzyme blend	Broiler	Unprocessed OP (50 g/kg, 100 g/kg, 50 g/kg with ENZ, 100 g/kg with ENZ), processed OP (50 g/kg, 100 g/kg, 50 g/kg with ENZ, 100 g/kg with ENZ), and control groups (without OP, and without OP with ENZ)	Sayehban et al. reported that carcass traits such as live BW, de-feathered BW, full abdomen carcass weight, empty abdomen carcass weight, eviscerated carcass weight, breast weight, thigh, and drumstick weight (legs), wing weight, and relative breast and wing weights were not different among dietary treatments. The 50 g OP/kg inclusion increased the eviscerated carcass, leg, and neck percentage values. Processing of OP increased breast percentages in broilers.	[136]
Olive meal (OM)	Broiler	0, 2, 4, 6, and 8% and enzymes	Sateri et al. indicated that BW and BW gain, feed intake, feed conversion efficiency, carcass traits, meat cuts (breast, drumsticks, and wings), the cecum microbiota, blood LDL and HDL cholesterol, triglycerides, total protein, albumin, glucose, and UA were not significantly different among the dietary groups. However, total cholesterol was higher in the 2% OM group (no enzyme supplementation) at 42 d compared to the 4% OM group (with enzyme addition). Birds fed 4% OM exhibited higher antibody titers after vaccinations against infectious bronchitis virus and Gumboro disease.	[142]
Olive pomace (OPO)	Broiler	0, 2.5, 5, 7.5%	Nasopoulou et al. reported higher growth rates at the 5 and 7.5% OPO groups, while the 5% OPO group had more potent in vitro antithrombotic properties compared to the control group. Grilled broiler meat of the 5% OPO group had acceptable sensory properties.	[139]
Olive pomace extract (OPE)	Broiler	Control (no additives), 100 ppm monensin, 500 or 1500 ppm OPE	Herrero-Encinas et al. found that OPE addition up to 1500 ppm did not affect daily gain, feed intake, and FCR. 500 ppm OPE supplementation decreased duodenal crypt depth, mannitol concentration, and ileal IL-8 expression in comparison with the control group.	[147]
		Control (no additives), 100 pm monensin, or 750 ppm OPE	Herrero-Encinas et al. reported that average daily gain was increased and FCR decreased in the experimental diets with no effect on feed intake. Bacterial composition at a family level in the caeca of broilers, plasma, and intestinal bile acid composition was not affected in the experimental groups. The OPE group showed a reduction of IL-8 expression in the ileum, while upregulation of the expression of TGF-β4 and Bu-1 in both experimental groups was observed.	[146]
Olive oil mill wastewater (OMWW) permeate or retentate	Broiler	-	Gerasopoulos et al. reported that broilers of the experimental groups had lower protein oxidation and lipid peroxidation levels and higher total antioxidant capacity in plasma and tissues. CAT activity in erythrocytes and tissues was significantly increased in the experimental groups. Erythrocytes in broilers with low glutathione (GSH) showed an increase in GSH levels with the inclusion of OMWW retentate, but in broilers with high GSH, it was reduced.	[143]

Table 2. *Cont.*

Agro-Industrial By-Product	Poultry Species	Examined Inclusion Levels	Key Findings as Reported by Authors	References
Pomegranate peel powder (PPP)	Broiler	Control group, or 0.5% colostin antibiotic, or 2, 3, 4 g PPP/kg, or 2, 3, 4 g PPP and 1 cm^3 probiotic	Abdel Baset et al. reported that live BW (5 wk) and BW gain (1–5 wk) were highest in the 2 and 4 g PPP, and positive control (PC, with antibiotic supplementation) groups compared to the negative control (NC, without additives) and other PPP groups. Feed intake was unaffected among dietary treatments but was affected during 1–3 wk of age. FCR was unaffected by the inclusion of PPP or additives in the diet. Daily feed conception and FCR were not affected by PPP dietary addition. The highest amounts of dressing and thigh output were recorded in the 3.0 g PPP plus 1 cm^3 probiotic and in the 4.0 g PPP plus 1 cm^3 probiotic/kg diet. The liver percentage was lower in the PPP-supplemented groups compared to the PC and NC groups (5 wk), while heart and gizzard were lower in the PC and NC groups compared to the PPP or additive-supplemented groups. AST was reduced in the 3 g PPP plus 1 cm^3 probiotics/kg diet and 4 g PPP/kg groups compared to the NC and PC groups. Alanine aminotransferase decreased in the 3 g and 4 g PPP plus 1 cm^3 probiotics/kg groups compared to the PC and NC groups. Urea and creatinine concentrations were lower in all treatments compared to the ones with no inclusion levels. Creatinine, total protein, and albumin concentrations were elevated in all treatments except for the NC group, and the PC group exhibited the lowest values. IgM and lysozyme were increased due to the incorporation of PPP in the diets. Reduced oxidative rancidity of meat was observed in the PPP groups.	[159]
	Laying hen	2 or 4%	Eid et al. reported that the negative effects of oxidative stress induced by dexamethasone on BW and egg production were alleviated in the PPP groups. Plasma cholesterol, triglyceride contents, and lipid peroxidation indicators (MDA) were reduced in the PPP groups, while the antioxidative enzymes (SOD, CAT, and GPx) and total antioxidant blood capacity were enhanced.	[169]
	Quail	2.5, 5.0, or 7.5%	Abbas et al. found that final BW was similar between treatments, while the 7.5% PPP group had the highest feed intake, and the feed intake in 2.5 and 5% PPP groups were not affected. FCR, egg production, egg numbers, egg weight, and egg mass were enhanced in the PPP groups. Serum cholesterol, triglyceride, glucose concentration, and GPT, 5% PPP) were reduced, and total protein increased, while the GOT to GPT ratio was not affected. PPP groups exhibited the highest relative weight in liver and heart, villus height, and crypt depth. 5 and 7.5% PPP groups recorded the highest ratio of villi length/villi depth. Liver weight, villus length, and crypt depth were higher in females than in males.	[171]
Pomegranate peel (PP)	Broiler	Control (no additives), vitamin E (100 mg/kg), or pomegranate peel (15,000 mg/kg), and others	Rajani et al. reported that the experimentally induced ascites mortality and MDA occurrence in meat (PP inclusion had the best effect) were decreased, and the right ventricular weight ratio was improved in the experimental groups. Growth performance was not affected, and meat shelf-life was extended in the experimental groups.	[161]

Table 2. *Cont.*

Agro-Industrial By-Product	Poultry Species	Examined Inclusion Levels	Key Findings as Reported by Authors	References
		2, 4, 6, or 8 g/kg PP or 0, 200 g/ton α-tocopherol acetate (vitamin E)	Akuru et al. indicated that the feed intake and FCR were increased in the vitamin E group at wk 3, but FCR was comparable to the control, 4, and 6 g/kg PP groups. The average final BW and average daily weight gain exhibited the highest values in the 2 and 4 g/kg PP groups, and the 2 g/kg PP groups had enhanced FCR and protein efficiency ratios in comparison to the vitamin E group. Thigh weight was the highest in the 4 g/kg PP group, while breast weight was the highest in the 8 g/kg PP group in comparison with the vitamin E group. The highest spleen and gizzard weights values were in the 4 g/kg PP group compared to the control (no additives) group, and nutrient digestibility was improved in comparison with the vitamin E group. The concentration of serum aspartate aminotransferase was reduced in the 4 g/kg PP group, while CAT enzyme activity in meat was the highest in the 8 g/kg PP group. The 4 g/kg PP group had better performance, digestibility, carcass, and organ indices compared to the vitamin E group.	[160]
Urea treated pomegranate peel (UTPP)	Broiler	0, 15, 30, or 50 g/kg	Hosseini-Vashan and Raei-Moghadam reported that BW gain increased during the start and overall experimental periods but decreased feed intake during starter and growing periods. FCR was enhanced in the experimental groups. Increasing levels of UTPP quadratically increased the breast yield, and the liver and abdominal fat decreased. The concentration of blood glucose, HDL, and globulin linearly increased with increasing UTPP dietary levels, while the plasma albumin, alkaline phosphatase, alanine aminotransferase, lactate dehydrogenase, cholesterol, LDL, and malondialdehyde concentrations were reduced at day 42 The bursa percentage increased with increasing levels of UTPP. The primary total, IgM, and IgG responses and the secondary total and IgG responses against sheep red blood cells were enhanced in the UTPP groups. The villus height, crypt depth, and villus height/crypt depth ratio, while decreasing the villus width, were increased, and the oxidative stability and water-holding capacity of breast meat was enhanced in the UTPP groups.	[165]
Pomegranate by-products (PB)	Broiler	0, 0.5, 1.0, or 2.0%	Ahmed et al. reported that crude protein and moisture contents were elevated, while ether extract in breast and thigh meat and cholesterol in breast meat were reduced. SFAs were reduced, and the sum of mono-unsaturated and n-3 fatty acids was increased in breast and thigh meat. n-6/n-3 ratio of breast and thigh meat was lower in 1 and 2% PB groups. The TBARS values and pH of breast and thigh meat were decreased in the BP groups.	[157]
Fermented pomegranate byproducts (FPB)	Broiler	0, 0.5, 1.0, or 2.0%	Bostami et al. reported that average daily weight gain during the finisher and overall period was increased in 1 and 2% FPB groups, while daily feed intake and FCR were not affected. Fecal pH tended to decrease in 0.5 and 2% FPB groups. Fecal ammonia emission was reduced in all the FPB groups, and hydrogen sulfide emission was decreased in 0.5 and 1% FPB groups. Feed cost per unit of weight gain was lower in the 1 and 2% FPB groups.	[158]

Table 2. *Cont.*

Agro-Industrial By-Product	Poultry Species	Examined Inclusion Levels	Key Findings as Reported by Authors	References
		0, 0.5, 1.0, and 2.0%	Ahmed et al. reported that increasing levels of FPB linearly increased weight gain and feed intake, and linearly reduced FCR. In the breast meat, crude protein, iron, magnesium, and sodium content were linearly higher, while cholesterol was linearly reduced. In thigh meat, ether extract and cholesterol were linearly lower with high moisture. The SFA % was linearly and quadratically reduced in breast and thigh meat, while MUFA of the breast (linear and quadratic) and n-3 fatty acids of breast and thigh (linear) increased in the FPB groups. The n-6/n-3 ratio of breast meat decreased in the FPB groups. The hypocholesterolaemic to the hypercholesterolaemic ratio of thigh meat increased with FPB inclusion. Breast and thigh meat had reduced TBARS and pH values in the FPB groups.	[162]
Raw (PPO) and fermented (FPPO) pomegranate pomace	Broiler	Control (no PP), 5 or 10 g/kg PPO, and 5 or 10 g/kg FPPO	Gungor et al. indicated that BW and FCR, serum GPx, SOD, and CAT levels were not affected in the experimental groups; however, malondialdehyde in breast meat was reduced. Caecal *C. perfringens* count and the villus height were reduced in the 10% PPO, 5% FPPO, and 10% FFPO groups compared to the control group. Ileum morphology was negatively affected by PPO and FPPO dietary inclusion. Crypt depth increased in the 5% PPO and 10% FPPO groups compared to the control and 10% PPO groups. The villus height-to-crypt depth ratio was reduced in the 5% PPO, 5% FPPO, and 10% FPPO groups.	[163]
Pomegranate pulp (PPU)	Broiler	0, 40, 70, or 100 g/kg	Hosseini-Vashan and Raei-Moghadam investigated the effect of PPU in thermoneutral and heat-stressed broilers in comparison with no added PPU diets. The concentration of uric acid, malondialdehyde, the enzyme activity of GPx, total antioxidant capacity, abdominal fat, and liver percentage were significantly affected by the inclusion of PPU in the heat-stressed broilers. Plasma protein and the enzyme activities of SOD were reduced in the PPU groups compared to the thermoneutral group. Plasma The plasma cholesterol and LDL concentrations were decreased compared to the control.	[164]
Pomegranate seed oil (PSO)	Broiler	0.0, 0.5, 1.0, 1.5% with or without 2% linseed oil (LO)	Manterys et al. reported that white blood cell levels were increased in 0.5 and 1% PSO supplemented with LO groups. Total cholesterol was increased with 1.5% PSO or with LO supplementation. PSO dietary treatment resulted in c9,t11 conjugated linoleic acid (CLA) concentration-dependent deposition in adipose tissue. ALA content was increased, and the n-6/n-3 ratio was reduced with LO addition. PSO and ALA influenced oleic acid proportion in adipose tissue. Liver parameters were not affected by PSO or LO incorporation. Health status was not affected by PSO dietary inclusion.	[166]
		0.0, 0.5, 1.0, 1.5% PSO with 0.0 or 2.0% LO	Szymczyk and Szczurek reported that the feed-to-gain ratio was enhanced with PSO inclusion in the diet of broilers (22–42 d). The abdominal fat percentage was higher in the 1.5% PSO group. Deposition of CLA in breast lipids increased with increasing levels of PSO. PUFA increased, MUFA increased, and SFA in breast lipids was not affected in the PSO groups. 2% LO incorporation increased total PUFA, decreased total MUFA proportions, and enhanced the n-6/n-3 ratio in breast meat compared to non-LO groups.	[167]

Table 2. *Cont.*

Agro-Industrial By-Product	Poultry Species	Examined Inclusion Levels	Key Findings as Reported by Authors	References
	Laying hen	2.5% sunflower oil (control), 0.5, 1.0, or 1.5% punicic acid (CLnA)	Kostogrys et al. reported that the color of the eggs' yolk was improved, while the hardness of hard-boiled egg yolks was not influenced. Dietary punicic acid incorporation resulted in an increase of CLnA and CLA levels in egg yolk. The Haugh units and pH values were similar between treatments. The egg albumen index was significantly higher in the 1% CLnA group. Feed consumption was the lowest in the 1.5% CLnA group. Egg yield increased in the CLnA groups. Eggs Shape Index of the CLnA was lower in the 0.5% CLnA group. SFA proportion increased, MUFA decreased in the experimental groups, while content and proportion of PUFA increased in the 0.5% CLnA and the lowest was observed in the 1.5% CLnA group.	[170]
Pomegranate and grape seed oil (GPO)	Broiler	2% replacement of soybean oil (5% in the diet)	Banaszkiewicz et al. indicated that the source of oil did not influence the slaughter yield, basic nutrients, and physical characteristics of the breast and thigh muscles. PSO inclusion enhanced the palatability of thigh muscles. GPO reduces the saturated fatty acids (palmitic) in muscles. The GPO group exhibited the deposition of a small amount of punicic acid and increased rumenic acid. The sum of the n-6 fatty acids and the n-6/n-3 ratio increased in the GPO group compared to the control group.	[168]
Tomato pomace (TP)	Broiler	0, 5, 10, or 15% substitution of SBM	Ghazi and Drakhshan reported that feed intake, weight gain, and FCR were similar between the treatments. No significant differences were recorded for breast, abdominal fat, liver, and gizzard weight.	[177]
		0% TP (rearing under thermoneutral zone), 0, 3 or 5% TP (heat-stressed broilers)	Hosseini-Vashan et al. indicated that BW and production index were elevated, and FCR was reduced in the 5% TP group (1–28 d), while reduced serum triglycerides and higher HDL cholesterol concentration were recorded in 28 d. The activities of GPx and SOD were elevated, and the concentration of MDA decreased in the 5% TP group (28 d). The adverse effects of heat stress on immune response were alleviated in the 5% TP group. The ash and Ca contents of the tibia were not significantly different between thermoneutral and heat-stressed broilers fed on 5% TP.	[180]
	Grower chicks	0, 5, 10, 15, or 20%	Yitbarek reported that TP groups recorded higher dry matter intake than the control group, and daily BW gain was highest in the 5% TP group. A significant difference in FCR was observed between the 5 and 20% TP groups. Economic efficiency was the highest in the 20% TP.	[179]
	Laying hen	0, 150, 170, or 190 g/kg	Salajegheh et al. found that BW, feed intake, egg production, FCR, egg weight, egg mass, eggshell weight, eggshell thickness, and Haugh unit were not affected by the dietary inclusion of TP. The yolk color score increased in the TP groups. Total serum protein, cholesterol, LDL, HDL, albumin, glucose, and triglyceride levels were not significantly different among treatments.	[186]
Tomato meal (TM)	Laying hen	0, 80, or 150 g/kg	Yannakopoulos et al. indicated that body weight gain, egg number, feed consumption, mortality, eggshell quality, and egg shape index were not influenced in the experimental groups. The yolk color score was improved, and the number of blood and meat spots decreased in the TM groups.	[183]

Table 2. *Cont.*

Agro-Industrial By-Product	Poultry Species	Examined Inclusion Levels	Key Findings as Reported by Authors	References
Tomato pulp (TPU)	Laying hen	Control (no additives), 28 g/ton carophyll, 40, 60, 80, or 120 g/kg TPU	Dotas et al. reported that egg production was not affected in the experimental groups, and food consumption, food efficiency, and egg weight were not different between the control and TPU groups. The number of broken eggs and the number of eggs without shells were not affected by dietary treatment. The yolk color was enhanced in the TPU groups, with the carophyll groups recording the same.	[182]
		0, 50, 100, or 150 kg/t	Jafari et al. reported that egg production and egg mass of hens (27–38 wk) were higher by the inclusion of up to 100 kg/t than the control, while final BW, egg weight, daily feed consumption, eggshell weight, eggshell thickness, Haugh units, and yolk color were similar to the control group. Conversely, lower egg production and egg mass, and increased feed efficiency were recorded in the 150 kg/t TPU group.	[176]
	Quail	0, 5, 10%	Nikolakakis et al. reported that final BW, daily feed consumption, FCR, and carcass weight, yield, and composition were similar among treatments, while ether extract content was lower in both TPU groups. Thigh and breast skin coloring was darker in the TPU groups, with the 10% TPU group exhibiting the darkest color. Carcass coloring, the fatty acid profile of carcasses, total SFA, MUFA, and PUFA were not different among treatments.	[189]
		0, 5, 10%	Botsoglou et al. indicated that MDA values in raw meat were higher after 6–9 d in the 10% TPU group and lower in the 5% TPU group. The oxidation profile of cooked meat was similar after 3, 6, and 9 d of storage among dietary treatments. MDA values of raw meat in the 5% TPU group were lower only at 100 and 150 min of iron-induced lipid oxidation. PUFA and USFA/SFA ratio (unsaturated fatty acids/SFA) was higher in the 10% TPU group.	[188]
Tomato powder (TPO)	Laying hen	0, 5, or 10 g/kg	Akdemir et al. indicated that feed intake, egg production, egg weight, and yolk color increased linearly, and feed conversion decreased linearly with increment dietary levels of TPO. Shell weight, shell thickness, and Haugh unit were not affected in the TPO groups. Concentrations of serum and egg yolk lycopene, β-carotene, lutein, and vitamin A were elevated in the TPO groups, while MDA decreased linearly with increasing levels of TPO in the diet.	[184]
	Quail	0, 2.5, 5%	Sahin et al. reported that increasing dietary levels of TPO linearly increased feed intake, live weight gain, and feed conversion under heat stress conditions but not under thermoneutral conditions. Serum lycopene and vitamin C, E, and A concentrations linearly increased with increasing levels of TPO. MDA in serum, liver, and muscles linearly decreased with increasing levels of TPO in both heat-stressed and thermoneutral groups.	[190]
Tomato pulp powder (TPP)	Quail	0, 2, 4, 6, 8%	Jouzi et al. reported that feed intake was similar between treatments, while BW and pre-slaughter weight was elevated in the 4% TPP group compared to the other groups. The feed coefficient was higher in the 6 and 8% TPP groups compared to the control. Wing weight was lower in the 2 and 4% TPP groups compared to the control. Breast, drumstick, and carcass yield, triglyceride, cholesterol, Zn, Cu, and Fe levels were reduced in the 2, 4, and 6% TPP groups.	[187]
Tomato waste juice (TWJ)	Broiler	0, 40, 80, 120 mL/d	Wahyuni et al. reported that the relative weight of the thymus, duodenum, jejunum, caecum, and liver was elevated in broilers (15–35 d) fed on TWJ. Cumulative feed intake, final BW, daily weight gain, and FCR were not affected by TWJ incorporation.	[181]

Table 2. *Cont.*

Agro-Industrial By-Product	Poultry Species	Examined Inclusion Levels	Key Findings as Reported by Authors	References
Tomato waste (TW)	Broiler	0, 5, 10, 15, or 20%	Lira et al. reported that feed intake was elevated during 1–7, 8–14, and 29–36 d, while gain weight and FCR deteriorated up to 29 d but not during 29–42 d. Carcass weight and weight of the noble parts, breast, drumstick, and thighs were reduced linearly with increasing TW levels (up to 28 d), while yield (%) was not affected except for heart and liver yields.	[178]
	Laying hen	5% flaxseed and 2.5, 5.0, or 7.5% TW	Panaite et al. reported that average daily feed intake and laying percentage were reduced in the 5 and 7.5% TW groups in comparison with the control. Yolk Roche color score was improved in the TW groups due to the enrichment of yolk with carotenoids, but transfer efficiency from feed to egg was reduced. In 5 and 7.5% TW groups (4 wk), lutein and zeaxanthin levels of egg yolk were elevated, and the color score was 3.5-fold compared to the control. The n-3 fatty acid content of egg yolk increased from 3.1 to 3.7-fold due to flaxseed addition compared to the control group, while the n-6/n-3 ratio decreased from 18.3 in the control to 4.1–5.4 in the flaxseed-supplemented groups.	[185]
Sugar beet pulp	Broiler	0, 30 g/kg oat hulls (OH) or SBP	Gonzalez-Alvarado et al. indicated that BW gain and feed-to-gain ratio were improved in the SBP diets compared to the control. Feed intake was reduced at 25–42 d, and the relative weight of the gastrointestinal tract and gizzard, the digesta content of the gizzard, and the total tract apparent digestibility of nutrients were improved in the SBP group.	[194]
		0, 25, 50, and 75 g/kg SBP or OH	Jimenez-Moreno et al. reported that feed intake or BW gain was not affected, while FCR was enhanced quadratically in the SBP and OH groups (1–18 d). Energy efficiency improved linearly in the SBP and OH groups (1–18 d). The coefficient of total tract apparent retention was improved by the incorporation of up to 50 g/kg SBP or OH.	[196]
		0, 25, 50, and 75 g of either OH or SBP	Jimenez-Moreno et al. reported that the relative weight of the gastrointestinal tract with digesta contents increased linearly with increasing levels of dietary fiber. The weight of the pancreas increased with increment levels of SBP, while the relative weight of the gizzard and its dry matter (DM) content was elevated, and gizzard pH was decreased in the experimental diets at all ages of broilers. Gizzards were heavier with higher DM content and gizzard pH in the OH group compared to the SBP one, while villus height (12 d) decreased in the SBP group. The pH of the digesta of the duodenum was elevated in the SBP and OH groups at 6 d and at 12 d in the SBP group.	[195]
		0%, 7.5% SBP, or 15% potato peel, with or without enzyme	Abdel-Hafeez et al. reported that the SBP or potato peel inclusion decreased BW, while feed intake, weight gain, and feed conversion were lower in the SBP group but were not different in the potato peel group compared to the control. Enzyme addition increased BW, feed intake, and feed conversion. The total cholesterol, LDL cholesterol serum levels, and carcass fat content were lower in the experimental groups, while carcass yield was not different. At the same time, SBP addition at greater levels (7.5%) decreased BW, weight gain, LDL, and total cholesterol serum levels.	[198]

Table 2. *Cont.*

Agro-Industrial By-Product	Poultry Species	Examined Inclusion Levels	Key Findings as Reported by Authors	References
		0%, 7.5% SBP, or 15% potato peel, with or without enzyme	Abdel-Daim et al. indicated that the digestibility of ether extract, crude fiber, or crude protein, physicochemical and sensory characteristics of the breast or thigh muscles, and intestinal morphology during starter and growing periods were not affected in the experimental groups. Villus height and villus height/crypt depth ratio decreased during the starting period but not in the grower period. Enzyme addition increased the digestibility of nutrients, enhanced the development of the small intestine, increased the crude protein content and the water-holding capacity, and reduced the ether extract of the meat and the cooking loss rate.	[197]
		23, 46, and 92 g/kg	Petterson and Razdan reported that the ileal digestibilities of organic matter, crude fat, and crude protein were reduced with increasing dietary levels of SBP. Total serum cholesterol levels decreased in the SBP groups. High triacylglycerol and total serum cholesterol concentrations of the restricted level-fed chickens exhibited a meal frequency factor. Growth performance was improved in the SBP groups, but with no statistical differences.	[199]
	Laying hen	0, 3, 5, 7%	Selim and Hussein reported that feed intake, egg production, egg weight and mass, and improved FCR, yolk color core, and Haugh unit linearly increased with SBP addition. Higher protein and lower ether extract in eggs of the SBP groups were observed, while serum total lipids, cholesterol, alanine aminotransferase, aspartate aminotransferase, and creatinine decreased. Egg yolk MDA, cholesterol, and triglyceride linearly decreased, and GPx increased with SBP dietary inclusion.	[201]
	Quail	0, 20, 40 g/kg and multi-enzyme 0, 1, or 2 g/kg	Alagawany and Attia reported that feed consumption, FCR, egg number, egg weight, egg mass, external and internal egg quality, N consumption, N in egg, N excretion, N fecal, N intake, and N retention were not affected by SBP inclusion. Increasing levels of SBP reduced final BW and fertility percentage. Hatchability percentages from fertile eggs increased with decreasing levels of SBP. Digestion coefficients of the nutrients excluding the N digestibility were significantly affected by SBP addition.	[191]
Sugar beet meal	Broiler	Control, 2.5% sugar beet meal, 2.5% neem leaf meal, 2.5% linseed meal, or 2.5% coriander seed meal	Kumari et al. reported enhanced BW, weight gain, feed conversion rate, performance index, and giblet relative weight in the sugar beet meal-fed group.	[193]
Aqueous methanolic extract of sugar beet	Broiler	100, 200, and 300 mg/kg BW, vitamin E 87 mg/kg, Baycox® 1 mL/L of water, PBS group (infected non medicated control group). Group served as non-infected, non-medicated group	Abbas et al. indicated that sugar beet exhibited good anticoccidial activity, which was evaluated based on the improvement of FCR, lesion score, oocyst score, and oocysts per g of feces. The serum profile of infected broilers was not significantly different by the inclusion of the sugar beet extract.	[200]

Table 2. *Cont.*

Agro-Industrial By-Product	Poultry Species	Examined Inclusion Levels	Key Findings as Reported by Authors	References
Brewers dried grains (BDG)	Broiler	Substitution of maize with 0, 25, 50, 75, or 100% BDG	Ironkwe and Bamgbose reported that the 50% BDG group exhibited the highest final live weight, daily weight gains, and the lowest FCR, followed by the 25, 75, and 100% BDG groups, with the lowest. Feed intake was the highest in the control group (0% BDG) and the lowest in the 100% BDG group, while cost per kg feed was the lowest in the latter group, which decreased with increasing BDG participation in the diet. Cost per kg weight gain decreased with increasing BDG levels.	[205]
		0, 15, 20, 25, 30, 35, or 40% BDG and sand 0 or 4%	Onwudike indicated that sand incorporation in diets improved digestibility, gain, and FCR, while the inclusion of higher than 15% to 20% BDG decreased feed conversion efficiency. At levels of 0, 15, 20, 25, and 30% BDG and 4% sand inclusion in the diet, no positive effects were observed in grower diets.	[206]
Brewers spent grains (BSG)	Broiler	Group I (whole BSG without xylanase), group II (whole BSG and xylanase top-dressed), group III (whole BSG and xylanase pre-treated), group IV (ground BSG without xylanase), group V (ground BSG and xylanase top-dressed), group VI (ground BSG and xylanase pre-treated)	Denstadli et al. reported that weight gain was not different among treatments, but feed intake increased by the xylanase supplementation and by the addition of coarse BSG. Feeding coarse BSG rather than ground BSG had a better effect on feed utilization, but no effect on ileal digestibility or apparent metabolizable energy was observed. Gizzard weight was elevated in the coarse-BSG groups compared to the ground-BSG groups. Jejunal viscosity decreased due to the enzyme supplementation in the diet. Higher concentrations of arabinose and xylose in caeca in the pre-treated diets compared with the untreated or top-dressed diets were observed. Enzyme supplementation in diets affected the caecal contents of rhamnose and mannose. Elevated ileal concentrations of mannose and glucose in groups fed on pre-treated diets compared with groups fed on top-dressed diets were recorded.	[208]
Fermented brewers spent grains, mineral sorbing complex, and probiotics	Quail	0 or 1.5%	Yurina et al. reported that an additive consisting of fermented brewers spent grains, mineral sorbing complex, and probiotics increased gross egg production by 3.8%, intensity of egg production by 2.3%, the intensity of egg production by 2.3%, decreased feed consumption for the production of 1 dozen eggs by 5.5%, and increased FCR in comparison with the control group. Livability (90%) of quails was not affected among the treatments.	[207]

ALA = α-linolenic acid; AO = antioxidants; AP = apple pomace; BDG = brewers dried grains; BL = *Bacillus licheniformis*; BSG = brewers spent grains; BW = body weight; CAT = catalase; CLA = conjugated linoleic acid; CLnA = punicic acid; CP = citrus pulp; CW = citrus waste; DCP = dried citrus pulp; DCSP = dried *Citrus sinensis* peel; DDGS = dried distillers' grains with solubles; DLP = dried lemon (*Citrus aurantifulia*) pulp; DM = dry matter; ENZ = enzyme; FCR = feed conversion ratio; FGP = fermented grape pomace; FPB = fermented pomegranate byproducts; FPPO = fermented pomegranate pomace; GOT = glutamic-oxaloacetic transaminase; GP = grape pomace; GPC = grape pomace concentrate; GPO = grape seed oil; GPT = glutamic-pyruvic transaminase; GPx = glutathione peroxidase; GSH = glutathione; HDL = high-density lipoprotein; HPSFM = high-protein sunflower meal; LDL = low-density lipoprotein; LFSFM = low-fiber sun flower meal; LO = linseed oil; LO-DDGS = low-oil distillers dried grains with solubles; MDA = malondialdehyde; NC = negative control; OC = olive cake; OH = oat hulls; OL = olive leaves; OM = olive meal; OMWW = olive oil mill wastewater; OP = olive pulp; OPE = olive pomace extract; OPO = olive pomace; OPU = orange pulp; PB = pomegranate by-products; PC = positive control; PP = pomegranate peel; PPO = raw pomegranate pomace; PPP = pomegranate peel powder; PPU = pomegranate pulp; PSO = pomegranate seed oil; SBM = soybean meal; SBP = sugar beet pulp; SC = sunflower cake; SFA = saturated fatty acids; SFM = sunflower meal; SOD = superoxide dismutase; TBA = thiobarbituric acid; TBARS = thiobarbituric acid reactive substances; TM = tomato meal; TP = tomato pomace; TPO = tomato powder; TPP = tomato pulp powder; TPU = tomato pulp; TW = tomato waste; TWJ = tomato waste juice; UA = uric acid; USFA = unsaturated fatty acids; UTPP = urea-treated pomegranate peel; WGPF = wine-grape pomace flour.

3. Potential Limitations of Utilizing Agro-Industrial By-Products in Poultry Nutrition

Potential limitations in using AIBP in poultry rations include nutrient variability, feed safety, sensitivity to peroxidation, the presence of anti-nutritional factors, their high content of fiber, and whether it is possible to achieve profit by this practice.

The nutrient composition of AIBPs may vary depending on the area, climate, and season, which limits their use in animal diets [10]. Nutrient variability can be reduced by the processing of AIBPs to produce a uniform feed with consistent nutrient composition while nutrient supplementation is necessary [60]. Furthermore, technological requirements are necessary to stabilize the final AIBP product and to abate seasonal variability [21] Seasonal availability of some by-products is a limit in their wide use, such as in the case of vegetables and fruit residues, for example, in cider and wine production, which takes place from September to October [10,209]. These by-products are mainly fed raw or after drying [210]. In DDGS, variability in the nutrient composition depends on different factors, including the origin of raw material, the processing methods applied, fermentation yeast properties, and year of production, and thus, chemical analysis of DDGS from different sources should be conducted regularly [83].

Considering a processing method to be applied to AIBPs, the good nutritional value of the final product and socioeconomic and ecological feasibility should be taken into account [211]. Drying of by-products with high moisture, often exceeding 80%, such as grape and tomato pomace and skins, is necessary to prevent microbial spoilage [60,212] which should take place by the producer soon after the generation of the by-products [18]. Water content higher than 20% in the by-product prior to processing could limit storage duration. Refrigeration or treatments with exogenous enzymes or fermentation permit a longer shelf life for the by-products. In the case of fruit pomaces, steam explosion, and amination may also be used [126]. In grape by-products, the above-mentioned processing methods and polyethylene glycol treatment release non-starch polysaccharides or linked tannins from cell walls and biomass resulting in an increase in digestibility and enhanced bioactive properties [104]. Moreover, transportation costs increase when the moisture content of AIBPs is high. The energy costs of drying of high-moisture AIBPs may be higher than the value of the feedstuff itself. Thus, mixing with other dry feedstuffs to reduce water content prior to processing can be applied [60]. In order to ensure the economic feasibility of using AIBPs in animal nutrition, the relative economic value should be low [210]. Solar drying is gaining much attention as an environmentally friendly and low-cost processing method. Methods of solar drying include open-air drying (sun drying) and drying with the use of sun dryers. In the case of apple and orange waste, treatment with greenhouse solar dryers is more suitable than sun drying in terms of time efficiency, minimization of microorganisms, and nutritional value [212]. However, drying of by-products may lead to a concentration of pesticide residues, while mycotoxins can be produced by molds in low water activity levels [25].

Food safety is a prerequisite to achieving food security and protecting the income of small-scale farmers. A circular economy may introduce safety hazards in the food supply chain. Although there are numerous scientific articles on the reuse of alternative foods and feeds, the safety of these products in a circular economy is sometimes neglected in the literature. The importance of safety in a circular economy has been reviewed by researchers, for instance, for insects, former food products [213], catering waste [211], and seaweeds [214], while the significance of emerging hazards has been indicated by EFSA et al. [215].

Although a circular economy is promoted in the EU through the EC Green Deal and Farm to Fork policy, no policies to monitor the safety of by-products exist [216]. The EU Regulation (EC) 178/2002 stipulates that the food business operators are responsible for the safety of the products put on the market [216,217]. Although several feed contamination episodes have occurred over the last decades, such as the carry-over of dioxins from feed by-products [218], limited data are available in the literature on the hazards of plant by-products and their carry-over in other parts of the food supply chain [216].

Different chemical hazards have been found in plant by-products, such as heavy metals, mycotoxins, pesticides, and plant toxins. In Table 3, hazards in each AIBP reported in the literature are presented. Mycotoxin occurrence varies depending on agronomic practices, the geographic location of the crop grown, and meteorological conditions [219] Heavy metals or other contaminants, such as antibiotic residues, may accumulate in by-products, such as sugar beet pulp, which can be taken up from the soil. The route of antibiotic residues contaminating the soil is via manure [220]. In the case of grape by-products, the level of heavy metals is variable and depends on soil composition and contamination and the grape variety [104].

DDGS and germ, rootlets, or brewer's spent grains may be contaminated by mycotoxins [216]. These by-products are of particular importance since mycotoxins are frequently found in cereals [218], and they accumulate primarily in the outer fractions of the grain, such as the fibers and husks [216]. In the case of DDGS, this co-product may contain three times the content of some mycotoxins compared to the raw material [216]. The primary mycotoxins that may be present in corn and can be found in DDGS include fumonisin, aflatoxin, deoxynivalenol (vomitoxin), zearalenone, and ochratoxin. The risk of contamination of corn DDGS is very low due to the control systems applied throughout the chain of farm-bioethanol industry-animal feed [82,221]. In sugar beet pulp silage, the mycotoxins ochratoxin A, zearalenone, mycophenolic acid, and roquefortine C were found in France [222]. Moreover, it is forecasted that a higher prevalence of aflatoxin producing species (i.e., *Aspergillus* spp.) will be evident due to climate change [223]. A reduction of mycotoxin levels in grains can be achieved by mitigation measures. These include good agricultural practices, plant breeding, use of less susceptible varieties, plant protection, crop rotation, drying, and storage. Appropriate sampling and testing are needed to identify potentially contaminated feed. Measures focusing on contaminated feed include visual/automated sorting, decontamination (e.g., ammoniation), the addition of binders, or the proper inclusion in the diet of less sensitive animal species [224]. Furthermore, risk assessment is necessary to evaluate the risk of mycotoxins to poultry and consumer health via potential carry-over. Antibiotic residues may also be found in DDGS as a result of using antibiotics in bioethanol production that function as inhibitors to microbial growth in order to enhance the fermentation process [216].

CP may be contaminated by pesticide residues and dioxins. For example, one sample of CP was found to contain the pesticide heptachlor at a level exceeding the maximum residue level (MRL) by the official control in Denmark from the results published between 1996 and 2008 [218]. In a risk assessment [218], it was concluded that if residue levels of the stobilurin pesticides azoxystrobin, pyroclostrobin, and the fungicides imazalil and thiabendazol in CP are lower than 0.5 mg/kg, inclusion levels of 20% or 23% in the poultry diet will not result in negative health effects for the consumer [218]. Distribution of pesticides in the by-products due to processing is another point to be considered. The concentration of pesticides in the by-product may occur if processing steps that include dehydration are used [25], while pesticides tend to concentrate in brewer's spent grains [220].

The transformation of pesticide residues in food processing by-products may take place depending on factors such as temperature and microbial activities resulting in contamination with other chemicals [25]. Polychlorinated dibenzo(p)dioxins and furans (PCDD/Fs) can be introduced in CP via contaminated lime (calcium hydroxide), which is used at a level of 2% to partly neutralize fruit acids and make it suitable as animal feed [225]. Another possible route of transmission of dioxin in CP was reported in Germany, The Netherlands, and Belgium, in 1998, via the use of certain types of waste as the fuel for the direct drying of CP [226]. Other persistent environmental pollutants include polycyclic aromatic hydrocarbons (PAHs), which may be present in by-products comprising fats and fatty acids, such as vegetable oils [216]. For example, PAHs have been identified in olive pomace oil [227].

The antinutritional factors that are usually present in AIBP may affect feed palatability, digestibility, and animal production performance [20]. The processing methods to deactivate the antinutritional factors include physical, chemical, and biological methods [20].

AIBPs may be a cheap source of feed, but to the best of our knowledge, economic analyses of the use of AIBPs in poultry nutrition are scarce. However, in practice, it is widely known that it may be cost-effective in comparison with using conventional feeds. In a study conducted on pomegranate by-products, feed cost per unit of weight gain was lower in experimental groups fed 1% or 2% fermented pomegranate byproducts in the diet compared to the basal diet containing solely conventional feeds [158]. In another study, the feeding cost in broilers fed a diet containing corn-silage was 1.95 –fold higher than the feeding cost in broilers fed a diet containing avocado and pomegranate by-products [228]. In a study evaluating the economic feasibility of OP in broiler diets, feed cost per kg live weight was statistically higher when 100 g/kg OP was fed compared to the inclusion level of 50 g/kg OP. In countries and regions that produce olive oil, the use of olive pomace may reduce poultry production costs, while in other regions, the cost of olive pomace can be a barrier for its use [141]. SFM inclusion level of 8% resulted in an improvement of the economic efficiency index compared to inclusion of 0, 16, and 24% SFM levels in broiler diets [68]. Increasing incorporation levels of sorghum DDGS in broiler diets decreased the cost of feed/kg, while at the level of 20% of DDGS, maximum financial returns were observed [229]. Another study found that 50% or 100% maize replacement with brewer's dried grain in the diet leads to the lowest production feed cost per kg weight gain of broilers [205]. One important factor that may reduce the cost is to maintain a small distance between livestock farms and the generation sites of by-products [230]. To calculate the true cost of AIBP feed, the Extension of the University of Georgia recommends taking into account the price of the feed delivered to the farm, the interest, the shrinkage and storage losses, and the extra handling cost [231]. It is encouraging that extensions of universities, such as in the case of the University of Georgia, provide recommendations about the incorporation of AIBP in animal diets. Economic incentives to alleviate the costs of using AIBPs in animal nutrition would foster the participation of the different stakeholders (e.g., farmers and feed manufacturers).

Table 3. Reported hazards found in agro-industrial by-products.

Agro-Industrial By-Product	Potential Hazards	References
Apple by-products	Amygdalin, pesticides (e.g., neonicotinoids and arsenic-based pesticides), patulin	[232]
Citrus pulp	PCBs and PCDD/Fs [1], ochratoxin A, pesticides (e.g., imidacloprid, abamectin, cypermethrin, and prochloraz	[218,225,233–235]
Sunflower meal	Alternariol, alternariol monomethyl ether and tenuazonic acid (*Alternaria* spp. toxins), *Fusarium* spp. toxins, aflatoxin B_1, heavy metals (e.g., Pb, Cd, Cr, As, Hg, Ni)	[236,237]
Wheat dried distillers' grain with solubles	Deoxynivalenol, enniatin B, ochratoxin, antibiotics Co-occurrence of deoxynivalenol with its acetylated and/or glycosylated derivatives, and DON with enniatins, beauvericin or zearalenone	[83,216,218,238–240]
Corn dried distillers' grain with solubles	Aflatoxins (e.g., AFB_1), deoxynivalenol, fumonisins, T-2 toxin, zearalenone, ochratoxin	[82,83,221]
Grape pomace	Heavy metals (e.g., Al, As, Pb, Cd, and Ni), toxins (e.g., ochratoxin A, biogenic amines)	[104,241]
Sugar beet pulp	Heavy metals (e.g., Al, As, Pb, Cd, and Ni)	[222]
Sugar beet pulp silage	ochratoxin A, zearalenone, mycophenolic acid and roquefortine C	
Brewery by-products	Aflatoxins (e.g., AFB_1), ochratoxin A, fumonisin B_1, acetyl-deoxynivalenols (ADONs), deoxynivalenol-3-glucoside (DON-3-Glc), HT-2, enniatins, patulin and gliotoxin, pesticides	[83,220,242]

[1] PCDD/F, polychlorinated dibenzo-p-dioxins and dibenzofurans; PCB, polychlorinated biphenyls.

4. Conclusions

AIBPs contain several bioactive compounds that may act as antimicrobial agents, antioxidants, and immune modulators. These properties may contribute to the role of AIBPs as functional feed ingredients in promoting the health, productivity, and well-being of poultry. Based on the findings reported in the present study, the following inclusion levels of agro-industrial by-products are proposed. Dried apple pomace at levels up to 6% in broiler diets and up to 25% in laying hens are recommended. Dried sweet orange peel up to 3% in broiler diets and dried citrus pulp up to 10% in broiler diets could be used, while favorable results were reported in ostrich and goose diets, while in hen diets variable results were found. Sunflower meal in broiler diets at levels up to 15%, especially when an enzyme mixture was added, provided favorable results, depending on the growing period. However, results were inconsistent among the studies. In the hen's diet, the inclusion of sunflower meal at levels up to 25% did not show negative results. In the case of dried distillers' grain with solubles, levels up to 24% in broiler diets and up to 25% in laying hen diets could be used, but variable results were reported depending on the growing period. Grape pomace can be included at levels up to 10% in broiler diets and up to 6% in hen diets. Olive cake could be incorporated at levels up to 20% in broiler and hen diets, with the addition of citric acid and olive pulp at levels up to 10%. Raw or fermented pomegranate pulp could be used at levels up to 2%. Sugar beet pulp could be incorporated at levels up to 5% in broiler diets. However, ways to reduce nutrient variability of ABIP across countries should be found.

Unfortunately, AIBPs also have shortcomings and limitations, such as the presence of anti-nutritional ingredients and chemical hazards. The importance of the control of potential hazards in AIBPs should be emphasized through proper legislation and knowledge of the different stakeholders involved. For example, the use of good agricultural practices and minimization measures of the antinutritional factors present in AIBPs. Furthermore, many of the studies reviewed herein presented notable differences in the characterization of extracts in terms of their biological properties when assessed. However, modern processing methods, new types/classifications, and appropriate developmental strategies are expanding the applications of AIBPs as animal feeds for poultry production.

Overall, given that the availability and price of the AIBPs may vary greatly across the regions, the use of such by-products as functional feed ingredients in poultry rations should therefore be adjusted according to the availability and cost of each by-product. Moreover, due to the nutrient variability of AIBPs, proximate analysis prior to feeding is necessary to manage animal diets and decrease costs. Future studies should confirm the efficacy of agro-industrial residues and their derivative products in substituting the use of conventional feedstuffs in poultry nutrition.

Author Contributions: Conceptualization, A.G., A.C.P., and E.Z.; methodology, A.G., A.C.P., and P.S.; software, A.G.; validation, A.G., E.G., M.G., and P.S.; formal analysis, A.G., A.C.P., M.G., and E.Z.; investigation, A.G., E.G., P.S., M.G., and A.C.P.; resources, A.C.P. and E.Z.; data curation, A.G., E.G., M.G., and P.S.; writing—original draft preparation, A.G., A.C.P., P.S., and E.Z.; writing—review and editing, A.G., E.G., M.G., A.C.P., E.Z., and P.S.; visualization, A.G.; supervision, A.C.P. and E.Z.; project administration, A.G., A.C.P., and P.S. All authors have read and agreed to the published version of the manuscript.

Funding: This research received no external funding.

Institutional Review Board Statement: Not applicable.

Informed Consent Statement: Not applicable.

Data Availability Statement: Not applicable.

Conflicts of Interest: The authors declare no conflict of interest.

References

1. United Nations Department of Economic and Social Affairs, Population Division. World Population Prospects 2022: Summary of Results. 2022. Available online: https://www.un.org/development/desa/pd/sites/www.un.org.development.desa.pd/files/wpp2022_summary_of_results.pdf (accessed on 1 December 2022).
2. Grumbine, R.E.; Xu, J.; Ma, L. An overview of the problems and prospects for circular agriculture in sustainable food systems in the anthropocene. *Circ. Agric. Syst.* **2021**, *1*, 3. [CrossRef]

3. WHO; FAO; IFAD; UNICEF; WFP. *The State of Food Security and Nutrition in the World 2022. Repurposing Food and Agricultural Policies to Make Healthy Diets More Affordable*; FAO: Rome, Italy, 2022. Available online: https://www.fao.org/3/cc0639en/cc0639en.pdf (accessed on 1 December 2022).
4. Brunetti, L.; Leuci, R.; Colonna, M.A.; Carrieri, R.; Celentano, F.E.; Bozzo, G.; Loiodice, F.; Selvaggi, M.; Tufarelli, V.; Piemontese L. Food industry byproducts as starting material for innovative, green feed formulation: A sustainable alternative for poultry feeding. *Molecules* **2022**, *27*, 4735. [CrossRef]
5. Van Zanten, H.H.E.; Herrero, M.; Van Hal, O.; Röös, E.; Muller, A.; Garnett, T.; Gerber, P.J.; Schader, C.; De Boer, I.J.M. Defining a land boundary for sustainable livestock consumption. *Glob. Change Biol.* **2018**, *24*, 4185–4194. [CrossRef]
6. Philippini, R.R.; Martiniano, S.E.; Ingle, A.P.; Marcelino, P.R.F.; Silva, G.M.; Barbosa, F.G.; dos Santos, J.C.; da Silva, S.S. Agroindustrial byproducts for the generation of biobased products: Alternatives for sustainable biorefineries. *Front. Energy Res* **2020**, *8*, 152. [CrossRef]
7. van Hal, O.; de Boer, I.J.M.; Muller, A.; de Vries, S.; Erb, K.H.; Schader, C.; Gerrits, W.J.J.; van Zanten, H.H.E. Upcycling food leftovers and grass resources through livestock: Impact of livestock system and productivity. *J. Clean. Prod.* **2019**, *219*, 485–496. [CrossRef]
8. Sonaiya, E.B.; Swan, S.E.J. *Small-Scale Poultry Production-Technical Guide*; Food and Agriculture Organization of the United Nations Rome, Italy, 2004.
9. Esposito, B.; Sessa, M.R.; Sica, D.; Malandrino, O. Towards circular economy in the agri-food sector. A systematic literature review. *Sustainability* **2020**, *12*, 7401. [CrossRef]
10. Correddu, F.; Lunesu, M.F.; Buffa, G.; Atzori, A.S.; Nudda, A.; Battacone, G.; Pulina, G. Can Agro-Industrial By-Products Rich in Polyphenols be Advantageously Used in the Feeding and Nutrition of Dairy Small Ruminants? *Animals* **2020**, *10*, 131. [CrossRef]
11. Thieme, O.; Makkar, H.P.S. Utilisation of loss and waste during the food-production cycle as livestock feed. *Anim. Prod. Sci.* **2017** *57*, 601–607. [CrossRef]
12. Angulo, J.; Mahecha, L.; Yepes, S.A.; Yepes, A.M.; Bustamante, G.; Jaramillo, H.; Valencia, E.; Villamil, T.; Gallo, J. Nutritional evaluation of fruit and vegetable waste as feedstuff for diets of lactating Holstein cows. *J. Environ. Manag.* **2012**, *95*, S210–S214 [CrossRef]
13. Sadh, P.K.; Duhan, S.; Duhan, J.S. Agro-industrial wastes and their utilization using solid state fermentation: A review. *Bioresour Bioprocess.* **2018**, *5*, 1. [CrossRef]
14. Makkar, H.P.S. Opinion paper: Food loss and waste to animal feed. *Animal* **2017**, *11*, 1093–1095. [CrossRef]
15. Leinonen, I.; Kyriazakis, I. How can we improve the environmental sustainability of poultry production? *Proc. Nutr. Soc.* **2016**, *75* 265–273. [CrossRef]
16. Eliopoulos, C.; Arapoglou, D.; Chorianopoulos, N.; Markou, G.; Haroutounian, S.A. Conversion of brewers' spent grain into proteinaceous animal feed using solid state fermentation. *Environ. Sci. Pollut. Res.* **2022**, *29*, 29562–29569. [CrossRef]
17. Azizi, M.; Seidavi, A.R.; Ragni, M.; Laudadio, V.; Tufarelli, V. Practical applications of agricultural wastes in poultry feeding in Mediterranean and Middle East regions. Part 1: Citrus, grape, pomegranate and apple wastes. *Worlds Poult. Sci. J.* **2018**, *74* 489–498. [CrossRef]
18. Chuang, W.-Y.; Lin, L.-J.; Shih, H.-D.; Shy, Y.-M.; Chang, S.-C.; Lee, T.-T. The Potential Utilization of High-Fiber Agricultural By-Products as Monogastric Animal Feed and Feed Additives: A Review. *Animals* **2021**, *11*, 2098. [CrossRef]
19. Abdel-Moneim, A.l.-M.E.; Shehata, A.M.; Alzahrani, S.O.; Shafi, M.E.; Mesalam, N.M.; Taha, A.E.; Swelum, A.A.; Arif, M.; Fayyaz, M.; Abd El-Hack, M.E. The role of polyphenols in poultry nutrition. *J. Anim. Physiol. Anim. Nutr.* **2020**, *104*, 1851–1866. [CrossRef]
20. Yang, K.; Qing, Y.; Yu, Q.; Tang, X.; Chen, G.; Fang, R.; Liu, H. By-product feeds: Current understanding and future perspectives. *Agriculture* **2021**, *11*, 207. [CrossRef]
21. Salami, S.A.; Luciano, G.; O'Grady, M.N.; Biondi, L.; Newbold, C.J.; Kerry, J.P.; Priolo, A. Sustainability of feeding plant by-products: A review of the implications for ruminant meat production. *Anim. Feed Sci. Technol.* **2019**, *251*, 37–55. [CrossRef]
22. AAFCO. *Official Feed Terms, Common or Usual Ingredient Names and Ingredient Definitions. Official Publication*; Eyck, R.T., Ed. AAFCO: Champaign, IL, USA, 2021.
23. FAOSTAT. Food and Agriculture Data. 2023. Available online: https://www.fao.org/faostat/en/#home (accessed on 1 December 2022).
24. Gołębiewska, E.; Kalinowska, M.; Yildiz, G. Sustainable Use of Apple Pomace (AP) in Different Industrial Sectors. *Materials* **2022** *15*, 1788. [CrossRef]
25. Rao, M.; Bast, A.; de Boer, A. Valorized Food Processing By-Products in the EU: Finding the Balance between Safety, Nutrition and Sustainability. *Sustainability* **2021**, *13*, 4428. [CrossRef]
26. Aghili, A.H.; Toghyani, M.; Tabeidian, S.A. Effect of incremental levels of apple pomace and multi enzyme on performance immune response, gut development and blood biochemical parameters of broiler chickens. *Int. J. Recycl. Org. Waste Agric.* **2019** *8*, 321–334. [CrossRef]
27. Gullón, B.; Yáñez, R.; Alonso, J.L.; Parajó, J.C. L-Lactic acid production from apple pomace by sequential hydrolysis and fermentation. *Bioresour. Technol.* **2008**, *99*, 308–319. [CrossRef]
28. Colombino, E.; Ferrocino, I.; Biasato, I.; Cocolin, L.S.; Prieto-Botella, D.; Zduńczyk, Z.; Jankowski, J.; Milala, J.; Kosmala, M. Fotschki, B.; et al. Dried fruit pomace inclusion in poultry diet: Growth performance, intestinal morphology and physiology. *J. Anim. Sci. Biotechnol.* **2020**, *11*, 63. [CrossRef]

29. Heidarisafar, Z.; Sadeghi, G.; Karimi, A.; Azizi, O. Apple Peel Waste as a Natural Antioxidant for Heat-Stressed Broiler Chickens. *Trop. Anim. Health Prod.* **2016**, *48*, 831–835. [CrossRef]
30. Ghaemi, H.; Nobakht, A.; Razzaghzadeh, S. The Effect of Apple Pulp and Multi Enzyme on Performance and Blood Parameters in Native Laying Hens. *J. Farm Anim. Physiol. Nutr.* **2014**, *9*, 10–21.
31. Akhlaghi, A.; Ahangari, Y.J.; Zhandi, M.; Peebles, E.D. Reproductive Performance, Semen Quality, and Fatty Acid Profile of Spermatozoa in Senescent Broiler Breeder Roosters as Enhanced by the Long-Term Feeding of Dried Apple Pomace. *Anim. Reprod. Sci.* **2014**, *147*, 64–73. [CrossRef]
32. Fiialovych, L.; Kyryliv, I. Laying Performance, Egg Quality and Hatching Results in Geese Fed with Dry Apple Pomaces. *Acta Sci. Pol. Zootech.* **2016**, *15*, 71–82. [CrossRef]
33. Bampidis, V.A.; Robinson, P.H. Citrus by-Products as Ruminant Feeds: A Review. *Anim. Feed Sci. Technol.* **2006**, *128*, 175–217. [CrossRef]
34. Diarra, S.S. Peel Meals as Feed Ingredients in Poultry Diets: Chemical Composition, Dietary Recommendations and Prospects. *J. Anim. Physiol. Anim. Nutr.* **2018**, *102*, 1284–1295. [CrossRef]
35. Abbasi, H.; Seidavi, A.; Liu, W.; Asadpour, L. Investigation on the Effect of Different Levels of Dried Sweet Orange (*Citrus sinensis*) Pulp on Performance, Carcass Characteristics and Physiological and Biochemical Parameters in Broiler Chicken. *Saudi J. Biol. Sci.* **2015**, *22*, 139–146. [CrossRef]
36. Alzawqari, M.H.; Al-Baddany, A.A.; Al-Baadani, H.H.; Alhidary, I.A.; Khan, R.U.; Aqil, G.M.; Abdurab, A. Effect of Feeding Dried Sweet Orange (*Citrus sinensis*) Peel and Lemon Grass (Cymbopogon Citratus) Leaves on Growth Performance, Carcass Traits, Serum Metabolites and Antioxidant Status in Broiler During the Finisher Phase. *Environ. Sci. Pollut. Res.* **2016**, *23*, 17077–17082. [CrossRef]
37. Ebrahimi, A.; Qotbi, A.A.A.; Seidavi, A. The Effects of Different Levels of Dried *Citrus sinensis* Peel on Broiler Carcass Quality. *Acta Sci. Vet.* **2013**, *41*, 1169.
38. Ebrahimi, A.; Qotbi, A.A.A.; Seidavi, A.; Edens, F.W.; Laudadio, V.; Tufarelli, V. Selected Plasma Constituents of Broiler Chickens Fed Different Levels of Dried Sweet Orange (*Citrus sinensis*) Peels. *J. Anim. Plant Sci.* **2016**, *26*, 949–955.
39. Behera, D.P.; Sethi, A.P.S.; Singh, C.; Singh, U.; Wadhwa, M. Effect of Citrus Waste on Blood Parameters of Broiler Birds with and without Cocktail of Enzymes. *Vet. World* **2019**, *12*, 483–488. [CrossRef]
40. Agu, P.N.; Oluremi, O.I.A.; Tuleun, C.D. Nutritional Evaluation of Sweet Orange (*Citrus sinensis*) Fruit Peel as a Feed Resource in Broiler Production. *Int. J. Poult. Sci.* **2010**, *9*, 684–688. [CrossRef]
41. Diaz-Vargas, M.; Murakami, A.E.; Pintro, P.T.M.; Ospina-Rojas, I.C.; de Souza, C.H.P.; Eyng, C. Dehydrated Citrus Pulp in Broiler Diets. *Can. J. Anim. Sci.* **2018**, *99*, 33–40. [CrossRef]
42. Zoidis, E.; Simitzis, P.; Kampantais, D.; Katsoulas, P.; Pappas, A.C.; Papadomichelakis, G.; Goliomytis, M. Dietary Orange Pulp and Organic Selenium Effects on Growth Performance, Meat Quality, Fatty Acid Profile, and Oxidative Stability Parameters of Broiler Chickens. *Sustainability* **2022**, *14*, 1534. [CrossRef]
43. Mourão, J.L.; Pinheiro, V.M.; Prates, J.A.M.; Bessa, R.J.B.; Ferreira, L.M.A.; Fontes, C.M.G.A.; Ponte, P.I.P. Effect of Dietary Dehydrated Pasture and Citrus Pulp on the Performance and Meat Quality of Broiler Chickens. *Poult. Sci.* **2008**, *87*, 733–743. [CrossRef]
44. Basir, R.; Toghyani, M. Effect of Dietary Graded Levels of Dried Lemon (*Citrus aurantifulia*) Pulp on Performance, Intestinal Morphology, and Humoral Immunity in Broiler Chickens. *Int. J. Recycl. Org. Waste Agric.* **2017**, *6*, 125–132. [CrossRef]
45. Lanza, M.; Fasone, V.; Galofaro, V.; Barbagallo, D.; Bella, M.; Pennisi, P. Citrus Pulp as an Ingredient in Ostrich Diet: Effects on Meat Quality. *Meat Sci.* **2004**, *68*, 269–275. [CrossRef]
46. Wang, C.; Gao, G.L.; Huang, J.X.; Zhang, K.S.; Zhong, H.; Wang, H.W.; Su, J.; Xie, M.; Wang, Q.G. Nutritive Value of Dry Citrus Pulp and Its Effect on Performance in Geese from 35 to 70 Days of Age. *J. Appl. Poult. Res.* **2017**, *26*, 253–259. [CrossRef]
47. Goliomytis, M.; Kostaki, A.; Avgoulas, G.; Lantzouraki, D.Z.; Siapi, E.; Zoumpoulakis, P.; Simitzis, P.; Deligeorgis, S.G. Dietary Supplementation with Orange Pulp (*Citrus sinensis*) Improves Egg Yolk Oxidative Stability in Laying Hens. *Anim. Feed Sci. Technol.* **2018**, *244*, 28–35. [CrossRef]
48. Nazok, A.; Rezaei, M.; Sayyahzadeh, H. Effect of Different Levels of Dried Citrus Pulp on Performance, Egg Quality, and Blood Parameters of Laying Hens in Early Phase of Production. *Trop. Anim. Health Prod.* **2010**, *42*, 737–742. [CrossRef]
49. Florou-Paneri, P.; Babidis, V.; Kufidis, D.; Christaki, E.; Spais, A.B. Effect of Feeding Dried Citrus Pulp on Quail Laying Performance and Some Egg Quality Characteristics. *Arch. Geflügelk.* **2001**, *65*, 178–181.
50. Schieber, A.; Stintzing, F.C.; Carle, R. By-Products of Plant Food Processing as a Source of Functional Compounds—Recent Developments. *Trends Food Sci. Technol.* **2001**, *12*, 401–413. [CrossRef]
51. Santana-Méridas, O.; González-Coloma, A.; Sánchez-Vioque, R. Agricultural Residues as a Source of Bioactive Natural Products. *Phytochem. Rev.* **2012**, *11*, 447–466. [CrossRef]
52. Fermoso, F.G.; Serrano, A.; Alonso-Fariñas, B.; Fernández-Bolaños, J.; Borja, R.; Rodríguez-Gutiérrez, G. Valuable Compound Extraction, Anaerobic Digestion, and Composting: A Leading Biorefinery Approach for Agricultural Wastes. *J. Agric. Food Chem.* **2018**, *66*, 8451–8468. [CrossRef]
53. Bocco, A.; Cuvelier, M.-E.; Richard, H.; Berset, C. Antioxidant Activity and Phenolic Composition of Citrus Peel and Seed Extracts. *J. Agric. Food Chem.* **1998**, *46*, 2123–2129. [CrossRef]

54. Coll, M.D.; Coll, L.; Laencina, J.; Tomás-Barberán, F.A. Recovery of Flavanones from Wastes of Industrially Processed Lemons. *Z. Naturforchung* **1998**, *206*, 404–407. [CrossRef]
55. Li, J.; Trinh, H.K.; Mirmajlessi, S.M.; Haesaert, G.; Xhaferi, R.; Delaere, I.; Höfte, M.; Raymaekers, K.; Cammue, B.P.A.; Jonckheere, W.; et al. Biopesticide and Plant Growth-Promoting Activity in Maize Distillers' Dried Grains with Solubles. *Ind. Crops Prod.* **2023**, *193*, 116175. [CrossRef]
56. Lu, Y.; Foo, L.Y. The Polyphenol Constituents of Grape Pomace. *Food Chem.* **1999**, *65*, 1–8. [CrossRef]
57. Trikas, E.D.; Melidou, M.; Papi, R.M.; Zachariadis, G.A.; Kyriakidis, D.A. Extraction, Separation and Identification of Anthocyanins from Red Wine by-Product and Their Biological Activities. *J. Funct. Foods* **2016**, *25*, 548–558. [CrossRef]
58. Obied, H.K.; Bedgood, D.R.J.; Prenzler, P.D.; Robards, K. Effect of Processing Conditions, Prestorage Treatment, and Storage Conditions on the Phenol Content and Antioxidant Activity of Olive Mill Waste. *J. Agric. Food Chem.* **2008**, *56*, 3925–3932. [CrossRef]
59. Kalogeropoulos, N.; Chiou, A.; Pyriochou, V.; Peristeraki, A.; Karathanos, V.T. Bioactive Phytochemicals in Industrial Tomatoes and Their Processing Byproducts. *LWT—Food Sci. Technol.* **2012**, *49*, 213–216. [CrossRef]
60. Čolović, D.; Rakita, S.; Banjac, V.; Đuragić, O.; Čabarkapa, I. Plant Food by-Products as Feed: Characteristics, Possibilities, Environmental Benefits, and Negative Sides. *Food Rev. Int.* **2019**, *35*, 363–389. [CrossRef]
61. Ditta, Y.A.; King, A.J. Recent Advances in Sunflower Seed Meal as an Alternate Source of Protein in Broilers. *Worlds Poult. Sci. J.* **2017**, *73*, 527–542. [CrossRef]
62. Golob, P.; Farrell, G.; Orchard, J.E. Crop Post-Harvest: Principles and Practice. In *Crop Post-Harvest: Science and Technology*; Farell Golob, G.P., Orchard, J.E., Eds.; John Wiley & Sons: Amsterdam, The Netherlands, 2002; Available online: https://docplayer.net/49528915-Crop-post-harvest-science-and-technology.html (accessed on 1 December 2022).
63. Pedrosa, M.M.; Muzquiz, M.; García-Vallejo, C.; Burbano, C.; Cuadrado, C.; Ayet, G.; Robredo, L.M. Determination of Caffeic and Chlorogenic Acids and Their Derivatives in Different Sunflower Seeds. *J. Sci. Food Agric.* **2000**, *80*, 459–464. [CrossRef]
64. Senkoylu, N.; Dale, N. Sunflower Meal in Poultry Diets: A Review. *Worlds Poult. Sci. J.* **2007**, *55*, 153–174. [CrossRef]
65. Sangsoponjit, S.; Suphalucksana, W.; Srikijkasemwat, K. Effect of Feeding Sunflower Meal on the Performance and Carcass Characteristics of Broiler Chickens. *Chem. Eng. Trans.* **2017**, *58*, 841–846.
66. Moghaddam, H.N.; Salari, S.; Arshami, J.; Golian, A.; Maleki, M. Evaluation of the Nutritional Value of Sunflower Meal and Its Effect on Performance, Digestive Enzyme Activity, Organ Weight, and Histological Alterations of the Intestinal Villi of Broiler Chickens. *J. Appl. Poult. Res.* **2012**, *21*, 293–304. [CrossRef]
67. Araújo, L.F.; da Silva Araújo, C.S.; Petroli, N.B.; de Laurentiz, A.C.; de Albuquerque, R.; de Trindade Neto, M.A. Sunflower Meal for Broilers of 22 to 42 Days of Age. *R. Bras. Zootec.* **2011**, *40*, 2142–2146. [CrossRef]
68. Araújo, W.A.G.; Albino, L.F.T.; Rostagno, H.S.; Hannas, M.I.; Pessoa, G.B.S.; Messias, R.K.G.; Lelis, G.R.; Ribeiro, V., Jr. Sunflower Meal and Enzyme Supplementation of the Diet of 21- to 42-D-Old Broilers. *Braz. J. Poult. Sci.* **2014**, *16*, 17–24. [CrossRef]
69. Tavernari, F.C.; Albino, L.F.T.; Morata, R.L. Inclusion of Sunflower Meal, with or without Enzyme Supplementation, in Broiler Diets. *Braz. J. Poult. Sci.* **2008**, *10*, 233–238. [CrossRef]
70. Gerzilov, V.; Petrov, P.B. Effects of Partial Substitution of Soybean Meal with High Protein Sunflower Meal in Broiler Diets. *Bulg. J. Agric. Sci.* **2022**, *28*, 151–157.
71. Kyrkelanov, N.; Chobanova, S.; Atanasoff, A. Investigation of Possible Use of Compound Feeds with Different Level of High-Protein Sunflower Meal in Broiler Chickens Nutrition. *Bulg. J. Agric. Sci.* **2020**, *26*, 121–125.
72. Chobanova, S. Effects of Compound Poultry Feed with Different Content of High-Protein Sunfl Ower Meal on Growth Performance of Broiler Chickens. *Bulg. J. Agric. Sci.* **2019**, *25*, 91–94.
73. Ciurescu, G.; Vasilachi, A.; Grigore, D.; Grosu, H. Growth Performance, Carcass Traits, and Blood Biochemistry of Broiler Chicks Fed with Low-Fibre Sunflower Meal and Phytase. *S. Afr. J. Anim. Sci.* **2019**, *49*, 4. [CrossRef]
74. Berwanger, E.; Nunes, R.V.; De Oliveira, T.M.M.; Bayerle, D.F.; Bruno, L.D.G. Performance and Carcass Yield of Broilers Fed Increasing Levels of Sunflower Cake. *Rev. Caatinga* **2017**, *30*, 201–212. [CrossRef]
75. De Oliveira, T.M.M.; Nunes, R.V.; Eyng, C.; Berwanger, E.; Bayerle, D.F. Sunflower Meal and Exogenous Enzymes in Initial Diets for Broilers. *Rev. Caatinga* **2016**, *29*, 996–1005. [CrossRef]
76. Horvatovic, M.P.; Glamocic, D.; Zikic, D.; Hadnadjev, T.D. Performance and Some Intestinal Functions of Broilers Fed Diets with Different Inclusion Levels of Sunflower Meal and Supplemented or Not with Enzymes. *Rev. Bras. Cienc. Avic.* **2015**, *17*, 25–30. [CrossRef]
77. Shi, S.R.; Lu, J.; Tong, H.B.; Zou, J.M.; Wang, K.H. Effects of Graded Replacement of Soybean Meal by Sunflower Seed Meal in Laying Hen Diets on Hen Performance, Egg Quality, Egg Fatty Acid Composition, and Cholesterol Content. *J. Appl. Poult. Res.* **2012**, *21*, 367–374. [CrossRef]
78. Das, S.K.; Biswas, A.; Neema, R.P.; Maity, B. Effect of Soybean Meal Substitution by Different Concentrations of Sunflower Meal on Egg Quality Traits of White and Coloured Dwarf Dam Lines. *Br. Poult. Sci.* **2010**, *51*, 427–433. [CrossRef]
79. Wadhwa, M.; Bakshi, M.P.S. Chapter 10—Application of Waste-Derived Proteins in the Animal Feed Industry. In *Protein Byproducts*; Dhillon, G.S., Ed.; Academic Press: Cambridge, MA, USA, 2016; pp. 161–192. [CrossRef]
80. Fan, W.; Sun, X.; Cui, G.; Li, Q.; Xu, Y.; Wang, L.; Li, X.; Hu, B.; Chi, Z. A Strategy of Co-Fermentation of Distillers Dried Grains with Solubles (Ddgs) and Lignocellulosic Feedstocks as Swine Feed. *Crit. Rev. Biotechnol.* **2022**, 1–15. [CrossRef]

81. Noblet, J.; Cozannet, P.; Skiba, F. Nutritional Value and Utilization of Wheat Dried Distillers Grain with Solubles in Pigs and Poultry. In *Biofuel Co-Products as Livestock Feed—Opportunities and Challenges*; Makkar, H.P.S., Ed.; FAO: Rome, Italy, 2012; pp. 163–174.
82. Salim, H.M.; Kruk, Z.A.; Lee, B.D. Nutritive Value of Corn Distillers Dried Grains with Solubles as an Ingredient of Poultry Diets: A Review. *Worlds Poult. Sci. J.* **2010**, *66*, 411–432. [CrossRef]
83. Pinotti, L.; Ottoboni, M.; Giromini, C.; Dell'Orto, V.; Cheli, F. Mycotoxin Contamination in the E. U. Feed Supply Chain: A Focus on Cereal Byproducts. *Toxins* **2016**, *8*, 45. [CrossRef]
84. Shim, M.Y.; Pesti, G.M.; Bakalli, R.I.; Tillman, P.B.; Payne, R.L. Evaluation of Corn Distillers Dried Grains with Solubles as an Alternative Ingredient for Broilers. *Poult. Sci.* **2011**, *90*, 369–376. [CrossRef]
85. Damasceno, J.L.; Rocha, C.S.; Eyng, C.; Broch, J.; Savaris, V.D.L.; Wachholz, L.; Tesser, G.L.S.; Avila, A.S.; Pacheco, W.J.; Nunes, R.V. Corn Distillers' Dried Grains with Solubles to Feed Broiler Chickens from 22 to 42 d of Age. *J. Appl. Poult. Res.* **2020**, *29*, 573–583. [CrossRef]
86. Loar, R.E.; Moritz, J.S.; Donaldson, J.R.; Corzo, A. Effects of Feeding Distillers Dried Grains with Solubles to Broilers from 0 to 28 Days Posthatch on Broiler Performance, Feed Manufacturing Efficiency, and Selected Intestinal Characteristics. *Poult. Sci.* **2010**, *89*, 2242–2250. [CrossRef]
87. Loar, R.E.; Donaldson, J.R.; Corzo, A. Effects of Feeding Distillers Dried Grains with Solubles to Broilers from 0 to 42 Days Posthatch on Broiler Performance, Carcass Characteristics, and Selected Intestinal Characteristics. *J. Appl. Poult. Res.* **2012**, *21*, 48–62. [CrossRef]
88. Min, Y.N.; Li, L.; Waldroup, P.W.; Niu, Z.Y.; Wang, Z.P.; Gao, Y.P.; Liu, F.Z. Effects of Dietary Distillers Dried Grains with Solubles Concentrations on Meat Quality and Antioxidant Status and Capacity of Broiler Chickens. *J. Appl. Poult. Res.* **2012**, *21*, 603–611. [CrossRef]
89. Schilling, M.W.; Battula, V.; Loar, R.E.; Jackson, V.; Kin, S.; Corzo, A. Dietary Inclusion Level Effects of Distillers Dried Grains with Solubles on Broiler Meat Quality. *Poult. Sci.* **2010**, *89*, 752–760. [CrossRef]
90. Fries-Craft, K.; Bobeck, E.A. Evaluation of a High-Protein Ddgs Product in Broiler Chickens: Performance, Nitrogen-Corrected Apparent Metabolisable Energy, and Standardised Ileal Amino Acid Digestibility. *Br. Poult. Sci.* **2019**, *60*, 749–756. [CrossRef]
91. Guney, A.C.; Shim, M.Y.; Batal, A.B.; Dale, N.M.; Pesti, G.M. Effect of Feeding Low-Oil Distillers Dried Grains with Solubles on the Performance of Broilers. *Poult. Sci.* **2013**, *92*, 2070–2076. [CrossRef]
92. Benavides, P.T.; Cai, H.; Wang, M.; Bajjalieh, N. Life-Cycle Analysis of Soybean Meal, Distiller-Dried Grains with Solubles, and Synthetic Amino Acid-Based Animal Feeds for Swine and Poultry Production. *Anim. Feed Sci. Technol.* **2020**, *268*, 114607. [CrossRef]
93. Loar, R.E.; Schilling, M.W.; McDaniel, C.D.; Coufal, C.D.; Rogers, S.F.; Karges, K.; Corzo, A. Effect of Dietary Inclusion Level of Distillers Dried Grains with Solubles on Layer Performance, Egg Characteristics, and Consumer Acceptability. *J. Appl. Poult. Res.* **2010**, *19*, 30–37. [CrossRef]
94. Masa'deh, M.K.; Purdum, S.E.; Hanford, K.J. Dried Distillers Grains with Solubles in Laying Hen Diets Phosphorus. *Poult. Sci.* **2011**, *90*, 1960–1966. [CrossRef]
95. El-Hack, M.E.A.; Mahrose, K.M.; Attia, F.A.M.; Swelum, A.A.; Taha, A.E.; Shewita, R.S.; Hussein, E.-S.O.S.; Alowaimer, A.N. Laying Performance, Physical, and Internal Egg Quality Criteria of Hens Fed Distillers Dried Grains with Solubles and Exogenous Enzyme Mixture. *Animals* **2019**, *9*, 150. [CrossRef]
96. Shalash, S.M.M.; El-Wafa, S.A.; Hassan, R.A.; Ramadan, N.A.; Mohamed, M.S.; El-Gabry, H.E. Evaluation of Distillers Dried Grains with Solubles as Feed Ingredient in Laying Hen Diets. *Int. J. Poult. Sci.* **2010**, *9*, 537–545. [CrossRef]
97. Sun, H.; Lee, E.J.; Samaraweera, H.; Persia, M.; Ahn, D.U. Effects of Increasing Concentrations of Corn Distillers Dried Grains with Solubles on Chemical Composition and Nutrient Content of Egg. *Poult. Sci.* **2013**, *92*, 233–242. [CrossRef]
98. Abd El-Hack, M.E.; El-Hindawy, M.M.; Attia, A.I.; Mahrose, K.M. Does the Use of Distiller's Dried Grains with Solubles (Ddgs) in Layer Diets Affect the Nutrients Digestibility and Manure Pollution by Nitrogen and Phosphorous? *Environ. Sci. Pollut. Res.* **2017**, *24*, 13335–13343. [CrossRef]
99. Wu-Haan, W.; Powers, W.; Angel, R.; Applegate, T.J. The Use of Distillers Dried Grains Plus Solubles as a Feed Ingredient on Air Emissions and Performance from Laying Hens. *Poult. Sci.* **2010**, *89*, 1355–1359. [CrossRef]
100. Ding, X.M.; Qi, Y.Y.; Zhang, K.Y.; Tian, G.; Bai, S.P.; Wang, J.P.; Peng, H.W.; Lv, L.; Xuan, Y.; Zeng, Q.F. Corn Distiller's Dried Grains with Solubles as an Alternative Ingredient to Corn and Soybean Meal in Pekin Duck Diets Based on Its Predicted Ame and the Evaluated Standardized Ileal Digestibility of Amino Acids. *Poult. Sci.* **2022**, *101*, 101974. [CrossRef]
101. Sirohi, R.; Tarafdar, A.; Singh, S.; Negi, T.; Gaur, V.K.; Gnansounou, E.; Bharathiraja, B. Green Processing and Biotechnological Potential of Grape Pomace: Current Trends and Opportunities for Sustainable Biorefinery. *Bioresour. Technol.* **2020**, *314*, 123771. [CrossRef]
102. Brenes, A.; Viveros, A.; Chamorro, S.; Arija, I. Use of Polyphenol-Rich Grape by-Products in Monogastric Nutrition. A Review. *Anim. Feed Sci. Technol.* **2016**, *211*, 1–17. [CrossRef]
103. Mavrommatis, A.; Giamouri, E.; Myrtsi, E.D.; Evergetis, E.; Filippi, K.; Papapostolou, H.; Koulocheri, S.D.; Zoidis, E.; Pappas, A.C.; Koutinas, A.; et al. Antioxidant Status of Broiler Chickens Fed Diets Supplemented with Vinification by-Products: A Valorization Approach. *Antioxidants* **2021**, *10*, 1250. [CrossRef]

104. Costa, M.M.; Alfaia, C.M.; Lopes, P.A.; Pestana, J.M.; Prates, J.A.M. Grape by-Products as Feedstuff for Pig and Poultry Production. *Animals* **2022**, *12*, 2239. [CrossRef]
105. Kumanda, C.; Mlambo, V.; Mnisi, C.M. Valorization of Red Grape Pomace Waste Using Polyethylene Glycol and Fibrolytic Enzymes: Physiological and Meat Quality Responses in Broilers. *Animals* **2019**, *9*, 779. [CrossRef]
106. Van Niekerk, R.F.; Mnisi, C.M.; Mlambo, V. Polyethylene Glycol Inactivates Red Grape Pomace Condensed Tannins for Broiler Chickens. *Br. Poult. Sci.* **2020**, *61*, 566–573. [CrossRef]
107. Aditya, S.; Ohh, S.-J.; Ahammed, M.; Lohakare, J. Supplementation of Grape Pomace (Vitis Vinifera) in Broiler Diets and Its Effect on Growth Performance, Apparent Total Tract Digestibility of Nutrients, Blood Profile, and Meat Quality. *Anim. Nutr.* **2018**, *4*, 210–214. [CrossRef]
108. Goñi, I.; Brenes, A.; Centeno, C.; Viveros, A.; Saura-Calixto, F.; Rebolé, A.; Arija, I.; Estevez, R. Effect of Dietary Grape Pomace and Vitamin E on Growth Performance, Nutrient Digestibility, and Susceptibility to Meat Lipid Oxidation in Chickens. *Poult. Sci.* **2007**, *86*, 508–516. [CrossRef]
109. Gungor, E.; Altop, A.; Erener, G. Effect of Raw and Fermented Grape Pomace on the Growth Performance, Antioxidant Status, Intestinal Morphology, and Selected Bacterial Species in Broiler Chicks. *Animals* **2021**, *11*, 364. [CrossRef]
110. Turcu, R.P.; Panaite, T.D.; Untea, A.E.; Șoica, C.; Iuga, M.; Mironeasa, S. Effects of Supplementing Grape Pomace to Broilers Fed Polyunsaturated Fatty Acids Enriched Diets on Meat Quality. *Animals* **2020**, *10*, 947. [CrossRef]
111. Hosseini-Vashan, S.J.; Safdari-Rostamabad, M.; Piray, A.H.; Sarir, H. The Growth Performance, Plasma Biochemistry Indices, Immune System, Antioxidant Status, and Intestinal Morphology of Heat-Stressed Broiler Chickens Fed Grape (Vitis Vinifera) Pomace. *Anim. Feed Sci. Technol.* **2020**, *259*, 114343. [CrossRef]
112. Kumanda, C.; Mlambo, V.; Mnisi, C.M. From Landfills to the Dinner Table: Red Grape Pomace Waste as a Nutraceutical for Broiler Chickens. *Sustainability* **2019**, *11*, 1931. [CrossRef]
113. Haščík, P.; Čech, M.; Čuboň, J.; Bobko, M.; Arpášová, H.; Pavelková, A.; Kačániová, M.; Tkáčová, J.; Čeryová, N. Effect of Grape Pomace Supplementation on Meat Performance of Broiler Chicken Ross 308. *J. Microbiol. Biotechnol. Food Sci.* **2020**, *10*, 140–144.
114. Jurčaga, L.; Bobko, M.; Haščík, P.; Bobková, A.; Demianová, A.; Belej, Ľ.; Kročko, M. Effect of Dietary Red Grape Pomace on Lipid Oxidation in Meat of Broiler Chickens. *J. Microbiol. Biotechnol. Food Sci.* **2021**, *10*, e3769. [CrossRef]
115. Bennato, F.; Di Luca, A.; Martino, C.; Ianni, A.; Marone, E.; Grotta, L.; Ramazzotti, S.; Cichelli, A.; Martino, G. Influence of Grape Pomace Intake on Nutritional Value, Lipid Oxidation and Volatile Profile of Poultry Meat. *Foods* **2020**, *9*, 508. [CrossRef]
116. Reyes, P.; Urquiaga, I.; Echeverría, G.; Durán, E.; Morales, M.S.; Valenzuela, C. Wine Grape Pomace Flour in Broiler Diets Effects Growth and Some Meat Characteristics. *Anim. Prod. Sci* **2020**, *60*, 1210–1216. [CrossRef]
117. Kasapidou, E.; Sossidou, E.N.; Zdragas, A.; Papadaki, C.; Vafeas, G.; Mitlianga, P. Effect of Grape Pomace Supplementation on Broiler Meat Quality Characteristics. *Eur. Poult. Sci.* **2016**, *80*, 135–142. [CrossRef]
118. Brenes, A.; Viveros, A.; Goñi, I.; Centeno, C.; Sáyago-Ayerdy, S.G.; Arija, I.; Saura-Calixto, F. Effect of Grape Pomace Concentrate and Vitamin E on Digestibility of Polyphenols and Antioxidant Activity in Chickens. *Poult. Sci.* **2008**, *87*, 307–316. [CrossRef]
119. Lichovnikova, M.; Kalhotka, L.; Adam, V.; Klejdus, B.; Anderle, V. The Effects of Red Grape Pomace Inclusion in Grower Diet on Amino Acid Digestibility, Intestinal Microflora, and Sera and Liver Antioxidant Activity in Broilers. *Turkish J. Vet. Anim. Sci.* **2015**, *39*, 406–412. [CrossRef]
120. Ebrahimzadeh, S.K.; Navidshad, B.; Farhoomand, P.; Aghjehgheshlagh, F.M. Effects of Grape Pomace and Vitamin E on Performance, Antioxidant Status, Immune Response, Gut Morphology and Histopathological Responses in Broiler Chickens. *S Afr. J. Anim. Sci.* **2018**, *48*, 324–336. [CrossRef]
121. Kara, K.; Kocaoğlu Güçlü, B.; Baytok, E.; Şentürk, M. Effects of Grape Pomace Supplementation to Laying Hen Diet on Performance, Egg Quality, Egg Lipid Peroxidation and Some Biochemical Parameters. *J. Appl. Anim. Res.* **2016**, *44*, 303–310. [CrossRef]
122. Reis, J.H.; Gebert, R.R.; Barreta, M.; Boiago, M.M.; Souza, C.F.; Baldissera, M.D.; Santos, I.D.; Wagner, R.; Laporta, L.V.; Stefani, L.M.; et al. Addition of Grape Pomace Flour in the Diet on Laying Hens in Heat Stress: Impacts on Health and Performance as Well as the Fatty Acid Profile and Total Antioxidant Capacity in the Egg. *J. Therm. Biol.* **2019**, *80*, 141–149. [CrossRef]
123. Fróes, H.G.; Jácome, I.M.T.D.; Tavares, R.A.; Garcia, R.G.; Domingues, C.H.F.; Bevilaqua, T.M.S.; Martinelli, M.; Naas, I.A.; Borille, R. Grape (Vitis Vinifera) Pomace Flour as Pigment Agent of Quail Eggs. *Braz. J. Poult. Sci.* **2018**, *20*, 183–188. [CrossRef]
124. Khdair, A.; Abu-Rumman, G. Sustainable Environmental Management and Valorization Options for Olive Mill Byproducts in the Middle East and North Africa (Mena) Region. *Processes* **2020**, *8*, 671. [CrossRef]
125. Khwaldia, K.; Attour, N.; Matthes, J.; Beck, L.; Schmid, M. Olive Byproducts and Their Bioactive Compounds as a Valuable Source for Food Packaging Applications. *Compr. Rev. Food Sci. Food Saf.* **2022**, *21*, 1218–1253. [CrossRef]
126. Erinle, T.J.; Adewole, D.I. Fruit Pomaces—Their Nutrient and Bioactive Components, Effects on Growth and Health of Poultry Species, and Possible Optimization Techniques. *Anim. Nutr.* **2022**, *9*, 357–377. [CrossRef]
127. Pečjak, M.; Levart, A.; Salobir, J.; Rezar, V. Effect of the Supplementation of Olive Leaves and Olive Cake on Growth Performance and Bone Mineralisation of Broiler Chickens. *Acta Fytotech. Zootech.* **2020**, *23*, 105–111. [CrossRef]
128. Al-Harthi, M.A. The Efficacy of Using Olive Cake as a by-Product in Broiler Feeding with or without Yeast. *Ital. J. Anim. Sci.* **2016**, *15*, 512–520. [CrossRef]
129. Al-Harthi, M. The Effect of Olive Cake, with or without Enzymes Supplementation, on Growth Performance, Carcass Characteristics, Lymphoid Organs and Lipid Metabolism of Broiler Chickens. *Braz. J. Poult. Sci.* **2017**, *19*, 83–90. [CrossRef]

130. Al-Harthi, M.A.; Attia, Y.A.; El-Shafey, A.S.; Elgandy, M.F. Impact of Phytase on Improving the Utilisation of Pelleted Broiler Diets Containing Olive by-Products. *Ital. J. Anim. Sci.* **2020**, *19*, 310–318. [CrossRef]
131. Al-Harthi, M.A.; Attia, Y.A. Effect of Citric Acid on the Nutritive Value of Olive Cake in Broiler Diets. *Eur. Poult. Sci.* **2016**, *80*, 153. [CrossRef]
132. Saleh, A.A.; Paray, B.A.; Dawood, M.A.O. Olive Cake Meal and Bacillus Licheniformis Impacted the Growth Performance, Muscle Fatty Acid Content, and Health Status of Broiler Chickens. *Animals* **2020**, *10*, 695. [CrossRef]
133. Rebollada-Merino, A.; Ugarte-Ruiz, M.; Hernández, M.; Miguela-Villoldo, P.; Abad, D.; Cuesta-Álvaro, P.; Rodríguez-Lázaro, D.; de Juan, L.; Domínguez, L.; Rodríguez-Bertos, A. Dietary Supplementation with Fermented Defatted "Alperujo" Induces Modifications of the Intestinal Mucosa and Cecal Microbiota of Broiler Chickens. *Poult. Sci.* **2020**, *99*, 5308–5315. [CrossRef]
134. Tufarelli, V.; Passantino, L.; Zupa, R.; Crupi, P.; Laudadio, V. Suitability of Dried Olive Pulp in Slow-Growing Broilers: Performance, Meat Quality, Oxidation Products and Intestinal Mucosa Features. *Poult. Sci.* **2022**, *101*, 102230. [CrossRef]
135. Papadomichelakis, G.; Pappas, A.C.; Tsiplakou, E.; Symeon, G.K.; Sotirakoglou, K.; Mpekelis, V.; Fegeros, K.; Zervas, G. Effects of Dietary Dried Olive Pulp Inclusion on Growth Performance and Meat Quality of Broiler Chickens. *Livest. Sci.* **2019**, *221*, 115–122. [CrossRef]
136. Sayehban, P.; Seidavi, A.; Dadashbeiki, M.; Ghorbani, A.; de Araújo, W.A.; Durazzo, A.; Lucarini, M.; Gabrielli, P.; Omri, B.; Teixeira Albino, L.F.; et al. Olive Pulp and Exogenous Enzymes Feed Supplementation Effect on the Carcass and Offal in Broilers: A Preliminary Study. *Agriculture* **2020**, *10*, 359. [CrossRef]
137. Sayehban, P.; Seidavi, A.; Dadashbeiki, M.; Ghorbani, A.; Araújo, W.A.G.; Albino, L.F.T. Effects of Different Dietary Levels of Two Types of Olive Pulp and Exogenous Enzyme Supplementation on the Gastrointestinal Tract Size, Immunology and Hematology of Broilers. *Braz. J. Poult. Sci.* **2015**, *17*, 73–85. [CrossRef]
138. Branciari, R.; Galarini, R.; Giusepponi, D.; Trabalza-Marinucci, M.; Forte, C.; Roila, R.; Miraglia, D.; Servili, M.; Acuti, G.; Valiani, A. Oxidative Status and Presence of Bioactive Compounds in Meat from Chickens Fed Polyphenols Extracted from Olive Oil Industry Waste. *Sustainability* **2017**, *9*, 1566. [CrossRef]
139. Nasopoulou, C.; Lytoudi, K.; Zabetakis, I. Evaluation of Olive Pomace in the Production of Novel Broilers with Enhanced in Vitro Antithrombotic Properties. *Eur. J. Lipid Sci. Technol.* **2018**, *120*, 1700290. [CrossRef]
140. Pappas, A.C.; Tsiplakou, E.; Papadomichelakis, G.; Mitsiopoulou, C.; Sotirakoglou, K.; Mpekelis, V.; Haroutounian, S.A.; Fegeros, K.; Zervas, G. Effects of Olive Pulp Addition to Broiler Diets on Performance, Selected Biochemical Parameters and Antioxidant Enzymes. *J. Hell. Vet. Med.* **2019**, *70*, 1687–1696. [CrossRef]
141. Sayehban, P.; Seidavi, A.; Dadashbeiki, M.; Ghorbani, A.; Araújo, W.A.G.; Albino, L.F.T. Effects of Different Levels of Two Types of Olive Pulp with or without Exogenous Enzyme Supplementation on Broiler Performance and Economic Parameters. *Braz. J. Poult. Sci.* **2016**, *18*, 489–500. [CrossRef]
142. Sateri, S.; Seidavi, A.; Bouyeh, M.; Kutzler, M.; Neumann, P.; Laudadio, V.; Loperfido, F.; Tufarelli, V. Effect of Olive Meal and Supplemental Enzymes on Performance Traits, Blood Biochemistry, Humoral Immunity Response and Caecal Microbiota of Broilers. *S. Afr. J. Anim. Sci.* **2017**, *47*, 804–812. [CrossRef]
143. Gerasopoulos, K.; Stagos, D.; Kokkas, S.; Petrotos, K.; Kantas, D.; Goulas, P.; Kouretas, D. Feed Supplemented with Byproducts from Olive Oil Mill Wastewater Processing Increases Antioxidant Capacity in Broiler Chickens. *Food Chem. Toxicol.* **2015**, *82*, 42–49. [CrossRef]
144. Bonos, E.; Skoufos, I.; Petrotos, K.; Giavasis, I.; Mitsagga, C.; Fotou, K.; Vasilopoulou, K.; Giannenas, I.; Gouva, E.; Tsinas, A.; et al. Innovative Use of Olive, Winery and Cheese Waste by-Products as Functional Ingredients in Broiler Nutrition. *Vet. Sci.* **2022**, *9*, 290. [CrossRef]
145. Petrotos, K.; Papaioannou, C.; Kokkas, S.; Gkoutsidis, P.; Skoufos, I.; Tzora, A.; Bonos, E.; Tsinas, A.; Giavasis, I.; Mitsagga, C. Optimization of the Composition of a Novel Bioactive Silage Produced by Mixing of Ground Maize Grains with Olive Mill Waste Waters, Grape Pomace and Feta Cheese Whey. *AgriEngineering* **2021**, *3*, 868–893. [CrossRef]
146. Herrero-Encinas, J.; Blanch, M.; Pastor, J.J.; Mereu, A.; Ipharraguerre, I.R.; Menoyo, D. Effects of a Bioactive Olive Pomace Extract from Olea Europaea on Growth Performance, Gut Function, and Intestinal Microbiota in Broiler Chickens. *Poult. Sci.* **2020**, *99*, 2–10. [CrossRef]
147. Herrero-Encinas, J.; Blanch, M.; Pastor, J.J.; Menoyo, D. Diet Supplementation with a Bioactive Pomace Extract from Olea Europaea Partially Mitigates Negative Effects on Gut Health Arising from a Short-Term Fasting Period in Broiler Chickens. *Animals* **2020**, *10*, 349. [CrossRef]
148. Afsari, M.; Mohebbifar, A.; Torki, M. Effects of Dietary Inclusion of Olive Pulp Supplemented with Probiotics on Productive Performance, Egg Quality and Blood Parameters of Laying Hens. *Annu. Res. Rev. Biol.* **2013**, *4*, 198–211. [CrossRef]
149. Zangeneh, S.; Torki, M. Effects of B-Mannanase Supplementing of Olive Pulp Included Diet on Performance of Laying Hens, Egg Quality Characteristics, Humoral and Cellular Immune Response and Blood Parameters. *Glob. Vet.* **2011**, *7*, 391–398.
150. Zarei, M.; Ehsani, M.; Torki, M. Productive Performance of Laying Hens Fed Wheat-Based Diets Included Olive Pulp with or without a Commercial Enzyme Product. *Afr. J. Biotechnol.* **2011**, *10*, 4303–4312.
151. Al-Harthi, M.A.; Attia, Y.A. Effect of Citric Acid on the Utilization of Olive Cake Diets for Laying Hens. *Ital. J. Anim. Sci.* **2015**, *14*, 3966. [CrossRef]
152. Ozcan, C.; Cimrin, T.; Yakar, Y.; Alasahan, S. Effects of Olive Cake Meal on Serum Constituents and Fatty Acid Levels in Breast Muscle of Japanese Quail. *S. Afr. J. Anim. Sci.* **2020**, *50*, 874–880. [CrossRef]

153. Abd El-Moneim, A.E.-M.E.; Sabic, E.M.; Abu-Taleb, A.M. Influence of Dietary Supplementation of Irradiated or Non-Irradiated Olive Pulp on Biochemical Profile, Antioxidant Status and Immune Response of Japanese Quails. *Biol. Rhythm Res.* **2019**, *53*, 519–534. [CrossRef]
154. Abd El-Moneim, A.E.; Sabic, E.M. Beneficial Effect of Feeding Olive Pulp and Aspergillus Awamori on Productive Performance, Egg Quality, Serum/Yolk Cholesterol and Oxidative Status in Laying Japanese Quails. *J. Anim. Feed. Sci.* **2019**, *28*, 52–61. [CrossRef]
155. Valenti, B.; Luciano, G.; Morbidini, L.; Rossetti, U.; Codini, M.; Avondo, M.; Priolo, A.; Bella, M.; Natalello, A.; Pauselli, M. Dietary Pomegranate Pulp: Effect on Ewe Milk Quality During Late Lactation. *Animals* **2019**, *9*, 283. [CrossRef]
156. Cano-Lamadrid, M.; Martínez-Zamora, L.; Castillejo, N.; Artés-Hernández, F. From Pomegranate Byproducts Waste to Worth: A Review of Extraction Techniques and Potential Applications for Their Revalorization. *Foods* **2022**, *11*, 2596. [CrossRef]
157. Ahmed, S.T.; Islam, M.M.; Bostami, A.B.M.R.; Mun, H.-S.; Kim, Y.-J.; Yang, C.-J. Meat Composition, Fatty Acid Profile and Oxidative Stability of Meat from Broilers Supplemented with Pomegranate (*Punica granatum* L.) by-Products. *Food Chem.* **2015**, *188*, 481–488. [CrossRef]
158. Bostami, A.B.M.R.; Ahmed, S.T.; Islam, M.M.; Mun, H.S.; Ko, S.Y.; Kim, S.S.; Yang, C.J. Growth Performance, Fecal Noxious Gas Emission and Economic Efficacy in Broilers Fed Fermented Pomegranate Byproducts as Residue of Fruit Industry. *Int. J. Adv. Res.* **2015**, *3*, 102–114.
159. Abdel Baset, S.; Ashour, E.A.; Abd El-Hack, M.E.; El-Mekkawy, M.M. Effect of Different Levels of Pomegranate Peel Powder and Probiotic Supplementation on Growth, Carcass Traits, Blood Serum Metabolites, Antioxidant Status and Meat Quality of Broilers. *Anim. Biotechnol.* **2020**, *33*, 690–700. [CrossRef] [PubMed]
160. Akuru, E.A.; Mpendulo, C.T.; Oyeagu, C.E.; Nantapo, C.W.T. Pomegranate (*Punica granatum* L.) Peel Powder Meal Supplementation in Broilers: Effect on Growth Performance, Digestibility, Carcase and Organ Weights, Serum and Some Meat Antioxidant Enzyme Biomarkers. *Ital. J. Anim. Sci.* **2021**, *20*, 119–131. [CrossRef]
161. Rajani, J.; Karimi Torshizi, M.A.; Rahimi, S. Control of Ascites Mortality and Improved Performance and Meat Shelf-Life in Broilers Using Feed Adjuncts with Presumed Antioxidant Activity. *Anim. Feed Sci. Technol.* **2011**, *170*, 239–245. [CrossRef]
162. Ahmed, S.T.; Ko, S.Y.; Yang, C.J. Improving the Nutritional Quality and Shelf Life of Broiler Meat by Feeding Diets Supplemented with Fermented Pomegranate (*Punica granatum* L.) by-Products. *Br. Poult. Sci.* **2017**, *58*, 694–703. [CrossRef] [PubMed]
163. Gungor, E.; Altop, A.; Erener, G.; Coskun, I. Effect of Raw and Fermented Pomegranate Pomace on Performance, Antioxidant Activity, Intestinal Microbiota and Morphology in Broiler Chickens. *Arch. Anim. Nutr.* **2021**, *75*, 137–152. [CrossRef]
164. Hosseini-Vashan, S.J.; Raei-Moghadam, M.S. Antioxidant and Immune System Status, Plasma Lipid, Abdominal Fat, and Growth Performance of Broilers Exposed to Heat Stress and Fed Diets Supplemented with Pomegranate Pulp (*Punica granatum* L.). *J. Appl. Anim. Res.* **2019**, *47*, 521–531. [CrossRef]
165. Hosseini-Vashan, S.J.; Sharifian, M.; Piray, A.H.; Fathi-Nasri, M.H. Growth Performance, Carcass and Blood Traits, Immunity, Jejunal Morphology and Meat Quality of Heat-Stressed Broiler Chickens Fed Urea-Treated Pomegranate (*Punica granatum* L.) Peel. *Anim. Feed Sci. Technol.* **2020**, *267*, 114553. [CrossRef]
166. Manterys, A.; Franczyk-Zarow, M.; Czyzynska-Cichon, I.; Drahun, A.; Kus, E.; Szymczyk, B.; Kostogrys, R.B. Haematological Parameters, Serum Lipid Profile, Liver Function and Fatty Acid Profile of Broiler Chickens Fed on Diets Supplemented with Pomegranate Seed Oil and Linseed Oil. *Br. Poult. Sci.* **2016**, *57*, 771–779. [CrossRef]
167. Szymczyk, B.; Szczurek, W. Effect of Dietary Pomegranate Seed Oil and Linseed Oil on Broiler Chicken's Performance and Meat Fatty Acid Profile. *J. Anim. Feed. Sci.* **2016**, *25*, 37–44. [CrossRef]
168. Banaszkiewicz, T.; Białek, A.; Tokarz, A.; Kaszperuk, K. Effect of Dietary Grape and Pomegranate Seed Oil on the Post-Slaughter Value and Physicochemical Properties of Muscles of Broiler Chickens. *Acta Sci. Pol. Technol. Aliment.* **2018**, *17*, 199–209.
169. Eid, Y.; Kirrella, A.A.; Tolba, A.; El-Deeb, M.; Sayed, S.; El-Sawy, H.B.; Shukry, M.; Dawood, M.A.O. Dietary Pomegranate by-Product Alleviated the Oxidative Stress Induced by Dexamethasone in Laying Hens in the Pre-Peak Period. *Animals* **2021**, *11*, 1022. [CrossRef]
170. Kostogrys, R.B.; Filipiak-Florkiewicz, A.; Dereń, K.; Drahun, A.; Czyżyńska-Cichoń, I.; Cieślik, E.; Szymczyk, B.; Franczyk-Żarów, M. Effect of Dietary Pomegranate Seed Oil on Laying Hen Performance and Physicochemical Properties of Eggs. *Food Chem.* **2017**, *221*, 1096–1103. [CrossRef] [PubMed]
171. Abbas, R.J.; Al-Salhie, K.C.K.; Al-Hummod, S.K. The Effect of Using Different Levels of Pomegranate (*Punica granatum*) Peel Powder on Productive and Physiological Performance of Japanese Quail (*Coturnix coturnix* Japonica). *Livest. Res. Rural Dev.* **2017**, *29*, 231.
172. Lu, Z.; Wang, J.; Gao, R.; Ye, F.; Zhao, G. Sustainable Valorisation of Tomato Pomace: A Comprehensive Review. *Trends Food Sci. Technol.* **2019**, *86*, 172–187. [CrossRef]
173. Seidavi, A.R.; Azizi, M.; Ragni, M.; Laudadio, V.; Tufarelli, V. Practical Applications of Agricultural Wastes in Poultry Feeding in Mediterranean an Middle East Regions. Part 2: Tomato, Olive, Date, Sunflower Wastes. *Worlds Poult. Sci. J.* **2018**, *74*, 443–452. [CrossRef]
174. Zuorro, A.; Fidaleo, M.; Lavecchia, R. Enzyme-Assisted Extraction of Lycopene from Tomato Processing Waste. *Enzyme Microb. Technol.* **2011**, *49*, 567–573. [CrossRef]
175. Del Valle, M.; Cámara, M.; Torija, M.-E. Chemical Characterization of Tomato Pomace. *J. Sci. Food Agric.* **2006**, *86*, 1232–1236. [CrossRef]

76. Jafari, M.; Pirmohammadi, R.; Bampidis, V. The Use of Dried Tomato Pulp in Diets of Laying Hens. *Int. J. Poult. Sci.* **2006**, *5*, 618–622.
77. Ghazi, S.; Drakhshan, A. The Effects of Different Levels of Tomato Pomace in Broilers Chick Performance. In Proceedings of the 12th European Poultry Conference, Verona, Italy, 10–14 September 2006.
78. Lira, R.C.; Rabello, C.B.; Mohaupt Marques Ludke, M.D.C.; Ferreira, P.V.; Lana, G.R.Q.; Valerio Lana, S.R. Productive Performance of Broiler Chickens Fed Tomato Waste. *Rev. Bras. Zootec.* **2010**, *39*, 1074–1081. [CrossRef]
79. Yitbarek, M.B. The Effect of Feeding Different Levels of Dried Tomato Pomace on the Performance of Rhode Island Red (Rir) Grower Chicks. *Int. J. Livest. Prod.* **2013**, *4*, 35–41. [CrossRef]
80. Hosseini-Vashan, S.J.; Golian, A.; Yaghobfar, A. Growth, Immune, Antioxidant, and Bone Responses of Heat Stress-Exposed Broilers Fed Diets Supplemented with Tomato Pomace. *Int. J. Biometeorol.* **2016**, *60*, 1183–1192. [CrossRef] [PubMed]
81. Wahyuni, H.I.; Yudiarti, T.; Widiastuti, E.; Sugiharto, S.; Isroli, I.; Sartono, T.A. The Use of Tomato Waste Juice as an Antioxidant Source for Broiler Chickens. *IOP Conf. Ser. Earth Environ. Sci.* **2020**, *518*, 012006. [CrossRef]
82. Dotas, D.; Zamanidis, S.; Balios, J. Effect of Dried Tomato Pulp on the Performance and Egg Traits of Laying Hens. *Br. Poult. Sci.* **1999**, *40*, 695–697. [CrossRef] [PubMed]
83. Yannakopoulos, A.L.; Tserveni-Gousi, A.S.; Christaki, E.V. Effect of Locally Produced Tomato Meal on the Performance and the Egg Quality of Laying Hens. *Anim. Feed Sci. Technol.* **1992**, *36*, 53–57. [CrossRef]
84. Akdemir, F.; Orhan, C.; Sahin, N.; Sahin, K.; Hayirli, A. Tomato Powder in Laying Hen Diets: Effects on Concentrations of Yolk Carotenoids and Lipid Peroxidation. *Br. Poult. Sci.* **2012**, *53*, 675–680. [CrossRef]
85. Panaite, T.D.; Nour, V.; Vlaicu, P.A.; Ropota, M.; Corbu, A.R.; Saracila, M. Flaxseed and Dried Tomato Waste Used Together in Laying Hens Diet. *Arch. Anim. Nutr.* **2019**, *73*, 222–238. [CrossRef]
86. Salajegheh, M.H.; Ghazi, S.; Mahdavi, R.; Mozafari, O. Effects of Different Levels of Dried Tomato Pomace on Performance, Egg Quality and Serum Metabolites of Laying Hens. *Afr. J. Biotechnol.* **2012**, *11*, 15373–15379.
87. Jouzi, H.; Vali, N.; Pourreza, J. The Effects of Tomato Pulp Powder Supplementation on Performance and Some Blood Parameters in Japanese Quail (*Coturnix japonica*). *J. Agric. Biol. Sci.* **2015**, *10*, 103–107.
88. Botsoglou, N.; Papageorgiou, G.; Nikolakakis, I.; Florou-Paneri, P.; Giannenas, I.; Dotas, V.; Sinapis, E. Effect of Dietary Dried Tomato Pulp on Oxidative Stability of Japanese Quail Meat. *J. Agric. Food Chem.* **2004**, *52*, 2982–2988. [CrossRef]
89. Nikolakakis, I.; Banakis, D.; Florou-Paneri, P.; Dotas, V.; Giannenas, I.; Botsoglou, N. Effect of Dried Tomato Pulp on Performance and Carcass Characteristics of Growing Quails. *Arch. Geflügelkd* **2004**, *68*, 34–38.
90. Sahin, N.; Orhan, C.; Tuzcu, M.; Sahin, K.; Kucuk, O. The Effects of Tomato Powder Supplementation on Performance and Lipid Peroxidation in Quail. *Poult. Sci.* **2008**, *87*, 276–283. [CrossRef] [PubMed]
91. Alagawany, M.; Attia, A. Effects of Feeding Sugar Beet Pulp and Avizyme Supplementation on Performance, Egg Quality, Nutrient Digestion and Nitrogen Balance of Laying Japanese Quail. *Avian Biol. Res.* **2015**, *8*, 79–88. [CrossRef]
92. Ajila, C.M.; Brar, S.K.; Verma, M.; Tyagi, R.D.; Godbout, S.; Valéro, J.R. Bio-Processing of Agro-Byproducts to Animal Feed. *Crit. Rev. Biotechnol.* **2012**, *32*, 382–400. [CrossRef] [PubMed]
93. Kumari, P.; Kumar, K.; Kumar, S. Effect of Dietary Supplement of Sugar Beet, Neem Leaf, Linseed and Coriander on Growth Performance and Carcass Trait of Vanaraja Chicken. *Vet. World* **2014**, *7*, 639–643. [CrossRef]
94. González-Alvarado, J.M.; Jiménez-Moreno, E.; González-Sánchez, D.; Lázaro, R.; Mateos, G.G. Effect of Inclusion of Oat Hulls and Sugar Beet Pulp in the Diet on Productive Performance and Digestive Traits of Broilers from 1 to 42 Days of Age. *Anim. Feed Sci. Technol.* **2010**, *162*, 37–46. [CrossRef]
95. Jiménez-Moreno, E.; Frikha, M.; de Coca-Sinova, A.; Lázaro, R.P.; Mateos, G.G. Oat Hulls and Sugar Beet Pulp in Diets for Broilers. 2. Effects on the Development of the Gastrointestinal Tract and on the Structure of the Jejunal Mucosa. *Anim. Feed Sci. Technol.* **2013**, *182*, 44–52. [CrossRef]
96. Jiménez-Moreno, E.; Frikha, M.; de Coca-Sinova, A.; García, J.; Mateos, G.G. Oat Hulls and Sugar Beet Pulp in Diets for Broilers 1. Effects on Growth Performance and Nutrient Digestibility. *Anim. Feed Sci. Technol.* **2013**, *182*, 33–43. [CrossRef]
97. Abdel-Daim, A.S.A.; Tawfeek, S.S.; El-Nahass, E.S.; Hassan, A.H.A.; Youssef, I.M.I. Effect of Feeding Potato Peels and Sugar Beet Pulp with or without Enzyme on Nutrient Digestibility, Intestinal Morphology, and Meat Quality of Broiler Chickens. *Poult. Sci. J.* **2020**, *8*, 189–199.
98. Abdel-Hafeez, H.M.; Saleh, E.S.E.; Tawfeek, S.S.; Youssef, I.M.I.; Abdel-Daim, A.S.A. Utilization of Potato Peels and Sugar Beet Pulp with and without Enzyme Supplementation in Broiler Chicken Diets: Effects on Performance, Serum Biochemical Indices and Carcass Traits. *J. Anim. Physiol. Anim. Nutr.* **2018**, *102*, 56–66. [CrossRef]
99. Pettersson, D.; Razdan, A. Effects of Increasing Levels of Sugar-Beet Pulp in Broiler Chicken Diets on Nutrient Digestion and Serum Lipids. *Br. J. Nutr.* **1993**, *70*, 127–137. [CrossRef]
100. Abbas, A.; Iqbal, Z.; Abbas, R.Z.; Khan, M.K.; Khan, J.A.; Sindhu, Z.D.; Mahmood, M.S.; Saleemi, M.K. In vivo Anticoccidial Effects of Beta Vulgaris (Sugar Beet) in Broiler Chickens. *Microb. Pathog.* **2017**, *111*, 139–144. [CrossRef] [PubMed]
101. Selim, S.; Hussein, E. Production Performance, Egg Quality, Blood Biochemical Constituents, Egg Yolk Lipid Profile and Lipid Peroxidation of Laying Hens Fed Sugar Beet Pulp. *Food Chem.* **2020**, *310*, 125864. [CrossRef] [PubMed]
102. Westendorf, M.L.; Wohlt, J.E. Brewing by-Products: Their Use as Animal Feeds. *Vet. Clin. N. Am.—Food Anim. Pract.* **2002**, *18*, 233–252. [CrossRef] [PubMed]

203. Lynch, K.M.; Steffen, E.J.; Arendt, E.K. Brewers' Spent Grain: A Review with an Emphasis on Food and Health. *J. Inst. Brew.* **2016** 122, 553–568. [CrossRef]
204. Stengel, G. Brewers Grains: The Industry. In Proceedings of the Alternative Feeds for Dairy and Beef Cattle Symposium; Jordan E.R., Ed.; University of Missouri: Columbia, MO, USA, 1991; pp. 86–89.
205. Ironkwe, M.O.; Bamgbose, A.M. Effect of Replacing Maize with Brewer's Dried Grain in Broiler Finisher Diet. *Int. J. Poult. Sci.* **2011**, *10*, 710–712. [CrossRef]
206. Onwudike, O.C. The Effects of Dietary Sand on the Usage of Diets Containing Brewer's Dried Grains by Growing Chicks. *Poult. Sci.* **1986**, *65*, 1129–1136. [CrossRef]
207. Yurina, N.A.; Labutina, N.D.; Danilova, A.A.; Vlasov, A.B.; Khorin, B.V.; Yurin, D.A. Feed Additive for Quails Based on Fermented Brewer's Spent Grain. *IOP Conf. Ser. Earth Environ. Sci.* **2021**, *938*, 012005. [CrossRef]
208. Denstadli, V.; Westereng, B.; Biniyam, H.G.; Ballance, S.; Knutsen, S.H.; Svihus, B. Effects of Structure and Xylanase Treatment of Brewers' Spent Grain on Performance and Nutrient Availability in Broiler Chickens. *Br. Poult. Sci.* **2010**, *51*, 419–426. [CrossRef]
209. San Martin, D.; Ramos, S.; Zufía, J. Valorisation of Food Waste to Produce New Raw Materials for Animal Feed. *Food Chem.* **2016** *198*, 68–74. [CrossRef]
210. Kasapidou, E.; Sossidou, E.; Mitlianga, P. Fruit and Vegetable Co-Products as Functional Feed Ingredients in Farm Animal Nutrition for Improved Product Quality. *Agriculture* **2015**, *5*, 1020–1034. [CrossRef]
211. Georganas, A.; Giamouri, E.; Pappas, A.C.; Papadomichelakis, G.; Fortatos, S.; Manios, T.; Lasaridi, K.; Fegeros, K.; Tsiplakou, E. Zervas, G. Redefining the Future of Catering Waste Application in Animal Diets—A Review on the Minimization of Potential Hazards in Catering Waste Prior to Application in Animal Diets. *Anim. Feed Sci. Technol.* **2022**, *289*, 115334. [CrossRef]
212. Badaoui, O.; Djebli, A.; Hanini, S. Solar Drying of Apple and Orange Waste: Evaluation of a New Thermodynamic Approach and Characterization Analysis. *Renew. Energy* **2022**, *199*, 1593–1605. [CrossRef]
213. Pinotti, L.; Giromini, C.; Ottoboni, M.; Tretola, M.; Marchis, D. Review: Insects and Former Foodstuffs for Upgrading Food Waste Biomasses/Streams to Feed Ingredients for Farm Animals. *Animal* **2019**, *13*, 1365–1375. [CrossRef] [PubMed]
214. Morais, T.; Inácio, A.; Coutinho, T.; Ministro, M.; Cotas, J.; Pereira, L.; Bahcevandziev, K. Seaweed Potential in the Animal Feed: A Review. *J. Mar. Sci. Eng.* **2020**, *8*, 559. [CrossRef]
215. EFSA; James, K.; Millington, A.; Randall, N. Food and Feed Safety Vulnerabilities in the Circular Economy. *EFSA Support Publ.* **2022**, *19*, 7226E.
216. van Asselt, E.D.; Arrizabalaga-Larrañaga, A.; Focker, M.; Berendsen, B.J.A.; van de Schans, M.G.M.; van der Fels-Klerx, H.J. Chemical Food Safety Hazards in Circular Food Systems: A Review. *Crit. Rev. Food Sci. Nutr.* **2022**, 1–13. [CrossRef]
217. EC. Regulation (Ec) No 178/2002 of the European Parliament and of the Council of 28 January 2002 Laying Down the General Principles and Requirements of Food Law, Establishing the European Food Safety Authority and Laying Down Procedures in Matters of Food Safety. *Off. J. Eur. Communities* **2002**, *L 31*, 1–24.
218. Granby, K.; Mortensen, A.; Broesboel-Jensen, B. 21—Potential Contamination Issues Arising from the Use of Biofuel and Food Industry by-Products in Animal Feed. In *Animal Feed Contamination*; Fink-Gremmels, J., Ed.; Woodhead Publishing: Sawston, UK 2012; pp. 514–539.
219. Battilani, P.; Logrieco, A.F. Global Risk Maps for Mycotoxins in Wheat and Maize. In *Mycotoxin Reduction in Grain Chains*; Leslie J.F., Logrieco, A.F., Eds.; John Wiley & Sons: Hoboken, NJ, USA, 2014; pp. 309–326.
220. Focker, M.; van Asselt, E.D.; Berendsen, B.J.A.; van de Schans, M.G.M.; van Leeuwen, S.P.J.; Visser, S.M.; van der Fels-Klerx, H.J. Review of Food Safety Hazards in Circular Food Systems in Europe. *Food Res. Int.* **2022**, *158*, 111505. [CrossRef]
221. Wu, F.; Munkvold, G.P. Mycotoxins in Ethanol Co-Products: Modeling Economic Impacts on the Livestock Industry and Management Strategies. *J. Agric. Food Chem.* **2008**, *56*, 3900–3911. [CrossRef]
222. Boudra, H.; Rouillé, B.; Lyan, B.; Morgavi, D.P. Presence of Mycotoxins in Sugar Beet Pulp Silage Collected in France. *Anim. Feed Sci. Technol.* **2015**, *205*, 131–135. [CrossRef]
223. van der Spiegel, M.; van der Fels-Klerx, H.J.; Marvin, H.J.P. Effects of Climate Change on Food Safety Hazards in the Dairy Production Chain. *Food Res. Int.* **2012**, *46*, 201–208. [CrossRef]
224. FAO; WHO. Hazards Associated with Animal Feed. Report of the Joint Fao/Who Expert Meeting—12–15 May 2015. In *FAO Animal Production and Health Report No. 13*; FAO: Rome, Italy, 2019.
225. Schmid, P.; Gujer, E.; Degen, S.; Zennegg, M.; Kuchen, A.; Wüthrich, C. Levels of Polychlorinated Dibenzo-P-Dioxins and Dibenzofurans in Food of Animal Origin. The Swiss Dioxin Monitoring Program. *J. Agric. Food Chem.* **2002**, *50*, 7482–7487 [CrossRef] [PubMed]
226. den Hartog, J. 26—The Gmp+ Feed Safety Assurance (Fsa) Scheme. In *Animal Feed Contamination*; Fink-Gremmels, J., Ed. Woodhead Publishing: Sawston, UK, 2012; pp. 625–649.
227. Guillén, M.D.; Sopelana, P.; Palencia, G. Polycyclic Aromatic Hydrocarbons and Olive Pomace Oil. *J. Agric. Food Chem.* **2004**, *52* 2123–2132. [CrossRef] [PubMed]
228. Leontopoulos, S.; Skenderidis, P.; Petrotos, K.; Giavasis, I. Corn Silage Supplemented with Pomegranate (*Punica granatum*) and Avocado (Persea Americana) Pulp and Seed Wastes for Improvement of Meat Characteristics in Poultry Production. *Molecules* **2021**, *26*, 5901. [CrossRef] [PubMed]
229. Adama, T.Z.; Ogunbajo, S.A.; Mambo, M. Feed Intake, Growth Performance and Nutrient Digestibility of Broiler Chicks Fed Diets Containing Varying Levels of Sorghum Dried Brewers' Grains. *Int. J. Poult. Sci.* **2007**, *6*, 592–598. [CrossRef]

230. Ominski, K.; McAllister, T.; Stanford, K.; Mengistu, G.; Kebebe, E.G.; Omonijo, F.; Cordeiro, M.; Legesse, G.; Wittenberg, K. Utilization of by-Products and Food Waste in Livestock Production Systems: A Canadian Perspective. *Anim. Front.* **2021**, *11*, 55–63. [CrossRef]
231. Bernard, J.K. Considerations for Using by-Product Feeds. University of Georgia EXTENSION. Available online: https://extension.uga.edu/publications/detail.html?number=B862#Economics (accessed on 20 October 2022).
232. Lyu, F.; Luiz, S.F.; Azeredo, D.R.; Cruz, A.G.; Ajlouni, S.R.; Chaminda, S. Apple Pomace as a Functional and Healthy Ingredient in Food Products: A Review. *Processes* **2020**, *8*, 319. [CrossRef]
233. Malisch, R. Increase of the Pcdd/F-Contamination of Milk, Butter and Meat Samples by Use of Contaminated Citrus Pulp. *Chemosphere* **2000**, *40*, 1041–1053. [CrossRef]
234. Li, Y.; Jiao, B.; Zhao, Q.; Wang, C.; Gong, Y.; Zhang, Y.; Chen, W. Effect of Commercial Processing on Pesticide Residues in Orange Products. *Eur. Food Res. Technol.* **2012**, *234*, 449–456. [CrossRef]
235. Veldman, A.; Borggreve, G.J.; Mulders, E.J.; Van De Lagemaat, D. Occurrence of the Mycotoxins Ochratoxin a, Zearalenone and Deoxynivalenol in Feed Components. *Food Addit. Contam.* **1992**, *9*, 647–655. [CrossRef]
236. Smeu, I.; Dobre, A.A.; Cucu, E.M.; Mustățea, G.; Belc, N.; Ungureanu, E.L. Byproducts from the Vegetable Oil Industry: The Challenges of Safety and Sustainability. *Sustainability* **2022**, *14*, 2039. [CrossRef]
237. Nawaz, S.; Scudamore, K.A.; Rainbird, S.C. Mycotoxins in Ingredients of Animal Feeding Stuffs: I. Determination of Alternaria Mycotoxins in Oilseed Rape Meal and Sunflower Seed Meal. *Food Addit. Contam.* **1997**, *14*, 249–262. [CrossRef] [PubMed]
238. Mortensen, A.; Granby, K.; Eriksen, F.D.; Cederberg, T.L.; Friis-Wandall, S.; Simonsen, Y.; Broesbøl-Jensen, B.; Bonnichsen, R. Levels and Risk Assessment of Chemical Contaminants in Byproducts for Animal Feed in Denmark. *J. Environ. Sci. Health B* **2014**, *49*, 797–810. [CrossRef]
239. Zachariasova, M.; Dzuman, Z.; Veprikova, Z.; Hajkova, K.; Jiru, M.; Vaclavikova, M.; Zachariasova, A.; Pospichalova, M.; Florian, M.; Hajslova, J. Occurrence of Multiple Mycotoxins in European Feedingstuffs, Assessment of Dietary Intake by Farm Animals. *Anim. Feed Sci. Technol.* **2014**, *193*, 124–140. [CrossRef]
240. Tolosa, J.; Rodríguez-Carrasco, Y.; Ruiz, M.J.; Vila-Donat, P. Multi-Mycotoxin Occurrence in Feed, Metabolism and Carry-over to Animal-Derived Food Products: A Review. *Food Chem. Toxicol.* **2021**, *158*, 112661. [CrossRef] [PubMed]
241. Lavelli, V. Circular Food Supply Chains—Impact on Value Addition and Safety. *Trends Food Sci. Technol.* **2021**, *114*, 323–332. [CrossRef]
242. Mastanjević, K.; Lukinac, J.; Jukić, M.; Šarkanj, B.; Krstanović, V.; Mastanjević, K. Multi-(Myco)Toxins in Malting and Brewing by-Products. *Toxins* **2019**, *11*, 30. [CrossRef]

sustainability

MDPI

Review

An Overview of Poultry Greenhouse Gas Emissions in the Mediterranean Area

Foivos Zisis [1], Elisavet Giamouri [1], Christina Mitsiopoulou [1], Christos Christodoulou [1], Charalampos Kamilaris [2], Alexandros Mavrommatis [1], Athanasios C. Pappas [1,*] and Eleni Tsiplakou [1]

[1] Laboratory of Nutritional Physiology and Feeding, Department of Animal Science, Agricultural University of Athens, Iera Odos 75, 11855 Athens, Greece

[2] Centre for Innovation Excellence in Livestock, Innovation Centre, York Science Park, Innovation Way, Heslington, York YO10 5DG, UK

* Correspondence: apappas@aua.gr

Abstract: The growing population and income drive the rapid increase in food demand. Greece and a few other Mediterranean countries are characterized as countries with a high proportion of mountains favoring goat and sheep breeding; however, poultry breeding is also important, and production is increasing rapidly. Poultry breeding is characterized by the millions of birds reared with increased quantities and prices of feedstuffs. There is a parallel increase in greenhouse gas (GHG) emissions., since poultry production generates a significant amount of GHG. The aim of the present study was to provide an overview of poultry GHG in the Mediterranean area. Emissions' sources and mitigation practices are presented. Future is promising given that sustainable practices are implemented.

Keywords: soybean replacement; manure management; Mediterranean; biogas production; novel feeding; poultry; sustainable farming

Citation: Zisis, F.; Giamouri, E.; Mitsiopoulou, C.; Christodoulou, C.; Kamilaris, C.; Mavrommatis, A.; Pappas, A.C.; Tsiplakou, E. An Overview of Poultry Greenhouse Gas Emissions in the Mediterranean Area. *Sustainability* **2023**, *15*, 1941. https://doi.org/10.3390/su15031941

Academic Editor: Roberto Mancinelli

Received: 21 December 2022
Revised: 16 January 2023
Accepted: 16 January 2023
Published: 19 January 2023

1. Introduction

The global population, estimated at nearly 8 billion people worldwide in 2022, tends to approach 9.7 billion by 2050 and 10.4 billion by 2100 [1]. Likewise, in 1950, 30% of the population lived in urban areas, in 2000 47%, while 68% is foreseen by 2050 [2]. This alteration led to a more fast-paced way of life and to a shift in food habits more western-based, abandoning the Mediterranean diet concept, as well as to considerable levels of animal-derived protein consumption [3]. Besides that, there is a trend in shift from red to white meat [4], especially towards poultry. Thus, to meet the demand, poultry farming grows constantly [5]. This leads to thriving environmental concerns about animal production [6], hence setting the development of sustainable animal diets as a top priority.

In terms of environmental concern, climate change severely affects livestock. On the other side, concurrently livestock farming is one of the main contributors to greenhouse gas (GHG) emissions [7], either through the animal physiological processes or the food supply chains. Thoroughly, the three so-called GHGs, which customarily contribute to this phenomenon, are carbon dioxide (CO_2), methane (CH_4), and nitrous oxide (N_2O) [8], while others such as ozone (O_3), hydrofluorocarbons (HFCs), and sulphur hexafluoride (SF_6) have a slighter contribution [9].

Nevertheless, the concerns over the impact of the livestock sector om climate change have led to a plethora of studies aimed at improving the scientific knowledge over this [10–14]. Moreover, there is a great need for further research, not only on the impact of poultry farming on GHG emissions, but more importantly on the mitigation practices that need to be established for lessening such environmental burden. It is important to note that even though the food systems have a major contribution to the GHG emissions (~30%), the opportunity for reducing them has received less attention, partly because it seems like an unavoidable burden for meeting the nutritional global demand [15]. Nonetheless, at the same time, consumers' demand for products with a low environmental footprint [16] and that promote animal welfare [17] appears to be increasing.

Greece and a few other Mediterranean countries in the European Union (EU) are characterized as countries with a high proportion of mountains favoring goat and sheep breeding; however, poultry breeding is also important, and production is increasing rapidly. Moreover, the production system established in countries of the Mediterranean region is generally based on intensive production, thus the application of GHG mitigation strategies could be more holistic and irreversible. As a result, it could be more efficient compared to ruminants' sector, where the variety of production systems make difficult the application of such strategies. In addition, the EU Mediterranean countries share not only similar climate conditions and farming systems, but more importantly share common policies under the auspices of the European Union, potentially establishing a more unified approach. Considering the above evidence and due to the specific socioeconomic aspects analyzed below, we chose to study poultry, aiming to explore options to help lessen the environmental impact associated with poultry production in EU Mediterranean countries.

2. Greenhouse Gas Emissions

2.1. Calculation of the GHG Emissions

A widely described term is that of "Carbon Footprint" (CF), a reference to the sum of GHG emissions generated and associated with any activity by a product or service system [18]. Contrary to what the word implies, the CF includes CO_2, N_2O, and CH_4 emissions expressed in carbon dioxide equivalents (CO_2-eq) [19].

Likewise, several methodologies have been applied to assess the environmental impact of livestock production systems [7]. The LCA (Life Cycle Assessment) is an ISO-standardized environmental accounting tool (14040 and 14044 ISO standards) used to evaluate the environmental impact generated through the different life-cycle stages of a product "from raw material acquisition, via production and use stages to waste management" [20]. The methodology has also been applied to livestock and food production systems [21,22]. However, for conducting a holistic approach, the keynote is the inclusion not only of the in-farm emissions but also the emissions contained in each stage of production as fertilizer, crop or feed production, animal facilities, processing, transportation, market distribution, product consumption, and waste management [23].

2.2. Mediterranean Statistics for Livestock GHG

The principal sources for GHGs involved in livestock are N_2O from fertilizer application, during cultivation, and manure management [24]; CH_4 originates from enteric fermentation and manure management plus the CO_2 mainly from fuel combustion on-farm, such as from heavy farm machinery operations, heating, electricity, fertilizer production [25], and land-use change (LUC) for feed production [26]. In detail, a FAO report [27] described that livestock accounts for 14.5% of total anthropogenic GHG emissions or 7.1 Gt CO_2-eq. Of this total, the share of feed production and processing is about 45% or 3.2 Gt CO_2-eq; enteric fermentation about 39% or 2.8 Gt CO_2-eq, being the second-largest source of emissions; and manure management about 10% or 0.71 Gt CO_2-eq. The remaining 6% or 0.42 Gt CO_2-eq is due to the processing and transportation of the animal products. The largest contributor of livestock emissions globally is the cattle production with a contribution of 4.6 Gt CO_2-eq or 61%, while other species generate much lower emissions, such as poultry 0.7 Gt CO_2-eq (8%) [19,27]. By accounting only for the direct CH_4 and N_2O emissions from enteric fermentation and total manure management, the contribution is estimated at 5.4 Gt CO_2-eq [19,28].

Specifically for the EU countries, agriculture constitutes 11.4% of total GHG emissions [29], from which CH_4 from enteric fermentation accounts for 42% while manure management 14% [30]. Combining CH_4 from enteric fermentation and N_2O emissions from soils, a sum of 81% of the total agricultural emissions results [30]. However, the quantities of GHG emissions differ between species, regions, or production systems, as organic systems tend to produce higher GHG emissions [31] (Figure 1 and Table S1 in the Supplementary Materials). Moreover, farm-gate emissions have increased by up to 11% from 2000 to 2019, mainly attributed to livestock (~55%) [32]. Therefore, each production system should be studied separately, requiring a different approach for mitigation options [33].

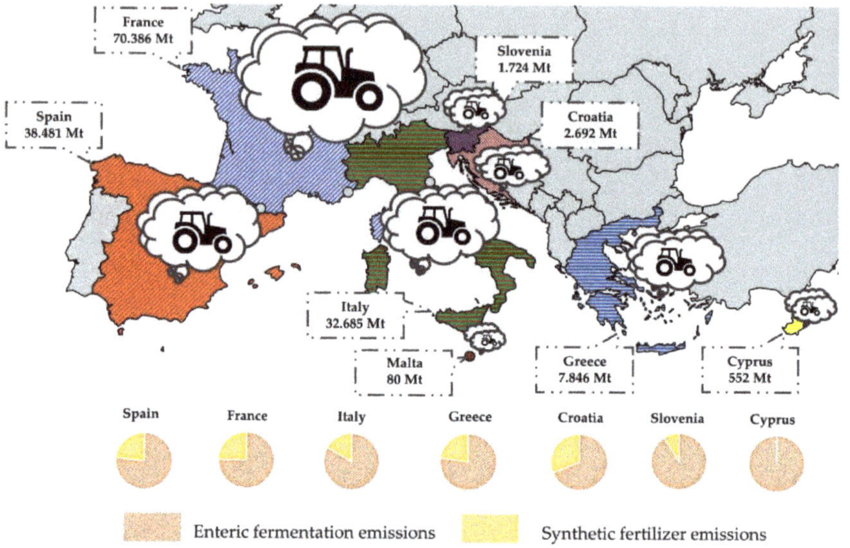

Figure 1. Emissions, in EU Mediterranean countries, originating from agriculture sector, expressed as total of CO_2-eq and as ratio of enteric fermentation to synthetic fertilizers emissions (only total emissions are presented for Malta, Supplementary Data at Table S1) [32].

Nevertheless, agriculture in Greece accounted for 8.84 million tonnes CO_2-eq in 2016 but 7.78 million tonnes CO_2-eq in 2019 [34], Croatia and Malta being among the countries presenting a substantial decrease within the EU [30]. Furthermore, emissions from agriculture accounted for 9.2% of total emissions in 2018 and decreased by 22.19% compared to 1990 levels [34]. This outcome is due to the reduction of N_2O emissions from soil because of the reduced use of synthetic nitrogen fertilizers and the animal population number. Furthermore, in Greece, CH_4 represents the main GHG from agriculture, in a range of 48% to 58% of total GHG [34].

Hence, greater interest should be given to research in mitigating GHG emissions. According to Gerber et al. [27], good practices and technological application in animal nutrition, health, as well as manure management can help to improve livestock production and reduce global GHG emissions by 30%. For this purpose, in turn are described some mitigation practices that can be adopted for poultry farming applied in the EU Mediterranean countries and include—but are not limited to—feed management, feed production options, manure management for fertilization, and energy mitigation on site.

3. Poultry

Primarily, poultry meat represents a major part of the total amount of meat produced [35] and is projected to represent 41% of all meat sources protein by 2030 [5]. Over the last 50 years, the average annual growth for poultry meat has been 5%, while it has been 3.1% for pork, only 1.5% for beef, and 1.7% for small ruminant meat [36], operating as a forerunner of the total meat production. Additionally, the rapid increase in poultry production is further supported by the increased poultry population from 4.2 billion birds in 1961 to more than 35 billion birds in 2020 [32]. Similarly, egg production was 15 million tonnes in 1961, while in 2018 it exceeded 92 million tonnes [32]. Likewise, poultry production is also a major agricultural sector in the Mediterranean area (Table 1), and a substantial source of GHG emissions for Greece [37], compared to other Mediterranean countries.

Table 1. Population and production (meat and eggs) of poultry in 2020, in EU Mediterranean countries.

Countries	Poultry	
	Meat Production (Thousand Tonnes)	**Eggs Production (Thousands of Tonnes)**
Croatia	54	30.1
Cyprus	27	10.5
France	1121	786.1
Greece	228	74.3
Italy	1055	717.4
Malta	4	5.5
Slovenia	64	24.8
Spain	1412	810.9

The data were collected by FAO [32].

The sources of the emissions derive from feed production, LUC, and the energy use during the operations in and out of the farms (international trade, slaughterhouse, feed production, hatchery). It has been vastly reported that feed production is the most important source of GHG emissions [33,38–40], and the rate varies widely in the literature from 45% to 93.7% [41–43]. For example, the major environmental cost from an intensive egg production system was feed production in Spain [44], while comparable results were found in Netherlands [45], Iran [46], and Canada [42], implying an independence for each country.

As for the broiler sector, the most is related to LUC and animal feed [6] while the rest to performance objectives, use of energy, the handling of litter, and the stocking density [47,48]. Energy combustion is an important source of GHG in poultry production, both for eggs and meat [45,49,50], required for the ventilation, feeding, lighting, egg collection, sorting, heating, and operation of the mechanical equipment.

As a result, it is essential to mention that the differences in emissions sources, for meat and eggs, result in different values for emissions, as well as in different ways of expressing the rate of GHG emissions. For instance, in a meta-analysis study, with protein as a basis of calculation, 100 g of poultry meat protein accounted for 5.7 kg CO_2-eq, while 100 g of egg protein accounted for a lower rate of 4.21 kg CO_2-eq [6]. On the other hand, Clune et al. [51] reported values based on the kg of product at a rate of 3.65 kg CO_2-eq/kg BFM (bone-free meat for chicken) and 3.46 kg CO_2-eq/kg eggs.

3.1. Greenhouse Gas Emissions

3.1.1. CO_2 Emissions

Carbon dioxide emissions originate mainly from fuel consumption [52] and electricity [53] and are attributed to the appropriate machinery, transportation, and ventilation equipment used in the animal houses and the crop production. Notwithstanding, respiration is not considered a source of emissions, because the emitted and absorbed quantities of CO_2 are equivalent [54]. As for the crop production, this turns on the cereal production, a major and necessary part in poultry feed and diets, as well as on the LUC and the transportation of other essential feedstuff. Nevertheless, the cereal production is of great concern and concurrently a potent area for further research, as has been extensively reviewed by Różewicz [55].

Moreover, the health of birds highly depends on the temperature in the poultry house, with emissions deriving from the need for optimal temperature conditions and the adjustments needed to achieve this. For instance, in poultry houses there is the need for heating in cold months of the year but also the need for cooling during hot months, to avoid heat stress [56]. This is even greater for small chicks that lack proper thermoregulation mechanisms. To properly establish such thermal systems, a significant amount of CO_2 emissions may be generated, due to energy consumption.

Besides, more CO_2 emissions are generated from hatchery processes due to extensive energy consumption. In detail, a range of 12–23% was reported by Usubharatana and Phungrassami [57].

Last, the international trade in poultry meat contributes to significant emissions of CO_2 by fuel use for the shipping and total transportation of poultry meat, estimated at 256,000 tonnes of CO_2 [58]. Emissions are proportional to the use of fuel and strictly related to the distance travelled [39]. Therefore, the potential lack of availability of local feed is a major challenge for producers due to the increased need for fuels.

3.1.2. CH_4 Emissions

Methane is the most important greenhouse gas related to animal agriculture with a global warming potential that is 28-fold that of CO_2 [59], mainly coming from ruminants' enteric fermentation and manure storage [60]. Conversely, poultry as monogastric do not pose a significant share from enteric fermentation, so the CH_4 originates mainly from the management of waste and excreta generated. The type of litter, moisture, and temperature affect the emissions and concentrations of gases [61]. Egg production includes a variety of housing systems and litter-processing practices. The waste which is produced from the various operations of poultry farms are liquids, the litter, dead birds, broken eggs, and the eggs discarded during the packaging process in larger production units.

Often, although the term chicken manure is used, it represents a general term including but not limited to chicken slurry, dry chicken excreta, chicken manure, and fresh chicken excreta [62]. In a biological production system, excreta in piles are composted, which increases aeration and reduces anaerobic CH_4 production.).

3.1.3. N_2O Emissions

The N_2O emissions are generally produced due to the high use of N fertilizers for the feed production, slightly more than 35 percent of emissions [27], and the spreading of the animal manure on the field. Under this context, it is necessary to divide the source of the fertilisers into artificial and natural manure, not only under the strict limits of GHG emissions but also based on a more sustainable production activity [63]. For fertilizer manufacturing, 0.7% of total GHG emissions or 0.41 Gt CO_2-eq were reported [64]. Excess of N can be converted to N_2O through the nitrification–denitrification process, where ammonium (NH_4^+) or organic N is converted to NO_3^- and NO_2^- during nitrification, and then via anaerobic treatment NO_3^- and NO_2^- are reduced to N_2, while through denitrification N_2O and nitric oxide (NO) are produced [9].

4. Mitigation Strategies

To lessen the environmental burden, a variety of strategies should be addressed. Mitigation strategy is a group of practices that need to be implemented for GHG emissions' reduction. In general, it is crucial to separate the mitigation practices in different sections according to their aim. In poultry, the practices are related, but not limited to, feed utilization, manure management through utilizing it for energy production or as a fertilizer, as well as energy management through cost effective solutions and renewable energy sources, and data collection for extra data analysis, a novel way of predicting the environmental impact. Moreover, feed efficiency and growth rate are related to the quantity and quality of generated excreta [65]. FAO [66] estimates that available improved-farming practices can lower emissions by up to 30%, including mitigation for feed production, nutrition, energy combustion, and faecal management. However, management practices can only lower GHG emissions up to a specific point. For any further mitigation, livestock reduction is needed or an increased efficiency.

4.1. Poultry Feed Formulation and Additives

A summary of the mitigation strategies related to poultry nutrition is presented in Table 2 and presented in detail in the following sections.

Table 2. Summary of the dietary practices applied to lessen the GHG emissions.

Dietary Strategy	Nutritional Practice	Effect-Impact	Supporting Evidence	References
Replacement of soybean	Peas	↓ 8.21% GHG	Local availability reduces the transportation emissions	[67]
	Semi-leafless peas			[68]
	Domestically peas and rapeseed			[49]
	Cottonseed meal			[69]
Replacement of palm oil	Cotton seed oil	↓ 22% CF		[70]
Alternative protein sources (insects)	Mealworm	↓ LUC	- Welfare, growth performance or any other physiological or morphological feature	[71,72]
	Black soldier larvae fat		No alteration on performance or meat and carcass quality	[73]
Balance low-protein diets	Amino acids	↓ Loss of nutrients	↓ Energy demand	[74]
Improve bioavailability of nutrients	*E. coli* phytase	↑ Bioavailability		[75]
	Zn and phytase		↑ Body weight and nutrient usage	[76]
	Decrease or replace the amount of soybean meal with protease and corn gluten meal			[77]
Waste valorization	Hotel food residues	↓ Loss of nutrients ↓ Energy for feed production ↑ Supply of bioactive compounds	No impact on FCR, mortality, carcass, or breast yield	[78,79]
	Vinification by-products (ground grape pomace, wine lees extract and grape stem extract)		- Feed intake, FCR, carcass yield, and the weight of the internal organs not affected	[80]
	Wine lees extract rich in yeast cell walls, and grape stem extracts		- Improvement of the broilers' oxidative status	[81]

↑ = Increase, ↓ = Decrease.

4.1.1. Unconventional Feed Formulation

Under this context, total or partial replacement of ingredients in diets can be applied. During volatile ingredient markets, when there is pressure to reduce diet costs, or when common ingredients become scarce, there is a need to use alternative feed ingredients considering potential limitations. Likewise, the high demand for protein supply for livestock led to the constantly increasing global production of soy products, which is severely related to deforestation and soil degradation. Poultry production uses 37% of soybean [82], the largest amount of any livestock sector in Europe. Therefore, great interest should be given to soybean that highly contributes to LUC [83]. It is important to note that a

1% soybean reduction can decrease more than 10% N excretion [84]. Most notably, soybean can be replaced by semi-leafless peas [68], domestically cultivated peas, and rapeseed [49] however, the local availability and presence of antinutritional factors should be taken intc account. A beneficial strategy for the poultry sector with a crucial impact is the cultivation of legumes, a practice under serious development. The locally produced legumes no only provide poultry with an efficient amount of protein but also have the potential fo substituting imported soybean. For instance, under the perspective of reducing the globa transportation emissions due to soy imports from south America, the cultivation of legumes such as faba bean or lupine, can be turned into a viable solution [85]. A decrease of 8.21% for total GHG was reported by Fatica et al. [67] by examining the total for the replacemen of soybean with peas, in broiler diets in Italy, suggesting it as a means for reducing the GHG emissions in the Mediterranean area. Moreover, Abín et al. [70] substituted palm oi with cotton seed oil, which reduced the CF by 22%, ensuing a CF of 2.3 kg CO_2-eq per kg o eggs. Furthermore, Ceylan et al. [86] tested sunflower, fish, linseed, and rapeseed oil while Yuan et al. [69] replaced soybean meal with expanded cottonseed meal in laying hen diets at the levels of 6, 8, and 10%. The benefits of these substitutions are not solely related tc the use of feed with lower CF but also to changes within the cropping systems [49]. Unde this scope, using good quality non-contaminated feed which does not contain an excess of minerals such as copper or zinc but those quantities that are required for animal health could be a beneficial strategy [58].

The prospect of using alternative protein-rich feed ingredients in poultry diets may benefit global food security and demand for protein. Apart from that, they do not compete with the use of arable land, a crucial benefit on the effect of climate change. Some suggestions with environmental potential for upcoming use in poultry nutrition include but are not limited to micro- and macro-algae, duckweed, yeast protein concentrate (YPC) bacterial protein meal (BPM), leaf protein concentrate (LPC), and insect meal [87]. However, their role in mitigating GHG emissions needs to be clearly evaluated, but it seems promising since, for example, insects can turn low-grade biowaste into proteins [88]. Under this context, insects are acknowledged as a natural feed for wild-grown chickens that car further be used in organic systems, so a modest hypothesis is whether the inclusion o insects in poultry diets would benefit, in terms of sustainable production, animal welfare and with reduced environmental impact [89]. Since insects are high in protein, fat, essentia amino acids, and micronutrients [90,91], a promising option for soybean replacement is the use of insects, such as black soldier fly larvae, maggot meal and mealworm, as protein sources [92–94]. For example, mealworms' crude protein content (~50.4%) is at the same level as or even higher than soybean (~49.5%) but less than fishmeal (~69%). In addition saturated and monosaturated fatty acids are also present in high levels [88]. As for the feed efficiency, the mealworm inclusion in poultry diets did not affect the welfare, growth performance, or any other physiological or morphological feature in free-range chickens [71] or broilers [72]. Mealworms are a better option than maggot and silkworm as they improve both broiler performance and meat quality [95]. However, Biasato et al. [96] suggested inclusion at a low level (~5%), due to risks for the intestinal morphology. Moreover, the substitution of 50% and 100% of soybean meal with the same level of black soldier larvae fa did not present any alteration on performance or meat and carcass quality [73], prompting an innovative and promising feed-supplementation method. Nevertheless, insects require special treatment for growth and development [87] for heating and drying, so the energy demand needs to be limited, as reported by de Boer et al. [97]. Tallentire et al. [87] reported that LPC had the lowest GHG emissions, followed by YPC and insect meal. However the potential of future growth in insect meals has some crucial limitations, since it is no only limited for the large volume the market needs but also in relation to the substantia ingredients for the poultry feed, mainly amino acids and trace elements [98], issues which need to be resolved. Nevertheless, more research is needed for the examination of the economic viability [99].

4.1.2. Feed Production Practices

Feed production is the biggest opportunity to reduce GHG emissions in poultry farming, since it constitutes a major emissions factor. In that part, the main target is to reduce the need for nitrogen fertilizer added on soil or use the least amounts that can be handled without compromising N emissions. That could be achieved through pruning and management systems that increase nitrogen yield in crops. Also, a good option is the use of crops that require less N per unit of yield compared to conventional crops. Moreover, soil management using suitable cultivation techniques (e.g., minimum slope), integrated pest management, and targeted use of fertilizers is a practice not only good for GHG emissions but also with advantages for soil and total crop production [58].

4.1.3. Amino Acids

It is essential to provide poultry diets that meet the nutritional requirements in their production and development stage to reduce the loss of nutrients and balance the direct consequences on profit [100] as well as preserve welfare. Furthermore, for low protein diets, it is crucial to supplement the ratios with additional essential amino acids or to modify them to balance amino acids and avoid reduced-feed intake resulting, in turn, in reduced production [101,102]. Although amino acids appear to have the greatest carbon intensity among other feed constituents, they have low impact due to the small quantities used [103]. Amongst others, the use of amino acids may result in a viable replacement of soybean in poultry meat production [104]. Moreover, feeding with the digestible amino acids tend to reduce not only the environmental impact but also the cost and the energy demand for their production [74]. That is why the use of amino acids is highly related to precision feeding as a technique.

4.1.4. Exogenous Enzymes

Another option is to improve the digestibility of feed and the bioavailability of nutrients using exogenous enzymes, such as phytases and proteases, and to ensure a balanced microflora (eubiosis) in the digestive system of the bird. The idea of including exogenous enzymes in poultry diets is to decrease the feed's protein level without affecting the growth performance and improve the environmental impact as a result. For example, Al-Harthi et al. [75] described some broiler feeding scenarios and reported that the diet consisted of olive cake supplemented with 500 FTU of *Escherichia coli* (*E. coli*) phytase was the most beneficial and economically competent scenario among those they examined without affecting performance. Also, improved body weight and nutrient usage was indicated in wheat–soybean meal-fed broilers supplemented with Zn and phytase [76]. Besides, Giannenas et al. [77] examined some scenarios, decreasing the amount of soybean meal with the addition of protease or replacing soybean meal by corn gluten meal and including a protease, providing potential environmental benefits.

4.1.5. Waste Valorization

Food waste has generally been considered as a great loss of nutrients, so a modest proposal could be the optimization of such novel feeding technique, especially for poultry. Thus, the incorporation of food waste or by-products could imply a potent material for further utilization [105], contributing to both the circular economy and upcycling. In experiments performed in broilers, the results from incorporating vinification waste by-products [80,81] and hotel food residues [78,79] depicted some promising results. Apart from the concept of sustainable production and the lack of contradiction with human food consumption, there are some vital points to emphasize on this promising strategy. For example, it is of utmost importance to preserve the presence of beneficial bioactive compounds in food waste by examining the proper method of transforming waste into animal feed [106]. Also, to abate the potential hazards of food waste as feed supply, several techniques have been studied and evaluated in recent research [107].

As far as the performance concerns, in an experiment conducted by supplementing 15% of the ration with hotel residues, in broiler diets there was no impact reported on the feed conservation ratio (FCR), mortality, carcass, or breast yield [78]. Moreover, there was no biochemical parameter examined implying any physiological malfunction due to the food-waste incorporation. In another study, Giamouri et al. [79] examined not only the supplementation of hotel food residues in broiler diets, but also the condition of supply, sterilized or non-sterilized. The broilers were supplied with four different treatments: a control, non-meat treatment (100 g dehydrated food residues without any meat), non-sterilized treatment (NS) (100 g non-sterilized dehydrated food residues/kg feed), and sterilized treatment (100 g sterilized dehydrated food residues/kg feed). The results from this study suggested the significance of such treatments as feedstuffs for broilers. In depth, the performance, body weight, FCR, mortality, carcass yield, meat quality, weight of organs, as well as a plethora of biochemical and hematological parameters examined, presented no effect from meat treatment, suggesting the sterilization as a good means for maintaining high levels of hygiene. In addition, the treatment with no meat even resulted in lighter broilers, and a worsening in FCR showed no impact on meat quality, indicating space for optimization.

Furthermore, in an experiment conducted by Giamouri et al. [80], broiler diets were supplemented with three kinds of vinification by-products such as ground grape pomace, wine-lees extract, and grape stem extract. The results depicted the feed intake, FCR, carcass yield, and the weight of the internal organs as not having any significant difference. However, a lighter color was observed for the treatment groups compared to the control. Lastly, the ground grape pomace group demonstrated higher levels of polyunsaturated fatty acids and lower saturated fatty acids compared to the other groups. Nevertheless, these results come in combination with another study by Mavrommatis et al. [81] where the incorporation of wine-lees extract rich in yeast cell walls and grape-stem extracts in broiler diets led to a considerable improvement in the broilers' oxidative status. The use of specific by-products can not only contribute to the purpose of a circular economy but also in supplying diets with significant bioactive compounds, which in another way would have been lost. Although the valorization of food waste and agro-industrial by-products is mostly related to an environmental approach, under the perspective of a circular economy, some parallel effects on GHG emissions can also be unveiled. More specifically, by applying this novel feeding practice, a significant decongestion of the environment using organic wastes can be achieved. Combined with that, the utilization of a plethora of bioactive nutrients, contained in waste, reveals the potential for improving the animal performance and concurrently to further minimize the CF of poultry meat products per unit of protein.

Finally, any process of transformation of food waste to poultry feed needs to produce low carbon emissions. In this regard, Georganas et al. [106] reported an energy-efficient and cost-effective transformation process of food waste from hotels into animal feed by applying solar energy for pasteurizing and drying of the food waste.

4.2. Manure Management

Manure management, including collection, storage, treatment, transportation, and the final utilization, can lessen the burden of environmental pollution and on public health through specific actions. Any efficient system aims to prevent manure and its constituents from accessing the environment, as well as being profitable. Moreover, by applying the manure under specific requirements to avoid CH_4 and N_2O formation [108] for fertilization and feed production, a diminished use of nitrogen fertilizer and, subsequently, nitrogen loss can be achieved. Specifically, poultry manure consists of many essential macronutrients and micronutrients, functioning as a rich organic nutrient for plant growth [109]. However, there is great concern over the overuse and improper application, resulting in severe environmental and health hazards [110]. The spread of manure on crops and fields, although representing the simplest and a low-cost method of manure management, and apart from the high emissions of N_2O and ammonia levels, reveals a high handicap over the unutilized energy coming from manure [111]. Therefore, the ease of collection combined with the prevention of evaporation, runoff, and leaching to prevent losses are keys to an efficient system.

Spreading broiler manure, without any treatment or any further process on the fields, is the simplest paradigm of manure handling. After cleaning the stable, the manure is stored for a limited amount of time and then distributed to the field. The main advantage of this approach is the relatively low costs. Apart from the space needed for manure storage (usually a concrete structure), only a small amount of equipment for transport and distribution (generally manure spreader) is required. Despite this, this kind of manure handling is vastly associated with NH_3 and N_2O emissions; hence, the energy potential of the manure utilization remains unused [111].

4.2.1. Storage

The frequent removal from animal houses is appropriate for avoiding manure fermentation and GHG emissions, as extended storage may increase CH_4 emissions [112]. Therefore, the area manure is intended to be stored in should meet the proper period boundaries. Additionally, the duration of storage is highly dependent on the climate conditions, and consequently different limitations are set due to the conditions, from 3 months in drier countries such as Greece to even 10 months in Finland [113].

An option that is not so expensive and can be widely used is the cooling of the slurry channels [114]. As for the appropriate temperature, the cooling of slurry below 10 °C tends to reduce CH_4 emissions at a rate of 30% to 46% compared with no cooling conditions [65] and can also mitigate NH_3 emissions from in-house manure storage [115]. The combination of frequent removal and cooling by taking it to an outside storage is based on the premise of significant temperature-difference conditions [116] and can be used in cold or mild climates.

4.2.2. Process

Anaerobic digestion is important. Since there is a great need for renewable fuel sources worldwide, the interest aims on efforts such as biogas produced under the anaerobic digestion (AD) process. Anaerobic digestion is the degradation of microorganisms in organic matter under anaerobic conditions producing CH_4, CO_2, and other gases as by-products. The output of this process is the utilization of biogas over heat and electricity on the last level of process, thus indicating one of the most efficient methods for emissions reduction both from manure and energy [117]. The biogas digestate may also replace the use of mineral fertilizers, especially if it is kept in closed tanks, with CH_4 at low rates [111]. The composition of raw biogases is generally 55–70% CH_4, 30–45% CO_2, nitrogen (0–15%), O_2 (0–3%), H_2O (1–5%), hydrocarbons (0–200 mg m^{-3}), hydrogen sulphide (0–10,000 ppmv), NH_3 (0–100 ppmv), and siloxanes (0–41 mg Si m^{-3}) and the heating value of the gas varies from 18 to 30 MJ/m^3 [118]. Biogas can be efficiently used for replacing electricity or fuel combustion on the farm, since poultry excreta produce 310 m^3 biogas/tonne on dry matter basis [119,120]. In an experiment conducted in Greece, biogas from chicken manure was implied to reduce the CF of the farm from 1.38 to 0.49 kg CO_2-eq/head, also demonstrating the economical aspect of the payback for the farm in around 8 years [121]. Production of biogas via anaerobic digestion of biodegradable materials such as biomass, manure, sewage, municipal waste, green waste, plant material, and energy crops [122–124], is a technology that can be implemented at the industrial, village and farm-household scales [120]. Moreover, this technique over waste can effectively reduce odor emission or contaminants making it a promising solution for the utilization of waste from slaughterhouses [125]. However, there is a limitation due to high levels of ammonium, which is toxic for biogas bacteria, and may contain sand and other materials; hence, it should be used in small quantities [120].

Composting provides several benefits related to manure handling, odor, manure moisture and pathogen control, organic matter stabilization, additional farm income [126], lower density, as well as easier and longer storage than raw chicken manure [127]. Nevertheless,

C and N losses are of major concern during composting; for example, N loss results eithe in ammonia emissions, nitrates leaching [128], or CO_2 evaporation [129]. The benefits, such as lessened odor and CH_4 emissions compared with anaerobically stored manure, make i a favorable option [126] to increase nutrient conservation and to produce stable organi fertilizer [130]. Thus, utilizing additives such as zeolite, biochar [129], vermiculite [131], o vinegar waste residues [132] have been examined.

Although on a smaller scale evaluated, the technique of gasification could be highlighted. Gasification functions a dual purpose, partially transforming carbonaceous conten into syngas as fuel and generating biochar as a by-product [133].

The combustion of poultry litter can be a sustainable form of manure management [134 Ogino et al. [135] reported a 42% reduction in GHG emissions by applying a low-protein diet combined with litter incineration compared to a conventional farming system.

A summary of the practices applied to lessen the manure emissions is presented in Table 3.

Table 3. Summary of the practices applied to lessen the manure emissions.

Manure Strategy	Manure Practice	Effect-Impact	References
Frequent removal of manure		↓ CH_4	[112,113]
Cooling manure	Cooling < 10 °C	↓ CH_4 30–46%, ↓ NH_3	[65,115]
Biogas for energy	Producing biogas from manure	↓ GHG emissions from energy ↓ CF 1.38 to 0.49 kg CO_2-eq/head	[117,121]
Biogas digestate for fertilizer	Biogas digestate stored in closed tanks	↓ GHG fertilizer production	[111]
Composting manure	Composting and use of additives (zeolite, biochar etc.)	↓ CH_4, fertilizer production ↑ nutrient conservation	[128–132]
Incineration	Litter incineration and low-protein diet	↓ 42% GHG	[135]

↑ = Increase, ↓ = Decrease.

4.3. Energy Management and Hatchery Practices

Energy management is an important factor for livestock production and mitigation policies. First and foremost, measures should be established to improve the heating requirements of the animal houses. Moreover, the use, selection and maintenance of high performance lighting systems, exhaust fans, and the improved building insulation appears to be of utmost importance. In addition, replacing fossil fuels with renewable energy sources, such as wind or solar energy, biomass, and the energy of biogas produced from manure, can provide immediate reduction in GHG emissions. Thus, the farm buildings should be supplied by renewable energy sources for heating, ventilation, air conditioning, and lighting [136]. In that area of interest, significant work has been made. In an experiment conducted in a cold region of Greece, the holistic approach of a heat pump for heating, ventilation, and air conditioning systems for broiler houses presented promising results [137]. In another experiment, the use of geothermal heat pumps not only reduced the cost and the fuel consumption compared to a conventional broiler house, but also lessened the CO_2 emissions, prompting a cleaner energy source [138].

As reported by Li et al. [139], solar thermal energy can be used in heating (water and space) and manure drying. Moreover, electricity through solar panels could be generated directly. Thus, a critical application is to maintain solar panels on the rooftops of the farm buildings, especially for countries with a lot of sun during the year, such as the Mediterranean ones. For this purpose, the use of combined photovoltaic panels and heat pumps could also be used, even for fulfilling the energy demand of the poultry houses [140,141]. In addition, wind turbines combined with fans can play a major part in the energy efficiency of a farm, generating electricity and ventilating the facilities [133]. Furthermore, a simple suggestion is the use of LED technology, as a more reliable, energy efficient option preserving lifetime, as well as being environmentally friendly [136].

In addition, poultry-related stakeholders should avail themselves of new hatchery techniques, including but not limited to on-tray feeding at hatchery and immediate access to feed [142,143], on-farm hatching of eggs with immediate access to feed [144], and in ovo supplementation of nutrients [145].

4.4. Data Management

Last but not least, smart technology should be included to help in precision livestock farming. The control of indoor climate conditions contributes to expressing the maximum genetic potential and increasing the productivity of the animals [146]. By installing a simple control system, with little cost, for example temperature sensors, microphones, and cameras, a farmer can preserve the appropriate conditions for the animals' welfare on a daily 24 h basis [147] and collect the proper data for extra analysis. To extend this, in the big-data world of the 21st century, the development of algorithmic models, software, image analysis tools and the collection and interpretation of such data can help the farmers to achieve the overall control of the farm and the scientists to improve their research and better analyze the animal productivity and sustainability [148–150] as well as lessen the environmental impact [10].

5. Future Prospects

Research in the accessible literature demonstrated the plethora of technical options for mitigating poultry emissions. Nutritional management, mainly through the replacement of established feed material (e.g., soybean), feed additives (e.g., amino acids), or novel feeding practices (e.g., food waste and by-products) have been cherished as the main options. Their effectiveness can be significantly increased when nutrition management is improved and productivity is increased. Diets also affect manure emissions, as they alter their content; the composition of the proportions and additives affect the form and amount of N manure. Furthermore, GHG emissions from manure can be effectively controlled by shortening storage life, ensuring proper conditions, or mainly by utilizing the unique characteristics to generate biogas. However, direct and indirect N_2O emissions are much more difficult to avoid once N is excreted. Techniques that block emissions during the early management stages retain N in the faeces that are often emitted at later stages. Another possibility from an environmental point of view could be the optimization of the productive life of the poultry species, such as broilers. In addition, various factors should be considered, such as precision feeding and the design of housing systems, to ensure that adverse effects are avoided. Another important contribution is the use of materials, equipment, and supplies with the lowest energy combustion. Not only that, but aiming for energy efficiency with a combination of renewable energy sources such as solar-wind energy, providing heating and electricity in farms, can effectively lessen greenhouse gas emissions and improve farm sustainable-energy independence. A characteristic of the Mediterranean countries is the natural gift of periods of sunshine year-round; thus, we should maximize the use of solar power to cover farms' needs. Nevertheless, the proposals mentioned in the previous sections address important technical elements that need to be implemented to achieve the goal of reduced GHG emissions. However, for all these techniques to be effective, they need to be carried out based on overall central planning, in agreement with all stakeholders, including but not limited to organizations and business entities. These agreements will concern new "green development" plans as formulated by the Paris and Kyoto agreements. One such example is the EU's "Green Agreement 2050", an agreement for achieving a sustainable economy with zero net GHG emissions by 2050. Livestock and food production should be developed with the least impact to the environment and at the same time should be safe, nutritional, and of high quality. In addition, further research is needed to quantify the economic aspects of mitigation, as well as mitigation practices that may have an impact on other environmental and broader development goals, such as food security and animal welfare.

Supplementary Materials: The following supporting information can be downloaded at: https://www.mdpi.com/article/10.3390/su15031941/s1, Table S1: Emissions from agriculture as a total, enteric fermentation, and synthetic fertilizer use in 2019, in EU Mediterranean countries [32].

Author Contributions: Conceptualization, F.Z. and A.C.P.; investigation, F.Z., E.G., C.M., C.C., A.M., E.T. and C.K.; writing—original draft preparation, F.Z., E.G., C.M., C.C., A.M., E.T., C.K. and A.C.P.; writing—review and editing, F.Z., E.G., C.M., C.C., A.M., C.K., E.T. and A.C.P.; visualization, A.M.; supervision, E.T. and A.C.P.; project administration, E.T., C.K. and A.C.P. All authors have read and agreed to the published version of the manuscript.

Funding: This research received no external funding.

Institutional Review Board Statement: Not applicable.

Data Availability Statement: Not applicable.

Conflicts of Interest: The authors declare no conflict of interest.

References

1. United Nations Department of Economic and Social Affairs. *United Nations World Population Prospects 2022: Summary of Results*; United Nations: New York, NY, USA, 2022.
2. United Nations, Department of Economic and Social Affairs. *World Urbanization Prospects: 2018: Highlights*; Population Division; United Nations: New York, NY, USA, 2019; p. 126.
3. Godfray, H.C.J.; Aveyard, P.; Garnett, T.; Hall, J.W.; Key, T.J.; Lorimer, J.; Pierrehumbert, R.T.; Scarborough, P.; Springmann, M.; Jebb, S.A. Meat Consumption, Health, and the Environment. *Science* **2018**, *361*, eaam5324. [CrossRef] [PubMed]
4. González, N.; Marquès, M.; Nadal, M.; Domingo, J.L. Meat Consumption: Which Are the Current Global Risks? A Review of Recent (2010–2020) Evidences. *Food Res. Int.* **2020**, *137*, 109341. [CrossRef] [PubMed]
5. OECD/FAO. *OECD-FAO Agricultural Outlook 2021–2030*; OECD Publishing: Paris, France, 2021. [CrossRef]
6. Poore, J.; Nemecek, T. Reducing Food's Environmental Impacts through Producers and Consumers. *Science* **2018**, *360*, 987–992. [CrossRef]
7. Grossi, G.; Goglio, P.; Vitali, A.; Williams, A.G. Livestock and Climate Change: Impact of Livestock on Climate and Mitigation Strategies. *Anim. Front.* **2019**, *9*, 69–76. [CrossRef] [PubMed]
8. Naser, H.M.; Nagata, O.; Sultana, S.; Hatano, R. Carbon Sequestration and Contribution of CO_2, CH_4 and N_2O Fluxes to Global Warming Potential from Paddy-Fallow Fields on Mineral Soil Beneath Peat in Central Hokkaido, Japan. *Agriculture* **2019**, *10*, 6. [CrossRef]
9. Environmental Protection Agency. *Inventory of U.S. Greenhouse Gas Emissions and Sinks: 1990–2018*; Inventory of U.S. Greenhouse Gas Emissions and Sinks; EPA: Washington, DC, USA, 2020; p. 733.
10. Tullo, E.; Finzi, A.; Guarino, M. Review: Environmental Impact of Livestock Farming and Precision Livestock Farming as a Mitigation Strategy. *Sci. Total Environ.* **2019**, *650*, 2751–2760. [CrossRef] [PubMed]
11. van Lingen, H.J.; Niu, M.; Kebreab, E.; Valadares Filho, S.C.; Rooke, J.A.; Duthie, C.-A.; Schwarm, A.; Kreuzer, M.; Hynd, P.I.; Caetano, M.; et al. Prediction of Enteric Methane Production, Yield and Intensity of Beef Cattle Using an Intercontinental Database. *Agric. Ecosyst. Environ.* **2019**, *283*, 106575. [CrossRef]
12. Arndt, C.; Hristov, A.N.; Price, W.J.; McClelland, S.C.; Pelaez, A.M.; Cueva, S.F.; Oh, J.; Bannink, A.; Bayat, A.R.; Crompton, L.A.; et al. Strategies to Mitigate Enteric Methane Emissions by Ruminants—A Way to Approach the 2.0 °C Target. CABI Agrirxiv. 2021. Available online: https://agrirxiv.org/search-details/?pan=20210085288 (accessed on 13 November 2022).
13. Christodoulou, C.; Mavrommatis, A.; Mitsiopoulou, C.; Symeon, G.; Dotas, V.; Sotirakoglou, K.; Kotsampasi, B.; Tsiplakou, E. Assessing the Optimum Level of Supplementation with Camelina Seeds in Ewes' Diets to Improve Milk Quality. *Foods* **2021**, *10*, 2076. [CrossRef]
14. Mostert, P.F.; Bos, A.P.; van Harn, J.; de Jong, I.C. The Impact of Changing toward Higher Welfare Broiler Production Systems on Greenhouse Gas Emissions: A Dutch Case Study Using Life Cycle Assessment. *Poult. Sci.* **2022**, *101*, 102151. [CrossRef]
15. Clark, M.A.; Domingo, N.G.G.; Colgan, K.; Thakrar, S.K.; Tilman, D.; Lynch, J.; Azevedo, I.L.; Hill, J.D. Global Food System Emissions Could Preclude Achieving the 1.5° and 2°C Climate Change Targets. *Science* **2020**, *370*, 705–708. [CrossRef]
16. Ravaglia, P.; Famiglietti, J.; Valentino, F. Certification and Added Value for Farm Productions. In *Advances in Chemical Pollution, Environmental Management and Protection*; Elsevier: Amsterdam, The Netherlands, 2018; Volume 2, pp. 63–108. ISBN 9780128128664.
17. Guarino Amato, M.; Castellini, C. Adaptability Challenges for Organic Broiler Chickens: A Commentary. *Animals* **2022**, *12*, 1354. [CrossRef] [PubMed]

8. Wiedmann, T.; Minx, J. A Definition of Carbon Footprint. In *Ecological Economics Research Trends*; Nova Publishers: New York, NY, USA, 2007; pp. 55–65. ISBN 9781600219412.
9. Opio, C.; Gerber, P.; Mottet, A.; Falcucci, A.; Tempio, G.; MacLeod, M.; Vellinga, T.; Henderson, B.; Steinfeld, H. *Greenhouse Gas Emmission from Ruminant Supply Chains: A Global Life Cycle Assessment*; Food and Agriculture Organization of the United Nations: Rome, Italy, 2013.
10. *ISO 14044:2006*; Environmental Management—Life Cycle Assessment—Requirements and Guidelines. International Organization for Standardization: Geneva, Switzerland. Available online: https://www.iso.org/cms/render/live/en/sites/isoorg/contents/data/standard/03/84/38498.html (accessed on 21 November 2022).
11. De Vries, M.; de Boer, I.J.M. Comparing Environmental Impacts for Livestock Products: A Review of Life Cycle Assessments. *Livest. Sci.* **2010**, *128*, 1–11. [CrossRef]
12. Roy, P.; Nei, D.; Orikasa, T.; Xu, Q.; Okadome, H.; Nakamura, N.; Shiina, T. A Review of Life Cycle Assessment (LCA) on Some Food Products. *J. Food Eng.* **2009**, *90*, 1–10. [CrossRef]
13. Dijkman, T.J.; Basset-Mens, C.; Antón, A.; Núñez, M. LCA of Food and Agriculture. In *Life Cycle Assessment*; Hauschild, M.Z., Rosenbaum, R.K., Olsen, S.I., Eds.; Springer International Publishing: Cham, Switzerland, 2018; pp. 723–754. ISBN 9783319564746.
14. Dangal, S.R.S.; Tian, H.; Pan, S.; Zhang, L.; Xu, R. Greenhouse Gas Balance in Global Pasturelands and Rangelands. *Environ. Res. Lett.* **2020**, *15*, 104006. [CrossRef]
15. Blandford, D.; Hassapoyannes, K. *The Role of Agriculture in Global GHG Mitigation*; Food, Agriculture and Fisheries Papers; OECD: Paris, France, 2018.
16. Crippa, M.; Solazzo, E.; Guizzardi, D.; Monforti-Ferrario, F.; Tubiello, F.N.; Leip, A. Food Systems Are Responsible for a Third of Global Anthropogenic GHG Emissions. *Nat. Food* **2021**, *2*, 198–209. [CrossRef]
17. Gerber, P.J.; Steinfeld, H.; Henderson, B.; Mottet, A.; Opio, C.; Dijkman, J.; Falcucci, A.; Tempio, G. *Tackling Climate Change through Livestock: A Global Assessment of Emissions and Mitigation Opportunities*; Food and Agriculture Organization of the United Nations: Rome, Italy, 2013.
18. Gerber, P.J.; Hristov, A.N.; Henderson, B.; Makkar, H.; Oh, J.; Lee, C.; Meinen, R.; Montes, F.; Ott, T.; Firkins, J.; et al. Technical Options for the Mitigation of Direct Methane and Nitrous Oxide Emissions from Livestock: A Review. *Animal* **2013**, *7*, 220–234. [CrossRef]
19. EUROSTAT. Climate Change—Driving Forces. Available online: https://ec.europa.eu/eurostat/statistics-explained/index.php?title=Climate_change_-_driving_forces (accessed on 5 December 2022).
20. European Environmental Agency (EEA). Greenhouse Gas Emissions from Agriculture in Europe. Available online: https://www.eea.europa.eu/ims/greenhouse-gas-emissions-from-agriculture (accessed on 21 November 2022).
21. Buratti, C.; Fantozzi, F.; Barbanera, M.; Lascaro, E.; Chiorri, M.; Cecchini, L. Carbon Footprint of Conventional and Organic Beef Production Systems: An Italian Case Study. *Sci. Total Environ.* **2017**, *576*, 129–137. [CrossRef]
22. Food and Agriculture Organization of the United Nations. *World Food and Agriculture—Statistical Yearbook 2021*; FAO: Rome, Italy, 2021; ISBN 9789251343326.
23. Costantini, M.; Ferrante, V.; Guarino, M.; Bacenetti, J. Environmental Sustainability Assessment of Poultry Productions through Life Cycle Approaches: A Critical Review. *Trends Food Sci. Technol.* **2021**, *110*, 201–212. [CrossRef]
24. Ministry of Environment and Energy. *Climate Change—National Inventory Report of Greece for Greenhouse and Other Gases for the Years 1990–2019*; Ministry of Environment and Energy: Athens, Greece, 2021.
25. OECD; FAO. *Agricultural Outlook 2020–2029*; OECD/FAO Agricultural Outlook; OECD Publishing: Paris, France, 2020.
26. Alexandratos, N.; Bruinsma, J. *World Agriculture towards 2030/2050: The 2012 Revision*; ESA Working Paper No. 12-03. June 2012; Food and Agriculture Organization of the United Nations: Italy, Rome, 2012; Available online: https://www.fao.org/3/ap106e/ap106e.pdf (accessed on 14 November 2022).
27. Lesschen, J.P.; van den Berg, M.; Westhoek, H.J.; Witzke, H.P.; Oenema, O. Greenhouse Gas Emission Profiles of European Livestock Sectors. *Anim. Feed. Sci. Technol.* **2011**, *166–167*, 16–28. [CrossRef]
28. Dunkley, C.S.; Fairchild, B.D.; Ritz, C.W.; Kiepper, B.H.; Lacy, M.P. Carbon Footprint of Poultry Production Farms in South Georgia: A Case Study. *J. Appl. Poult. Res.* **2015**, *24*, 73–79. [CrossRef]
29. Vetter, S.H.; Malin, D.; Smith, P.; Hillier, J. The Potential to Reduce GHG Emissions in Egg Production Using a GHG Calculator—A Cool Farm Tool Case Study. *J. Clean. Prod.* **2018**, *202*, 1068–1076. [CrossRef]
30. Andretta, I.; Hickmann, F.M.W.; Remus, A.; Franceschi, C.H.; Mariani, A.B.; Orso, C.; Kipper, M.; Létourneau-Montminy, M.-P.; Pomar, C. Environmental Impacts of Pig and Poultry Production: Insights from a Systematic Review. *Front. Vet. Sci.* **2021**, *8*, 750733. [CrossRef] [PubMed]
31. Pelletier, N. Environmental Performance in the US Broiler Poultry Sector: Life Cycle Energy Use and Greenhouse Gas, Ozone Depleting, Acidifying and Eutrophying Emissions. *Agric. Syst.* **2008**, *98*, 67–73. [CrossRef]
32. Pelletier, N. Life Cycle Assessment of Canadian Egg Products, with Differentiation by Hen Housing System Type. *J. Clean. Prod.* **2017**, *152*, 167–180. [CrossRef]
33. Skunca, D.; Tomasevic, I.; Djekic, I. Environmental Performance of the Poultry Meat Chain—LCA Approach. *Procedia Food Sci.* **2015**, *5*, 258–261. [CrossRef]
34. Dekker, S.E.M.; de Boer, I.J.M.; Vermeij, I.; Aarnink, A.J.A.; Koerkamp, P.W.G.G. Ecological and Economic Evaluation of Dutch Egg Production Systems. *Livest. Sci.* **2011**, *139*, 109–121. [CrossRef]

45. Leinonen, I.; Williams, A.G.; Wiseman, J.; Guy, J.; Kyriazakis, I. Predicting the Environmental Impacts of Chicken Systems in th United Kingdom through a Life Cycle Assessment: Egg Production Systems. *Poult. Sci.* **2012**, *91*, 26–40. [CrossRef]
46. Ghasempour, A.; Ahmadi, E. Assessment of Environment Impacts of Egg Production Chain Using Life Cycle Assessmen *J. Environ. Manag.* **2016**, *183*, 980–987. [CrossRef]
47. Bengtsson, J.; Seddon, J. Cradle to Retailer or Quick Service Restaurant Gate Life Cycle Assessment of Chicken Products i Australia. *J. Clean. Prod.* **2013**, *41*, 291–300. [CrossRef]
48. Prudêncio da Silva, V.; van der Werf, H.M.G.; Soares, S.R.; Corson, M.S. Environmental Impacts of French and Brazilian Broile Chicken Production Scenarios: An LCA Approach. *J. Environ. Manag.* **2014**, *133*, 222–231. [CrossRef]
49. Cederberg, C.; Wivstad, M.; Bergkvist, P.; Mattsson, B.; Ivarsson, K. Environmental Assessment of Plant Protection Strategie Using Scenarios for Pig Feed Production. *AMBIO A J. Hum. Environ.* **2005**, *34*, 408–413. [CrossRef]
50. Xin, H.; Gates, R.S.; Green, A.R.; Mitloehner, F.M.; Moore, P.A.; Wathes, C.M. Environmental Impacts and Sustainability of Eg Production Systems. *Poult. Sci.* **2011**, *90*, 263–277. [CrossRef] [PubMed]
51. Clune, S.; Crossin, E.; Verghese, K. Systematic Review of Greenhouse Gas Emissions for Different Fresh Food Categorie *J. Clean. Prod.* **2017**, *140*, 766–783. [CrossRef]
52. Chianese, D.S.; Rotz, C.A.; Richard, T.L. Richard Whole-Farm Greenhouse Gas Emissions: A Review with Application to Pennsylvania Dairy Farm. *Appl. Eng. Agric.* **2009**, *25*, 431–442. [CrossRef]
53. Olesen, J.E.; Schelde, K.; Weiske, A.; Weisbjerg, M.R.; Asman, W.A.H.; Djurhuus, J. Modelling Greenhouse Gas Emissions fron European Conventional and Organic Dairy Farms. *Agric. Ecosyst. Environ.* **2006**, *112*, 207–220. [CrossRef]
54. Zervas, G.; Tsiplakou, E. Life Cycle Assessment of Animal Origin Products. *Adv. Anim. Biosci.* **2016**, *7*, 191–195. [CrossRef]
55. Różewicz, M. Production, Use and Efficiency of Utilising Grains of Various Cereal Species as Feed Resources for Poultr Production. *Pol. J. Agron.* **2019**, *38*, 66–74. [CrossRef]
56. Kapica, J.; Pawlak, H.; Ścibisz, M. Carbon Dioxide Emission Reduction by Heating Poultry Houses from Renewable Energ Sources in Central Europe. *Agric. Syst.* **2015**, *139*, 238–249. [CrossRef]
57. Usubharatana, P.; Phungrassami, H. Greenhouse Gas Emissions of One-Day-Old Chick Production. *Pol. J. Environ. Stud.* **2017**, 2 1269–1277. [CrossRef]
58. Gerber, P.; Opio, C.; Steinfeld, H. Poultry production and the environment—A review. Available online: https://www semanticscholar.org/paper/Poultry-production-and-the-environment-%E2%80%93-a-review-Gerber-Opio/11b57f4788910bc 263f7eebbe74c58c3eaff779 (accessed on 20 December 2022).
59. Tapio, I.; Snelling, T.J.; Strozzi, F.; Wallace, R.J. The Ruminal Microbiome Associated with Methane Emissions from Ruminan Livestock. *J. Anim. Sci. Biotechnol.* **2017**, *8*, 7. [CrossRef]
60. Wolf, J.; Asrar, G.R.; West, T.O. Revised Methane Emissions Factors and Spatially Distributed Annual Carbon Fluxes for Globa Livestock. *Carbon Balance Manag.* **2017**, *12*, 16. [CrossRef] [PubMed]
61. Redding, M.R. Bentonite Can Decrease Ammonia Volatilisation Losses from Poultry Litter: Laboratory Studies. *Anim. Prod. Sc* **2013**, *53*, 1115. [CrossRef]
62. Wedwitschka, H.; Gallegos Ibanez, D.; Schäfer, F.; Jenson, E.; Nelles, M. Material Characterization and Substrate Suitabilit Assessment of Chicken Manure for Dry Batch Anaerobic Digestion Processes. *Bioengineering* **2020**, *7*, 106. [CrossRef] [PubMed]
63. Gržinić, G.; Piotrowicz-Cieślak, A.; Klimkowicz-Pawlas, A.; Górny, R.L.; Ławniczek-Wałczyk, A.; Piechowicz, L.; Olkowska, E Potrykus, M.; Tankiewicz, M.; Krupka, M.; et al. Intensive Poultry Farming: A Review of the Impact on the Environment an Human Health. *Sci. Total Environ.* **2022**, *858*, 160014. [CrossRef]
64. Food and Agriculture Organization FAOSTAT. Available online: https://fenix.fao.org/faostat/internal/en/ (accessed o 9 December 2022).
65. Monteny, G.-J.; Bannink, A.; Chadwick, D. Greenhouse Gas Abatement Strategies for Animal Husbandry. *Agric. Ecosyst. Environ* **2006**, *112*, 163–170. [CrossRef]
66. Food and Agriculture Organization of the United Nations. *FAO's Work on Climate Change United Nations Climate Change Conferenc 2017*; Food and Agriculture Organization of the United Nations: Rome, Italy, 2017; p. 40.
67. Fatica, A.; Fantuz, F.; Wu, M.; Tavaniello, S.; Maiorano, G.; Salimei, E. Soybean vs. Pea Bean in the Diet of Medium-Growin Broiler Chickens Raised under Semi-Intensive Conditions of Inner Mediterranean Areas: Growth Performance and Environmenta Impact. *Animals* **2022**, *12*, 649. [CrossRef]
68. Koivunen, E.; Tuunainen, P.; Valkonen, E.; Valaja, J. Use of Semi-Leafless Peas (*Pisum sativum* L.) in Laying Hen Diets *Agric. Food Sci.* **2015**, *24*, 84–91. [CrossRef]
69. Yuan, C.; Song, H.; Zhang, X.; Jiang, Y.; Zhang, A.; Azzam, M.M.; Zou, X. Effect of Expanded Cottonseed Meal on Layin Performance, Egg Quality, Concentrations of Free Gossypol in Tissue, Serum and Egg of Laying Hens: Laying Hens Fed Expande Cottonseed Meal. *Anim. Sci. J.* **2014**, *85*, 549–554. [CrossRef]
70. Abín, R.; Laca, A.; Laca, A.; Díaz, M. Environssessmentsesment of Intensive Egg Production: A Spanish Case Study. *J. Clean. Prod* **2018**, *179*, 160–168. [CrossRef]
71. Biasato, I.; De Marco, M.; Rotolo, L.; Renna, M.; Lussiana, C.; Dabbou, S.; Capucchio, M.T.; Biasibetti, E.; Costa, P.; Gai, F.; et a Effects of Dietary *Tenebrio Molitor* Meal Inclusion in Free-Range Chickens. *J. Anim. Physiol. Anim. Nutr.* **2016**, *100*, 1104–1112 [CrossRef]

72. Bovera, F.; Loponte, R.; Marono, S.; Piccolo, G.; Parisi, G.; Iaconisi, V.; Gasco, L.; Nizza, A. Use of Tenebrio Molitor Larvae Meal as Protein Source in Broiler Diet: Effect on Growth Performance, Nutrient Digestibility, and Carcass and Meat Traits. *J. Anim. Sci.* **2016**, *94*, 639–647. [CrossRef] [PubMed]
73. Schiavone, A.; Cullere, M.; De Marco, M.; Meneguz, M.; Biasato, I.; Bergagna, S.; Dezzutto, D.; Gai, F.; Dabbou, S.; Gasco, L.; et al. Partial or Total Replacement of Soybean Oil by Black Soldier Fly Larvae (*Hermetia Illucens* L.) Fat in Broiler Diets: Effect on Growth Performances, Feed-Choice, Blood Traits, Carcass Characteristics and Meat Quality. *Ital. J. Anim. Sci.* **2017**, *16*, 93–100. [CrossRef]
74. Mosnier, E.; van der Werf, H.M.G.; Boissy, J.; Dourmad, J.-Y. Evaluation of the Environmental Implications of the Incorporation of Feed-Use Amino Acids in the Manufacturing of Pig and Broiler Feeds Using Life Cycle Assessment. *Animal* **2011**, *5*, 1972–1983. [CrossRef]
75. Al-Harthi, M.A.; Attia, Y.A.; El-Shafey, A.S.; Elgandy, M.F. Impact of Phytase on Improving the Utilisation of Pelleted Broiler Diets Containing Olive By-Products. *Ital. J. Anim. Sci.* **2020**, *19*, 310–318. [CrossRef]
76. Zarghi, H.; Golian, A.; Hassanabadi, A.; Khaligh, F. Effect of Zinc and Phytase Supplementation on Performance, Immune Response, Digestibility and Intestinal Features in Broilers Fed a Wheat-Soybean Meal Diet. *Ital. J. Anim. Sci.* **2022**, *21*, 430–444. [CrossRef]
77. Giannenas, I.; Bonos, E.; Anestis, V.; Filioussis, G.; Papanastasiou, D.K.; Bartzanas, T.; Papaioannou, N.; Tzora, A.; Skoufos, I. Effects of Protease Addition and Replacement of Soybean Meal by Corn Gluten Meal on the Growth of Broilers and on the Environmental Performances of a Broiler Production System in Greece. *PLoS ONE* **2017**, *12*, e0169511. [CrossRef]
78. Giamouri, E.; Pappas, A.C.; Papadomichelakis, G.; Tsiplakou, E.; Sotirakoglou, K.; Markakis, N.; Galliou, F.; Manios, T.; Zentek, J.; Lasaridi, K.; et al. The Food for Feed Concept. Performance of Broilers Fed Hotel Food Residues. *Br. Poult. Sci.* **2021**, *62*, 452–458. [CrossRef] [PubMed]
79. Giamouri, E.; Pappas, A.C.; Papadomichelakis, G.; Simitzis, P.E.; Manios, T.; Zentek, J.; Lasaridi, K.; Tsiplakou, E.; Zervas, G. The Food for Feed Concept: Redefining the Use of Hotel Food Residues in Broiler Diets. *Sustainability* **2022**, *14*, 3659. [CrossRef]
80. Giamouri, E.; Mavrommatis, A.; Simitzis, P.E.; Mitsiopoulou, C.; Haroutounian, S.A.; Koutinas, A.; Pappas, A.C.; Tsiplakou, E. Redefining the Use of Vinification Waste By-Products in Broiler Diets. *Sustainability* **2022**, *14*, 15714. [CrossRef]
81. Mavrommatis, A.; Giamouri, E.; Myrtsi, E.D.; Evergetis, E.; Filippi, K.; Papapostolou, H.; Koulocheri, S.D.; Zoidis, E.; Pappas, A.C.; Koutinas, A.; et al. Antioxidant Status of Broiler Chickens Fed Diets Supplemented with Vinification By-Products: A Valorization Approach. *Antioxidants* **2021**, *10*, 1250. [CrossRef]
82. Fraanje, W.; Garnett, T. *Soy: Food, Feed, and Land Use Change (Foodsource: Building Blocks)*; Technical Report; Food Climate Research Network, University of Oxford: Oxford, UK, 2020.
83. Song, X.-P.; Hansen, M.C.; Potapov, P.; Adusei, B.; Pickering, J.; Adami, M.; Lima, A.; Zalles, V.; Stehman, S.V.; Di Bella, C.M.; et al. Massive Soybean Expansion in South America since 2000 and Implications for Conservation. *Nat. Sustain.* **2021**, *4*, 784–792. [CrossRef] [PubMed]
84. Alfonso-Avila, A.R.; Cirot, O.; Lambert, W.; Létourneau-Montminy, M.P. Effect of Low-Protein Corn and Soybean Meal-Based Diets on Nitrogen Utilization, Litter Quality, and Water Consumption in Broiler Chicken Production: Insight from Meta-Analysis. *Animal* **2022**, *16*, 100458. [CrossRef] [PubMed]
85. Różewicz, M.; Grabiński, J.; Sułek, A. Possibilities and Limitations in the Use of Legumes from Domestic Cultivation in Poultry Feed in the Context of Fodder Protein Deficit. *Pol. J. Agron.* **2018**, *35*, 32–44. [CrossRef]
86. Ceylan, N.; Ciftçi, I.; Mızrak, C.; Kahraman, Z.; Efil, H. Influence of different dietary oil sources on performance and fatty acid profile of egg yolk in laying hens. *J. Anim. Feed Sci.* **2011**, *20*, 71–83. [CrossRef]
87. Tallentire, C.W.; Mackenzie, S.G.; Kyriazakis, I. Can Novel Ingredients Replace Soybeans and Reduce the Environmental Burdens of European Livestock Systems in the Future? *J. Clean. Prod.* **2018**, *187*, 338–347. [CrossRef]
88. Veldkamp, T.; Bosch, G. Insects: A Protein-Rich Feed Ingredient in Pig and Poultry Diets. *Anim. Front.* **2015**, *5*, 45–50. [CrossRef]
89. Vauterin, A.; Steiner, B.; Sillman, J.; Kahiluoto, H. The Potential of Insect Protein to Reduce Food-Based Carbon Footprints in Europe: The Case of Broiler Meat Production. *J. Clean. Prod.* **2021**, *320*, 128799. [CrossRef]
90. Rumpold, B.A.; Schlüter, O.K. Nutritional Composition and Safety Aspects of Edible Insects. *Mol. Nutr. Food Res.* **2013**, *57*, 802–823. [CrossRef]
91. Hong, J.; Han, T.; Kim, Y.Y. Mealworm (Tenebrio Molitor Larvae) as an Alternative Protein Source for Monogastric Animal: A Review. *Animals* **2020**, *10*, 2068. [CrossRef]
92. van Huis, A. *Edible Insects: Future Prospects for Food and Feed Security*; FAO Forestry Paper; Food and Agriculture Organization of the United Nations: Rome, Italy, 2013; ISBN 9789251075951.
93. Khan, S.; Khan, R.U.; Alam, W.; Sultan, A. Evaluating the Nutritive Profile of Three Insect Meals and Their Effects to Replace Soya Bean in Broiler Diet. *J. Anim. Physiol. Anim. Nutr.* **2018**, *102*, e662–e668. [CrossRef]
94. Selaledi, L.; Mbajiorgu, C.A.; Mabelebele, M. The Use of Yellow Mealworm (T. Molitor) as Alternative Source of Protein in Poultry Diets: A Review. *Trop. Anim. Health Prod.* **2020**, *52*, 7–16. [CrossRef] [PubMed]
95. Khan, S.; Naz, S.; Sultan, A.; Alhidary, I.A.; Abdelrahman, M.M.; Khan, R.U.; Khan, N.A.; Khan, M.A.; Ahmad, S. Worm Meal: A Potential Source of Alternative Protein in Poultry Feed. *World's Poult. Sci. J.* **2016**, *72*, 93–102. [CrossRef]

96. Biasato, I.; Gasco, L.; De Marco, M.; Renna, M.; Rotolo, L.; Dabbou, S.; Capucchio, M.T.; Biasibetti, E.; Tarantola, M.; Sterpone, L.; et al. Yellow Mealworm Larvae (Tenebrio Molitor) Inclusion in Diets for Male Broiler Chickens: Effects on Growth Performance, Gut Morphology, and Histological Findings. *Poult. Sci.* **2018**, *97*, 540–548. [CrossRef] [PubMed]
97. De Boer, H.C.; van Krimpen, M.M.; Blonk, H.; Tyszler, M. *Replacement of Soybean Meal in Compound Feed by European Protein Sources—Effects on Carbon Footprint*; Livestock Research Report; Wageningen UR (University & Research Centre) Livestock Research: Wageningen, The Netherlands, 2014.
98. Gasco, L.; Acuti, G.; Bani, P.; Dalle Zotte, A.; Danieli, P.P.; De Angelis, A.; Fortina, R.; Marino, R.; Parisi, G.; Piccolo, G.; et al. Insect and Fish By-Products as Sustainable Alternatives to Conventional Animal Proteins in Animal Nutrition. *Ital. J. Anim. Sci.* **2020**, *19*, 360–372. [CrossRef]
99. Tavares, M.N.; Pereira, R.T.; Silva, A.L.; Lemes, L.R.; Menten, J.F.M.; Gameiro, A.H. Economic Viability of Insect Meal as a Novel Ingredient in Diets for Broiler Chickens. *J. Insects Food Feed* **2022**, *8*, 1015–1025. [CrossRef]
100. Maharjan, P.; Martinez, D.A.; Weil, J.; Suesuttajit, N.; Umberson, C.; Mullenix, G.; Hilton, K.M.; Beitia, A.; Coon, C.N. Review: Physiological Growth Trend of Current Meat Broilers and Dietary Protein and Energy Management Approaches for Sustainable Broiler Production. *Animal* **2021**, *15*, 100284. [CrossRef]
101. Belloir, P.; Méda, B.; Lambert, W.; Corrent, E.; Juin, H.; Lessire, M.; Tesseraud, S. Reducing the CP Content in Broiler Feeds: Impact on Animal Performance, Meat Quality and Nitrogen Utilization. *Animal* **2017**, *11*, 1881–1889. [CrossRef]
102. Hilliar, M.; Hargreave, G.; Girish, C.K.; Barekatain, R.; Wu, S.-B.; Swick, R.A. Using Crystalline Amino Acids to Supplement Broiler Chicken Requirements in Reduced Protein Diets. *Poult. Sci.* **2020**, *99*, 1551–1563. [CrossRef]
103. Benavides, P.T.; Cai, H.; Wang, M.; Bajjalieh, N. Life-Cycle Analysis of Soybean Meal, Distiller-Dried Grains with Solubles, and Synthetic Amino Acid-Based Animal Feeds for Swine and Poultry Production. *Anim. Feed. Sci. Technol.* **2020**, *268*, 114607. [CrossRef]
104. Selle, P.H.; de Paula Dorigam, J.C.; Lemme, A.; Chrystal, P.V.; Liu, S.Y. Synthetic and Crystalline Amino Acids: Alternatives to Soybean Meal in Chicken-Meat Production. *Animals* **2020**, *10*, 729. [CrossRef]
105. Giamouri, E.; Papadomichelakis, G.; Pappas, A.C.; Simitzis, P.E.; Galliou, F.; Paßlack, N.; Zentek, J.; Lasaridi, K.; Fegeros, K.; Manios, T.; et al. Meat Quality Traits as Affected by the Dietary Inclusion of Food Waste in Finishing Pigs. *Sustainability* **2022**, *14*, 6593. [CrossRef]
106. Georganas, A.; Giamouri, E.; Pappas, A.C.; Papadomichelakis, G.; Galliou, F.; Manios, T.; Tsiplakou, E.; Fegeros, K.; Zervas, G. Bioactive Compounds in Food Waste: A Review on the Transformation of Food Waste to Animal Feed. *Foods* **2020**, *9*, 291. [CrossRef] [PubMed]
107. Georganas, A.; Giamouri, E.; Pappas, A.C.; Papadomichelakis, G.; Fortatos, S.; Manios, T.; Lasaridi, K.; Fegeros, K.; Tsiplakou, E.; Zervas, G. Redefining the Future of Catering Waste Application in Animal Diets—A Review on the Minimization of Potential Hazards in Catering Waste Prior to Application in Animal Diets. *Anim. Feed. Sci. Technol.* **2022**, *289*, 115334. [CrossRef]
108. Bouwman, A.F.; Boumans, L.J.M.; Batjes, N.H. Emissions of N_2O and NO from Fertilized Fields: Summary of Available Measurement Data: Summary of NO and N_2O measurement data. *Glob. Biogeochem. Cycles* **2002**, *16*, 6-1–6-13. [CrossRef]
109. Awasthi, M.K.; Duan, Y.; Awasthi, S.K.; Liu, T.; Zhang, Z. Influence of Bamboo Biochar on Mitigating Greenhouse Gas Emissions and Nitrogen Loss during Poultry Manure Composting. *Bioresour. Technol.* **2020**, *303*, 122952. [CrossRef]
110. USEPA. Quantitative Microbial Risk Assessment to Estimate Illness in Freshwater Impacted by Agricultural Animal Sources of Fecal Contamination. Report EPA 822-R-10-005; U.S. Environmental Protection Agency Office of Water: Washington, DC, USA, 2010.
111. Kreidenweis, U.; Breier, J.; Herrmann, C.; Libra, J.; Prochnow, A. Greenhouse Gas Emissions from Broiler Manure Treatment Options Are Lowest in Well-Managed Biogas Production. *J. Clean. Prod.* **2021**, *280*, 124969. [CrossRef]
112. Mohankumar Sajeev, E.P.; Winiwarter, W.; Amon, B. Greenhouse Gas and Ammonia Emissions from Different Stages of Liquid Manure Management Chains: Abatement Options and Emission Interactions. *J. Environ. Qual.* **2018**, *47*, 30–41. [CrossRef]
113. Burton, C.H.; Turner, C. *Manure Management: Treatment Strategies for Sustainable Agriculture*, 2nd ed.; Burton, C.H., Turner, C., Eds.; Silsoe Research Institute: Silsoe, UK, 2003; ISBN 9780953128266.
114. Sommer, S.G.; Petersen, S.O.; Møller, H.B. Algorithms for Calculating Methane and Nitrous Oxide Emissions from Manure Management. *Nutr. Cycl. Agroecosystems* **2004**, *69*, 143–154. [CrossRef]
115. Groenestein, C.M.; Smits, M.C.J.; Huijsmans, J.F.M.; Oenema, O. *Measures to Reduce Ammonia Emissions from Livestock Manures; Now, Soon and Later*; Wageningen UR (University & Research Centre) Livestock Research: Wageningen, The Netherlands, 2011.
116. Sommer, S.G.; Olesen, J.E.; Petersen, S.O.; Weisbjerg, M.R.; Valli, L.; Rodhe, L.; Béline, F. Region-Specific Assessment of Greenhouse Gas Mitigation with Different Manure Management Strategies in Four Agroecological Zones: Region-specific assessment of greenhouse gas mitigation. *Glob. Chang. Biol.* **2009**, *15*, 2825–2837. [CrossRef]
117. Aguirre-Villegas, H.A.; Larson, R.A. Evaluating Greenhouse Gas Emissions from Dairy Manure Management Practices Using Survey Data and Lifecycle Tools. *J. Clean. Prod.* **2017**, *143*, 169–179. [CrossRef]
118. Awe, O.W.; Zhao, Y.; Nzihou, A.; Minh, D.P.; Lyczko, N. A Review of Biogas Utilisation, Purification and Upgrading Technologies. *Waste Biomass Valorization* **2017**, *8*, 267–283. [CrossRef]
119. Teenstra, E.; De Buisonjé, F.; Ndambi, A.; Pelster, D. *Manure Management in the (Sub-)Tropics: Training Manual for Extension Workers*; Livestock Research Report; Wageningen UR (University & Research Centre) Livestock Research: Rome, Italy; Wageningen, The Netherlands, 2015.

20. Teenstra, E.; Vellinga, T.; Aektasaeng, N.; Amatayakul, W.; Ndambi, A.; Pelster, D.; Germer, L.; Jenet, A.; Opio, C.; Andeweg, K. *Global Assessment of Manure Management Policies and Practices*; Livestock Research Report; Wageningen UR (University & Research Centre) Livestock Research: Wageningen, The Netherlands, 2014.
21. Kalogiannis, A.; Vasiliadou, I.A.; Spyridonidis, A.; Diamantis, V.; Stamatelatou, K. Biogas Production from Chicken Manure Wastes Using an LBR-CSTR Two-stage System: Process Efficiency, Economic Feasibility, and Carbon Dioxide Footprint. *J. Chem. Technol. Biotechnol.* **2022**, *97*, 2952–2961. [CrossRef]
22. Da Costa Gomez, C. Biogas as an Energy Option: An Overview. In *The Biogas Handbook*; Elsevier: Amsterdam, The Netherlands, 2013; pp. 1–16. ISBN 9780857094988.
23. Bhatia, S.C. Biogas. In *Advanced Renewable Energy Systems*; WPI Publishing: New York, NY, USA, 2015; p. 47. ISBN 9780429091575.
24. Teng, Z.; Hua, J.; Wang, C.; Lu, X. Design and Optimization Principles of Biogas Reactors in Large Scale Applications. In *Reactor and Process Design in Sustainable Energy Technology*; Elsevier: Amsterdam, The Netherlands, 2014; pp. 99–134. ISBN 9780444595669.
25. Siddiki, S.Y.A.; Uddin, M.N.; Mofijur, M.; Fattah, I.M.R.; Ong, H.C.; Lam, S.S.; Kumar, P.S.; Ahmed, S.F. Theoretical Calculation of Biogas Production and Greenhouse Gas Emission Reduction Potential of Livestock, Poultry and Slaughterhouse Waste in Bangladesh. *J. Environ. Chem. Eng.* **2021**, *9*, 105204. [CrossRef]
26. Hristov, A.N.; Oh, J.; Firkins, J.L.; Dijkstra, J.; Kebreab, E.; Waghorn, G.; Makkar, H.P.S.; Adesogan, A.T.; Yang, W.; Lee, C.; et al. SPECIAL TOPICS—Mitigation of Methane and Nitrous Oxide Emissions from Animal Operations: I. A Review of Enteric Methane Mitigation Options1. *J. Anim. Sci.* **2013**, *91*, 5045–5069. [CrossRef]
27. Vandecasteele, B.; Reubens, B.; Willekens, K.; De Neve, S. Composting for Increasing the Fertilizer Value of Chicken Manure: Effects of Feedstock on P Availability. *Waste Biomass Valorization* **2014**, *5*, 491–503. [CrossRef]
28. Steiner, C.; Das, K.C.; Melear, N.; Lakly, D. Reducing Nitrogen Loss during Poultry Litter Composting Using Biochar. *J. Environ. Qual.* **2010**, *39*, 1236–1242. [CrossRef]
29. Janczak, D.; Malińska, K.; Czekała, W.; Cáceres, R.; Lewicki, A.; Dach, J. Biochar to Reduce Ammonia Emissions in Gaseous and Liquid Phase during Composting of Poultry Manure with Wheat Straw. *Waste Manag.* **2017**, *66*, 36–45. [CrossRef]
30. Godlewska, P.; Schmidt, H.P.; Ok, Y.S.; Oleszczuk, P. Biochar for Composting Improvement and Contaminants Reduction. A Review. *Bioresour. Technol.* **2017**, *246*, 193–202. [CrossRef]
31. He, Z.; Lin, H.; Hao, J.; Kong, X.; Tian, K.; Bei, Z.; Tian, X. Impact of Vermiculite on Ammonia Emissions and Organic Matter Decomposition of Food Waste during Composting. *Bioresour. Technol.* **2018**, *263*, 548–554. [CrossRef]
32. Liu, C.; Zhang, X.; Zhang, W.; Wang, S.; Fan, Y.; Xie, J.; Liao, W.; Gao, Z. Mitigating Gas Emissions from Poultry Litter Composting with Waste Vinegar Residue. *Sci. Total Environ.* **2022**, *842*, 156957. [CrossRef]
33. He, X.; Hu, Q.; Chen, J.; Leong, W.Q.; Dai, Y.; Wang, C.-H. Energy and Environmental Risk Assessments of Poultry Manure Sustainable Solution: An Industrial Case Study in Singapore. *J. Clean. Prod.* **2022**, *339*, 130787. [CrossRef]
34. Choudhury, A.; Felton, G.; Moyle, J.; Lansing, S. Fluidized Bed Combustion of Poultry Litter at Farm-Scale: Environmental Impacts Using a Life Cycle Approach. *J. Clean. Prod.* **2020**, *276*, 124231. [CrossRef]
35. Ogino, A.; Oishi, K.; Setoguchi, A.; Osada, T. Life Cycle Assessment of Sustainable Broiler Production Systems: Effects of Low-Protein Diet and Litter Incineration. *Agriculture* **2021**, *11*, 921. [CrossRef]
36. Cui, Y.; Theo, E.; Gurler, T.; Su, Y.; Saffa, R. A Comprehensive Review on Renewable and Sustainable Heating Systems for Poultry Farming. *Int. J. Low-Carbon Technol.* **2020**, *15*, 121–142. [CrossRef]
37. Manolakos, D.; Panagakis, P.; Bartzanas, T.; Bouzianas, K. Use of Heat Pumps in HVAC Systems for Precise Environment Control in Broiler Houses: System's Modeling and Calculation of the Basic Design Parameters. *Comput. Electron. Agric.* **2019**, *163*, 104876. [CrossRef]
38. Choi, H.C.; Salim, H.M.; Akter, N.; Na, J.C.; Kang, H.K.; Kim, M.J.; Kim, D.W.; Bang, H.T.; Chae, H.S.; Suh, O.S. Effect of Heating System Using a Geothermal Heat Pump on the Production Performance and Housing Environment of Broiler Chickens. *Poult. Sci.* **2012**, *91*, 275–281. [CrossRef]
39. Li, Y.; Arulnathan, V.; Heidari, M.D.; Pelletier, N. Design Considerations for Net Zero Energy Buildings for Intensive, Confined Poultry Production: A Review of Current Insights, Knowledge Gaps, and Future Directions. *Renew. Sustain. Energy Rev.* **2022**, *154*, 111874. [CrossRef]
40. Gurler, T.; Elmer, T.; Cui, Y.; Omer, S.; Riffat, S. Experimental Investigation of a Novel PVt/Heat Pump System for Energy-Efficient Poultry Houses. *Int. J. Low-Carbon Technol.* **2018**, *13*, 404–413. [CrossRef]
41. Cui, Y.; Theo, E.; Gurler, T.; Su, Y.; Saffa, R. Feasibility of Hybrid Renewable Heating System Application in Poultry House: A Case Study of East Midlands, UK. *Int. J. Low-Carbon Technol.* **2021**, *16*, 73–88. [CrossRef]
42. Wijnen, H.J.; Molenaar, R.; Kemp, B.; van Roovert-Reijrink, I.A.M.; van den Brand, H.; van der Pol, C.W. Effects of Late Incubation Temperature and Moment of First Post-Hatch Feed Access on Neonatal Broiler Development, Temperature Preference, and Stress Response. *Poult. Sci.* **2022**, *101*, 102088. [CrossRef]
43. Gaweł, A.; Madej, J.P.; Kozak, B.; Bobrek, K. Early Post-Hatch Nutrition Influences Performance and Muscle Growth in Broiler Chickens. *Animals* **2022**, *12*, 3281. [CrossRef]
44. Witjes, V.L.; Bruckmaier, R.M.; Gebhardt-Henrich, S.G.; Toscano, M.J. Effects of On-Farm Hatching on Short Term Stress Indicators, Weight Gain, and Cognitive Ability in Layer Chicks. *Appl. Anim. Behav. Sci.* **2022**, *254*, 105692. [CrossRef]

145. Arain, M.A.; Nabi, F.; Marghazani, I.B.; Hassan, F.U.; Soomro, H.; Kalhoro, H.; Soomro, F.; Buzdar, J.A. *In Ovo* Delivery of Nutraceuticals Improves Health Status and Production Performance of Poultry Birds: A Review. *World's Poult. Sci. J.* **2022**, *7* 765–788. [CrossRef]
146. Costantino, A.; Fabrizio, E.; Calvet, S. The Role of Climate Control in Monogastric Animal Farming: The Effects on Animal Welfare, Air Emissions, Productivity, Health, and Energy Use. *Appl. Sci.* **2021**, *11*, 9549. [CrossRef]
147. Berckmans, D. General Introduction to Precision Livestock Farming. *Anim. Front.* **2017**, *7*, 6–11. [CrossRef]
148. Fournel, S.; Rousseau, A.N.; Laberge, B. Rethinking Environment Control Strategy of Confined Animal Housing Systems through Precision Livestock Farming. *Biosyst. Eng.* **2017**, *155*, 96–123. [CrossRef]
149. Rowe, E.; Dawkins, M.S.; Gebhardt-Henrich, S.G. A Systematic Review of Precision Livestock Farming in the Poultry Sector: Is Technology Focussed on Improving Bird Welfare? *Animals* **2019**, *9*, 614. [CrossRef]
150. Li, N.; Ren, Z.; Li, D.; Zeng, L. Review: Automated Techniques for Monitoring the Behaviour and Welfare of Broilers and Laying Hens: Towards the Goal of Precision Livestock Farming. *Animal* **2020**, *14*, 617–625. [CrossRef]

MDPI
St. Alban-Anlage 66
4052 Basel
Switzerland
www.mdpi.com

Sustainability Editorial Office
E-mail: sustainability@mdpi.com
www.mdpi.com/journal/sustainability

www.ingramcontent.com/pod-product-compliance
Lightning Source LLC
Chambersburg PA
CBHW061604220926
51850CB00005B/417
9783036596068